Charge Exchange and the Theory of Ion–Atom Collisions

B. H. BRANSDEN

and

M. R. C. McDOWELL

CLARENDON PRESS · OXFORD

1992

Oxford University Press, Walton Street, Oxford OX2 6DP
Oxford New York Toronto
Delhi Bombay Calcutta Madras Karachi
Petaling Jaya Singapore Hong Kong Tokyo
Nairobi Dar es Salaam Cape Town
Melbourne Auckland
and associated companies in
Berlin Ibadan

Oxford is a trade mark of Oxford University Press

Published in the United States
by Oxford University Press, New York

A catalogue record for this book is available from the British Library

Library of Congress Cataloging in Publication Data
Bransden, B. H., 1926–
Charge exchange and the theory of ion–atom collisions / B. H. Bransden and M. R. C. McDowell
(The International series of monographs on physics; 82)
Includes bibliographical references and index.
1. Ion–atom collisions. 2. Charge exchange. I. McDowell, M. R. C. II. Title. III. Series: International series of monographs on physics (Oxford, England); 82.
QC794.6.C6B72 1992 539.7'57—dc20 91-36597

ISBN 0 19 852020 4

Typeset by Integral Typesetting, Gorleston, Norfolk
Printed in Great Britain by
Bookcraft (Bath) Ltd.
Midsomer Norton, Avon

To Brenda and Pauline

PREFACE

During the last two decades the theory of ion–atom collisions, and particularly of charge exchange reactions, has advanced rapidly to the point where existing texts are no longer suitable as an introduction to the subject. This book aims to remedy this situation by providing an account of modern theoretical methods used to study the interaction of positive ions with atoms (or ions), concentrating particularly on charge exchange reactions. These reactions cannot be studied in isolation, and we shall have to consider to some extent the whole range of ion–atom collisions leading to elastic scattering, excitation, and ionization, so that in many respects this book can be considered as a replacement for *Introduction to the theory of ion–atom collisions* written by J. P. Coleman and one of the present authors (M. R. C. McDowell), and which is now out of print.

The material of this book is presented at a level suitable for beginning research students and is self-contained, except that a knowledge of undergraduate quantum mechanics and atomic physics is assumed. It should also be useful for experimentalists who wish to assess the status of theoretical treatments of those collision processes in which they are interested. We have made no attempt to be encyclopaedic in our treatment, and we have excluded discussion of chemical reactions, ion–molecule scattering, and the scattering of ions by surfaces, although the methods developed can often be extended to these situations. In particular, we have not surveyed the large body of experimental data available in the literature, but have concentrated on comparing experiment with theory for a limited number of representative reactions.

Following a general introduction (Chapter 1) and a short account of the semi-classical theory of elastic scattering (Chapter 2), the theory of non-relativistic ion–atom collisions is presented in both quantum and semi-classical versions in Chapter 3. This is followed in Chapter 4 by an account of some applications of the quantum coupled channel model. In Chapters 5 and 6 the semi-classical impact parameter coupled channel method is further developed and applied to a number of illustrative systems. Chapter 7 contains an account of some approximations to the coupled channel equations, which can be employed in certain restrictive circumstances, and in Chapter 8 further approximate methods are introduced which can be employed at intermediate energies. The problems arising when applying the theory to many electron systems are discussed briefly in Chapter 9, while in Chapters 10 and 11 perturbation methods suitable for application to high-energy collisions are considered, together with the corrections that result from satisfying the requirement of relativistic invariance. Certain special topics and a note on units are relegated to the appendices.

It is very sad to record that during the writing of this book my co-author and friend of long standing, Coulter McDowell, became seriously ill and he was unable to continue to work on the preparation of the text. Coulter and I had worked together on many occasions over a period of more than thirty years until his tragic and disabling illness caused his withdrawal from active life and I am deeply aware of how greatly his warmth, enthusiasm, and zest for life have been missed by his friends and colleagues. Although Coulter could not take part in the later stages of writing this book, it is my profound hope that something of his inspiration and deep knowledge of atomic physics will be conveyed to the reader through these pages.

Our thanks are due to many people who have contributed to the writing of this book, particularly to J. McCann, R. Potvliege, and H. Slim who have read and commented on the draft text, and to D. P. Dewangan, A. M. Ermolaev, D. R. Flower, R. K. Janev, C. J. Noble, R. Shingal, and other colleagues from whom we have benefited through many discussions. It is a pleasure to acknowledge our debt to Miss Penny Carse for her patient and careful typing of the manuscript and to Mrs Pauline Russell for drawing the figures.

Durham B.H.B.
January 1992

ACKNOWLEDGEMENT

Certain diagrams in this book are based on figures appearing in published research articles, identified in the captions. Permission for the use of this material, granted by the publishers and by (at least one) of the authors of the articles concerned, is gratefully acknowledged.

CONTENTS

1

INTRODUCTION

A charge exchange reaction in atomic physics is one in which an electron, or several electrons, are transferred from one atomic system to another. This book is about the theory of charge exchange in two-body ion–atom collisions. Other kinds of charge exchange, for instance those involving molecules, or three-body collisions, will not be discussed; however, the general ideas which will be developed can be extended to cover these more complicated situations, and also to rearrangement collisions involving other particles such as positrons or muons.

Collisions between atoms (or ions) A and B are classified, according to the final products of the reaction, into three broad categories. First there is elastic scattering in which the internal energies of A and B do not change, but the incident particles are deflected through a certain angle:

$$\mathrm{A} + \mathrm{B} \rightarrow \mathrm{A} + \mathrm{B}. \tag{1.1}$$

Then there are inelastic reactions in which either atom A or B, or both, are excited to a different energy level:

$$\begin{aligned} \mathrm{A} + \mathrm{B} &\rightarrow \mathrm{A}^* + \mathrm{B} \\ &\rightarrow \mathrm{A} + \mathrm{B}^* \\ &\rightarrow \mathrm{A}^* + \mathrm{B}^*. \end{aligned} \tag{1.2}$$

Included in this category are those reactions in which one or both atoms are ionized and one or more electrons are ejected, leaving a positive ion, or ions:

$$\begin{aligned} \mathrm{A} + \mathrm{B} &\rightarrow \mathrm{A} + \mathrm{B}^+ + \mathrm{e}^- \\ &\rightarrow \mathrm{A}^+ + \mathrm{B} + \mathrm{e}^- \\ &\rightarrow \cdots. \end{aligned} \tag{1.3}$$

Finally, there are reactions in which one or more electrons are transferred between A and B. There are the charge exchange, or electron capture, reactions with which we are chiefly concerned:

$$\mathrm{A} + \mathrm{B} \rightarrow \mathrm{A}^+ + \mathrm{B}^-. \tag{1.4}$$

As we shall see, although some theoretical models can be developed in which charge exchange is considered by itself, in general, since the processes (1.1)–(1.4) are coupled, we shall be concerned in some degree with all types of collision.

Charge exchange was discovered by Henderson (1923) during the course of experiments in which α-particles were passed through absorbing screens of mica. Numbers of singly charged He^+ ions and also neutral He atoms were observed to emerge from the far side of the screens, and these were attributed to the capture of electrons by the α-particles. This interpretation of the experiments was confirmed, and the measurements extended to electron capture by α-particles from air and hydrogen, by Rutherford (1924), Henderson (1925), and Jacobsen (1926). The first quantitative measurements of cross sections for inelastic ion–atom scattering came with the development of ion accelerators in the 1930s, but systematic measurements of total cross sections over a wide range of energies and processes were not carried out until the 1960s, followed slightly later by measurements of the angular distribution of the scattered particles. Much more recently, new types of experiment have been devised providing more detailed information about the collision mechanism. An example is the detection in coincidence of the photons emitted from an excited state formed by electron capture together with the scattered particles. From the results of such an experiment the relative population of the different magnetic substates of the excited atomic level can be deduced. If the excited level is hydrogenic and hence degenerate, all the elements of the density matrix can be found, giving a complete description of the collision, by measuring the variation of the polarization of the emitted light when an electric field is applied to the excited atoms.

1.1 GENERAL CHARACTERISTICS

To obtain an initial orientation, it is instructive to examine the total cross sections for electron capture by protons from the two lightest atoms, hydrogen and helium. We shall start with a discussion of the $H^+ + H$ system.

$H^+ + H$

In Fig. 1.1, the total cross sections are shown for the inelastic processes

$$H^+ + H(1s) \rightarrow H^+ + H \quad (n = 2, 3) \tag{1.5a}$$

$$\rightarrow H^+ + H + e^- \tag{1.5b}$$

$$\rightarrow H + H^+ \tag{1.5c}$$

as a function of the kinetic energy of the incident proton. Although collision experiments can be performed in various ways, using colliding or merging beams of particles, we shall express collision energies in terms of the energy

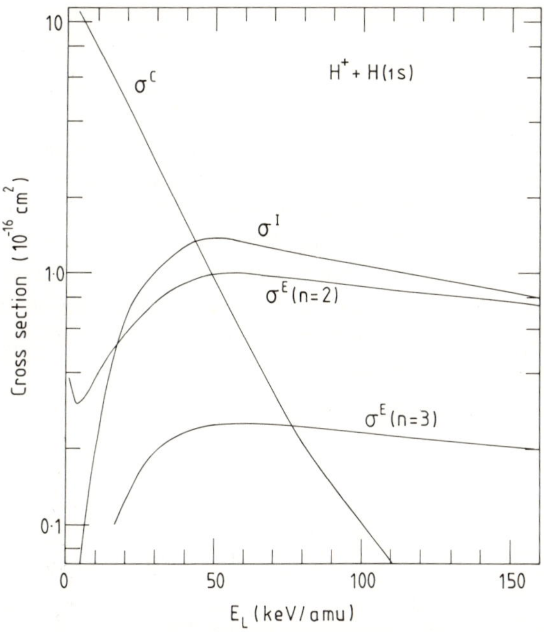

Fig. 1.1. Cross sections for the inelastic scattering of protons by atomic hydrogen in the ground state: σ^{C} is the cross section for capture into all bound states; σ^{I} is the cross section for ionization; σ^{E} ($n = 2, 3$) are the cross sections for excitation of the $n = 2, 3$ levels. The curves are based on the data reviewed by Park (1983).

divided by the mass of a projectile A incident on a stationary target B. This is the *laboratory system*, and the energy of A, expressed in keV divided by its mass expressed in atomic mass units (a.m.u.), will be denoted by E_{L}. If v_0 is the initial relative velocity of A and B *expressed in atomic units* (a.u.) (see Appendix A), then at non-relativistic velocities $E_{\mathrm{L}} = 25v_0^2$ keV/a.m.u.

From Fig. 1.1, it is seen that at low energies charge exchange is the dominant reaction. At small values of v_0, the charge exchange cross section σ^{C} behaves like $\sigma^{\mathrm{C}} = A|B - \log v_0|^2$, where A and B are constants; but as v_0 increases, σ^{C} decreases rapidly. For $v_0 \gg 1$ a.u., the non-relativistic theory shows that σ^{C} is at first proportional to E_{L}^{-6}, and then to $E_{\mathrm{L}}^{-5.5}$. This second asymptotic region is in fact only reached at energies for which relativistic corrections become important. As will be seen in Chapter 11, in the relativistic region at energies above several hundred MeV, the energy variation of the cross section is changed completely, the cross section ultimately becoming proportional to E_{L}^{-1}.

At low velocities ($v_0 \lesssim 1$ a.u.), the principal charge exchange channel in the reaction (1.5c) is one in which the electron is captured into the ground state of atomic hydrogen:

$$H^+ + H(1s) \rightarrow H(1s) + H^+. \tag{1.6}$$

This is a case of *symmetrical resonance* in which the initial and final systems are identical. Only in the case of symmetrical resonance does the charge exchange cross section increase as the velocity decreases; in other cases σ^C decreases to zero as v_0 decreases to zero. In contrast, the excitation and ionization cross sections (1.5a) and (1.5b), which are small at low energies, each increase to a maximum near 50 keV and thereafter decrease, ultimately like $(\log E_L)/E_L$. At high energies, inelastic scattering is dominated by ionization and, to a lesser extent, by discrete excitation.

In reactions such as (1.5), or in the reaction

$$He^+(1s) + He(1s^2) \rightarrow He(1s, nlm) + He^+(n'l') \tag{1.7}$$

where n, l, and m denote the principal, angular momentum, and magnetic quantum numbers, the nuclei of the colliding atoms are identical. In such *homonuclear reactions* there is, at first sight, no experimental distinction between elastic scattering (or excitation) on the one hand, and charge exchange on the other. However, this is not the case, except at very low energies, as the following argument shows. The strength and range of the interaction between a pair of atoms A and B are such that, at all energies greater than a few electron volts, the incident atom is only scattered through a very small angle. It follows that in reaction (1.5c), hydrogen atoms appearing close to the direction of the beam of incident protons (the forward direction) can be considered to be the products of charge exchange, while those appearing at large angles, close to 90° in the laboratory system† are hydrogen atoms recoiling after elastic scattering or excitation. In the centre-of-mass system of coordinates, in which the centre of mass of the colliding particles is at rest, and treating the two protons A and B as distinguishable, as in Fig. 1.2, separate amplitudes $f(\theta)$ for elastic scattering (or excitation) and $g(\bar{\theta})$ for electron capture can be defined. The angles θ and $\bar{\theta}$ are the angles at which H^+ ions and neutral hydrogen atoms, respectively, emerge from the scattering region measured with respect to the direction of incidence. In the centre-of-mass system we have that $\bar{\theta} = \pi - \theta$. Except at very low energies, the amplitude $f(\theta)$ is very small except near $\theta = 0$ and similarly $g(\bar{\theta})$ is very small except near $\bar{\theta} = 0$ (corresponding to $\theta = \pi$).

When A and B are treated as being indistinguishable, the differential cross section for the emergence of H^+ ions at angle θ depends on $|f(\theta) \pm g(\pi - \theta)|^2$, however, since the ranges of θ in which $f(\theta)$ and $g(\pi - \theta)$ are non-zero do

† The transformation between the laboratory and centre-of-mass system of coordinates is discussed in Appendix B. For particles of equal mass the angle of scattering in the centre of mass system θ is twice that in the laboratory system θ_L.

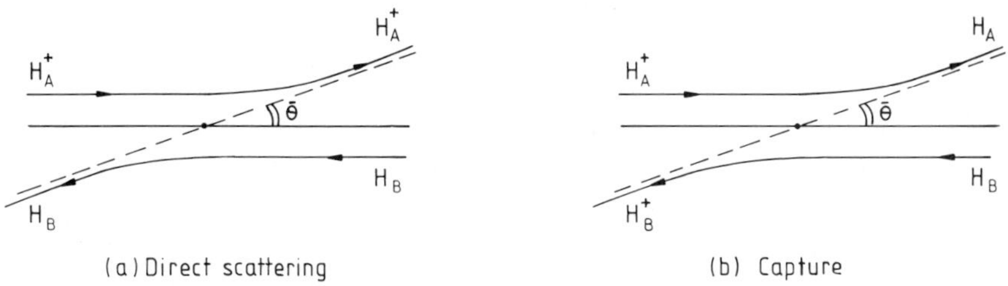

Fig. 1.2. Direct scattering and capture in the centre-of-mass system for the system $H^+ + H$, treating the two protons A and B as distinguishable.

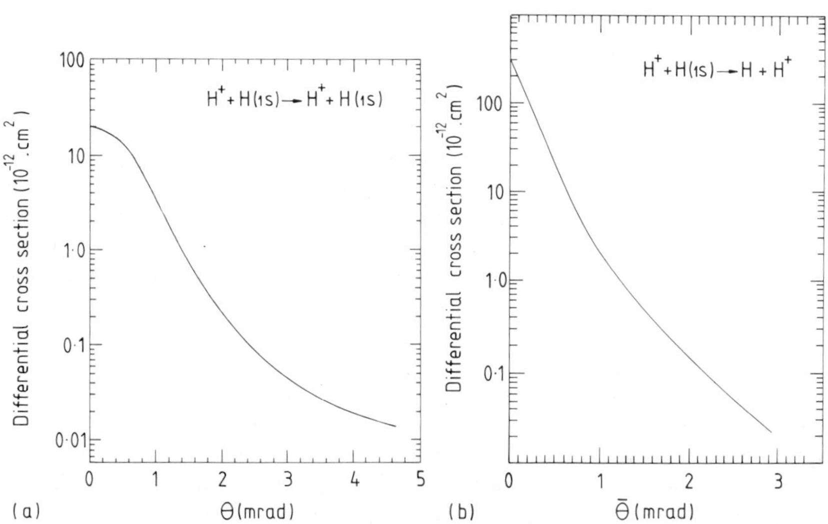

Fig. 1.3. Differential cross sections for the scattering of 60 keV/a.m.u. protons by atomic hydrogen; (a) elastic scattering as a function of θ, the angle of scattering in the centre-of-mass system; (b) charge exchange as a function of $\bar{\theta} = \pi - \theta$. The curves are based on the data of Park (1982) and Martin *et al.* (1981).

not overlap, separate elastic (or excitation) and capture cross sections can be defined as $|f(\theta)|^2$ and $|g(\bar{\theta})|^2$, respectively. For atoms with nuclei that are distinct no ambiguity arises, so that direct and exchange reactions can be separately identified at all energies. The peaking of the cross sections for elastic scattering about $\theta = 0$ and for charge exchange scattering about $\bar{\theta} = 0$ is illustrated in Fig. 1.3 for the interaction of 60 keV protons with atomic hydrogen.

$H^+ + He$

As a second example, the total cross sections for charge exchange, excitation to the 2^1S and 2^1P levels, and ionization, for the *heteronuclear* system

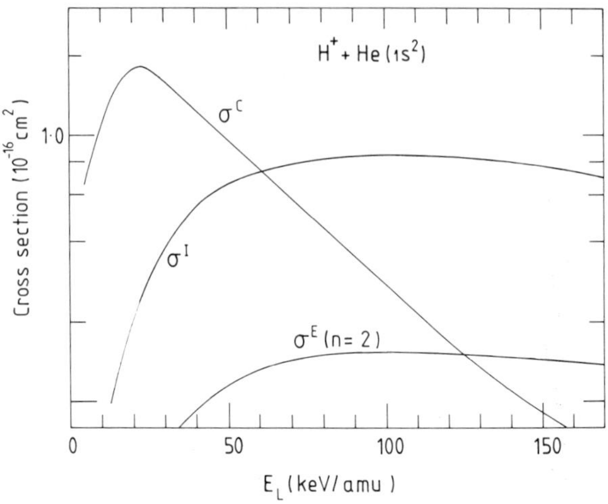

Fig. 1.4. Cross sections for the inelastic scattering of protons by helium in the ground state: σ^{C} is the cross section for capture into all bound states of atomic hydrogen; σ^{I} is the cross section for single ionization; σ^{E} is the cross section for excitation of the $(2^1\mathrm{S} + 2^1\mathrm{P})$ states. σ^{C} and σ^{I} are based on the measurements of Shah *et al.* (1989) and Shah and Gilbody (1985); σ^{E} is based on the measurements of Park and Schowengerdt (1969).

$H^+ + He$ are illustrated in Fig. 1.4. The cross section for charge exchange rises to a broad maximum, near $E_{\mathrm{L}} = 25$ keV/a.m.u., thereafter decreasing rapidly. Characteristically, charge exchange is the most important inelastic reaction at low energies, ionization and direct excitation dominating at higher energies. For this system the ionization cross section exceeds that for charge exchange for energies above ~60 keV/a.m.u. The general shape of the charge exchange cross section as a function of energy shown in Fig. 1.4 is typical of heteronuclear reactions, although the magnitude and position of the maximum in the cross section can vary widely. Kinematically it might be expected that electrons in the target with the same velocity as the incident ion would be captured most easily, and it is often the case that the maximum of the capture cross section occurs at a relative velocity v_0 which is comparable in magnitude, but slightly smaller than, the Bohr velocity v_e of the target electron. It is useful to make distinctions between low-, intermediate-, and high-energy regions according to whether $v_0 \ll v_e$, $v_0 \approx v_e$, or $v_0 \gg v_e$, respectively. In the low-energy region the relative motion of the colliding atoms is very slow compared with the electronic motion and the system can be viewed as a quasi-molecule which is changing its state adiabatically, the perturbation arising from the relative motion of the nuclei. At high, but non-relativistic, energies, the cross section for charge exchange is small (decreasing rapidly like v_0^{-12} or v_0^{-11}) and the interaction between

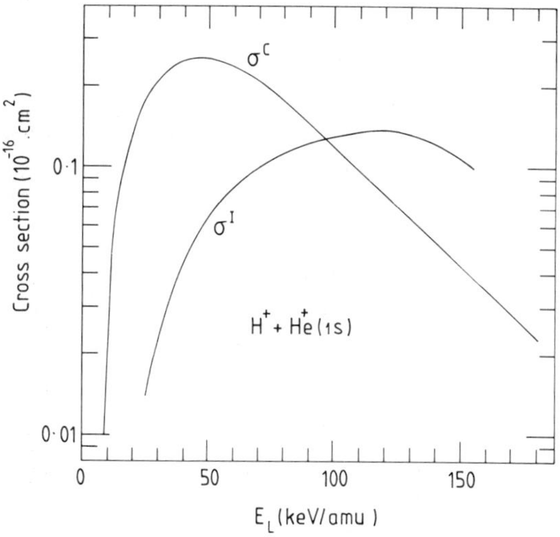

Fig. 1.5. Cross sections for the inelastic scattering of protons by He^+ ions in the ground state: σ^C is the cross section for capture into all bound states of atomic hydrogen; σ^I is the cross section for ionization. σ^C and σ^I are based on the data reviewed by Salzborn (1989).

the atoms can be considered as a small perturbation. In the intermediate-energy region, the cross section for charge exchange is often large and the theoretical simplifications implied by either an adiabatic or a perturbative treatment do not hold.

$H^+ + He^+$

As a third example, the total cross section for electron capture and ionization by H^+ from He^+ is shown in Fig. 1.5. The maximum in the cross section for charge exchange occurs at a higher energy than in the $H^+ + He$ reaction, and it is characteristic that the greater the binding energy of the electron before capture, the greater the energy at which the maximum in the cross section occurs. For example, in electron capture of a valence electron from Li by He^+ ions, the maximum is at $E_L \approx 2.5$ keV/a.m.u., but for capture from the K-shell of Li the maximum is at $E_L \approx 125$ keV/a.m.u.

Another important point to notice is that the magnitude of the electron capture cross section for $H^+ + He^+(1s)$ at the maximum is $\sim 2.5 \times 10^{-17}$ cm^2, considerably less than at the corresponding maximum cross section of $\sim 2 \times 10^{-16}$ cm^2 for capture by protons from neutral helium. In both cases the ground state of the hydrogen atom is the most important final state. The capture reaction in $H^+ + He^+(1s)$ is endothermic:

$$H^+ + He^+(1s) \rightarrow H(1s) + He^{++} + \Delta E, \qquad (1.8)$$

with $\Delta E = -41.5$ eV. The capture reaction in helium is also endothermic:

$$H^+ + He(1s^2) \rightarrow H(1s) + He^+ + \Delta E, \quad (1.9)$$

with $\Delta E = -11$ eV. Generally the greater the change in internal energy $|\Delta E|$, the smaller the cross section, and this is consistent with the cross section for reaction (1.8) being nearly an order of magnitude smaller than that for (1.9).

State selectivity

A feature of charge exchange, which in part arises from the sensitivity of the magnitude of the cross section to the energy defect ΔE, is that many reactions lead selectively to one, or to a small number, of final states. For instance, in the case of the reaction

$$He^{2+} + H(1s) \rightarrow He^+(nl) + H^+, \quad (1.10)$$

at low energies ($E_L \lesssim 15$ keV/a.m.u.) the dominant process is that leading to the $n = 2$ level of He^+. This is a *resonant reaction* with $\Delta E = 0$. For example, at $E_L = 12.5$ keV/a.m.u., capture into the $n = 2$ level accounts for 90 per cent of the total cross section. As the energy increases, final He^+ states with $n > 2$ become relatively more important and the reaction becomes gradually less state selective. In the more general case,

$$A^{Z+} + H(1s) \rightarrow A^{(Z-1)+}(nl) + H^+, \quad (1.11)$$

where A^{Z+} is a fully stripped ion of charge Z, the distribution of final states is also selective, the most important level of the hydrogenic ion $A^{(Z-1)}(nl)$ can be expressed by the semi-empirical formula $n \approx Z^{0.77}$.

1.2 APPLICATIONS

The study of charge exchange is important, not only in order to obtain an understanding of the basic mechanisms of rearrangement collisions, but also because charge exchange is a process which plays a vital role in the formation and decay of both astrophysical and laboratory plasmas. In addition, the state selective nature of charge exchange can be employed in diagnostic systems for laboratory plasmas, in pumping atomic levels which exhibit laser action, and in other applications.

A particular case, of astrophysical interest, concerns the problem of determining the physical state of planetary nebulae. These objects are low-density clouds of hydrogen surrounding a central star. The hydrogen is almost completely ionized under the radiation from the star and forms a plasma of temperature $\sim 10^4$ K, with a typical electron density of 10^{10} particles/m^3. In addition to the ionized hydrogen, there are small concentrations of neutral hydrogen and also of other atomic species such as C, N, O, and Ne in various states of ionization. Spectral lines originating from these ions can be analysed to obtain information about the physical state of a nebula. This is achieved by constructing a computer model taking into account the various reactions between the constituents of the nebula which control parameters such as the degree of ionization of each atomic species. Prominent amongst these reactions are charge exchange processes between the positive ions and neutral hydrogen, and the inverse reactions such as

$$O^{2+} + H \rightleftharpoons O^{+} + H^{+} \tag{1.12}$$

and

$$O^{+} + H \rightleftharpoons O + H^{+}. \tag{1.13}$$

The subject may be pursued by consulting the books edited by Burke *et al.* (1983) and Balian *et al.* (1974).

In thermonuclear research, the aim is to produce a hot plasma containing isotopes of hydrogen at a temperature of $\sim 10^8$ K. In forming a hot plasma, interactions necessarily take place between the plasma and the containing walls which introduce impurity ions of many kinds, particularly of common species such as O, C, and Fe. Although the concentration of impurity ions may be small, charge exchange between the ions and neutral hydrogen is rapid. As a consequence, impurity ions which are initially fully stripped capture electrons into very specific excited levels. For example, fully stripped ions of atomic oxygen, O^{8+}, capture electrons into excited levels of the hydrogen-like ion $O^{7+}(nlm)$ through the reaction:

$$O^{8+} + H(1s) \rightarrow O^{7+}(nlm) + H^{+}, \tag{1.14}$$

and the most probable value of n is $n = 5$. The subsequent line radiation from the excited states of O^{7+}, to which the plasma is transparent, is a major source of energy loss and plays a vital role in determining the maximum temperature that can be reached through any particular heating mechanism. In fact this effective cooling mechanism was responsible for the failure of the early attempts to heat plasmas to a significant temperature.

As an example of the use of charge exchange in diagnostics, we can instance the measurement of the concentration of impurity ions in a tokamak plasma. This can be achieved by injecting a neutral beam of hydrogen atoms into the plasma which interacts with the impurity ions. As mentioned in the last paragraph, a common impurity ion is C^{6+}. These ions interact with neutral hydrogen atoms, according to the charge exchange reaction

$$H(1s) + C^{6+} \rightarrow H^{+} + C^{5+}(nlm). \tag{1.15}$$

Since the C^{5+} ion is formed selectively in the level with $n = 4$, the subsequent radiation can be identified, and from a knowledge of the charge exchange cross section the impurity density can be calculated. Other impurity ions can be detected in a similar way, as can the α-particles produced by nuclear fusion. Further applications in the field of plasma diagnostics, and to neutral beam heating, are discussed in the volumes edited by McDowell and Ferendeci (1980) and by Joachain and Post (1983).

In other developments, it has been proposed to make use of the state selective nature of charge exchange to populate levels which radiate X-rays, and this may form a possible pumping mechanism for an X-ray laser. A particularly case which has been investigated extensively is the reaction

$$He^{2+} + Li(2s) \rightarrow He^{+}(nlm) + Li^{+}, \tag{1.16}$$

for which the dominant final level of the helium ion, at energies below $E_L = 25$ keV/a.m.u., is that with $n = 3$.

1.3 DEVELOPMENT OF THE THEORY OF CHARGE EXCHANGE

Shortly after the discovery of charge exchange in ion–atom collisions by Henderson, a theoretical model of such reactions was constructed by Thomas (1927), based on classical mechanics. It was assumed that the electrons in the target were at rest and were uniformly distributed over the surface of a sphere. Kinematically the reaction requires two successive collisions to take place. In the first the incident ion A of velocity v_0 strikes an electron in the target. Since the ion A is much heavier than the electron it is essentially undeflected. On the other hand it is possible for the electron to acquire the speed v_0 and to make a collision with the target nucleus B in such a way that it is deflected to move in the same direction as A. Those electrons which emerge moving with the same velocity as A and in the same direction are considered to be captured. This mechanism produced a capture cross section decreasing with velocity like v_0^{-11}, which is, in fact, found from quantum theory to be the correct asymptotic velocity variation of the non-relativistic cross section.

The first applications of quantum mechanics to charge exchange were perturbation treatments by Oppenheimer (1928) and by Brinkman and Kramers (1930). The latter authors were the first to make use of the fact that, at all except the very lowest energies, the wavelength associated with the relative motion of the heavy particle is very small compared with the distance over which interaction takes place, with the consequence that the heavy-particle motion follows a classical Newtonian trajectory; only the electronic motion needs to be treated by quantum mechanics. At low energies the charge exchange cross section is large and a perturbation treatment is not adequate. When the relative velocity of the heavy particles is small, the interaction time is long compared with the characteristic times of electrons in bound state orbits. Under these circumstances the electron to be captured is shared between the two ions during the collision, forming a quasi-molecule. As the ions separate, the probabilities of capture by the projectile or of remaining associated with the target ion are comparable. The first theoretical description of this mechanism was given by Massey and Smith (1933) who developed a theory based on the expansion of the electronic wave function in a series of molecular orbitals. In the original form this model contained some serious difficulties, and it was not until some twenty-five years later that Bates and McCarroll (1958) corrected these and presented the theory in its modern form, although extensive application had to await the huge increase in computing power that is now available.

1.4 A SIMPLE MODEL

In general we can see that the mechanism of charge exchange at high impact velocities $v_0 \gg v_e$ is quite different from that at low velocities $v_0 \ll v_e$. In the latter regime, a collision complex is formed and the transition probability for rearrangement is high, while in the former, charge exchange can only take place by a mechanism which imparts the incident velocity v_0 to the captured electron; this kinematical condition is very restrictive and increasingly difficult to satisfy with increasing velocity. Some insight into the transition from a low- to high-energy regime is given by a simple classical model introduced by Bohr and Lindhardt (1954) and elaborated by Knudsen *et al.* (1981).

The Bohr–Lindhardt model

Let us consider a fully stripped ion A interacting with a hydrogenic ion composed of a nucleus B and a single electron. Electron capture is considered to take place if two conditions are satisfied, the first being a condition for the release of the electron from the orbit about B, and the second being concerned with the capture of the released electron by the nucleus A.

The release condition can be derived as follows (Briggs 1985). If Z_A and Z_B are the charges, measured in atomic units,† on the nuclei A and B respectively, then the initial energy of the electron bound to B in an orbit with principal quantum number n, is

$$E_n = -\frac{1}{2}\frac{Z_B^2}{n^2}. \tag{1.17}$$

As the nucleus A approaches the atom $(B + e^-)$, the electron gains an additional electrostatic energy which may be taken to be, on the average, $-Z_A/R$, where R is the internuclear distance between A and B. The total

† Atomic units, as defined in Appendix A, are employed throughout this book, except where stated.

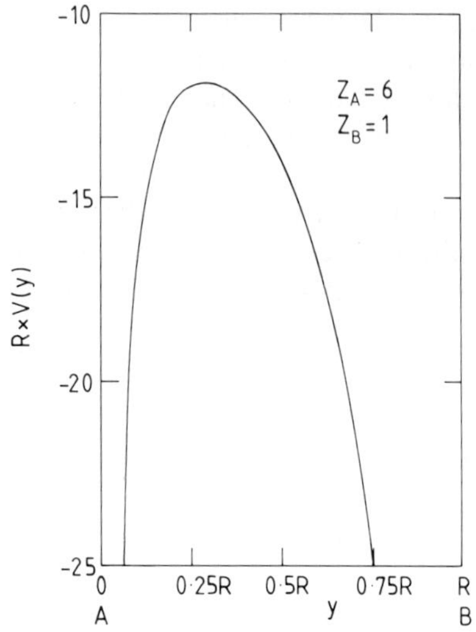

Fig. 1.6. The electrostatic potential $V(y)$ at a point on the internuclear line a distance y from the nucleus B, for the case $Z_B = 1$, $Z_A = 6$.

energy of the electron before capture is then

$$E = -\frac{1}{2}\frac{Z_B^2}{n^2} - \frac{Z_A}{R}. \tag{1.18}$$

At a point at distance y from the target nucleus B, along the internuclear line, the total potential energy acting on the electron becomes

$$V(y) = -\frac{Z_B}{y} - \frac{Z_A}{R-y}. \tag{1.19}$$

This potential (shown in Fig. 1.6) presents a barrier as the position of the electron changes from being in the vicinity of B to being close to A. The value of y at the maximum of V is given by the condition

$$\frac{\partial V(y)}{\partial y} = 0, \tag{1.20}$$

with the solution $y = y_m$ where

$$y_m = \frac{Z_B^{\frac{1}{2}}R}{Z_A^{\frac{1}{2}} + Z_B^{\frac{1}{2}}}. \tag{1.21}$$

At $y = y_m$, the height of the barrier is

$$V(y_m) = -\frac{1}{R}(Z_A^{\frac{1}{2}} + Z_B^{\frac{1}{2}})^2. \tag{1.22}$$

The *release* condition is taken to be that the initial energy of the electron given by (1.18) should be equal to the height of the barrier (1.22). This occurs at an internuclear separation $R = R_1$, where

$$R_1 = \frac{2n^2(Z_B + 2Z_A^{\frac{1}{2}}Z_B^{\frac{1}{2}})}{Z_B^2}. \tag{1.23}$$

For the case $Z_A \gg Z_B$, R_1 is given approximately by the expression

$$R_1 \approx \frac{4n^2 Z_A^{\frac{1}{2}}}{Z_B^{\frac{3}{2}}}. \tag{1.24}$$

The second, or *capture*, condition is derived as follows. With respect to the incident nucleus A the electron possesses a kinetic energy $v_0^2/2$ where v_0 is the relative velocity of A and B. Capture can only take place if this kinetic energy is balanced by the potential energy $-Z_A/R$ which binds the electron to A in the final state. This occurs at $R = R_2$ where

$$-\frac{Z_A}{R_2} + \tfrac{1}{2}v_0^2 = 0, \tag{1.25}$$

that is

$$R_2 = \frac{2Z_A}{v_0^2}. \tag{1.26}$$

$R_2 < R_1$ *(high velocities)*

If $R_2 < R_1$, which happens for sufficiently large values of v_0, the cross section could be taken to be the geometrical cross section πR_2^2. However this is not quite good enough since, when $R_2 < R < R_1$, the electron will be released, but not captured, which corresponds to ionization. To allow for this effect, the cross section πR_2^2 should be weighted by the probability of ionizing the electron before capture takes place. This probability is taken to be equal to the ratio of the collision time (R_2/v_0) to the periodic time of the electron in the initial Bohr orbit (a_n/u_n), where a_n is the radius of the orbit and u_n the velocity of the electron in the orbit. Thus for $R_2 < R_1$ the capture cross section becomes

$$\sigma_1^{C} = \pi R_2^2\left[\frac{R_2}{v_0}\frac{u_n}{a_n}\right] = 8\pi Z_A^3\left[\frac{u_n}{a_n}\right]v_0^{-7}. \tag{1.27}$$

$R_1 < R_2$ *(low velocities)*

Bohr and Lindhardt then argue that if $R_1 < R_2$, which happens at low velocities, capture occurs with 100 per cent probability whenever the nuclei A and B approach to within the distance R_1. The cross section for capture is in this case geometrical, with the value

$$\sigma_2^{\mathrm{C}} = \pi R_1^2$$

$$\approx \frac{16\pi n^4 Z_{\mathrm{A}}}{Z_{\mathrm{B}}^3}. \tag{1.28}$$

The capture cross section can be written in terms of the Bohr radius a_n and the Bohr velocity u_n of the orbit where (see Bransden and Joachain 1983)

$$a_n = n^2/Z_{\mathrm{B}}, \qquad u_n = Z_{\mathrm{B}}/n. \tag{1.29}$$

It is found that

$$\sigma_2^{\mathrm{C}} = 16\pi Z_{\mathrm{A}} \frac{a_n}{u_n^2}. \tag{1.30}$$

The condition $R_1 < R_2$ is satisfied at low incident velocities

$$v_0^2 < \tfrac{1}{2}\left(\frac{Z_{\mathrm{A}}}{Z_{\mathrm{B}}}\right)^{\frac{1}{2}} u_n^2. \tag{1.31}$$

In this region the cross section provided by the model (1.30) is constant, and is proportional to the charge on the incident nucleus A.

The overall picture provided by the model is that a low-velocity region, in which the cross section is independent of velocity, is succeeded by a region in which the cross section decreases rapidly with increasing velocity. This

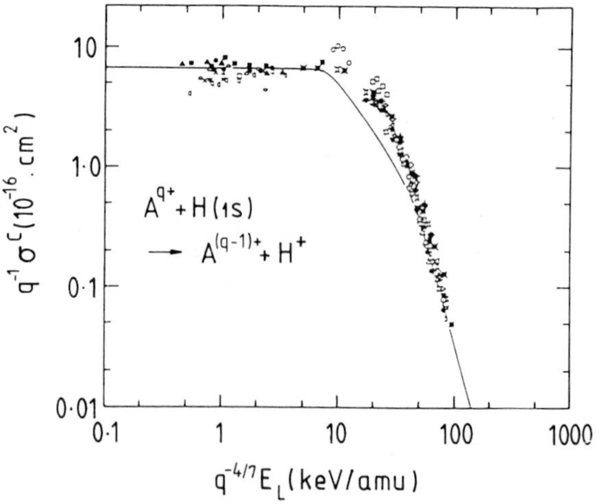

Fig. 1.7. Comparison between the predictions of the Bohr–Lindhardt model (the solid curve) and experimental data for the cross section for capture by ions of overall charge q from H(1s). The data points refer to various fully and partially stripped ions with $4 \leq q \leq 13$ (after Knudsen *et al.* (1981)).

prediction is qualitatively correct, although the model is too simple to hold in detail, and does not provide the correct velocity variation of the cross section at very high energies. In Fig. 1.7, taken from the work of Knudsen *et al.* (1981), the model capture cross section is shown compared with the experimental data relating to partially and fully stripped ions of overall charge q incident on atomic hydrogen in the ground state. The general tendency of the data follows the predictions of the model remarkably well.

Other simple classical models for capture have been developed, for example, the 'overbarrier' model of Grozdanov (1980) and Bárány *et al.* (1985), but we shall not pursue them here. To obtain a detailed and quantitative description of charge exchange, semi-classical or fully quantal methods must be employed to obtain approximate solutions of the Schrödinger equation for ion–atom systems, and an account of these methods is the major topic of this book.

2

ELASTIC SCATTERING OF IONS BY ATOMS

2.1 INTRODUCTION

In this chapter a short account of the theory of ion–atom elastic scattering will be given on the assumption that the interaction between two atoms (or ions) A and B can be represented by a real central energy-independent potential $W(R)$, where $\mathbf{R} = \mathbf{R}_A - \mathbf{R}_B$ is the position vector of A relative to B. This model can be applied to describe the relative motion of A and B approximately even if the collision is inelastic (capture, excitation, or ionization), provided the impact energy is large compared with changes in the internal electronic energies of the atoms. This is usually the case when the impact energy is of the order of 100 eV or greater.

Laboratory and centre-of-mass frames

We shall define the laboratory frame as one in which the target atom B is initially at rest and the projectile A is moving in a direction parallel to the Z-axis with velocity $\mathbf{v}_0$. In the laboratory frame the centre of mass of A and B moves with the constant velocity

$$\mathbf{V} = \frac{M_A \mathbf{v}_0}{M_A + M_B}, \tag{2.1}$$

where M_A and M_B are the masses of the ions A and B, respectively. The dynamics of the collision between A and B is most conveniently expressed in the centre-of-mass frame, the frame in which the centre of mass is at rest. In the centre-of-mass frame the relative motion of A and B can be described by an equivalent one-body system in which a particle of mass μ, where

$$\mu = \frac{M_A M_B}{M_A + M_B}, \tag{2.2}$$

is scattered by the potential $W(R)$. The mass μ is known as the reduced mass of A and B and is derived in Appendix B.

For an elastic collision the angles of scattering θ in the centre-of-mass system and θ_L in the laboratory system are related kinematically by the expression derived in Appendix B:

$$\tan \theta_L = \frac{\sin \theta}{\cos \theta + \tau}, \qquad \tau = \frac{M_A}{M_B}. \tag{2.3}$$

2.2 THE INTERACTION POTENTIAL BETWEEN ATOMIC SYSTEMS

Formal expressions can be obtained for the effective interaction potential between two atoms, starting from the Schrödinger equation for the system. Such *optical potentials* are energy dependent, non-local, and complex. However, provided coupling to the inelastic channels is weak, the interaction in the elastic scattering channel can be represented by a real, local, and energy-independent potential $W(\mathbf{R})$. If the two atoms are in spherically symmetrical ground states, the potential is central, $W(\mathbf{R}) = W(R)$. At low energies an *ab initio* calculation of $W(R)$ can be attempted using the Born–Oppenheimer separation in which the electronic Schrödinger equation is solved (approximately) for fixed values of the internuclear distance R to obtain electronic energy eigenvalues $E_n(R)$. The relative motion of the atomic nuclei is then determined as a potential scattering problem taking $W(R) = E_{el}(R)$, where $E_{el}(R)$ is the particular eigenenergy for the state in which the system separates into two ground-state atoms in the limit $R \to \infty$. The characteristics of the potential energies $E_n(R)$ are discussed further in Chapter 4 and in Appendix D and elementary accounts of calculational methods have been given by McDowell and Coleman (1970) and by Bransden and Joachain (1983). An alternative approach is the development of semi-empirical potentials depending on a number of parameters which can be determined by comparison with experimental scattering data.

The typical behaviour of the interaction potential is shown in Fig. 2.1, for the interaction between two spherically symmetrical atoms in the ground state. At small distances the potential is strongly repulsive, due both to the Coulomb repulsion between the nuclei and to the exchange interaction between the electronic distributions. At large distances, the interaction is attractive and is due to the multipole moments induced in each atom (or ion) by the other. The depth of the potential well formed by the inner repulsion and outer attraction varies greatly between different atomic systems but typically is between 10^{-3} and 5 eV. The position of the minimum of the well also varies, but not so widely, typical values being between 1 and 10 a.u.

The long-range part of the potential is readily calculated from second-order perturbation theory (Bransden and Joachain 1983; McDowell and Coleman 1970). The leading terms are of the form:

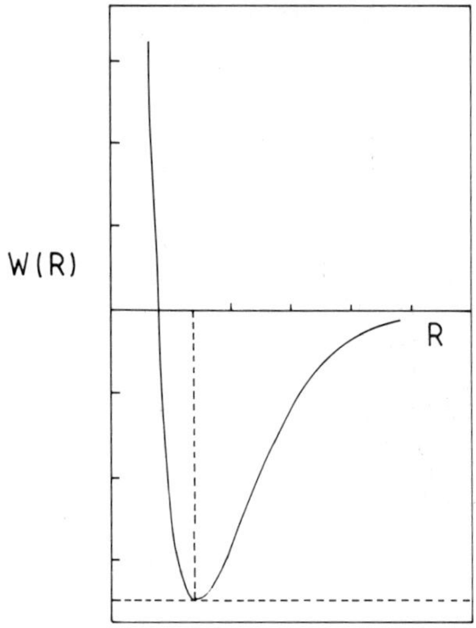

Fig. 2.1. Typical form of the ion–atom potential $W(R)$.

(1) both atoms neutral

$$W(R) \sim -\frac{C_{VW}}{R^6}; \tag{2.4}$$

(2) atom A neutral and atom B ionized with overall charge q_B

$$W(R) \sim -\frac{\alpha_A q_B^2}{R^4}; \tag{2.5}$$

(3) both atoms ionized with charges q_A, q_B

$$W(R) \sim \frac{q_A q_B}{R}. \tag{2.6}$$

Here C_{VW} is the Van der Waals coefficient and α_A is the dipole polarizability of atom A. Both coefficients can be calculated, or estimated, readily.

A semi-empirical form of potential for the interaction between two neutral atoms containing the correct long-range attraction is

$$W(R) = f(R) \exp(-\beta R) - \frac{C_{VW}}{R^6}, \tag{2.7}$$

where β is a constant. The function $f(R)$ is chosen so that, for small R, $W(R) \approx \varepsilon + Z_A Z_B/R$, where ε is the energy of the appropriate united atom state, and $Z_A Z_B/R$ is the Coulomb repulsion between the nuclei.

Another form of potential which has been widely employed is the Lennard–Jones (n, m) potential, defined as

$$W(R) = \frac{m\varepsilon}{n-m}\left[\left(\frac{R_0}{R}\right)^n - \frac{n}{m}\left(\frac{R_0}{R}\right)^m\right], \tag{2.8}$$

where $m = 6$ for two neutral atoms or $m = 4$ for an ion–atom system. A typical value for n is $n = 12$. The parameters ε and R_0 are the depth of the well and the position of the minimum, respectively. Although the Lennard–Jones potential is highly singular at $R = 0$, this is unimportant for the analysis of low-energy scattering.

Yet another widely used potential is the Buckingham (6-exp) potential for two neutral atoms:

$$W(R) = \begin{cases} \infty & R < R_A \\ \dfrac{\varepsilon}{B}\left[\dfrac{6A}{\gamma}\exp\left(-\dfrac{\gamma R}{R_0}\right) - \left(\dfrac{R_0}{R}\right)^6\right] & R > R_A, \end{cases} \tag{2.9}$$

where $A = \exp\gamma$, $B = 1 - 6/\gamma$, and R_A is such that $\max_{R \geq R_A} W(R) = W(R_A)$.

Although these potentials are, perhaps, the most frequently encountered, many others have been proposed and a convenient list has been given by Pauly (1979).

2.3 THE CLASSICAL APPROXIMATION

The quantum partial wave theory of potential scattering† is always applicable, but over a wide energy range it is often possible to describe atom–atom elastic scattering by classical mechanics modified, where necessary, to take into account quantum interference effects. This arises because, except at very low energies, the wavelength λ associated with the motion of two atoms is very small compared with the size of the interaction region. In terms of the incident velocity, we have that (in atomic units)

$$\lambda = \frac{2\pi}{\mu v_0}. \tag{2.10}$$

The interaction between the atoms, $W(R)$, is smoothly varying and only changes slowly over distances of the order a_0 ($a_0 = 1$ a.u.). It follows that if $\lambda < 1$ a localized wave packet can be formed, the centre of which will follow a classical trajectory in the potential $W(R)$. Since the lightest atom is hydrogen, and the mass of a proton is $M_P = 1{,}836$ atomic units, the smallest

† It is assumed that readers are familiar with the elementary theory of potential scattering as described, for example, in the texts of Bransden and Joachain (1983, 1989). More advanced accounts have been given by Bransden (1983) and Joachain (1975).

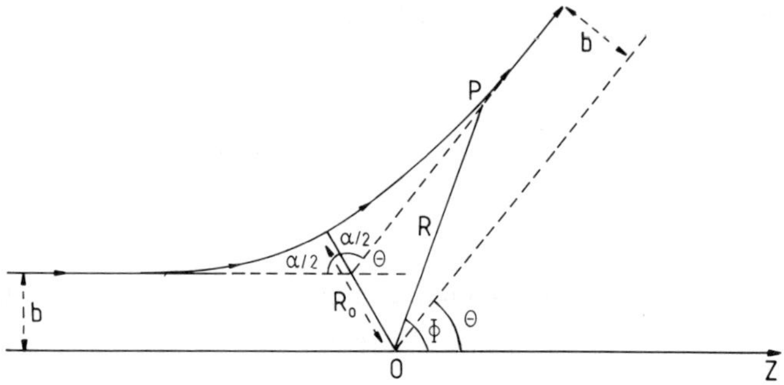

Fig. 2.2. The scattering of a particle in the field of a central potential whose origin is O, for the case of a repulsion.

reduced mass is $\mu = 918$. Hence $\lambda < 1$ for velocities greater than $2\pi/918$ a.u. which corresponds to $v_0 = 4.4 \times 10^6$ m s^{-1}, or an energy of 1.2 eV. Thus classical conditions are expected to apply for a system with reduced mass μ, when the energy $E \gtrsim 1.2\ (M_P/2\mu)$ eV.

A further condition, which must be satisfied if classical mechanics is to hold, is that the uncertainty in the angle through which the scattered atom is deflected should be small compared with the angle of scattering. Consider a particular classical trajectory defined by the initial velocity v_0 and an impact parameter b, where b is the perpendicular distance between the Z-axis and the initial straight line path. This situation is illustrated in Fig. 2.2, where the origin of coordinates O is taken to be the centre of force. Since the potential $W(R)$ is central the angular momentum $\mathbf{L}$, where $L = b\mu v_0$, is conserved and the motion is confined to one plane, taken to be the XZ-plane. Under classical conditions the wave function of the scattered particle is represented by a narrow wave packet of width Δx in the X-direction. The corresponding uncertainty in the x-component of the momentum p_x is, from the Heisenberg uncertainty principle, (in atomic units)

$$\Delta x \Delta p_x \gtrsim 1. \tag{2.11}$$

The uncertainty Δb in the impact parameter b must be of the same order as Δx so that

$$\Delta b \approx 1/\Delta p_x. \tag{2.12}$$

The uncertainty Δp_x in the transverse momentum corresponds to an uncertainty $\Delta\Phi$ in the angular coordinate Φ of the particle, such that

$$\Delta\Phi \approx \frac{\Delta p_x}{p} = \frac{1}{p\Delta b}, \tag{2.13}$$

where $p = \mu v_0$. The angle of scattering θ must also be uncertain by $\Delta\Phi$, and for θ to be well defined the condition

$$\theta \gg \Delta\Phi \tag{2.14}$$

must be satisfied. Since we must also ensure $b \gg \Delta b$, we have from (2.13) and (2.14)

$$\theta \gg \frac{1}{p\Delta b} \gg \frac{1}{pb}. \tag{2.15}$$

It follows that classical conditions will apply provided θ is greater than a critical angle θ_c where

$$\theta_c = 1/L, \tag{2.16}$$

where L is the classical angular momentum, $L = b\mu v_0 = bp$.

Even at thermal velocities ($v_0 \approx 10^5$ cm/s), the important values of L may be as large as 10^3 a.u., corresponding to a critical angle of the order of a few milliradians. The fraction of particles scattered through angles greater than θ_c may be quite small, since almost all elastic scattering is confined to a narrow cone about the forward direction at all energies greater than thermal. However, in actual experiments, measurements are usually confined to the region $\theta > \theta_c$, for which a classical description is well founded. Exceptional cases arise when more than one classical trajectory corresponds to a given angle of scattering. When this happens, a semi-classical theory can be developed to allow for the quantum interference effects that then occur.

2.3.1 The classical trajectory and the deflection function

To determine the classical trajectory in the central potential $W(R)$, plane polar coordinates (R, Φ) can be introduced such that

$$\begin{aligned} R^2 &= X^2 + Z^2, \\ \cos\Phi &= Z/R, \end{aligned} \tag{2.17}$$

as in Fig. 2.2. The component of the velocity of the atom perpendicular to $\mathbf{R}$ is $R\dot{\Phi}$ so that the angular momentum about the centre of force is†

$$L = \mu R^2 \dot{\Phi}, \tag{2.18}$$

and this is constant along a given trajectory. Thus

$$\dot{\Phi} = \frac{L}{\mu R^2} = \frac{v_0 b}{R^2}. \tag{2.19}$$

† The notation $\dot{A} \equiv dA/dt$, where A is a time-dependent quantity, is employed here and elsewhere in this book.

The total energy of the system E is the sum of the kinetic and potential energies and is also constant, being equal to $\mu v_0^2/2$:

$$E = \tfrac{1}{2}\mu v_0^2 = \frac{\mu}{2}(\dot{R}^2 + R^2\dot{\Phi}^2) + W(R)$$

$$= \frac{\mu}{2}\left(\dot{R}^2 + \frac{v_0^2 b^2}{R^2}\right) + W(R). \tag{2.20}$$

This relation provides an expression for $\dot{R}$, the radial velocity,

$$\dot{R} = \pm\left[\frac{2}{\mu}\{E - W(R)\} - \frac{v_0^2 b^2}{R^2}\right]^{\frac{1}{2}}, \tag{2.21}$$

from which an equation for the trajectory can be found:

$$\frac{\mathrm{d}\Phi}{\mathrm{d}R} = \frac{\dot{\Phi}}{\dot{R}} = \pm\frac{v_0 b}{R^2}\left[\frac{2}{\mu}\{E - W(R)\} - \frac{v_0^2 b^2}{R^2}\right]^{-\frac{1}{2}}$$

$$= \pm\frac{b}{R^2}\left[1 - \frac{W(R)}{E} - \frac{b^2}{R^2}\right]^{-\frac{1}{2}}. \tag{2.22}$$

At the point R_0 of closest approach $\dot{R} = 0$, so that, from (2.21), R_0 is the largest positive root of the equation

$$f(R) = 0, \tag{2.23}$$

where

$$f(R) = \left[1 - \frac{W(R)}{E} - \frac{b^2}{R^2}\right]^{\frac{1}{2}}. \tag{2.24}$$

The trajectory is clearly symmetrical about the point of closest approach, so that if 2α is the total change in the angle Φ during the collision, α can be obtained by integrating (2.22) along the trajectory from $R = R_0$ to $R = \infty$. Thus

$$\alpha = b\int_{R_0}^{\infty} [R^2 f(R)]^{-1}\,\mathrm{d}R. \tag{2.25}$$

The *classical deflection function* Θ is defined as

$$\Theta = \pi - 2\alpha \tag{2.26}$$

and is a single-valued function of b and E for a given potential:

$$\Theta = \Theta(b, E). \tag{2.27}$$

The inverse relation on the other hand may not be single valued and several values of b may correspond to a given value of Θ. The deflection angle can be of any magnitude and of either sign.

2.3.2 The cross section

If we consider the scattering of a uniform beam of atoms by the centre of force, the intensity of the scattered beam will be axially symmetrical about the Z-axis (for a central potential) and will depend only on the angle of scattering θ, which is defined so that $0 \le \theta \le \pi$. To obtain θ from the deflection function, we first form the quantity

$$\beta = |\Theta| - 2\pi n, \tag{2.28}$$

where either $n = 0$, or n is an integer chosen so that $0 \le \beta \le 2\pi$. Then

$$\theta = \begin{cases} \beta & \text{if } \beta \le \pi \\ 2\pi - \beta & \text{if } \pi \le \beta \le 2\pi. \end{cases} \tag{2.29}$$

In view of (2.27), atoms of energy E having impact parameters between b and $b + \mathrm{d}b$ are scattered between angles θ and $\theta + \mathrm{d}\theta$. The differential cross section $I(\theta)$ is defined to be the number of atoms scattered in the direction θ into an element of solid angle $\mathrm{d}\Omega = 2\pi \sin\theta \, \mathrm{d}\theta$, per unit time, per unit incident flux. If the incident flux is F, the number of particles scattered per unit time having impact parameters between b and $b + \mathrm{d}b$ is $F(2\pi b \, \mathrm{d}b)$, and if only a single value of b corresponds to the angle of scattering θ, it follows that

$$I(\theta) = \left| \frac{F(2\pi b \, \mathrm{d}b)}{F \, \mathrm{d}\Omega} \right| = \frac{b}{\sin\theta} \left| \frac{\mathrm{d}b}{\mathrm{d}\Theta} \right|. \tag{2.30}$$

If several values of b, say $b_1, b_2, b_3, \ldots$, correspond to the scattering angle θ, the differential cross section is the sum of contributions from each trajectory

$$I(\theta) = \sum_i \frac{b_i}{\sin\theta} \left| \frac{\mathrm{d}b_i}{\mathrm{d}\Theta} \right|. \tag{2.31}$$

In general a monotonic repulsive potential produces a positive deflection function Θ decreasing from $\Theta = \pi$ at $b = 0$ to $\Theta = 0$ in the limit $b \to \infty$, while a monotonic attractive potential leads to a negative deflection function for which $\Theta = 0$ at $b = 0$ and $\Theta \to 0$ as $b \to \infty$ as shown in Fig. 2.3. As we have seen, the characteristic potentials which arise in ion–atom scattering are not monotonic, but possess an inner repulsive region and an outer region of attraction. This type of potential produces a deflection function which changes sign as illustrated schematically in Fig. 2.4. It is seen that, in this particular case, for $\theta > \theta_{\mathrm{m}}$ only one value of b contributes, while for $\theta > \theta_{\mathrm{m}}$ three values contribute. When more than one value of b contributes to a given θ, as we shall see below, a phase of quantum mechanical origin must

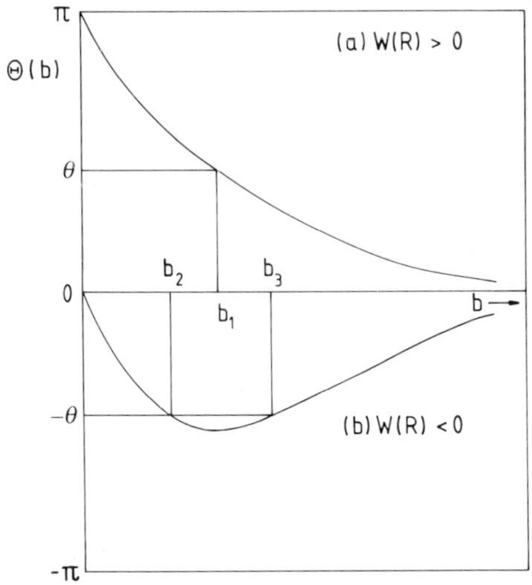

Fig. 2.3. Deflection functions corresponding to scattering by: (a) a monotonic repulsive potential, upper curve; (b) a monotonic attractive potential, lower curve. In the case of a repulsion, only one value of b at $b = b_1$ contributes to the scattering at angle θ, while in the case of an attraction two values of b, at $b = b_2$ and $b = b_3$, contribute to scattering at the angle θ.

be associated with each path and the formula (2.31) in which intensities are added must be replaced by one in which probability amplitudes are summed.

Referring to (2.30) and (2.31), it is seen that the classical scattering cross section may possess singularities at certain values of θ and v_0. A *glory singularity* arises if $\mathrm{d}\theta/\mathrm{d}b \neq 0$ at the points $\theta = 0$ or $\theta = \pi$, because of the factor $1/\sin\theta$. Another case arises if, for some value of b, $b = b_\mathrm{m}$, corresponding to $\theta = \theta_\mathrm{m}$ the classical deflection function has a minimum or maximum (as in the case illustrated in Fig. 2.4). At such a point $|\mathrm{d}\theta/\mathrm{d}b| = |\mathrm{d}\Theta/\mathrm{d}b| = 0$ and $I(\theta)$ becomes singular. This is known as a *rainbow singularity* and the angle θ_m is known as the *rainbow angle*. Finally, with an attractive potential it is possible for $|\Theta|$ to exceed π, corresponding to the scattered particle traversing one or more complete orbits round the centre of force. A particular case arises if the effective potential W_eff,

$$W_\mathrm{eff}(R) = W(R) + \frac{b^2 E}{R^2}, \tag{2.32}$$

possesses a maximum at a certain value of b, $b = \bar{b}$ at $R = R_0$. If this occurs, the equation (2.23) possesses repeated roots at $R = R_0$, and the deflection angle Θ becomes logarithmically infinite as $b \to \bar{b}$. This singularity in $\Theta(b, v_0)$ is called an *orbiting singularity*. The effect on the classical differential cross section has been analysed by Ford and Wheeler (1959). In the vicinity of

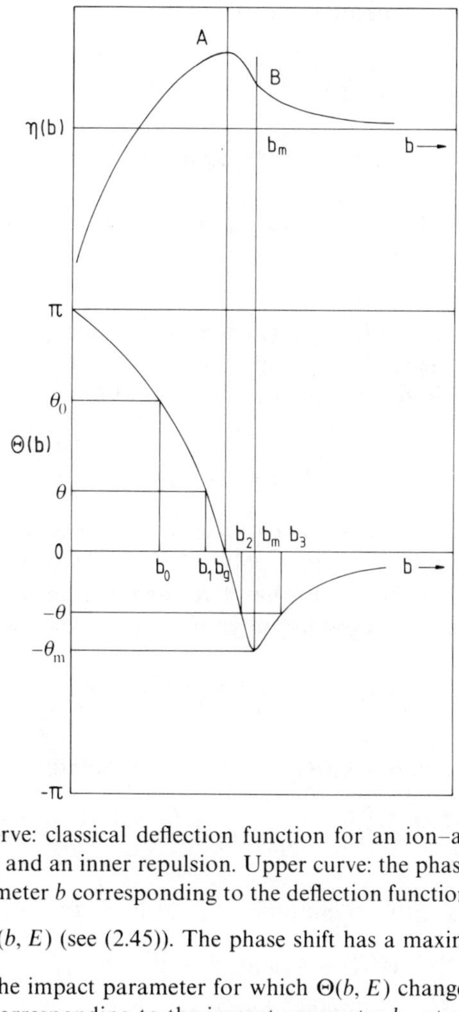

Fig. 2.4. Lower curve: classical deflection function for an ion–atom potential with an outer attraction and an inner repulsion. Upper curve: the phase shift as a function of the impact parameter b corresponding to the deflection function of the lower curve where $\dfrac{\partial \eta(b)}{\partial b} = \dfrac{k}{2}\,\Theta(b, E)$ (see (2.45)). The phase shift has a maximum at the point A corresponding to the impact parameter for which $\Theta(b, E)$ changes sign, and a point of inflection at B corresponding to the impact parameter b_m at which $\Theta(b, E)$ has a minimum.

these singularities a semi-classical rather than a purely classical theory is needed and this will be outlined in the following section.

2.4 THE SEMI-CLASSICAL APPROXIMATION

A full quantum mechanical treatment of elastic scattering from the central potential $W(R)$ starts from the time-independent Schrödinger equation (in the centre-of-mass coordinate system)

$$\left[-\frac{1}{2\mu}\nabla_R^2 + W(R) - E\right]\psi(\mathbf{R}) = 0, \tag{2.33}$$

which has to be solved subject to the boundary condition

$$\psi(\mathbf{R}) \underset{R\to\infty}{\sim} \exp(ikZ) + R^{-1}f(\theta)\exp(ikR), \tag{2.34}$$

where $E = \frac{1}{2}\mu v_0^2$ and $k = \mu v_0$.

The first term in (2.34) represents a plane wave incident parallel to the Z-axis and the second outgoing scattered waves. The amplitude of the scattered wave, $f(\theta)$, determines the differential cross section through the relation

$$I(\theta) = |f(\theta)|^2. \tag{2.35}$$

In the partial-wave treatment of Faxén and Holtsmark (1927), $\psi(\mathbf{R})$ is developed in a series of Legendre polynomials. It is assumed that $W(R)$ is not more singular at the origin that $1/R^2$ and vanishes, as $R \to \infty$, faster than $1/R$. If $P_l(x)$ is the Legendre polynomial of order l, we have

$$\psi(\mathbf{R}) = \sum_{l=0}^{\infty} R^{-1}U_l(R)P_l(\cos\theta), \tag{2.36}$$

where the radial functions $U_l(R)$ are the regular solutions of

$$\frac{d^2U_l(R)}{dR^2} + \left\{k^2 - 2\mu W(R) - \frac{l(l+1)}{R^2}\right\}U_l(R) = 0. \tag{2.37}$$

These solutions have the asymptotic form at large R:

$$U_l(R) \sim C_l \sin(kR - \tfrac{1}{2}l\pi + \eta_l), \tag{2.38}$$

where C_l and η_l are independent of R. The constants η_l are known as phase shifts and determine the scattering amplitude which has the series expansion:

$$f(\theta) = \frac{1}{2ik}\sum_{l=0}^{\infty}(2l+1)(e^{2i\eta_l} - 1)P_l(\cos\theta). \tag{2.39}$$

The expression (2.39) for $f(\theta)$ is always convergent for a central potential satisfying the specified conditions, but the evaluation of the sum is often impractical, or inconvenient, in atom–atom scattering because of the large number of l values of importance. Because of this difficulty, semi-classical methods have been developed which have much of the simplicity of the purely classical approach, but which allow for the quantum interference effects which are produced in scattering by realistic potentials.

In the semi-classical approximation (Ford and Wheeler 1959; Berry and Mount 1972; Child 1974), the leading term is retained in the expansion of the phase shift η_l in powers of $\hbar$, using the WKB (Wentzel–Kramers–Brillouin) method. An effective one-dimensional potential $W_{\text{eff}}(R)$ can be defined such that (cf. (2.32))

$$W_{\text{eff}}(R) \approx W(R) + \frac{(l+\frac{1}{2})^2}{2\mu R^2}, \tag{2.40}$$

where the second term represents the centrifugal barrier. Provided this potential possesses a single classical turning point, the following WKB approximation is obtained for the phase shifts (see Appendix C).

$$\eta_l = \frac{\pi(l+\frac{1}{2})}{2} - kR_0 + \int_{R_0}^{\infty} [\kappa(R') - k]\, \mathrm{d}R', \tag{2.41}$$

where

$$\kappa(R) = \left\{2\mu[E - W(R)] - \frac{(l+\frac{1}{2})^2}{R^2}\right\}^{\frac{1}{2}}. \tag{2.42}$$

For a strong attraction, the variation of W_{eff} with l is shown in Fig. 2.5. It is seen that for larger values of E and l there is a single classical turning point but for smaller values there are three turning points at $R = R_0$, R_1, and R_2. The classical orbiting singularity occurs when the energy E coincides with the maximum in the effective potential where the turning points at R_0

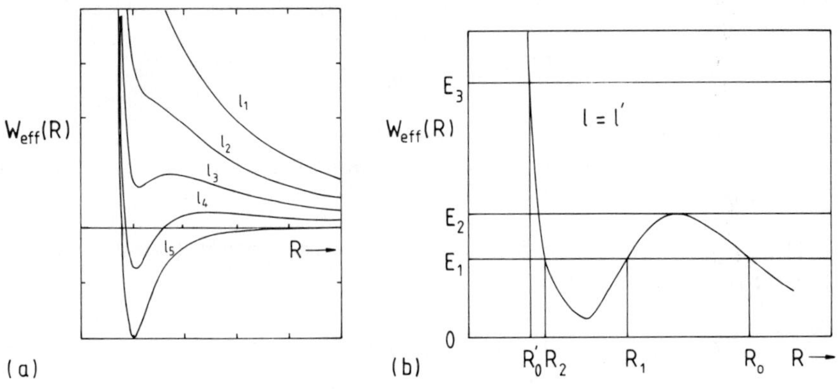

Fig. 2.5. (a) The variation of the effective potential $W_{\text{eff}}(R)$ defined by (2.40) for different values of the angular momentum quantum number l, with $l_1 > l_2 > l_3 > l_4 > l_5$. (b) This is illustrated on a larger scale than (a) and shows the variation of $W_{\text{eff}}(R)$ with R for a given value of $l = l'$, where l' is small enough for $W_{\text{eff}}(R)$ to exhibit a maximum and a minimum. For large energies such as $E = E_3$ there is one classical turning point at $R = R_0'$, but for lower energies such as $E = E_1$ there are three turning points at R_0, R_1, and R_2. For the energy $E = E_2$ which coincides with the maximum in $W_{\text{eff}}(R)$ the turning points R_0 and R_1 coalesce.

and R_1 coalesce (Fig. 2.5(b)). The modification to the semi-classical phase shift for three turning points will be discussed later.

The condition for the expression (2.41) to be a good approximation to the exact phase shift is that the effective potential should not vary appreciably over distances comparable to the de Broglie wavelength, except in the immediate vicinity of the classical turning point. This condition can be expressed as

$$\left|\frac{1}{\kappa(R)}\frac{\mathrm{d}W_{\text{eff}}}{\mathrm{d}R}\right| < 1; \qquad R \gg R_0 \quad \text{and} \quad R \ll R_0. \tag{2.43}$$

In general, under semi-classical conditions, the important values of l are large, in which case l can be related to the classical impact parameter b by setting

$$b = L/k \simeq \frac{l+\frac{1}{2}}{k}, \tag{2.44}$$

where L is the classical angular momentum.

Making this identification and treating l as a continuous variable it is seen that $\kappa(R)$ and $f(R)$, defined by (2.42) and (2.24), respectively, differ only by a constant:

$$\kappa(R) = (2\mu E)^{\frac{1}{2}} f(R),$$

and that, from (2.26), (2.25), and (2.41)

$$\Theta(b, E) = \frac{2}{k}\frac{\partial \eta_l}{\partial b} = 2\frac{\partial \eta_l}{\partial l}. \tag{2.45}$$

This is known as the *classical equivalence relation.*

The development of the semi-classical approximation proceeds by two further steps.

1. The summation in the expression (2.39) for the scattering amplitude $f(\theta)$ is replaced by an integration

$$f(\theta) = \frac{1}{ik}\int_0^\infty (l+\tfrac{1}{2})[\exp(2i\eta_l) - 1]P_l(\cos\theta)\,\mathrm{d}l. \tag{2.46}$$

2. The Legendre polynomials are replaced by the leading term of their asymptotic expansions:

$$P_l(\cos\theta) \simeq \begin{cases} [\tfrac{1}{2}(l+\tfrac{1}{2})\pi\sin\theta]^{-\frac{1}{2}}\sin[(l+\tfrac{1}{2})\theta+\pi/4] & \sin\theta > 1/l \qquad (2.47\text{a}) \\ (\cos\theta)^l J_0[(l+\tfrac{1}{2}(\theta] & \sin\theta \leq 1/l, \qquad (2.47\text{b}) \end{cases}$$

where $J_0(x)$ is the zero-order Bessel function of the first kind. These approximations for $P_l(\cos\theta)$ can be employed for all $l \geq 1$.

Using the approximations (2.47a and b) in (2.46), the integral over l can be evaluated by the stationary phase method or a similar approximation.

2.5 SCATTERING OUTSIDE THE SMALL-ANGLE REGION

We shall first consider scattering at angles for which approximation (2.47a) holds, that is at angles not too close to the forward ($\theta = 0$) or backward ($\theta = \pi$) directions. Using the relationship that for $\theta \neq 0$

$$\sum_{l=0}^{\infty} (2l + 1P_l(\cos\theta) = 0, \tag{2.48}$$

it is seen that the term $[\exp(2i\eta_l) - 1]$ in (2.46) can be replaced by $\exp(2i\eta_l)$. Using (2.47a), we then find:

$$f(\theta) = -\frac{1}{2k}\left(\frac{2}{\pi \sin\theta}\right)^{\frac{1}{2}} \int_0^{\infty} (l + \tfrac{1}{2})^{\frac{1}{2}}[e^{i\phi_+} - e^{i\phi_-}]\, dl, \tag{2.49a}$$

where

$$\phi_{\pm} = 2\eta_l \pm (l + \tfrac{1}{2})\theta \pm \frac{\pi}{4}. \tag{2.49b}$$

For most values of l, the variation of ϕ_+ and ϕ_- with l is fast and the real and imaginary parts of the integrand in (2.49a) oscillate rapidly about zero. The major contributions to the integral therefore come from regions in which either $\partial\phi_+/\partial l$ or $\partial\phi_-/\partial l$ is zero, and these are regions of *stationary phase*. Using the classical equivalence relation (2.45) we see that if $\partial\phi_{\pm}/\partial l = 0$, for $l = \bar{l}$

$$\theta = \mp\Theta(\bar{b}, E), \qquad \text{where } k\bar{b} = (\bar{l} + \tfrac{1}{2}). \tag{2.50}$$

This relation shows that the point of stationary phase, at $l = \bar{l}$, for a given angle of scattering θ, is determined by the classical impact parameter corresponding to the same angle of scattering. The relationships (2.45) and (2.50) are correct in the case that there is only one classical turning point in the effective potential $W_{\text{eff}}(R)$, in which case $|\Theta| < \pi$. If orbiting occurs for which $|\Theta| \geq \pi$, a modification must be made, which will be considered in Section 2.7. If $\Theta(b, E)$ arises from a repulsive trajectory, $\Theta > 0$ and the point of stationary phase is in ϕ_-, while if $\Theta(b, E)$ arises from an attractive trajectory, $\Theta < 0$, and the points of stationary phase are in ϕ_+.

Repulsive potentials

Let us first consider a repulsive potential $W(R) > 0$, for which $\Theta(b, E)$ decreases monotonically from $\Theta = \pi$ at $b = 0$ to zero as $b \to \infty$. There is one point of stationary phase for a given angle of scattering θ at $l = \bar{l}$, that is at $b = \bar{b}$, and expanding ϕ_- about this point we find

$$\phi_-(l) \approx \phi_-(\bar{l}) + \tfrac{1}{2}\phi''_-(\bar{l})(l - \bar{l})^2. \tag{2.51}$$

Since

$$\phi''_-(\bar{l}) = 2\frac{\partial^2 \eta_l}{\partial l}\bigg|_{l=\bar{l}} = \frac{\partial \Theta}{\partial l}\bigg|_{l=\bar{l}} \tag{2.52}$$

we see that ϕ''_- is negative and

$$\phi''_-(\bar{l}) = -\left|\frac{\partial \Theta}{\partial l}\right|_{l=\bar{l}} = -\frac{1}{k}\left|\frac{\mathrm{d}\theta}{\mathrm{d}b}\right|. \tag{2.53}$$

Using (2.53) in (2.49), it is found that

$$f(\theta) = \frac{1}{k}\left(\frac{1}{2\pi \sin\theta}\right)^{\frac{1}{2}} (\bar{l} + \tfrac{1}{2})^{\frac{1}{2}} \exp[\mathrm{i}\phi_-(\bar{l})] \int_{-\infty}^{\infty} \exp\left[-\frac{\mathrm{i}}{2k}\left|\frac{\mathrm{d}\theta}{\mathrm{d}b}\right|(l - \bar{l})^2\right] \mathrm{d}l, \tag{2.54}$$

where, since the only important contribution to the integral comes from the region of $l = \bar{l}$, the slowly varying term $(l + \frac{1}{2})^{\frac{1}{2}}$ has been taken outside the integral, and the lower limit has been extended from $l = 0$ to $l = -\infty$.

Using the standard result

$$\int_{-\infty}^{\infty} \exp(\pm \mathrm{i}\alpha x^2)\,\mathrm{d}x = \left(\frac{\pi}{\alpha}\right)^{\frac{1}{2}} \exp\left(\pm \mathrm{i}\frac{\pi}{4}\right), \qquad \alpha > 0, \tag{2.55}$$

we obtain

$$f(\theta) = \left[\frac{b}{\sin\theta}\left|\frac{\mathrm{d}b}{\mathrm{d}\theta}\right|\right]^{\frac{1}{2}} \exp(\mathrm{i}\gamma)$$
$$= [I^{\mathrm{cl}}(\theta)]^{\frac{1}{2}} \exp(\mathrm{i}\gamma), \tag{2.56}$$

where

$$\gamma = \phi_-(\bar{l}) - \frac{\pi}{4} = 2\eta_{\bar{l}} - (\bar{l} + \tfrac{1}{2})\theta - \frac{\pi}{2}, \tag{2.57}$$

and where $I^{\mathrm{cl}}(\theta)$ is the classical differential cross section (2.30). It follows that for a purely repulsive potential the semi-classical and classical differential cross sections are identical. This result is not correct if only a few low-order phase shifts contribute to the sum (2.39) and requires that the region of stationary phase, of width $\Delta l \approx |\phi''|^{-\frac{1}{2}}$, is large compared with unity. It is clearly necessary that the potential $W(R)$ should vary smoothly with R in order that η_l should also vary smoothly with l.

Inner repulsion – outer attraction

Let us consider the typical form of deflection functions for ion–atom scattering, in the absence of orbiting, shown in Fig. 2.4, corresponding to a potential $W(R)$ with an inner repulsive region and an outer attractive region. It is seen that the most negative value of Θ is equal to $-\theta_m$, where θ_m is the rainbow angle and

$$\text{if} \quad \theta < \theta_m \begin{cases} \phi_+ \text{ has two regions of stationary phase at } b = b_2 \text{ and } b = b_3 \\ \qquad\qquad \text{corresponding to } l = \bar{l}_2 \text{ and } l = \bar{l}_3 \\ \text{and} \\ \phi_- \text{ has one region of stationary phase at } b = b_1 \\ \qquad\qquad \text{corresponding to } l = \bar{l}_1; \end{cases}$$

$$\text{or if } \theta > \theta_m \begin{cases} \phi_+ \text{ has no regions of stationary phase} \\ \text{and} \\ \phi_- \text{ has one region of stationary phase at } b = b_0 \\ \qquad\qquad \text{corresponding to } l = \bar{l}_0. \end{cases}$$

Provided the regions of stationary phase at $b = b_i$ are well separated the scattering amplitude $f(\theta)$ is the sum of contributions $f_i(\theta)$ from each region, so that

$$f(\theta) = \sum_i f_i(\theta). \tag{2.58}$$

The contributions from the three points of stationary phase, for $\theta < \theta_m$, can be evaluated along the same lines as for the repulsive potential. It is found that

$$f_i(\theta) = [I_i^{\text{cl}}(\theta)]^{\frac{1}{2}} \exp(\mathrm{i}\gamma_i), \tag{2.59}$$

where $I_i^{\text{cl}}(\theta)$ is the classical differential cross section for $b = b_i$ and

$$\begin{aligned} \gamma_1 &= \phi_-(\bar{l}_1) = 2\eta_{\bar{l}_1} - (\bar{l}_1 + \tfrac{1}{2})\theta - \frac{\pi}{2}, \\ \gamma_2 &= \phi_+(\bar{l}_2) = 2\eta_{\bar{l}_2} + (\bar{l}_2 + \tfrac{1}{2})\theta - \pi, \\ \gamma_3 &= \phi_+(\bar{l}_3) = 2\eta_{\bar{l}_3} + (\bar{l}_3 + \tfrac{1}{2})\theta - \frac{\pi}{2}. \end{aligned} \tag{2.60}$$

In obtaining (2.60) the fact has been used that $\partial\Theta/\partial l$ is negative at $l = \bar{l}_1$ and $l = \bar{l}_2$ (that is $b = b_1$ and $b = b_2$) but positive at $l = \bar{l}_3$ ($b = b_3$).

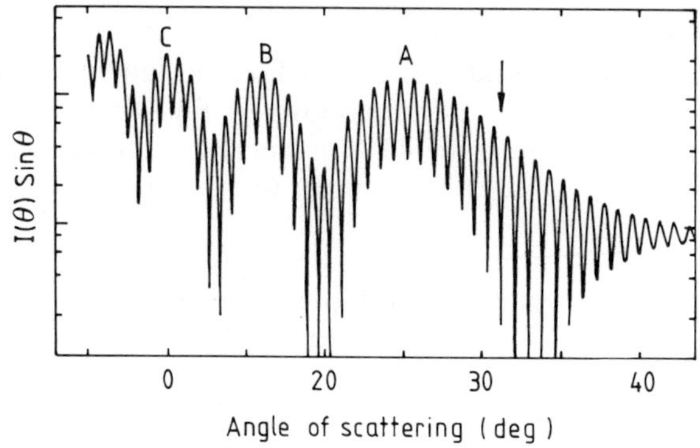

Fig. 2.6. Typical form of the differential cross section, weighted by $\sin\theta$, for a deflection function of the form shown in Fig. 2.4. The primary and supernumerary rainbow maxima are at A, B, C . . . and the rainbow angle at $\theta = \theta_m$ is marked by an arrow. The rapid oscillations are due to interference between the amplitude for rainbow scattering $f_m(\theta)$ with the amplitude $f_0(\theta)$, $\theta > \theta_m$ or $f_1(\theta)$, $\theta < \theta_m$ (see (2.61) and (2.63)).

The differential cross section

$$I(\theta) = \left|\sum_{i=1}^{3} f_i(\theta)\right|^2 \tag{2.61}$$

shows two types of oscillations, as illustrated in Fig. 2.6.

There are slow oscillations in $I(\theta)$ as a function of θ, due to interference between $f_2(\theta)$ and $f_3(\theta)$, controlled by the phase difference $\gamma_3 - \gamma_2 \approx$ constant $+ (\bar{l}_3 - \bar{l}_2)\theta$, so that the angular separation between successive maxima is

$$\Delta\theta = \frac{2\pi}{\bar{l}_3 - \bar{l}_2}. \tag{2.62}$$

The interference of f_1 with f_2 and f_1 with f_3 superimposes further rapid oscillations on this supernumerary rainbow pattern.

The vicinity of the rainbow angle $\theta = \theta_m$ requires special treatment, as the points of stationary phase at $\bar{l}_2$ and $\bar{l}_3$ coalesce and $f_2(\theta)$ and $f_3(\theta)$ cannot be treated separately. We discuss this in the next paragraph. As we shall see the amplitude for rainbow scattering, $f_m(\theta)$, extends into the region $\theta > \theta_m$, so that although there is only one point of stationary phase in that region, at $l = \bar{l}_0$ corresponding to an amplitude $f_0(\theta)$, the differential cross section

$$I(\theta) = |f_0(\theta) + f_m(\theta)|^2 \tag{2.63}$$

also shows interference oscillations, dying out as θ increases.

2.6 RAINBOW SCATTERING

At the point $l = \bar{l}_m$ ($b = b_m$), the classical deflection function is at a minimum and can be expressed as

$$\Theta(l) = \Theta_m + q(l - \bar{l}_m)^2, \qquad q = \frac{1}{2}\frac{d^2\Theta}{dl^2}\bigg|_{l=\bar{l}_m}. \tag{2.64}$$

Using the equivalence relation (2.45) we see that for l close to $\bar{l}_m$

$$\begin{aligned} 2\eta_l &\approx 2\eta_{\bar{l}_m} + \int_{\bar{l}_m}^{l} \Theta(l)\,dl \\ &= 2\eta_{\bar{l}_m} + \Theta_m(l - l_m) + \tfrac{1}{3}q(l - \bar{l}_m)^3, \end{aligned} \tag{2.65}$$

from which it is found that (with $\theta_m = -\Theta_m$)

$$\phi_+(l) = \phi_+(l_m) + (\theta - \theta_m)(l - \bar{l}_m) + \tfrac{1}{3}q(l - \bar{l}_m)^3, \tag{2.66a}$$

where

$$\phi_+(l_m) = 2\eta_{\bar{l}_m} + (\bar{l}_m + \tfrac{1}{2})\theta - \pi/4. \tag{2.66b}$$

Thus from (2.49)

$$f_m(\theta) = -\frac{1}{k}\left[\frac{\bar{l}_m + \frac{1}{2}}{2\pi \sin\theta}\right]^{\frac{1}{2}} e^{i\delta} \int_{-\infty}^{\infty} \exp\{i[(\theta - \theta_m)(l - \bar{l}_m) + \tfrac{1}{3}q(l - \bar{l}_m)^3]\}\,dl, \tag{2.67a}$$

with

$$\delta = \phi_+(\bar{l}_m). \tag{2.67b}$$

The integral in (2.67a) can be expressed in terms of the Airy function Ai(x), where

$$\mathrm{Ai}(x) = \frac{1}{2\pi}\int_{-\infty}^{\infty} \exp(ixt + t^3/3)\,dt. \tag{2.68}$$

This function decays exponentially for $x > 0$ and oscillates for $x < 0$, with the asymptotic forms

$$\mathrm{Ai}(x) \sim \begin{cases} \tfrac{1}{2}\pi^{-\frac{1}{2}}x^{-\frac{1}{4}}\exp(-\tfrac{2}{3}x^{\frac{3}{2}}) & x > 0 \quad (2.69a) \\ \pi^{-\frac{1}{2}}(-x)^{-\frac{1}{4}}\sin\left[\tfrac{2}{3}(-x)^{\frac{3}{2}} + \dfrac{\pi}{4}\right] & x < -1. \quad (2.69b) \end{cases}$$

From (2.67a) and (2.68)

$$f_m(\theta) = \frac{1}{k}\left[\frac{2\pi(\bar{l}_m + \frac{1}{2})}{\sin\theta}\right]^{\frac{1}{2}} e^{i\delta} q^{-\frac{1}{3}} \mathrm{Ai}\left[\frac{\theta - \theta_m}{q^{\frac{1}{3}}}\right]. \tag{2.70}$$

For $\theta > \theta_m$, the interference between $f_m(\theta)$ and $f_0(\theta)$ produces oscillations, the amplitude of which decreases exponentially with increasing angle (see Fig. 2.6). For $\theta < \theta_m$ the amplitude $f_m(\theta)$ oscillates and reproduces the supernumerary rainbows, which can be described alternatively as being due to interference between $f_2(\theta)$ and $f_3(\theta)$. In fact an improved approximation for $f_m(\theta)$ due to Berry (1966) exactly reproduces the result $f_m(\theta) = f_2(\theta) + f_3(\theta)$, for values of θ less than θ_m.

2.7 ORBITING

We now turn to the situation shown in Fig. 2.5(b) for which three classical turning points are associated with the effective potential (2.40). The semi-classical WKB phase shift (described in Appendix C) becomes, apart from a small correction,

$$\bar{\eta}_l = \eta_l + \eta_l^{(r)}, \tag{2.71}$$

where η_l is again given by (2.41), and

$$\tan \eta_l^{(r)} = -\tfrac{1}{4} \exp(-2\gamma) \cot\left(\phi - \frac{\pi}{2}\right). \tag{2.72}$$

The factor $\exp(-2\gamma)$ allows for tunnelling through the barrier from R_0 to R_1, and γ can be expressed as

$$\gamma = \int_{R_1}^{R_0} |k^2 - 2\mu W_{\text{eff}}(R)|^{\frac{1}{2}}\, \mathrm{d}R. \tag{2.73}$$

The parameter ϕ is the change in phase across the well as R changes from R_1 to R_2 and is

$$\phi = \int_{R_2}^{R_1} |k^2 - 2\mu W_{\text{eff}}(R)|^{\frac{1}{2}}\, \mathrm{d}R. \tag{2.74}$$

In the semi-classical approximation the energy levels of the well between R_1 and R_2 are the energies $E = E_n$ at which ϕ obeys the Bohr quantization condition (Bransden and Joachain 1989, p. 396)

$$\phi = (n + \tfrac{1}{2})\pi. \tag{2.75}$$

Near these energies, the phase shift $\eta_l^{(r)}$ has a typical resonance behaviour

$$\tan \delta_l^{(r)} = \frac{\Gamma_n}{2(E_n - E)}, \qquad \Gamma_n = \frac{\exp(-2\gamma E_n)}{2(\partial\phi/\partial E)_{E=E_n}}, \tag{2.76}$$

where Γ_n is the width of the resonance. For atom–atom scattering under semi-classical conditions it can be shown that usually $\Gamma_n \ll 1$ eV and the experimental energy resolution is such that no resonance contributions can be observed in the differential cross section. When the energy coincides with the top of the barrier, the classical orbiting singularity occurs. In contrast to the classical case, the semi-classical deflection function, defined as

$$\Theta(b, E) = 2\frac{\partial \bar{\eta}_l}{\partial l}, \tag{2.77}$$

is not logarithmically infinite and the associated effects in the differential cross section disappear. However, under orbiting conditions $\delta_l^{(r)}$ and Γ_n may be large in which case a correspondingly large contribution to the total cross section of order $4\pi R_0^2$ will arise.

2.8 SMALL-ANGLE SCATTERING

Using the small-angle approximation (2.47b) for $P_l(\cos\theta)$ in (2.46), the scattering amplitude becomes

$$f(\theta) = \frac{1}{\mathrm{i}k}\int_0^\infty (l+\tfrac{1}{2})[\mathrm{e}^{2\mathrm{i}\eta_l} - 1]J_0((l+\tfrac{1}{2})\theta)\,\mathrm{d}l, \tag{2.78}$$

where the approximation $(\cos\theta)^l \approx 1$ has been made. Referring to Fig. 2.4, it is seen that near $\theta = 0$ there are two contributions to the scattering amplitude. The first, and larger, contribution $f^{\mathrm{d}}(\theta)$ arises from large values of angular momentum and represents diffraction scattering from the outer part of the potential. The second, or 'glory' contribution arises from the confluence of the two points of stationary phase at $l = \bar{l}_1$ $(b = \bar{b}_1)$ and $l = \bar{l}_2$ $(b = \bar{b}_2)$ associated with positive and negative values of the deflection function, respectively (see Fig. 2.4).

To calculate $f^{\mathrm{d}}(\theta)$, η_l given by (2.41) is expanded for large $b = (l+\tfrac{1}{2})/k$, with $R_0 \approx b$, to obtain (see Appendix C),

$$\eta_l \approx \eta(b) = -\frac{\mu}{k}\int_b^\infty \frac{RW(R)}{\sqrt{R^2 - b^2}}\,\mathrm{d}R. \tag{2.79}$$

Since the potential, and hence the phase shifts, are small in regions of large b, the approximation $\exp(2\mathrm{i}\eta_l) - 1 \approx 2\mathrm{i}\eta_l$ can be made to obtain from (2.78) the result

$$f^{\mathrm{d}}(\theta) = 2k\int_{b_1}^\infty b\eta(b)J_0(kb\theta)\,\mathrm{d}b, \tag{2.80}$$

where b_1 is chosen to be large enough to ensure that the integral in (2.79) defining $\eta(b)$ originates from the tail of the potential. If, for large R, the potential is of the form

$$W(R) = -c/R^s, \qquad s > 5, \tag{2.81}$$

where c is a constant, it can be shown (McDowell and Coleman (1970)) that

$$\eta(b) = \frac{cf(s)k}{2Eb^{s-1}}, \tag{2.82}$$

where

$$f(s) = \tfrac{1}{2}\sqrt{\pi}\,\frac{\Gamma[\tfrac{1}{2}(s-1)]}{\Gamma(s/2)}. \tag{2.83}$$

Expanding $J_0(kb\theta)$ we can obtain the approximation, accurate for small θ,

$$J_0(kb\theta) \approx 1 - (\tfrac{1}{2}kb\theta)^2 + \cdots. \tag{2.84}$$

It is then found from (2.80) that

$$f^{\mathrm{d}}(\theta) = \tfrac{1}{2}\pi k\left[\frac{b_1^2}{s-3} - \frac{b_1^4}{4(s-5)}\,k^2\theta^2 + \cdots\right]. \tag{2.85}$$

To $f^{\mathrm{d}}(\theta)$ must be added the 'glory' amplitude $f^{\mathrm{g}}(\theta)$ which originates from points of stationary phase at $l = b_1$ and $l = b_2$. At the point $l = l_{\mathrm{g}}$, at which Θ changes sign, the phase shift η_l has a maximum. Expanding about this point,

$$\eta_l = \eta_{\mathrm{g}} + \tfrac{1}{2}\left.\frac{\mathrm{d}^2\eta_l}{\mathrm{d}l^2}\right|_{l=l_{\mathrm{g}}}(l - l_{\mathrm{g}})^2 + \cdots. \tag{2.86}$$

Writing (2.86) in terms of b, and using the classical equivalence relation, we find that

$$\eta(b) = \eta_{\mathrm{g}} + \tfrac{1}{4}\left(\frac{\mathrm{d}\Theta}{\mathrm{d}b}\right)_{b=b_{\mathrm{g}}} k(b - b_{\mathrm{g}})^2 + \cdots. \tag{2.87}$$

Then, from (2.78) and replacing $\exp(2\mathrm{i}\eta_l) - 1$ by $\exp(2\mathrm{i}\eta_l)$ which is permissible except at $\theta = 0$ (or $\theta = \pi$), we have

$$f^{\mathrm{g}}(\theta) = -\mathrm{i}kb_{\mathrm{g}}J_0(kb_{\mathrm{g}}\theta)\int_0^\infty \exp[\mathrm{i}\alpha(b - b_{\mathrm{g}})^2]\exp(2\mathrm{i}\eta_{\mathrm{g}})\,\mathrm{d}b, \tag{2.88}$$

where

$$\alpha = \frac{k}{2}\left(\frac{\mathrm{d}\Theta}{\mathrm{d}b}\right)_{b=b_{\mathrm{g}}}.$$

Since $\alpha < 0$ (see Fig. 2.4), on evaluation of the integral (see (2.56)) we obtain

$$f^{\mathrm{g}}(\theta) = -\mathrm{i}kb_{\mathrm{g}}J_0(kb_{\mathrm{g}}\theta)\,\mathrm{e}^{2\mathrm{i}\eta_{\mathrm{g}}}\sqrt{\frac{\pi}{-\alpha}}\,\mathrm{e}^{-\mathrm{i}\pi/4}. \tag{2.89}$$

The differential cross section near $\theta = 0$ is

$$I(\theta) = |f^{d}(\theta) + f^{g}(\theta)|^2. \tag{2.90}$$

Since both $f^{g}(\theta)$ and $f^{d}(\theta)$ behave like $a + b\theta^2$, we have that

$$I(\theta) \approx a_1 + a_2\theta^2 + \cdots, \tag{2.91}$$

and $I(\theta)$ is finite at $\theta = 0$, in contrast to the classical differential cross section which exhibits the glory singularity.

A high-energy approximation

In general, $f^{d}(\theta) > f^{g}(\theta)$ and at high energies $f^{g}(\theta)$ can be ignored. Since the approximation (2.79) is valid both for large k and large l, a useful high-energy approximation, for small-angle scattering, called the eikonal approximation, is

$$f(\theta) \approx \frac{k}{l}\int_0^\infty b[e^{i\chi(b)} - 1]J_0(kb\theta)\,db, \tag{2.92}$$

where

$$\begin{aligned}\chi(b) &= -2\frac{\mu}{k}\int_b^\infty \frac{RW(R)}{\sqrt{R^2 - b^2}}\,dR\\ &= -\frac{1}{v_0}\int_{-\infty}^{\infty} W(R)\,dZ; \qquad Z \equiv (R^2 - b^2)^{\frac{1}{2}}.\end{aligned} \tag{2.93}$$

This simple approximation is frequently accurate even at large angles and over a wide energy range. Details of applications and a different derivation have been given in a review by Byron and Joachain (1977).

2.9 IDENTICAL PARTICLES

In the theory developed so far, it has been tacitly assumed that we have been dealing with the scattering of one atom by another atom which is distinguishable from the first. If the two atoms are identical and not distinguishable, the expression given earlier must be modified to take this into account. If classical mechanics is employed, then the flux of particles in a given direction in the centre-of-mass system is the sum of both projectile particles scattered through an angle θ and recoil particles scattered through the angle $\pi - \theta$. The differential cross section is thus $\mathcal{I}(\theta)$ where

$$\mathcal{I}(\theta) = I(\theta) + I(\pi - \theta), \tag{2.94}$$

and $I(\theta)$ is the cross section calculated without regard to the symmetry.

In the semi-classical theory or in the full quantum theory, a wave function must be formed of appropriate symmetry. For example, if both atoms are in states of zero internal orbital angular momentum and have zero spin, the wave function is symmetrical under the operation $\mathbf{R} \to -\mathbf{R}$. In this case the scattering amplitude $\mathscr{F}(\theta)$ is

$$\mathscr{F}(\theta) = f(\theta) + f(\pi - \theta), \tag{2.95}$$

where $f(\theta)$ is obtained by ignoring the identity of the particles. The differential cross section

$$I(\theta) = |\mathscr{F}(\theta)|^2 = |f(\theta) + f(\pi - \theta)|^2 \tag{2.96}$$

shows oscillations due to interference between $f(\theta)$ and $f(\pi - \theta)$, in addition to the oscillations considered earlier.

Finally it should be noted that symmetry effects arise when considering the scattering of an ion X^+ by the corresponding neutral atom X, because of the identity of the two nuclei concerned. A discussion has been given by McDowell and Coleman (1970).

2.10 THE INVERSE PROBLEM

Many experiments to determine the elastic scattering of atoms by atoms or of ions by atoms have been carried out, with a view to determining the parameters in some assumed ion–atom potential – the inverse problem. This work lies outside the scope of the present book, but those interested in both the experimental methods and in the determination of potentials can refer to the book by Massey (1971) or the article of Pauly (1979).

3

INELASTIC ION–ATOM COLLISIONS – COUPLED CHANNEL MODELS

In this chapter we shall begin to develop the quantum theory of inelastic ion–atom collisions. In order to discuss the basic concepts without inessential complication, the theory will be developed for a simple system composed of a single electron and two inert singly charged ions A^+ and B^+. This will allow us to discuss charge exchange reactions like $H^+ + H(1s) \to H(nlm) + H^+$ and $H^+ + Li^0 \to H(nlm) + Li^+$ where, in the second example, Li^+ is treated as an inert core interacting with the electron through an effective potential. The restriction to singly charged cores and one-electron systems will be removed later.

In general we shall suppose the ion A^+ is initially incident on a neutral atom composed of the electron and the ionic core B^+, in which case the possible collision channels are:

$$A^+ + (B^+ + e^-) \to A^+ + (B^+ + e^-), \qquad \text{elastic scattering} \qquad (3.1a)$$

$$\to A^+ + (B^+ + e^-)^*, \qquad \text{excitation} \qquad (3.1b)$$

$$\to (A^+ + e^-) + B^+, \qquad \text{charge exchange} \qquad (3.1c)$$

$$\to A^+ + B^+ + e^-, \qquad \text{ionization.} \qquad (3.1d)$$

Although the charge exchange reaction (3.1c) is the topic of primary interest to us in this book, all the channels (3.1a) to (3.1d) are coupled, and in a detailed theory it is usually not possible to calculate the cross section for a particular reaction in isolation.

3.1 THE SCHRÖDINGER EQUATION

The time-independent, non-relativistic, Schrödinger equation for the three-body system A^+, B^+, and e^- is

$$H\Psi = E\Psi, \qquad (3.2)$$

where

$$H = T + V,$$

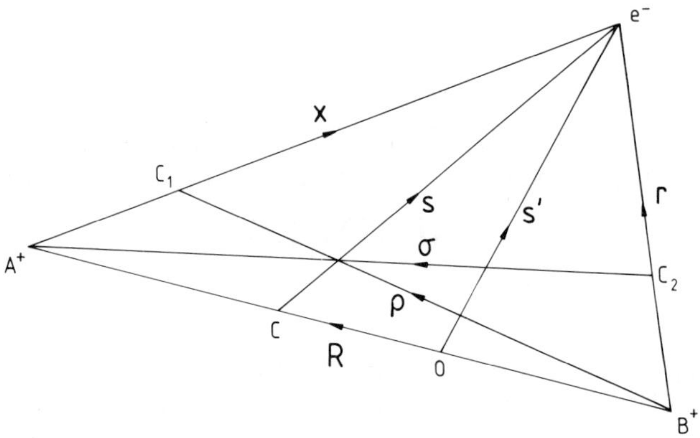

Fig. 3.1. Relative coordinates for a one-electron system. A and B are nuclei, or ionic cores, with masses M_A and M_B, respectively. C is the centre of mass of A and B, C_1 is the centre of mass of A and the electron e, and C_2 is the centre of mass of B and the electron. O is an arbitrary point on the line AB with AO $= pR$ and BO $= qR$ with $(p + q) = 1$.

and T and V are the kinetic and potential energy operators, E being the total energy.

It is convenient to work in the position representation and in the centre-of-mass frame. Two independent position vectors are required to describe the relative motion of the particles in the centre-of-mass frame and these can be chosen in various ways. The kinetic energy operator T takes a simple form in each of three sets of Jacobi coordinates defined as follows and illustrated in Fig. 3.1.

Coordinate set A

In the first set, which is the most appropriate for the description of the initial arrangement $A^+ + (B^+ + e^-)$, the position vectors are $(\mathbf{r}, \boldsymbol{\sigma})$, where $\mathbf{r}$ is the relative position vector of the electron (e^-) with respect to the ion core B^+ and $\boldsymbol{\sigma}$ is the position vector of A^+ with respect to the centre of mass of the subsystem $(B^+ + e^-)$. In this system of coordinates

$$T = -\frac{1}{2m_B}\nabla_r^2 - \frac{1}{2\mu_A}\nabla_\sigma^2, \tag{3.3}$$

where the reduced masses m_B and μ_A are

$$m_B = \frac{M_B}{M_B + 1}, \qquad \mu_A = \frac{M_A(M_B + 1)}{M_A + M_B + 1}. \tag{3.4}$$

In (3.4) M_A and M_B are the masses of the ions A^+ and B^+, while, since we are working in atomic units, the mass of the electron has been set equal to unity.

Coordinate set B

In the second set of coordinates $(\mathbf{x}, \boldsymbol{\rho})$, $\mathbf{x}$ is defined as the position vector of e^- relative to A^+, while $(-\boldsymbol{\rho})$ is the position vector of B^+ relative to the centre of mass of the subsystem $(A^+ + e^-)$. This set of coordinates provides the simplest description of the arrangement $B^+ + (A^+ + e^-)$ in the charge exchange channel. The kinetic energy operator is expressed as

$$T = -\frac{1}{2m_A}\nabla_x^2 - \frac{1}{2\mu_B}\nabla_\rho^2, \tag{3.5}$$

where the reduced masses are

$$m_A = \frac{M_A}{M_A + 1}, \qquad \mu_B = \frac{M_B(M_A + 1)}{M_A + M_B + 1}. \tag{3.6}$$

Coordinate set C

In the third coordinate set $(\mathbf{s}, \mathbf{R})$, which is particularly suitable for describing the pseudomolecule formed when all three particles are interacting at short distances, $\mathbf{s}$ is the position vector of the electron with respect to the centre of mass of A^+ and B^+, while $\mathbf{R}$ is the relative position vector of A^+ with respect to B^+. In this case T is expressed as

$$T = -\frac{1}{2m}\nabla_s^2 - \frac{1}{2\mu}\nabla_R^2, \tag{3.7}$$

where the reduced masses are

$$m = \frac{(M_A + M_B)}{M_A + M_B + 1}, \qquad \mu = \frac{M_A M_B}{M_A + M_B}. \tag{3.8}$$

Coordinate set D

Finally it is useful to introduce a fourth independent set of coordinates $(\mathbf{s}', \mathbf{R})$, where $\mathbf{s}'$ is the position vector of the electron relative to an arbitrary point, O, on the line AB (as shown in Fig. 3.1) and $\mathbf{R}$ is, as before, the relative position vector of A^+ relative to B^+. If the point O is taken so that $OA = pR$ and $OB = qR$ where $p + q = 1$, we have that

$$\mathbf{s}' = \mathbf{s} + \Delta\mathbf{R}, \qquad \text{where } \Delta = \frac{pM_A - qM_B}{M_A + M_B}. \tag{3.9}$$

Using (3.7) and (3.9) the kinetic energy operator can be transformed to

$$T = -\frac{1}{2\mu}\nabla_R^2 - \frac{1}{2m'}\nabla_{s'}^2 - \frac{1}{M}\nabla_R \cdot \nabla_{s'}, \tag{3.10}$$

where

$$m' = \frac{m\mu}{m\Delta^2 + \mu}, \qquad M = \frac{\mu}{\Delta}, \tag{3.11}$$

m and μ being defined by (3.8). In contrast to the three sets of Jacobi coordinates, T is not separable in this system, because of the presence of the cross-term in $\nabla_R \cdot \nabla_{s'}$.

As the mass of the lightest ion (H^+) is 1,836 in atomic units, it is a good approximation to neglect the mass of the electron with respect to the mass of the ion cores in the definitions of the reduced masses. In this approximation, which we shall employ unless otherwise stated, we have

$$m_A = m_B = m = m' = 1; \qquad \mu_A = \mu_B = \mu. \tag{3.12}$$

It should be emphasized that while it is permissible to neglect the mass of the electron in the reduced masses, it is not, in general, allowable to make the 'zero-mass approximation' in which $\boldsymbol{\sigma}$ and $\boldsymbol{\rho}$ are replaced by $\mathbf{R}$, without making a phase correction to the asymptotic wave functions. This will be discussed below.

The potential energy

The potential energy operator is the sum of the interparticle interaction energies:

$$V = V_{Ae}(x) + V_{Be}(r) + V_{AB}(R), \tag{3.13a}$$

where x, r, and R are to be expressed in whichever set of independent variables, (A), (B), (C), or (D), is chosen. In the case of proton scattering by atomic hydrogen, both A^+ and B^+ are protons, and the interactions are Coulombic (spin–orbit coupling being neglected) with

$$V_{Ae} = -\frac{1}{x}, \qquad V_{Be} = -\frac{1}{r}, \qquad V_{AB} = \frac{1}{R}. \tag{3.13b}$$

More generally, V_{Ae}, V_{Be}, and V_{AB} are effective potentials chosen to represent the interaction between the 'active' electron and the ion cores A and B, and the interaction between the cores. It will be assumed that these effective potentials are central. For example, if A^+ and B^+ represent Na^+ and Li^+ ions, V_{Ae} and V_{Be} represent effective central potentials between the electron and Na^+ or Li^+, respectively, and V_{AB} represents an effective Na^+–Li^+ interaction. Some examples of effective potentials are given in Chapter 4. Usually, a specific form is chosen for a potential such as $V_{Ae}(x)$ depending on parameters which are adjusted so that on diagonalizing the Hamiltonian H_A of the atom ($A^+ + e^-$), where $H_A = -\frac{1}{2}\nabla_x^2 + V_{Ae}(x)$, the energy spectrum obtained represents the observed energy spectrum of the atom as accurately as possible. The problems that arise in constructing effective potentials have been discussed by, among others, Bardsley (1974), Daniele (1979), Peach (1982), and Ermolaev (1984, 1985).

3.2 ASYMPTOTIC WAVE FUNCTIONS AND BOUNDARY CONDITIONS

We shall first treat the case in which the ions A^+ and B^+ are singly charged so that the atoms $(A^+ + e^-)$ and $(B^+ + e^-)$ are neutral. At large separations, $\sigma \gg r$, the interaction between the incident ion A^+ and the target atom $(B^+ + e^-)$ decreases faster than σ^{-1} and can be neglected. In the region $\sigma \gg r$, the wave function $\Psi \to \Psi_A$, where the asymptotic wave function Ψ_A satisfies the Schrödinger equation

$$\left[-\frac{1}{2\mu}\nabla_\sigma^2 - \tfrac{1}{2}\nabla_r^2 + V_{\mathrm{Be}}(r) - E\right]\Psi_A = 0, \tag{3.14}$$

where the approximate reduced masses of (3.12) have been employed, and E is the total energy, in the centre-of-mass frame.

The solutions of (3.14) can be written in terms of the wave functions $\phi_j(\mathbf{r})$ of the subsystem $(B^+ + e^-)$ which satisfy the Schrödinger equation

$$[-\tfrac{1}{2}\nabla_r^2 + V_{\mathrm{Be}}(r) - \varepsilon_j]\phi_j(\mathbf{r}) = 0. \tag{3.15}$$

The index j stands collectively for the set of quantum numbers nlm – the principal, angular momentum, and magnetic quantum numbers, and ε_j is the eigenenergy corresponding to the level j. We shall assume a 'box' normalization of the continuum solutions of (3.15), so that all eigenvalues can be labelled by a discrete index. The wave functions ϕ_j can be taken to be orthonormal:

$$\langle\phi_j|\phi_k\rangle \equiv \int \phi_j^*(\mathbf{r})\phi_k(\mathbf{r})\,d\mathbf{r} = \delta_{jk}. \tag{3.16}$$

It is easy to verify that solutions of (3.14) are of the form $\mathscr{F}_j(\boldsymbol{\sigma})\phi_j(\mathbf{r})$ and the general solution is

$$\Psi_A = \sum_j \mathscr{F}_j(\boldsymbol{\sigma})\phi_j(\mathbf{r}), \tag{3.17}$$

where the sum is over the complete set of functions ϕ_j and where $\mathscr{F}_j(\boldsymbol{\sigma})$ satisfies the free-particle equation

$$\left[-\frac{1}{2\mu}\nabla_\sigma^2 - \frac{1}{2\mu}k_j^2\right]\mathscr{F}_j(\boldsymbol{\sigma}) = 0. \tag{3.18}$$

In (3.18) $k_j^2/2\mu$ is the kinetic energy of relative motion in the channel j and

$$\frac{1}{2\mu}k_j^2 \equiv E - \varepsilon_j. \tag{3.19}$$

If the initial state of the target atom is labelled i, the boundary condition to be satisfied by the complete wave function Ψ corresponds to an incident plane wave in channel i together with outgoing spherical waves in all energetically accessible channels (Bransden 1983, McDowell and Coleman 1970). The appropriate solution of (3.18) is, for 'open' channels with $k_j^2 > 0$,

$$\mathscr{F}_j(\boldsymbol{\sigma}) = \delta_{ij}\exp(\mathrm{i}\mathbf{k}_i\cdot\boldsymbol{\sigma}) + \sigma^{-1}f_{ji}(\theta,\phi)\exp(\mathrm{i}k_j\sigma), \tag{3.20a}$$

and for 'closed' channels with $k_j^2 < 0$

$$\mathscr{F}_j(\boldsymbol{\sigma}) \to 0 \qquad \text{as } \sigma \to \infty. \tag{3.20b}$$

In (3.20a), the vector $\mathbf{k}_i$ is of magnitude k_i and is in a direction parallel to the direction of incidence, which we take to be the Z-axis, and (θ, ϕ) are the polar angles of $\boldsymbol{\sigma}$.

The scattering amplitudes $f_{ji}(\theta, \phi)$ define the differential cross sections $(\mathrm{d}\sigma_{ji}^E/\mathrm{d}\Omega)$, in the centre-of-mass frame, for elastic scattering or excitation into an element of solid angle $\mathrm{d}\Omega$, in the direction (θ, ϕ):

$$\frac{\mathrm{d}\sigma_{ji}^E}{\mathrm{d}\Omega} = \frac{k_j}{k_i}|f_{ji}(\theta,\phi)|^2. \tag{3.21}$$

The total cross sections are determined by integration over all solid angles:

$$\sigma_{ji}^E = \frac{k_j}{k_i}\int_{(4\pi)} |f_{ji}(\theta,\phi)|^2\,\mathrm{d}\Omega. \tag{3.22}$$

Exchange scattering

To determine the asymptotic form of the wave function in the charge exchange channels (3.1c), coordinate set B is used. At large separations of the atom $(A^+ + e^-)$ and the residual ion B^+, $\rho \gg x$, and $\Psi \to \Psi_B$, where the asymptotic wave function Ψ_B satisfies the equation

$$\left[-\frac{1}{2\mu}\nabla_\rho^2 - \tfrac{1}{2}\nabla_x^2 + V_{Ae}(x) - E\right]\Psi_B = 0. \tag{3.23}$$

Proceeding as before, the wave function $\chi_n(\mathbf{x})$ and eigenenergies η_n of the atom $(A^+ + e^-)$ are introduced, where n labels all necessary quantum numbers, satisfying

$$[-\tfrac{1}{2}\nabla_x^2 + V_{Ae}(x) - \eta_n]\chi_n(\mathbf{x}) = 0, \tag{3.24}$$

with

$$\langle \chi_n | \chi_m \rangle = \int \chi_n^*(\mathbf{x})\chi_m(\mathbf{x})\, \mathrm{d}r = \delta_{nm}. \tag{3.25}$$

Solutions of (3.23) are of the form $\mathscr{G}_n(\boldsymbol{\rho})\chi_n(\mathbf{x})$ and the general solution is

$$\Psi_\mathrm{B} = \sum_n \mathscr{G}_n(\boldsymbol{\rho})\chi_n(\mathbf{x}), \tag{3.26}$$

where the sum is over the whole set of functions $\chi_n(x)$ and where $\mathscr{G}_n(\boldsymbol{\rho})$ is a solution of the free-particle equation

$$\left[-\frac{1}{2\mu}\nabla_\rho^2 - \frac{1}{2\mu}\kappa_n^2\right]\mathscr{G}_n(\boldsymbol{\rho}) = 0, \tag{3.27}$$

with

$$\frac{1}{2\mu}\kappa_n^2 \equiv (E - \eta_n). \tag{3.28}$$

Since there is no incident plane wave in any rearranged channel, for $\kappa_n^2 > 0$,

$$\mathscr{G}_n(\boldsymbol{\rho}) = \rho^{-1} g_{ni}(\bar{\theta}, \bar{\phi}) \exp \mathrm{i}\kappa_n \rho, \tag{3.29a}$$

and in the closed channels with $\kappa_n^2 < 0$,

$$\mathscr{G}_n(\boldsymbol{\rho}) \to 0 \qquad \text{as } \rho \to \infty, \tag{3.29b}$$

where $(\bar{\theta}, \bar{\phi})$ are the polar angles of $\boldsymbol{\rho}$. The differential cross sections for charge exchange from the incident channel i into the channel n are determined by the scattering amplitudes g_{ni} by

$$\frac{\mathrm{d}\sigma_{ni}^\mathrm{C}}{\mathrm{d}\Omega} = \frac{\kappa_n}{k_i}|g_{ni}(\bar{\theta}, \bar{\phi})|^2. \tag{3.30}$$

The total charge exchange cross section for capture into the final state n is

$$\sigma_{ni}^\mathrm{C} = \frac{\kappa_n}{k_i}\int_{(4\pi)} |g_{ni}(\bar{\theta}, \bar{\phi})|^2\, \mathrm{d}\Omega. \tag{3.31}$$

Overall Coulomb scattering

We now discuss briefly the modifications to the asymptotic forms that arise if the ions A and B are not singly charged, but possess charges Z_A and Z_B, respectively (in atomic units). In the direct channels the overall interaction in the region $\sigma \gg r$ no longer vanishes faster than $1/\sigma$, but becomes equal to the Coulomb interaction $Z_\mathrm{A}(Z_\mathrm{B} - 1)/\sigma$. Correspondingly equation (3.18) is modified to become

$$\left[-\frac{1}{2\mu}\nabla_\sigma^2 + \frac{Z_\mathrm{A}(Z_\mathrm{B} - 1)}{\sigma} - \frac{1}{2\mu}k_j^2\right]\mathscr{F}_j^\mathrm{c}(\boldsymbol{\sigma}) = 0. \tag{3.32}$$

The Coulomb interaction modifies the incident and outgoing waves by logarithmic factors (see Bransden 1983, Chapter 3), and (3.20a) is replaced by

$$\mathscr{F}_j^{c}(\boldsymbol{\sigma}) = \delta_{ij} \exp\left\{ \mathrm{i}\left[\mathbf{k}_i \cdot \boldsymbol{\sigma} + \frac{\alpha}{2k_i} \log(k_i\sigma - \mathbf{k}_i \cdot \boldsymbol{\sigma}) \right] \right\}$$

$$+ [\delta_{ij} f_c(\theta) + f_{ji}(\theta, \phi)]\sigma^{-1} \exp\left\{ \mathrm{i}\left[k_j\sigma - \frac{\alpha}{2k_j} \log(2k_j\sigma) \right] \right\}, \quad (3.33)$$

where $\alpha = 2\mu Z_A(Z_B - 1)$. The amplitude $f_c(\theta)$ is the amplitude for pure Coulomb scattering and is given by

$$f_c(\theta) = -\frac{\alpha}{4k_i^2 \sin^2\theta/2} \frac{\Gamma(1 + \mathrm{i}\alpha/2k_i)}{\Gamma(1 - \mathrm{i}\alpha/2k_i)} \mathrm{e}^{-\mathrm{i}\alpha/2k_i \log(\sin^2\theta/2)}. \quad (3.34)$$

The differential cross sections become

$$\frac{\mathrm{d}\sigma_{ji}^E}{\mathrm{d}\Omega} = \frac{k_j}{k_i} |f_c(\theta)\delta_{ij} + f_{ji}(\theta, \phi)|^2. \quad (3.35)$$

Similarly, in the rearrangement (charge exchange) channel, there is an overall Coulomb potential $Z_B(Z_A - 1)/\rho$ and (3.29) is modified so that

$$\mathscr{G}_n^{c}(\boldsymbol{\rho}) \to \rho^{-1} g_{ni}(\bar{\theta}, \bar{\phi}) \exp \mathrm{i}\left[\kappa_n\rho - \frac{\beta}{2\kappa_n} \log 2\kappa_n\rho \right], \quad (3.36)$$

where $\beta = 2\mu Z_B(Z_A - 1)$. In this case the relationship (3.30) between the amplitude g_{ni} and the differential cross section is unaltered.

3.3 THE VARIATIONAL METHOD AND THE DERIVATION OF COUPLED CHANNEL EQUATIONS

The direct solution of the three-body Schrödinger equation, subject to the boundary conditions discussed in the last section is not yet feasible. However, by taking a suitable form of trial function to represent Ψ, the variational method can be used to obtain coupled integro-differential equations which represent the Schrödinger equation approximately and which are tractable. Trial functions can be taken of the form

$$\Psi = \sum_{j=1}^{N} \Phi_j(\mathbf{r}, \boldsymbol{\sigma})F_j(\boldsymbol{\sigma}) + \sum_{n=N+1}^{M+N} \mathrm{X}_n(\mathbf{x}, \boldsymbol{\rho})G_n(\boldsymbol{\rho}), \quad (3.37)$$

where the functions Φ_j and X_n are linearly independent, and have the property that at large separations of A and B

$$\Phi_j(\mathbf{r}, \boldsymbol{\sigma}) \underset{\sigma \to \infty}{\longrightarrow} \phi_j(\mathbf{r}), \qquad X_n(\mathbf{x}, \boldsymbol{\rho}) \underset{\rho \to \infty}{\longrightarrow} \chi_n(\mathbf{x}). \tag{3.38}$$

At present only the case in which ϕ_j and χ_n represent bound states will be considered. It should be noted that only two position vectors are independent and $(\mathbf{r}, \boldsymbol{\sigma})$, $(\mathbf{x}, \boldsymbol{\rho})$ are to be expressed in any one of the sets of coordinates $(\mathbf{r}, \boldsymbol{\sigma})$, $(\mathbf{x}, \boldsymbol{\rho})$, or $(\mathbf{s}, \mathbf{R})$ as discussed above. The boundary conditions (3.17) and (3.26) are satisfied provided, for large separations of A and B,

$$F_j(\boldsymbol{\sigma}) \underset{\sigma \to \infty}{\sim} \mathscr{F}_j(\boldsymbol{\sigma}), \qquad G_n(\boldsymbol{\rho}) \underset{\rho \to \infty}{\sim} \mathscr{G}_n(\boldsymbol{\rho}), \tag{3.39}$$

where $\mathscr{F}_j$ and $\mathscr{G}_n$ are given by (3.20) and (3.29), respectively.

The choice of the basis functions Φ_j and X_n will be discussed in detail a little later, but without loss of generality, the normalization and orthogonality conditions can be imposed that:

$$\begin{aligned} \int \Phi_j^*(\mathbf{r}, \boldsymbol{\sigma})\Phi_l(\mathbf{r}, \boldsymbol{\sigma})\, d\mathbf{r} &= \delta_{jl}, \qquad \text{(for all } \boldsymbol{\sigma}) \\ \int X_n^*(\mathbf{x}, \boldsymbol{\rho})X_m(\mathbf{x}, \boldsymbol{\rho})\, d\mathbf{x} &= \delta_{nm}, \qquad \text{(for all } \boldsymbol{\rho}). \end{aligned} \tag{3.40}$$

The functions Φ_j and X_n which belong to different arrangements are not, in general, orthogonal, but the overlap integrals,

$$S_{jn}(\boldsymbol{\sigma}) = \int \Phi_j^*(\mathbf{r}, \boldsymbol{\sigma})X_n(\mathbf{x}, \boldsymbol{\rho})\, d\mathbf{r}; \qquad \bar{S}_{nj}(\boldsymbol{\rho}) = \int X_n^*(\mathbf{x}, \rho)\Phi_j(\mathbf{r}, \boldsymbol{\sigma})\, d\mathbf{x}, \tag{3.41}$$

vanish for $\sigma \to \infty$ and $\rho \to \infty$, since the atomic bound state functions ϕ_j and χ_n centred on B and A respectively, do not overlap at large separations.

Because the functions Φ_j and X_n are asymptotic to bound state wave functions, the trial function (3.37) does not allow for ionization, and ways to overcome this will be discussed later.

The variational principle

To find the equations satisfied by the channel functions F_j and G_n for given sets of basis functions Φ_j and X_n we can use the Kohn variational principle, which may be formulated as follows. Consider the variation of the functional I, where

$$I[\Psi^*, \Psi] \equiv \iint \Psi^*(H - E)\Psi\, d\mathbf{r}\, d\boldsymbol{\sigma}, \tag{3.42}$$

under *independent* variations of Ψ^* and Ψ: $\Psi^* \to \Psi^* + \delta\Psi^*$ and $\Psi \to \Psi + \delta\Psi$. Provided the first-order quantities $\delta\Psi^*$ and $\delta\Psi$ vanish in the asymptotic regions $\rho \gg x$, $\rho \gg r$, it is easily shown that

$$\begin{aligned}\delta I &\equiv I[\Psi^* + \delta\Psi^*, \Psi + \delta\Psi] - I[\Psi^*, \Psi] \\ &= 0\end{aligned} \tag{3.43}$$

up to terms of the second order in $\delta\psi$ and $\delta\Psi^*$. (See Bransden (1983) p. 200 for a detailed treatment.)

Expressing Ψ by the trial function (3.37) and considering independent variations in each of the channel functions F_j, G_n and F_j^*, G_n^*

$$\begin{aligned} F_j &\to F_j + \delta F_j, \qquad & F_j^* &\to F_j^* + \delta F_j^* \\ G_n &\to G_n + \delta G_n, \qquad & G_n^* &\to G_n^* + \delta G_n^*, \end{aligned} \tag{3.44}$$

the condition $\delta I = 0$ is seen to be satisfied provided

$$\int \Phi_j^*(\mathbf{r}, \boldsymbol{\sigma})[H - E]\Psi(\mathbf{r}, \boldsymbol{\sigma})\, d\mathbf{r} = 0, \qquad j = 1, 2, \ldots, N,$$

and (3.45)

$$\int \mathrm{X}_n^*(\mathbf{x}, \boldsymbol{\rho})[H - E]\Psi(\mathbf{x}, \boldsymbol{\rho})\, d\mathbf{x} = 0, \qquad n = N+1, N+2, \ldots, N+M.$$

In the first set of equations all quantities are to be expressed in terms of the variables $(\mathbf{r}, \boldsymbol{\sigma})$ and in the second set, in terms of $(\mathbf{x}, \boldsymbol{\rho})$. Equations (3.45) also ensure that $I = 0$, a condition satisfied by the exact wave function.

The coupled equations

Using the orthogonality and normalization conditions expressed by (3.40), the equations (3.45) can be put into the form of coupled integro-differential equations for the channel functions F_j and G_n:

$$\begin{aligned}[\nabla_\sigma^2 + k_j^2]F_j(\boldsymbol{\sigma}) = 2\mu \sum_{l=1}^{N} [V_{jl}(\boldsymbol{\sigma}) + \mathbf{W}_{jl}(\boldsymbol{\sigma})\cdot\nabla_\sigma]F_l(\boldsymbol{\sigma}) \\ + 2\mu \sum_{m=N+1}^{M+N} \int \Phi_j^*(\mathbf{r}, \boldsymbol{\sigma})[H - E]\mathrm{X}_m(\mathbf{x}, \boldsymbol{\rho})\mathbf{G}_m(\boldsymbol{\rho})\, d\mathbf{r}, \\ j = 1, 2, 3, \ldots, N\end{aligned} \tag{3.46a}$$

$$\begin{aligned}[\nabla_\rho^2 + \kappa_n^2]G_n(\boldsymbol{\rho}) = 2\mu \sum_{m=N+1}^{M+N} [\bar{V}_{nm}(\boldsymbol{\rho}) + \overline{\mathbf{W}}_{nm}(\boldsymbol{\rho})\cdot\nabla_\rho]G_m(\boldsymbol{\rho}) \\ + 2\mu \sum_{l=1}^{N} \int \mathrm{X}_n^*(\mathbf{x}, \boldsymbol{\rho})[H - E]\Phi_l(\mathbf{r}, \boldsymbol{\sigma})F_l(\boldsymbol{\sigma})\, d\mathbf{x}, \\ n = N+1, N+2, \ldots, N+M.\end{aligned} \tag{3.46b}$$

The direct potentials V_{jl}, $\mathbf{W}_{jl}$, $\bar{V}_{jm}$, $\overline{\mathbf{W}}_{nm}$ which link channels belonging to the same arrangement are defined by the expressions

$$V_{jl}(\boldsymbol{\sigma}) = \int \Phi_j^*(\mathbf{r}, \boldsymbol{\sigma})[H - \delta_{jl}\varepsilon_j]\Phi_l(\mathbf{r}, \boldsymbol{\sigma})\, d\mathbf{r} \tag{3.47a}$$

$$\mathbf{W}_{jl}(\boldsymbol{\sigma}) = -\frac{1}{\mu}\int \Phi_j^*(\mathbf{r}, \boldsymbol{\sigma})\nabla_\sigma \Phi_l(\mathbf{r}, \boldsymbol{\sigma})\, d\mathbf{r} \tag{3.47b}$$

$$\bar{V}_{nm}(\boldsymbol{\rho}) = \int X_n^*(\mathbf{x}, \boldsymbol{\rho})[H - \delta_{nm}\eta_n]X_m(\mathbf{x}, \boldsymbol{\rho})\, d\mathbf{x} \tag{3.48a}$$

$$\overline{\mathbf{W}}_{nm}(\boldsymbol{\rho}) = -\frac{1}{\mu}\int X_n^*(\mathbf{x}, \boldsymbol{\rho})\nabla_\rho X_m(\mathbf{x}, \boldsymbol{\rho})\, d\mathbf{x}. \tag{3.48b}$$

In (3.46a), the independent variables are $(\mathbf{r}, \boldsymbol{\sigma})$ and in the integral term on the right hand side $\mathbf{x}$ and $\boldsymbol{\rho}$ must be expressed as

$$\begin{aligned} \mathbf{x} &= -\boldsymbol{\sigma} + \frac{M_B}{1 + M_B}\mathbf{r} \approx -\boldsymbol{\sigma} + \mathbf{r} \\ \boldsymbol{\rho} &= \frac{M_A}{M_A + 1}\boldsymbol{\sigma} + \frac{M_A + M_B + 1}{(M_B + 1)(M_B + 1)}\mathbf{r} \approx \boldsymbol{\sigma} + \frac{\mathbf{r}}{\mu}. \end{aligned} \tag{3.49}$$

In the same way, in (3.46b) the independent variables are $(\mathbf{x}, \boldsymbol{\rho})$ and under the integral on the right hand side $\mathbf{r}$ and $\boldsymbol{\sigma}$ must be expressed as

$$\begin{aligned} \mathbf{r} &= \boldsymbol{\rho} + \frac{M_A}{M_A + 1}\mathbf{x} \approx \boldsymbol{\rho} + \mathbf{x} \\ \boldsymbol{\sigma} &= \frac{M_B}{M_B + 1}\boldsymbol{\rho} - \frac{M_A + M_B + 1}{(M_A + 1)(M_B + 1)}\mathbf{x} \approx \boldsymbol{\rho} - \frac{\mathbf{x}}{\mu}. \end{aligned} \tag{3.50}$$

Equations (3.46) are of the same general form found in coupled channel models of other rearrangement collisions, for example the resonating group model of light nuclear reactions, or positronium formation reactions. However in the present case advantage can be taken of the small ratio of the electron to the nuclear mass to reduce the equations to *differential* rather than *integro-differential* form.

Consider first the set of equations (3.46a), and write

$$G_m(\boldsymbol{\rho}) = G_m\left(\boldsymbol{\sigma} + \frac{\mathbf{r}}{\mu}\right) = T\left(\frac{\mathbf{r}}{\mu}\right)G_m(\boldsymbol{\sigma}), \tag{3.51}$$

where $T(\mathbf{a})$ is the translation operator

$$T(\mathbf{a}) = \exp(i\mathbf{p}_\sigma \cdot \mathbf{a}), \qquad \mathbf{p}_\sigma = -i\nabla_\sigma. \tag{3.52}$$

To proceed further some approximation is required for $T(\mathbf{r}/\mu)$. We first notice that in the asymptotic region where σ is large $G_m(\boldsymbol{\sigma}) \to \mathscr{G}_m(\boldsymbol{\sigma})$, where $\mathscr{G}_m(\boldsymbol{\sigma})$ is obtained from (3.29a) by replacing $\boldsymbol{\rho}$ by $\boldsymbol{\sigma}$. We then have

$$\nabla_\sigma G_m(\boldsymbol{\sigma}) \approx \nabla_\sigma G_m(\boldsymbol{\sigma}) \approx \kappa_m \hat{\boldsymbol{\sigma}} G_m(\boldsymbol{\sigma}), \tag{3.53}$$

where $\hat{\boldsymbol{\sigma}}$ is a unit vector in the direction of $\boldsymbol{\sigma}$. For incident energies of the order of a few hundred eV per a.m.u. or greater, corresponding to velocities (in atomic units) $v_0 \geq 0.1$, the scattering amplitude is only large in the direction of incidence, which we shall take to be the Z-axis. It follows that $\mathscr{G}_m(\boldsymbol{\sigma})$ is small except when $\boldsymbol{\sigma}$ is parallel to the Z-axis. Defining $\mathbf{v}_m$ as a vector parallel to the Z-axis of magnitude κ_m/μ, in the asymptotic region we see that

$$G_m(\boldsymbol{\rho}) = T\left(\frac{\mathbf{r}}{\mu}\right) G_m(\boldsymbol{\sigma}) \approx e^{i\mathbf{v}_m\cdot\mathbf{r}} G_m(\boldsymbol{\sigma}). \tag{3.54}$$

A further approximation, accurate for fast collisions in which the wave function retains much of its asymptotic form throughout the collision, is to use (3.54) not just at large values but for all values of σ. Factors such as $\exp(i\mathbf{v}_m\cdot\mathbf{r})$ are called *translation factors* and were originally introduced by Bates and McCarroll (1958). To overcome the difficulty that (3.54) is strictly accurate only in the asymptotic region, more flexible translation factors can be employed of the form:

$$\exp[i\mathbf{v}_m\cdot\mathbf{r} f_m(\mathbf{r}, \boldsymbol{\sigma})], \tag{3.55}$$

where the functions f_m must satisfy the condition

$$f_m(\mathbf{r}, \boldsymbol{\sigma}) \to 1 \qquad \text{for } \sigma \to \infty \quad \text{and} \quad \sigma \gg r, \tag{3.56}$$

and we shall discuss later how such functions can be chosen.

Using the Bates–McCarroll plane-wave translation factors the integral terms on the right-hand side of (3.46a) can be reduced as follows:

$$2\mu \int \Phi_j^*(\mathbf{r}, \boldsymbol{\sigma})(H - E)X_m(\mathbf{x}, \boldsymbol{\rho})G_m(\boldsymbol{\rho})\,d\mathbf{r}$$

$$\approx S_{jm}(\boldsymbol{\sigma})[\nabla_\sigma^2 + \kappa_m^2]G_m(\boldsymbol{\sigma}) + 2\mu K_{jm}(\boldsymbol{\sigma})G_m(\boldsymbol{\sigma}) + 2\mu \mathbf{L}_{jm}(\boldsymbol{\sigma})\cdot\nabla_\sigma G_m(\boldsymbol{\sigma}), \tag{3.57}$$

where $S_{jm}(\boldsymbol{\sigma})$ is an overlap integral

$$S_{jm}(\boldsymbol{\sigma}) = \int \Phi_j^*(\mathbf{r}, \boldsymbol{\sigma})\, e^{i\mathbf{v}_m\cdot\mathbf{r}} X_m(\mathbf{x}, \boldsymbol{\rho})\,d\mathbf{r}, \tag{3.58}$$

and the interaction terms are

$$K_{jm}(\boldsymbol{\sigma}) = \int \Phi_j^*(\mathbf{r}, \boldsymbol{\sigma})\, e^{i\mathbf{v}_m\cdot\mathbf{r}}[H - \tfrac{1}{2}v_m^2 - \eta_m]X_m(\mathbf{x}, \boldsymbol{\rho})\,d\mathbf{r} \tag{3.59a}$$

$$\mathbf{L}_{jm}(\boldsymbol{\sigma}) = -\frac{1}{\mu}\int \Phi_j^*(\mathbf{r}, \boldsymbol{\sigma})\, e^{i\mathbf{v}_m\cdot\mathbf{r}} \nabla_\sigma X_m(\mathbf{x}, \boldsymbol{\rho})\,d\mathbf{r}. \tag{3.59b}$$

The second set of coupled equations, (3.46b), can be treated in the same way by setting

$$F_l(\boldsymbol{\sigma}) = F_l\left(\boldsymbol{\rho} - \frac{\mathbf{x}}{\mu}\right) = T\left(-\frac{\mathbf{x}}{\mu}\right)F_l(\boldsymbol{\rho}). \tag{3.60}$$

Making similar approximations to those of (3.53) and (3.54), we find

$$T\left(-\frac{\mathbf{x}}{\mu}\right)F_l(\boldsymbol{\rho}) \approx \exp(-\mathrm{i}\mathbf{v}_l\cdot\mathbf{x})F_l(\boldsymbol{\rho}), \tag{3.61}$$

where $\mathbf{v}_l$ is parallel to the Z-axis and is of magnitude κ_l/μ. More general translation factors can, of course, be introduced as in (3.55). The integral terms on the right-hand side of (3.46b) can now be reduced to obtain

$$2\mu\int \mathrm{X}_n^*(\mathbf{x}, \boldsymbol{\rho})(H - E)\Phi_l(\mathbf{r}, \boldsymbol{\sigma})F_l(\boldsymbol{\sigma})\,\mathrm{d}\mathbf{x}$$

$$\approx \bar{S}_{nl}(\boldsymbol{\rho})[\nabla_l^2 + k_l^2]F_l(\boldsymbol{\rho}) + 2\mu\bar{K}_{nl}(\boldsymbol{\rho})F_l(\boldsymbol{\rho}) + 2\mu\bar{\mathbf{L}}_{nl}(\boldsymbol{\rho})\cdot\boldsymbol{\nabla}_l F_l(\boldsymbol{\rho}), \tag{3.62}$$

where $\bar{S}_{nl}$, $\bar{K}_{nl}$, and $\bar{\mathbf{L}}_{nl}$ are given by

$$\bar{S}_{nl}(\boldsymbol{\rho}) = \int \mathrm{X}_n^*(\mathbf{x}, \boldsymbol{\rho})\,\mathrm{e}^{-\mathrm{i}\mathbf{v}_l\cdot\mathbf{x}}\Phi_l(\mathbf{r}, \boldsymbol{\sigma})\,\mathrm{d}\mathbf{x} \tag{3.63a}$$

$$\bar{K}_{nl}(\boldsymbol{\rho}) = \int \mathrm{X}_n^*(\mathbf{x}, \boldsymbol{\rho})\,\mathrm{e}^{-\mathrm{i}\mathbf{v}_l\cdot\mathbf{x}}[H - \tfrac{1}{2}v_l^2 - \varepsilon_l]\Phi_l(\mathbf{r}, \boldsymbol{\sigma})\,\mathrm{d}\mathbf{x} \tag{3.63b}$$

$$\bar{\mathbf{L}}_{nl}(\boldsymbol{\rho}) = -\frac{1}{\mu}\int \mathrm{X}_n^*(\mathbf{x}, \boldsymbol{\rho})\,\mathrm{e}^{-\mathrm{i}\mathbf{v}_l\cdot\mathbf{x}}\boldsymbol{\nabla}_l\Phi_l(\mathbf{r}, \boldsymbol{\sigma})\,\mathrm{d}\mathbf{x}. \tag{3.63c}$$

We can now see clearly why the approximation in which the electron mass is neglected completely (so that $\boldsymbol{\sigma} = \boldsymbol{\rho} = \mathbf{R}$) is inadmissible.

1. As the energy increases the oscillating factors $\exp(\mathrm{i}\mathbf{v}_m\cdot\mathbf{r})$, or $\exp(-\mathrm{i}\mathbf{v}_l\cdot\mathbf{x})$, dramatically decrease the importance of the exchange interaction terms compared with the direct interactions, and this does not occur in the zero-electron-mass model.

2. The individual terms such as $\mathrm{X}_m(\mathbf{x}, \mathbf{R})\,G_m(\mathbf{R})$ or $\Phi_j(\mathbf{r}, \mathbf{R})\,F_j(\mathbf{R})$ do not satisfy the asymptotic equation in the region of large R, unless the translational factors are included. This in turn leads to non-zero coupling between states at infinite separations and to difficulties in identifying the cross sections for capture into individual levels. Despite these difficulties, as the impact velocity tends to zero the translation factors become less important and have often been omitted in applications, although it is not clear that this is a valid procedure.

While it is not unreasonable to use the formalism given above, either with plane-wave translation factors, or, better still with the more flexible factors (3.55), a more exact approach has been developed by Delos (1981*a*, *b*), and also by Green (1981*a*, *b*), in the context of somewhat more general models using curvilinear coordinates in place of $\boldsymbol{\sigma}$ and $\boldsymbol{\rho}$. We shall discuss these later.

A low-velocity approximation

Since $\mu^{-1}\mathbf{p}_\sigma G_m(\boldsymbol{\sigma})$ is approximately equal to $\mathbf{v}_m G_m(\boldsymbol{\sigma})$, the translation operator (3.52) can be expanded, when the incident velocity is small, as

$$T\left(\frac{\mathbf{r}}{\mu}\right) \approx (1 + \mu^{-1}\mathbf{r}\cdot\boldsymbol{\nabla}_\sigma). \tag{3.64}$$

If this is done some extra terms in $\boldsymbol{\nabla}_\sigma G_m$ appear in the coupled equations, which cause no difficulty. There is however an apparent complication in that the equations are now of third order, containing $\nabla_\sigma^2[(\mathbf{r}\cdot\boldsymbol{\nabla}_\sigma)G_m]$. Delos (1981*a*, *b*) has shown how this can be overcome, demonstrating that if terms of order v_m^2 are neglected, the third-order terms can be eliminated in favour of one extra term depending on the gradient of $G_m(\boldsymbol{\sigma})$. The transformation of the function $F_l(\boldsymbol{\sigma})$ to $(1 - \mu^{-1}\mathbf{x}\cdot\boldsymbol{\nabla}_\rho)F_l(\boldsymbol{\rho})$ can be dealt with similarly. These expansions of $T(\mathbf{r}/\mu)$ and of $T(-\mathbf{x}/\mu)$ lead to some difficulties in that probability conservation (unitarity) is violated if only the lowest-order terms are retained. This problem does not arise with plane-wave translation factors, or more general factors like (3.55), where f is a real function.

The coupled differential equations

It is convenient to combine the coupled differential equations for F_j and G_n and, using a matrix notation, we write

$$\mathbf{N}(\mathbf{y})(\nabla_y^2 + \mathcal{K}^2)\mathbf{F}(\mathbf{y}) = 2\mu[\mathbf{M}(\mathbf{y})F(\mathbf{y}) + \mathbf{P}(\mathbf{y})\cdot\boldsymbol{\nabla}_y\mathbf{F}(\mathbf{y})], \tag{3.65}$$

where $\mathbf{F}(\mathbf{y})$ is an $(N + M)$-component column vector

$$\mathbf{F} = \begin{vmatrix} F_1 \\ F_2 \\ \vdots \\ F_N \\ G_1 \\ G_2 \\ \vdots \\ G_M \end{vmatrix} \tag{3.66}$$

and the dummy variable $\mathbf{y}$ stands for either of $\boldsymbol{\sigma}$ or $\boldsymbol{\rho}$. If plane-wave translation factors are employed, the generalized overlap matrix N is the $(N + M) \times (N + M)$ matrix

$$\mathbf{N} = \begin{pmatrix} \mathbf{1} & \mathbf{S} \\ \bar{\mathbf{S}} & \mathbf{1} \end{pmatrix}, \tag{3.67}$$

where the submatrices $\mathbf{S}$ and $\bar{\mathbf{S}}$ have elements given by (3.58) and (3.63a). The potential matrices are correspondingly

$$\mathbf{M} = \begin{pmatrix} \mathbf{V} & \mathbf{K} \\ \bar{\mathbf{K}} & \bar{\mathbf{V}} \end{pmatrix} \quad \text{and} \quad \mathbf{P} = \begin{pmatrix} \mathbf{W} & \mathbf{L} \\ \bar{\mathbf{L}} & \bar{\mathbf{W}} \end{pmatrix}, \tag{3.68}$$

again the elements of the submatrices being given by (3.47), (3.48), (3.59), and (3.63), and finally the matrix $\mathscr{K}^2$ is the diagonal matrix with elements K_j^2, where $K_j^2 = k_j^2$ for $j \leq N$ and $K_j^2 = \kappa_m^2$ where $j = m + N$, $m \leq M$.

The left-hand side of (3.65) can be diagonalized by inverting the overlap matrix, to obtain

$$(\nabla_y^2 + \mathscr{K}^2)\mathbf{F}(\mathbf{y}) = 2\mu\mathbf{N}^{-1}(\mathbf{y})[\mathbf{M}(\mathbf{y})\mathbf{F}(\mathbf{y}) + \mathbf{P}(\mathbf{y})\cdot\nabla_y\mathbf{F}(\mathbf{y})]. \tag{3.69}$$

Equations (3.65) or (3.69) remain too complicated for direct numerical solution, and the next step is either to make a partial-wave expansion to obtain radial equations in a single variable, or to make a semi-classical approximation. Both of these approaches will be outlined in due course, but we next turn to the question of obtaining similar coupled equations in the (**s**, **R**) set of coordinates – set C introduced in Section 3.1.

3.4 COUPLED EQUATIONS USING THE MOLECULAR COORDINATE SYSTEM (s, R)

Basis functions

Among the possible choices of basis functions Φ_j and X_n which can be employed in the trial function (3.37) are those of purely atomic form centred on A and B. That is, we take a set

$$\begin{aligned} \Phi_j(\mathbf{r}, \boldsymbol{\sigma}) &= \phi_j(\mathbf{r}), \qquad j = 1, 2, \ldots, N \\ \mathrm{X}_n(\mathbf{x}, \boldsymbol{\rho}) &= \chi_n(\mathbf{x}), \qquad n = 1, 2, \ldots, M, \end{aligned} \tag{3.70}$$

thus satisfying the asymptotic conditions (3.38) trivially. This choice causes no difficulty, and is appropriate for higher energies at which the electron can be associated with one or other centres A and B through the collision. In contrast, at low incident velocities for which $v_0 \ll v_e$ where v_e is the Bohr velocity of the bound target electron, the electron is expected to be shared between A and B for much of the duration of the collision. In the extreme limit $v_0 \to 0$, the Born–Oppenheimer separation becomes accurate and the electronic wave function becomes equal to the adiabatic molecular orbital $\psi_n(\mathbf{s}, \mathbf{R})$ satisfying the electronic Schrödinger equation

$$[-\tfrac{1}{2}\nabla_s^2 + V - E_n(R)]\psi_n(\mathbf{s}, \mathbf{R}) = 0. \tag{3.71}$$

In this equation the interionic distance $\mathbf{R}$ enters parametrically, and the eigenfunctions ψ_n and eigenvalues E_n are obtained by solving (3.71) for each value of R.

With this in mind it seems reasonable to set

$$\Phi_j(\mathbf{r}, \boldsymbol{\sigma}) = \psi_j^{\mathrm{B}}(\mathbf{s}, \mathbf{R}), \qquad j = 1, 2, \ldots, N \tag{3.72a}$$

$$\mathrm{X}_n(\mathbf{x}, \boldsymbol{\rho}) = \psi_n^{\mathrm{A}}(\mathbf{s}, \mathbf{R}), \qquad n = N+1, N+2, \ldots, N+M, \tag{3.72b}$$

where the functions $\psi_j^{\mathrm{B}}(\mathbf{s}, \mathbf{R})$ on the right-hand side of (3.72a) are to be expressed in terms of $(\mathbf{r}, \boldsymbol{\sigma})$, and the functions $\psi_n^{\mathrm{A}}(\mathbf{s}, \mathbf{R})$ on the right-hand side of (3.72b) are to be expressed in terms of $(\mathbf{x}, \boldsymbol{\rho})$. The functions ψ_j^{B} are those molecular orbitals (or combination of orbitals) which are asymptotic to atomic orbitals centred on B, and ψ_n^{A} are those asymptotic to atomic orbitals centred on A:

$$\begin{aligned} \psi_j^{\mathrm{B}}(\mathbf{s}, \mathbf{R}) &\underset{R\to\infty}{\sim} \phi_j(\mathbf{r}) \\ \psi_n^{\mathrm{A}}(\mathbf{s}, \mathbf{R}) &\underset{R\to\infty}{\simeq} \chi_n(\mathbf{x}). \end{aligned} \tag{3.73}$$

Although the choice of molecular orbitals which are functions of $(\mathbf{s}, \mathbf{R})$ as the basis functions seems very natural, and indeed this choice has been extensively employed, it does not fit very handily into the natural coordinates for the asymptotic states, which are the Jacobi coordinates $(\mathbf{r}, \boldsymbol{\sigma})$ or $(\mathbf{x}, \boldsymbol{\rho})$. One way out of this difficulty is to introduce new basis functions $\psi_n^{\mathrm{A,B}}$ by mapping the molecular orbitals, which are functions of $\mathbf{s}$ at constant values of $\mathbf{R}$, on to functions of $\mathbf{s}$ at constant values of $\boldsymbol{\sigma}$ or of $\boldsymbol{\rho}$. In this approach, discussed in detail by Delos (1981*a*, *b*), we set

$$\begin{aligned} \Phi_j(\mathbf{r}, \boldsymbol{\sigma}) &= \psi_j^{\mathrm{B}}(\mathbf{s}, \boldsymbol{\sigma}) \\ \mathrm{X}_n(\mathbf{x}, \boldsymbol{\rho}) &= \psi_n^{\mathrm{A}}(\mathbf{s}, \boldsymbol{\rho}), \end{aligned} \tag{3.74}$$

where $\psi_j^{\mathrm{B}}(\mathbf{s}, \boldsymbol{\sigma})$ is obtained by replacing $\mathbf{R}$ in $\psi_j^{\mathrm{B}}(\mathbf{s}, \mathbf{R})$ by $\boldsymbol{\sigma}$, and $\psi_n^{\mathrm{A}}(\mathbf{s}, \boldsymbol{\rho})$ is obtained by replacing $\mathbf{R}$ in $\psi_n^{\mathrm{A}}(\mathbf{s}, \mathbf{R})$ by $\boldsymbol{\rho}$. (The difference in the electron coordinates between $\mathbf{s}$ and $\mathbf{r}$ or $\mathbf{x}$ causes no difficulty since, for constant $\boldsymbol{\sigma}$ or $\boldsymbol{\rho}$, derivatives with respect to $\mathbf{r}$ or $\mathbf{x}$ can be expressed as derivatives with respect to $\mathbf{s}$.) A generalization of this approach has also been discussed by Delos (1981*a*, *b*), following earlier work by Mittleman (1969), Thorson and Delos (1978*a*, *b*), and Green (1981*a*, *b*). Rather than starting with the trial function (3.37), equations are developed for trial functions of the form

$$\Psi = \sum_{j=1}^{N} F_j(\xi_j)\psi_j^{B}(\mathbf{s}, \xi_j) + \sum_{n=N+1}^{N+M} G_n(\xi_n)\psi_n^{A}(\mathbf{s}, \xi_n), \tag{3.75}$$

where ξ_j and ξ_n are generalized *reaction coordinates* which become equal to $\boldsymbol{\sigma}$ in the asymptotic region of large R and $r \ll R$ (for which the electron is attached to the target B), and which become equal to $\boldsymbol{\rho}$ in the asymptotic region of large R and $x \ll R$ (for which the electron is attached to the projectile A). A suitable form for the coordinates ξ_j and ξ_n is

$$\xi_j = \mathbf{R} - \frac{\mathbf{w}_j}{\mu}, \tag{3.76a}$$

where

$$\mathbf{w}_j = \tfrac{1}{2}(f_j(\mathbf{s}, \mathbf{R}) + \lambda)\mathbf{s}_g + \tfrac{1}{8}(1 - \lambda^2)\mathbf{R}. \tag{3.76b}$$

The parameter λ and the vector $\mathbf{s}_g$ are defined as

$$\lambda = \frac{M_A - M_B}{M_A + M_B}, \tag{3.76c}$$

and

$$\mathbf{s}_g = \tfrac{1}{2}(\mathbf{x} + \mathbf{r}).$$

The vector $\mathbf{s}_g$ represents the position of the electron relative to the mid-point of AB. The functions $f_j(\mathbf{s}, \mathbf{R})$ are defined so that

$$f_j(\mathbf{s}, \mathbf{R}) \to \begin{cases} 1 & \text{if } R \to \infty \quad \text{and} \quad r \ll x \\ -1 & \text{if } R \to \infty \quad \text{and} \quad x \ll r \\ 0 & \text{if } R \to 0. \end{cases} \tag{3.77}$$

These functions, introduced originally by Schneiderman and Russek (1969) and called *switching functions*, can be chosen either on the grounds of physical reasonableness or optimized by the variational method. If translations akin to (3.51) and (3.60) are performed to introduce a common reaction coordinate, the coupled integro-differential equations can be converted to coupled differential equations, and in this case generalized translation factors similar to (3.55) arise naturally. This will be shown in Section 4.6.

A simplied approach

In the reaction coordinate approach the transition from integro-differential equations to differential equations, made possible because of the small ratio of the electronic to the ionic mass, m/M, is made at a late stage. A simpler approach, which was the one originally followed by Bates and McCarroll (1958), in turn patterned on the perturbed stationary state model due to

Massey and Smith (1933), is to work from the outset in terms of the molecular coordinate system $(\mathbf{s}, \mathbf{R})$, which we have labelled in Section 3.1 as coordinate set C.

Since the asymptotic forms (3.17) and (3.26) together with the boundary conditions (3.20) and (3.29) must be satisfied in order to identify the particular solutions of the Schrödinger equation which are required, the first task is to express the boundary conditions in terms of $(\mathbf{s}, \mathbf{R})$. Introducing the translation operator $T(\mathbf{a})$ as in the last section, and writing to a sufficient approximation $\boldsymbol{\sigma} \approx \mathbf{R} - \mathbf{s}/M_{\mathrm{B}}$, we have

$$\Psi_{\mathrm{A}} = \sum_j \phi_j(\mathbf{r})\mathscr{F}_j(\boldsymbol{\sigma}) \tag{3.78a}$$

$$= \sum_j \phi_j(\mathbf{r})T\left(-\frac{\mathbf{s}}{M_{\mathrm{B}}}\right)\mathscr{F}_j(\mathbf{R}) \tag{3.78b}$$

$$\simeq \sum_j \phi_j(\mathbf{r}) \exp(-\mathrm{i}\mathbf{v}_j^{\mathrm{c}}\cdot\mathbf{s})\mathscr{F}_j(\mathbf{R}), \tag{3.78c}$$

where in the last line the plane-wave approximation discussed in the previous section (see 3.54 and 3.61) has been introduced. The vector $\mathbf{v}_j^{\mathrm{c}}$ is parallel to the Z-axis and

$$v_j^{\mathrm{c}} \equiv \frac{k_j}{M_{\mathrm{B}}}. \tag{3.79a}$$

The vector $-\mathbf{v}_j^{\mathrm{c}}$ represents the velocity of B relative to C, the centre of mass of A and B, in the channel j and $\mathbf{v}_j^{\mathrm{c}}$ is taken to be in the positive Z-direction. If $\mathbf{v}_j$ is the velocity of A relative to B in channel j, $v_j \equiv k_j/\mu$ and

$$\mathbf{v}_j^{\mathrm{c}} = q_{\mathrm{c}}\mathbf{v}_j, \qquad \text{where } q_{\mathrm{c}} = \frac{M_{\mathrm{A}}}{M_{\mathrm{A}} + M_{\mathrm{B}}}. \tag{3.79b}$$

In the same way the asymptotic wave function in the exchange channels, (3.26), becomes, using $\boldsymbol{\rho} \approx \mathbf{R} + \mathbf{s}/M_{\mathrm{A}}$

$$\Psi_{\mathrm{B}} = \sum_n X_n(\mathbf{x})\mathscr{G}_n(\boldsymbol{\rho}) \tag{3.80a}$$

$$= \sum_n X_n(\mathbf{x})T\left(\frac{\mathbf{s}}{M_{\mathrm{A}}}\right)\mathscr{G}_n(\mathbf{R}) \tag{3.80b}$$

$$\simeq \sum_n X_n(\mathbf{x}) \exp(\mathrm{i}\bar{\mathbf{v}}_n^{\mathrm{c}}\cdot\mathbf{s})\mathscr{G}_n(\mathbf{R}) \tag{3.80c}$$

where

$$\bar{v}_n^{\mathrm{c}} = \kappa_n/M_{\mathrm{A}}. \tag{3.81a}$$

The vector $\bar{v}_n^{\mathrm{c}}$ is the velocity of A relative to the centre of mass A and B in the channel n, and $\bar{\mathbf{v}}_n^{\mathrm{c}}$ is also taken to be in the positive Z-direction. In terms of the velocity of A relative to B in channel n, $v_n \equiv \kappa_n/\mu$,

$$\bar{\mathbf{v}}_n^{\mathrm{c}} = p_c \mathbf{v}_n \qquad \text{where } p_{\mathrm{c}} = \frac{M_{\mathrm{B}}}{M_{\mathrm{A}} + M_{\mathrm{B}}}. \tag{3.81b}$$

It is easy to verify that the Ψ_{A} and Ψ_{B} satisfy the Schrödinger equation in the asymptotic regions $R \to \infty$ and $r \ll R$, or $x \ll R$, respectively. For example, in the initial arrangement, the asymptotic equation is (writing the kinetic energy in the form (3.7) with $m \simeq 1$) for $R \gg r$:

$$\left[-\frac{1}{2\mu}\nabla_R^2 - \tfrac{1}{2}\nabla_s^2 + V_{\mathrm{Be}}(r) - E\right]\Psi_{\mathrm{A}} = 0. \tag{3.82}$$

Since $\mathbf{r} = \mathbf{s} + \mu\mathbf{R}/M_{\mathrm{B}}$, we have that

$$\begin{aligned}\left[-\frac{1}{2\mu}\nabla_R^2 - \tfrac{1}{2}\nabla_s^2 + V_{\mathrm{Be}}(r) - E\right]&\phi_j(\mathbf{r})\exp(-\mathrm{i}\mathbf{v}_j^{\mathrm{c}}\cdot\mathbf{s})\mathscr{F}_j(\mathbf{R})\\ &= \phi_j(\mathbf{r})\exp(-\mathrm{i}\mathbf{v}_j^{\mathrm{c}}\cdot\mathbf{s})\left[-\frac{1}{2\mu}\nabla_R^2 - E + \tfrac{1}{2}v_j^{\mathrm{c}^2} + \varepsilon_j\right]\mathscr{F}_j(\mathbf{R})\\ &\quad - \frac{1}{M_{\mathrm{B}}}\exp(-\mathrm{i}\mathbf{v}_j\cdot\mathbf{s})\nabla_r\phi_j(\mathbf{r})\cdot\nabla_R\mathscr{F}_j(\mathbf{R})\\ &\quad + \mathrm{i}\exp(-\mathrm{i}\mathbf{v}_j^{\mathrm{c}}\cdot\mathbf{s})\mathbf{v}_j^{\mathrm{c}}\cdot\nabla_r\phi_j(\mathbf{r})\mathscr{F}_j(\mathbf{R}),\end{aligned} \tag{3.83}$$

where the Schrödinger equation for $\phi_j(\mathbf{r})$ has been employed and a small term $(1/2\mu)\nabla_R^2\phi_j = \frac{1}{2}M_{\mathrm{A}}/[M_{\mathrm{B}}(M_{\mathrm{A}} + M_{\mathrm{B}})]\nabla_r^2\phi_j$ has been neglected in comparison to the term $\frac{1}{2}\nabla_r^2$. Since

$$\nabla_R\mathscr{F}_j(\mathbf{R}) \simeq \mathrm{i}\mathbf{k}_j\mathscr{F}_j(\mathbf{R}), \tag{3.84}$$

the last two terms in (3.83) cancel. It follows that each term in (3.78c), and hence Ψ_{A}, is a solution of the wave equation for large R, $R \gg r$, provided $\mathscr{F}_j(\mathbf{R})$ satisfies the unperturbed equation

$$\left[-\frac{1}{2\mu}\nabla_R^2 - \frac{k_j^2}{2\mu}\right]\mathscr{F}_j(\mathbf{R}) = 0, \tag{3.85}$$

with

$$\frac{k_j^2}{2\mu} = E - \varepsilon_j - \tfrac{1}{2}v_j^{\mathrm{c}^2}. \tag{3.86}$$

The term $v_j^{\mathrm{c}^2}/2$ represents the kinetic energy of the electron, which shares the translational motion of B relative to the chosen origin C.

In the exchange channels it is again straightforward to show that the function (3.80c) satisfies the wave equation in the asymptotic region, provided $\mathscr{G}_n(\mathbf{R})$ satisfies the free-particle equation:

$$\left(-\frac{1}{2\mu}\nabla_R^2 - \frac{\kappa_n^2}{2\mu}\right)\mathscr{G}_n(\mathbf{R}) = 0, \tag{3.87a}$$

where

$$\frac{\kappa_n^2}{2\mu} = E - \eta_n - \tfrac{1}{2}\bar{v}_n^{c^2}. \tag{3.87b}$$

Trial functions

Parallelling the development in Section 3.3, trial functions are introduced similar in form to (3.37):

$$\Psi = \sum_{j=1}^{N} \Phi_j(\mathbf{s}, \mathbf{R})F_j(\mathbf{R}) + \sum_{n=N+1}^{N+M} \mathrm{X}_n(\mathbf{s}, \mathbf{R})G_n(\mathbf{R}). \tag{3.88}$$

In order to satisfy the asymptotic boundary conditions, $\phi_j(\mathbf{s}, \mathbf{R})$ and $\mathrm{X}_n(\mathbf{s}, \mathbf{R})$ should obey (cf. 3.38)

$$\begin{aligned} \Phi_j(\mathbf{s}, \mathbf{R}) &\to \phi_j(\mathbf{r}) \exp(-\mathrm{i}\mathbf{v}_j^c \cdot \mathbf{r}), \qquad R \to \infty, \quad r \ll R \\ \mathrm{X}_n(\mathbf{s}, \mathbf{R}) &\to \chi_n(\mathbf{x}) \exp(\mathrm{i}\bar{\mathbf{v}}_j^c \cdot \mathbf{x}), \qquad R \to \infty, \quad x \ll R. \end{aligned} \tag{3.89}$$

The channel functions $F_j(\mathbf{R})$ and $G_j(\mathbf{R})$ are asymptotic as $R \to \infty$ to the functions $\mathscr{F}_j(\mathbf{R})$ and $\mathscr{G}_j(\mathbf{R})$, which in turn are related to the scattering amplitudes through (3.20) (with $\boldsymbol{\sigma}$ set equal to $\mathbf{R}$) and (3.27) (with $\boldsymbol{\rho}$ set equal to $\mathbf{R}$).

As in (3.40) and (3.41), orthonormality conditions can be imposed and overlap matrices defined:

$$\begin{aligned} \int \Phi_j^*(\mathbf{s}, \mathbf{R})\Phi_l(\mathbf{s}, \mathbf{R})\, \mathrm{d}\mathbf{s} &= \delta_{jl} \\ \int \mathrm{X}_n^*(\mathbf{s}, \mathbf{R})\mathrm{X}_m(\mathbf{s}, \mathbf{R})\, \mathrm{d}\mathbf{s} &= \delta_{nm} \end{aligned} \tag{3.90}$$

$$S_{jn}(\mathbf{R}) = \int \Phi_j^*(\mathbf{s}, \mathbf{R})\mathrm{X}_n(\mathbf{s}, \mathbf{R})\, \mathrm{d}\mathbf{s}, \qquad \bar{S}_{nj}(\mathbf{R}) = S_{jn}^*(\mathbf{R}). \tag{3.91}$$

The coupled equations for F_j and G_n obtained from the variational method are

$$\begin{aligned} \int \Phi_j^*(\mathbf{s}, \mathbf{R})[H - E]\Psi(\mathbf{s}, \mathbf{R})\, \mathrm{d}\mathbf{s} &= 0, \qquad j = 1, 2, \ldots, N \\ \int \mathrm{X}_n^*(\mathbf{s}, \mathbf{R})[H - E]\Psi(\mathbf{s}, \mathbf{R})\, \mathrm{d}\mathbf{s} &= 0, \qquad n = N+1, N+2, \ldots, N+M, \end{aligned} \tag{3.92}$$

and these are *differential* equations rather than integro-differential equations. Using the orthonormality conditions (3.90), equations (3.92) are easily put in the form

$$[\nabla_R^2 + k_j^2]F_j(\mathbf{R}) + \sum_{m=1}^{M} S_{jm}(\mathbf{R})[\nabla_R^2 + \kappa_m^2]G_m(\mathbf{R})$$
$$= 2\mu \sum_{l=1}^{N} [V_{jl}(\mathbf{R}) + \mathbf{W}_{jl}(\mathbf{R})\cdot\nabla_{\mathrm{R}}]F_l(\mathbf{R})$$
$$+ 2\mu \sum_{m=N+1}^{N+M} [K_{jm}(\mathbf{R}) + \mathbf{L}_{jm}(\mathbf{R})\cdot\nabla_R]G_m(\mathbf{R}),$$
$$j = 1, 2, \ldots, N \qquad (3.93)$$
$$[\nabla_R^2 + \kappa_n^2]G_n(\mathbf{R}) + \sum_{l=1}^{N} \bar{S}_{nl}(\mathbf{R})[\nabla_R^2 + k_l^2]F_j(\mathbf{R})$$
$$= 2\mu \sum_{n=N+1}^{N+M} [\bar{V}_{nm}(\mathbf{R}) + \overline{\mathbf{W}}_{nm}(\mathbf{R})\cdot\nabla_{\mathrm{R}}]G_m(\mathbf{R})$$
$$+ 2\mu \sum_{l=1}^{N} [\bar{K}_{nl}(\mathbf{R}) + \bar{\mathbf{L}}_{nl}(\mathbf{R})\cdot\nabla_{\mathrm{R}}]F_l(\mathbf{R}),$$
$$n = N+1, N+2, \ldots, N+M.$$

The various potentials V_{jl}, $\mathbf{W}_{jl}$, K_{jm}, $\mathbf{L}_{jm}$ and corresponding barred quantities are essentially the same as those defined in (3.47), (3.48), (3.59), and (3.63) with the exception that $\mathbf{R}$ replaces $\boldsymbol{\sigma}$ or $\boldsymbol{\rho}$. To be quite specific the definitions are listed below.

$$V_{jl}(\mathbf{R}) = \int \Phi_j^*(\mathbf{s}, \mathbf{R})[H - \delta_{lj}\varepsilon_j]\Phi_l(\mathbf{s}, \mathbf{R})\, \mathrm{d}s$$
$$\bar{V}_{nm}(\mathbf{R}) = \int \mathrm{X}_n^*(\mathbf{s}, \mathbf{R})[H - \delta_{nm}\eta_n]\mathrm{X}_m(\mathbf{s}, \mathbf{R})\, \mathrm{d}\mathbf{s}$$
$$\mathbf{W}_{jl}(\mathbf{R}) = -\frac{1}{\mu}\int \Phi_j^*(\mathbf{s}, \mathbf{R})\nabla_R\Phi_l(\mathbf{s}, \mathbf{R})\, \mathrm{d}\mathbf{s}$$
$$\overline{\mathbf{W}}_{nm}(\mathbf{R}) = -\frac{1}{\mu}\int \mathrm{X}_n^*(\mathbf{s}, \mathbf{R})\nabla_R\mathrm{X}_m(\mathbf{s}, \mathbf{R})\, \mathrm{d}\mathbf{s} \qquad (3.94)$$
$$K_{jm}(\mathbf{R}) = \int \Phi_j^*(\mathbf{s}, \mathbf{R})[H - \eta_m]\mathrm{X}_m(\mathbf{s}, \mathbf{R})\, \mathrm{d}\mathbf{s}$$
$$\bar{K}_{nl}(\mathbf{R}) = \int \mathrm{X}_n^*(\mathbf{s}, \mathbf{R})[H - \varepsilon_l]\Phi_l(\mathbf{s}, \mathbf{R})\, \mathrm{d}\mathbf{s}$$
$$\mathbf{L}_{jm}(\mathbf{R}) = -\frac{1}{\mu}\int \Phi_j^*(\mathbf{s}, \mathbf{R})\nabla_R\mathrm{X}_m(\mathbf{s}, \mathbf{R})\, \mathrm{d}\mathbf{s}$$
$$\bar{\mathbf{L}}_{nl}(\mathbf{R}) = -\frac{1}{\mu}\int \mathrm{X}_n^*(\mathbf{s}, \mathbf{R})\nabla_R\Phi_l(\mathbf{s}, \mathbf{R})\, \mathrm{d}\mathbf{s}.$$

Once again the coupled equations can be expressed in the standard form (3.65), where the matrices **N**, **M**, and **P** are defined by (3.67) and (3.68).

Choice of Φ_j *and* X_n Although more general forms are possible, and indeed have been employed, the usual choice of basis functions is made between those of atomic and those of molecular character. For those of atomic character, we write analogously to (3.70)

$$\begin{aligned} \Phi_j(\mathbf{s}, \mathbf{R}) &= \phi_j(\mathbf{r}) \exp[i\gamma_j^B(\mathbf{s}, \mathbf{R})] \\ X_n(\mathbf{s}, \mathbf{R}) &= \chi_n(\mathbf{x}) \exp[i\gamma_n^A(\mathbf{s}, \mathbf{R})]. \end{aligned} \tag{3.95}$$

The simplest choice of the phases γ_j^B and γ_n^A is to use the asymptotic forms

$$\gamma_j^B = -\mathbf{v}_j^c \cdot \mathbf{s}, \qquad \gamma_n^A = \mathbf{v}_n^c \cdot \mathbf{s}; \tag{3.96}$$

but more flexible forms can be employed, such as

$$\begin{aligned} \gamma_j^B &= -\tfrac{1}{2}(f_j(\mathbf{s}, \mathbf{R}) + \lambda)(\mathbf{v}_j \cdot \mathbf{s}), \\ \gamma_n^A &= -\tfrac{1}{2}(f_n(\mathbf{s}, \mathbf{R}) + \lambda)(\mathbf{v}_n \cdot \mathbf{s}), \end{aligned} \tag{3.97}$$

where λ is defined by (3.76c) and $\mathbf{v}_j$ (or $\mathbf{v}_n$) is the relative velocity of A and B in the channel j (or n), and $f_j(\mathbf{s}, \mathbf{R})$ is a switching function

$$f_j(\mathbf{s}, \mathbf{R}) \to \begin{cases} +1 & r \ll x \quad \text{and} \quad R \to \infty \\ -1 & x \gg r \quad \text{and} \quad R \to \infty. \end{cases} \tag{3.98}$$

Similarly, if a basis which is molecular in character is employed, (3.72) is replaced by

$$\begin{aligned} \Phi_j(\mathbf{s}, \mathbf{R}) &= \psi_j^B(\mathbf{s}, \mathbf{R}) \exp[i\gamma_j^B(\mathbf{s}, \mathbf{R})] \\ X_n(\mathbf{s}, \mathbf{R}) &= \psi_n^A(\mathbf{s}, \mathbf{R}) \exp[i\gamma_n^A(\mathbf{s}, \mathbf{R})]. \end{aligned} \tag{3.99}$$

Finally we notice that since $(v_j^c)^2 - (v_l^c)^2$ is small and of the order $(\varepsilon_l - \varepsilon_j)/M_B$, it is usually permissible to set $\mathbf{v}_j^c = \mathbf{v}^c \equiv \mathbf{v}_i^c$, where i denotes the incident channel, and similarly to set $\mathbf{v}_m^c = \bar{\mathbf{v}}^c \equiv \bar{\mathbf{v}}_i^c$.

Summary

In this section, the derivation of suitable coupled differential equations for the channel wave functions F_j, G_m has been discussed at considerable length and the question immediately arises as to which formulation is to be preferred. We first recall that the integro-differential equations (3.46) contain no approximations other than that the basis is of finite size – a limitation common to all coupled channels models. However while equations of the form (3.46) must be used for light particle reactions, it would not be sensible to ignore the simplifications which arise from the small size of the electron–nucleon mass ratio. In converting from the integro-differential equations to coupled differential equations, the most accurate approach would appear to be that of introducing reaction coordinates as in (3.75)–(3.77). The main objections might be some degree of complication and also the arbitrariness of the choice of reaction coordinates.

At very low velocities, the coupled equations based on the expansion (3.64) may be expected to be accurate, but this method violates unitarity, and these errors will become significant as v_0 increases.

Finally we are left with the translation factor approach, leading to the expressions (3.57) and (3.62), which is entirely equivalent to the simplified approach leading to (3.93). This formulation suffers from two defects. The first is that we encounter an arbitrariness in the choice of the factors f_j, f_n introduced in (3.97). The second is the approximation in which the vectors $\mathbf{v}_m$ (see (3.79b) and (3.81b)) are taken to be parallel to the Z-axis. It is interesting to note that this approximation is removed in the impact parameter model employing non-linear trajectories, which will be described later in this chapter. Considered from the viewpoint of the variational method, it is clear that despite these defects, methods based on the Bates–McCarroll equations will become accurate as the basis size is increased. Furthermore, the use of plane wave, rather than more complicated translation factors, should be a good approximation at higher velocities ($v_0 \gtrsim v_e$). For these reasons, calculations based on (3.93) are capable of providing an accurate description of ion–atom interactions, particularly at the higher energies.

3.5 PARTIAL-WAVE EXPANSIONS

As we have already noted the coupled equations can be put into the general form (3.69), which in terms of **R** becomes

$$(\nabla_R^2 + \mathscr{K}^2)\mathbf{F}(\mathbf{R}) = 2\mu\mathbf{N}^{-1}(\mathbf{R})[\mathbf{M}(\mathbf{R}) + \mathbf{P}(\mathbf{R})\cdot\nabla_\mathbf{R}]\mathbf{F}(\mathbf{R}), \qquad (3.100)$$

To obtain a first orientation to the partial-wave expansions of these equations, let us consider the simplified situation in which a basis is chosen such that the coupling matrix **P** is small (or zero). We shall also suppose that the basis states are spherically symmetrical so that **N**(**R**) and **M**(**R**) are functions of the radial coordinate R only. The coupled equations in this model reduce to

$$(\nabla_R^2 + \mathscr{K}^2)\mathbf{F}(\mathbf{R}) = 2\mu\mathbf{U}(R)\mathbf{F}(\mathbf{R}), \qquad (3.101)$$

where

$$\mathbf{U}(R) = \mathbf{N}^{-1}(R)\mathbf{M}(R). \qquad (3.102)$$

Since **U**(**R**) is spherically symmetrical, the system is cylindrically symmetrical about the direction of incidence, the Z-axis, and the channel functions $\mathbf{F}_i(\mathbf{R})$

can be expanded in Legendre polynomials

$$F_j(\mathbf{R}) = \sum_{L=0}^{\infty} R^{-1} f_{L,j}(R) P_L(\cos\Theta), \tag{3.103}$$

where (R, Θ, Φ) are the polar coordinates of $\mathbf{R}$. Combining (3.101) with (3.103), the column vector of radial functions $\mathbf{f}_L(R)$ with elements $f_{L,j}(R)$ satisfies

$$\left[\frac{\mathrm{d}^2}{\mathrm{d}R^2} - \frac{L(L+1)}{R^2} + \mathscr{K}^2\right]\mathbf{f}_L(R) = 2\mu\mathbf{U}(R)\mathbf{f}_L(R). \tag{3.104}$$

Consistently with (3.20) and (3.29), the boundary conditions on $\mathbf{f}_L(R)$ (see Bransden (1983) for details) are, for 'open' channels with $K_j^2 > 0$,

$$\mathbf{f}_L(0) = 0, \qquad f_{L,j}(R) \underset{R\to\infty}{\sim} \delta_{ji} R \mathrm{i}^L j_L(K_j R) + a_{ji}^L \exp(\mathrm{i}K_j R), \tag{3.105a}$$

where i denotes the incident channel, and for the 'closed' channels with $K_j^2 < 0$

$$f_{L,j}(R) \underset{R\to\infty}{\rightarrow} 0. \tag{3.105b}$$

In the direct channels the amplitudes a_{ji}^L are related to the scattering amplitudes $f_{ji}(\Theta, \Phi)$ by

$$f_{ji}(\Theta, \Phi) = \sum_{L=0}^{\infty} (2L+1) a_{ji}^L P_L(\cos\Theta), \tag{3.106}$$

and in the exchange channels the a_{ji}^L are similarly related to the scattering amplitudes $g_{ni}(\Theta, \Phi)$.†

The $\mathscr{S}$-matrix can be obtained from the scattering amplitude matrix with elements a_{ji}^L, using the relation

$$\mathscr{S}_{ji}^L = 2\mathrm{i}\sqrt{K_i K_j}\, a_{ji} + \delta_{ij}. \tag{3.107}$$

Alternatively the $\mathscr{S}$-matrix can be found directly by solving the radial equations (3.104) with the boundary conditions

$$f_{L,j}(0) = 0,$$

$$f_{L,j}(R) \underset{R\to\infty}{\sim} k_j^{-\frac{1}{2}} \exp[-\mathrm{i}(K_i R - \tfrac{1}{2}L\pi)\delta_{ji} - k_j^{-\frac{1}{2}} \mathscr{S}_{ji}^L \exp[\mathrm{i}(K_j R - \tfrac{1}{2}L\pi)], \tag{3.108a}$$

† It should be noted that, as in the limit of large R, $\boldsymbol{\sigma} \approx \boldsymbol{\rho} \approx \mathbf{R}$, we have that $f_{ji}(\Theta, \Phi) \approx f_{ji}(\theta, \phi)$ and $g_{ni}(\Theta, \Phi) \approx g_{ni}(\bar{\theta}, \bar{\phi})$ where $f_{ji}(\theta, \phi)$ and $g_{ni}(\bar{\theta}, \bar{\phi})$ were introduced in (3.20a) and (3.29a), respectively.

for open channels, and

$$f_{L,j}(R) \xrightarrow[R \to \infty]{} 0, \tag{3.108b}$$

for closed channels. The exact matrix, $\mathscr{S}^L$, is symmetrical and unitary:

$$\mathscr{S}^L_{ji} = \mathscr{S}^L_{ij}; \qquad (\mathscr{S}^L)^{-1} = \mathscr{S}^{L\dagger}, \tag{3.109}$$

and the approximate $\mathscr{S}$-matrix obtained from the coupled equations (3.69) also possesses this property. The coupled equations for each L can be solved numerically, with the stated boundary conditions, using standard computational methods.

More generally, when the potentials are non-central and depend on the orientation of the interionic line **R**, the partial-wave analysis becomes more complicated and partial waves for different values of L are coupled. We shall return to this point subsequently in a discussion of the perturbed stationary state model, but for the present it is readily seen that the partial-wave method becomes impractical if the number of important partial waves becomes too large, and our next task is to provide an estimate of the number of partial waves required.

In Chapter 1, we noted that E_L, the incident laboratory energy divided by the mass of the projectile A, can be expressed in terms of the initial relative velocity of A and B, v_0, by the relation

$$E_L = 25 v_0^2 \text{ keV/a.m.u.}, \tag{3.110}$$

where v_0 is given in atomic units. It follows that the momentum K_i in the centre-of-mass frame is

$$K_i = \mu v_0 = \frac{\mu}{5}\sqrt{E_L}, \tag{3.111}$$

where μ is the reduced mass of A and B. Taking as an example the collision between the lightest ions, $H^+ + H$, for which $\mu = M_P/2$, we have

$$K_i \approx 183.6\sqrt{E_L}, \tag{3.112}$$

where K_i is in atomic units and E_L in keV/a.m.u.

If the interaction region is of extent a, we may expect to require at least $(K_i a)$ partial waves. The potential matrix elements linking the direct and exchange channels are usually of short range and, depending on the system, a can be from a few to a few tens of atomic units. Taking $a = 10$, we can draw up Table 3.1, showing the number, $L_{\max}$, of partial waves expected in terms of E_L and K_i.

We see in this particular example the number of partial waves required becomes quite large, even at low impact energies. For this reason the

Table 3.1. Possible number of partial waves, L_{max}, for an interaction of range 10 a.u.

v_0(a.u.)	E_L(keV/a.m.u.)	K_i(a.u.)	L_{max}
0.0022	10^{-4}	2	20
0.0065	10^{-3}	6	60
0.020	10^{-2}	18	180
0.063	10^{-1}	58	580
0.20	1.0	184	1840
0.63	10.0	580	5800

numerical solution of the second-order radial equations is not usually attempted at energies greater than a few eV (for a projectile of mass M_P). Fortunately, even at these energies, the reduced wavelength of the projectile, $\lambda\!\!\!^- = K_i^{-1}$ is small compared with the range of the interaction and this is the condition for the heavy particle motion to be described classically, or semi-classically. Semi-classical approximations are either based on a WKB approximation to the radial equations along the lines discussed in Chapter 2 for a single channel, or on an alternative eikonal approximation of the coupled equations. At energies E $(=M_A E_L)$ greater than a few hundred eV, the heavy particle motion can be approximated by an undeviated straight line, still further simplifying the problem. We shall outline each of these methods in turn, reserving details of application to later chapters.

3.6 SEMI-CLASSICAL APPROXIMATIONS TO THE COUPLED CHANNEL EQUATIONS

3.6.1 The WKB approximation to the radial equations

For simplicity, consider the case of a spherically symmetrical potential coupling matrix, as in (3.102). The radial equations are given by (3.104) which we express as

$$\left[\frac{d^2}{dR^2} - \frac{L(L+1)}{R^2} + K_j^2\right] f_{L,j}(R) = 2\mu \sum_{n=1}^{N+M} U_{jn}(R) f_{L,n}(R),$$

$$j = 1, 2, \ldots, N+M \quad (3.113)$$

Let us suppose the heavy particle motion in the jth channel can be described by an effective potential $W_j(R)$. Although expressions can be obtained for the effective potentials $W_j(R)$ (also called optical potentials) in the form of a perturbation series, it is often sufficient to use the approximation $W_j(R) \approx U_{jj}(R)$. This would be exact in the absence of coupling between channels

($U_{jn} = 0$, $j \neq n$) and should be accurate if the coupling is small. In general the most important contribution to $W_j(R)$ is the Coulomb repulsion between the ionic cores (contained in the diagonal potentials $U_{jj}(R)$), which reduces the probability of charge exchange or excitation at close separations.

In terms of the potentials $W_j(R)$, equation (3.113) can be rewritten as

$$\left[\frac{\mathrm{d}^2}{\mathrm{d}R^2} + p_j^2(R)\right] f_{L,j}(R) = 2\mu \sum_{n=1}^{N+M} [U_{jn}(R) - W_j(R)\delta_{jn}] f_{L,n}(R),$$

$$j = 1, 2, \ldots, N + M \quad (3.114)$$

where $p_j^2(R)$ is defined by

$$p_j^2(R) = K_j^2 - \frac{L(L+1)}{R^2} - 2\mu W_j(R). \quad (3.115)$$

The *uncoupled* equations,

$$\left[\frac{\mathrm{d}^2}{\mathrm{d}R^2} + p_j^2(R)\right] y_j(R) = 0, \quad (3.116)$$

have the primitive WKB solutions (see Appendix C) for $R > R_j$,

$$y_j^{\pm}(R) \approx p_j^{-\frac{1}{2}} \exp(\pm \mathrm{i} S_j(R)), \quad (3.117\text{a})$$

with

$$S_j(R) = \int_{R_j}^{R} p_j(R)\,\mathrm{d}R, \quad (3.117\text{b})$$

where R_j is the classical turning point for motion in the potential W_j. At the turning point, $p_j(R) = 0$ so that R_j is the largest real root of the equation

$$p_j^2(R) = 0. \quad (3.118)$$

The approximate solutions (3.117) are accurate provided the condition

$$\left|\frac{1}{p_j^2(R)} \frac{\mathrm{d}p_j(R)}{\mathrm{d}R}\right| \ll 1 \quad (3.119)$$

is satisfied in each channel j. This condition can be obtained from equation (C.9) of Appendix C by replacing $F(x)$ by $p_j^2(R)$. The solutions can be improved by making the Langer modification discussed in Appendix C in which $L(L+1)$ is replaced by $(L+\frac{1}{2})^2$.

The solutions of the coupled equations (3.114) can be expressed in terms of the WKB solutions of the uncoupled equations by writing

$$f_{L,j}(R) = p_j^{-\frac{1}{2}}[a_j^+(R) \exp(\mathrm{i}S_j(R)) + a_j^-(R) \exp(-\mathrm{i}S_j(R))]. \quad (3.120)$$

Since there are two functions a_j^+ and a_j^- in this representation of the single function $f_{L,j}(R)$, an extra condition can be imposed, which can be taken to be (Fröman and Fröman 1965)

$$\frac{\mathrm{d}a_j^+(R)}{\mathrm{d}R}\exp(\mathrm{i}S_j(R)) + \frac{\mathrm{d}a_j^-(R)}{\mathrm{d}R}\exp(-\mathrm{i}S_j(R)) = 0. \qquad (3.121)$$

If (3.120) is substituted into the coupled equations (3.114), the condition (3.121) has the effect of removing second-order derivatives of $a_j^{\pm}(R)$ and the following first-order coupled differential equations for $a_j^{\pm}(R)$ are obtained:

$$2\mathrm{i}p_j^{\frac{1}{2}}\frac{\mathrm{d}}{\mathrm{d}R}a_j^+(R) + \frac{\mathrm{d}^2}{\mathrm{d}R^2}(p_j^{-\frac{1}{2}})\{a_j^+(R) + a_j^-(R)\exp(-2\mathrm{i}S_j(R))\}$$
$$= 2\mu\sum_n \exp(-\mathrm{i}S_j(R))[U_{jn}(R) - \delta_{jn}W_j(R)]\exp(\mathrm{i}S_n(R))\,p_n^{-\frac{1}{2}}$$
$$\times\{a_n^+(R) + a_n^-(R)\exp(-2\mathrm{i}S_n(R))\},$$

and (3.122)

$$-2\mathrm{i}p_j^{\frac{1}{2}}\frac{\mathrm{d}}{\mathrm{d}R}a_j^-(R) + \frac{\mathrm{d}^2}{\mathrm{d}R^2}(p_j^{-\frac{1}{2}})\{a_j^-(R) + a_j^+(R)\exp(2\mathrm{i}S_j(R))\}$$
$$= 2\mu\sum_n \exp(\mathrm{i}S_j(R))[U_{jn}(R) - \delta_{jn}W_j(R)]\exp(-\mathrm{i}S_n(R))\,p_n^{-\frac{1}{2}}$$
$$\times\{a_n^-(R) + a_n^+(R)\exp(2\mathrm{i}S_n(R))\}, \qquad j = 1, 2, 3, \ldots, N+M$$

These equations are exact, but the approximation is now made in which the terms in $\mathrm{d}^2/\mathrm{d}R^2\,(p_j^{-\frac{1}{2}})$ are neglected, and this is accurate provided K_j is large and R is not too close to the classical turning point. Provided the coupling is weak in the sense that $|U_{jn}| \ll K_j^2$, the coefficients $a_j^{\pm}(R)$ vary slowly compared with $S_j(R)$. It follows that the rapidly oscillating terms containing $a_j^{\pm}\exp(\pm 2\mathrm{i}S_j)$ can be neglected compared with $a_j^{\pm}$, in which case the equations decouple into separate sets for a_j^+ and a_j^-. We obtain, in the region well away from the classical turning points $R > R_j$,

$$\pm\mathrm{i}\frac{\mathrm{d}}{\mathrm{d}R}a_j^{\pm}(R) = \sum_{n=1}^{N+M} 2\mu(p_j p_n)^{-\frac{1}{2}}[U_{jn}(R) - \delta_{jn}W_j(R)]$$
$$\times\exp[\pm\mathrm{i}(S_n(R) - S_j(R))]a_j^{\pm}(R), \qquad j = 1, 2, 3, \ldots, N+M$$
$$(3.123)$$

To investigate the boundary conditions placed on $a_j^{\pm}(R)$, we first note that the WKB phase shift of order L, $\eta_{L,j}$ (see Appendix C), for elastic scattering by a potential $W_j(R)$ is

$$\eta_{L,j} = \int_{R_j}^{\infty}\{p_j(R) - K_j\}\,\mathrm{d}R - K_jR_j + (L + \tfrac{1}{2})\frac{\pi}{2}. \qquad (3.124)$$

Now comparing (3.120) with the asymptotic form of $f_{L,j}(R)$, (3.108), the boundary conditions on $a_j^{\pm}$ are found to be, for large R,

$$a_j^-(R) \underset{R\to\infty}{\longrightarrow} \delta_{ji} \exp \mathrm{i}\left(\eta_{L,i} - \frac{\pi}{4}\right), \tag{3.125a}$$

$$a_j^+(R) \underset{R\to\infty}{\longrightarrow} -\mathscr{S}^L_{ji} \exp -\mathrm{i}\left(\eta_{L,j} - \frac{\pi}{4}\right). \tag{3.125b}$$

A further condition is required to relate a_j^+ and a_j^-. This is easily found, provided the coupling between the channels is weak, and can be neglected in the vicinity of classical turning points at $R = R_j$. It then follows that the radial wave function in each channel for R close to R_j will be that for elastic scattering in the potential $W_j(R)$. In the WKB approximation (see Appendix C, equation (C.28)) these radial functions are $f_{Lj}(R) = C_j p_j^{-\frac{1}{2}}(R) \times \cos\{S_j(R) - \pi/4\}$, where C_j are constants. From (3.120), we see that this will be the case provided the conditions

$$a_j^+(R_j) = -\mathrm{i}a_j^-(R_j) \tag{3.126}$$

are satisfied.

Common trajectory approximation

The numerical solution of the coupled equations (3.123) subject to (3.126) with different turning points and potentials in each channel, is complicated. In the case of a two-channel system Bates and Crothers (1970) used different potentials in each channel, but forced a common turning point. In contrast, most calculations have introduced an average potential $W(R)$ in place of the channel potentials $W_j(R)$ and correspondingly employed a *common trajectory* $\mathbf{R}(t)$ to describe the heavy particle motion at a constant angular momentum $(L + \frac{1}{2})$ and at a mean kinetic energy of $K^2/2\mu$ (Berson 1968; Riley 1973; Riley and Green 1971; Gaussorgues *et al.* 1975*a*).

The radial coordinate R at a point on the common classical trajectory is related to the time t by

$$t = \pm \int_{R_0}^{R} \mu[K^2 - 2\mu W(R) - (L + \tfrac{1}{2})^2/R^2]^{-\frac{1}{2}}\, \mathrm{d}R, \tag{3.127}$$

which is obtained from (2.21) of Chapter 2, with $E = K^2/2\mu$ and $\mu v_0 b = L + \frac{1}{2}$. The $+$ sign corresponds to $\mathrm{d}R/\mathrm{d}t > 0$ and outward motion after the turning point is reached, and the $-$ sign to inward motion. The turning point R_0 is the largest root of

$$K^2 - 2\mu W(R) - \frac{(L + \frac{1}{2})^2}{R^2} = 0, \tag{3.128}$$

and the turning point is reached at the time $t = 0$.

The radial action functions $S_n(R)$ defined by (3.117b) can be written approximately as

$$S_n(R) \approx \int_{R_0}^{R} \left[K^2 - 2\mu W(R) - \frac{(L+\frac{1}{2})^2}{R^2} \right]^{\frac{1}{2}} \mathrm{d}R$$
$$+ \frac{1}{2} \int_{R_0}^{R} [K_n^2 - 2\mu W_n(R) - K^2 + 2\mu W(R)]$$
$$\times \left[K^2 - 2\mu W(R) - \frac{(L+\frac{1}{2})^2}{R^2} \right]^{-\frac{1}{2}} \mathrm{d}R. \qquad (3.129a)$$

where the Langer modification $L(L+1) \to (L+\frac{1}{2})^2$ has been made. Similarly the factors $(p_j p_n)^{-\frac{1}{2}}$ in (3.123) can be approximated by

$$(p_j p_n)^{-\frac{1}{2}} \approx p^{-1} = \left[K^2 - 2\mu W(R) - \frac{(L+\frac{1}{2})}{R^2} \right]^{-\frac{1}{2}}. \qquad (3.129b)$$

The condition for the mean trajectory approximations (3.129a) and (3.129b) to be accurate is that

$$|K_n^2 - K^2 - 2\mu W_n + 2\mu W| \ll K^2, \qquad \text{for all } n. \qquad (3.130)$$

Making these approximations and transforming to the variable t by using (3.127), the coupled equations (3.123) become

$$\mathrm{i}\frac{\mathrm{d}}{\mathrm{d}t} a_j^{\mp}(t) = \sum_{n=1}^{N+M} [U_{jn}(t) - \delta_{jn} W_j(t)] \exp\left\{\mathrm{i}\int_0^t [\Delta\varepsilon_{jn} + W_n - W_j]\right\} a_n^{\mp}(t),$$
$$j = 1, 2, \ldots, N+M$$

where

$$\Delta\varepsilon_{ji} = \frac{1}{2\mu}(K_j^2 - K_n^2). \qquad (3.131)$$

The equation for $a_j^-(t)$ is integrated from $t = -\infty$ to the turning point at $t = 0$, using the boundary condition (3.125a). At $t = 0$, $a_j^+(0) = -\mathrm{i}a_j^-(0)$ and for $t > 0$ the equation for $a_j^+(t)$ is integrated from $t = 0$ to $t = \infty$, $\mathscr{S}_{ji}^L$ being identified by boundary conditions (3.125b). If a new function $c_j^L(t)$ is introduced by setting

$$c_j^L(t) = \begin{cases} a_j^-(t) \exp\left(-\mathrm{i}\left(\eta_{L,j} - \dfrac{\pi}{4}\right)\right) & t < 0 \\ -a_j^+(t) \exp\left(\mathrm{i}\left(\eta_{L,j} - \dfrac{\pi}{4}\right)\right) & t > 0, \end{cases} \qquad (3.132)$$

$c_j^L(t)$ satisfies the same equations as $a_j^{\mp}(t)$, (3.131), with the boundary condition $c_j^L(-\infty) = \delta_{ij}$, the subscript i indicating the incident channel. The $\mathscr{S}$-matrix is now obtained by integrating the equations from $t = -\infty$ to $t = +\infty$, and setting

$$\mathscr{S}_{ji}^L = c_j^L(\infty) \exp \mathrm{i}(\eta_{L,j} + \eta_{L,i}). \tag{3.133}$$

Equations (3.131) have been derived under the assumption that the potentials $U_{jn}(R)$ are spherically symmetrical. Rather than give the appropriate derivation for non-spherical potentials, we shall show below (in (3.153)) that equations of the same form can be obtained directly from the coupled channel equations *before* making an angular momentum decomposition, provided that the number of important partial waves is large enough to treat L as a continuous variable. The new derivation does not depend on the assumption of spherical symmetry. In Chapter 4, we shall discuss the very low energy situation in which both the number of partial waves is limited and the potentials are not spherically symmetrical.

The differential cross section

The scattering amplitude from which the differential cross section is computed is, for the case of spherically symmetrical potentials, expressed by (cf. (3.106) and (3.107))

$$f_{ji}(\Theta) = \sum_{L=0}^{\infty} (2L+1) \frac{1}{2\mathrm{i}} \left(\frac{1}{K_i K_j}\right)^{\frac{1}{2}} (\mathscr{S}_{ji}^L - \delta_{ji}) P_L(\cos\Theta). \tag{3.134}$$

The term in δ_{ji} only contributes in the forward direction since

$$\tfrac{1}{2} \sum_{L=0}^{\infty} (2L+1) P_L(\cos\Theta) = \delta(\cos\Theta - 1), \tag{3.135}$$

and can be omitted. In the common trajectory approximation with $K_i K_j \approx K^2$ and $\eta_{L,i} \approx \eta_{L,j} = \eta_L$, the sum can be replaced by an integral over L, provided a large number of partial waves contribute:

$$f_{ji}(\Theta) \simeq \frac{1}{2\mathrm{i}K} \int_0^{\infty} (2L+1) c_j^L(\infty) \exp(2\mathrm{i}\eta_L) P_L(\cos\Theta)\, \mathrm{d}L. \tag{3.136}$$

If the variation of the phase shifts with L is smooth and if c_j^L is slowly varying the integral may be evaluated by the stationary phase approximation (see Chapter 2) using the asymptotic expression valid for $\sin\Theta > 1/L$

$$P_L(\cos\Theta) \approx \left(\frac{2}{\pi(L+\frac{1}{2})\sin\Theta}\right)^{\frac{1}{2}} \sin\left((L+\tfrac{1}{2})\Theta + \frac{\pi}{4}\right). \tag{3.137a}$$

It is found that

$$\frac{\mathrm{d}\sigma_{ji}}{\mathrm{d}\sigma} = \frac{K_j}{K_i} |f_{ji}(\Theta)|^2 \approx \left(\frac{\mathrm{d}\sigma}{\mathrm{d}\Omega}\right)_{\mathrm{class}} |c_j^{\bar{L}}(\infty)|^2, \tag{3.137b}$$

where $\bar{L}$ is the classical angular momentum appropriate to the angle of scattering Θ in the average potential $W(R)$, and $(d\sigma/d\Omega)_{\text{class}}$ is the corresponding *classical* differential cross section. Although widely used, the stationary phase approximation (3.137b) is not always very accurate. This has been illustrated by Gaussorgues *et al.* (1975*b*), who have calculated cross sections for the scattering of H^+ by H(1s) using both (3.137) and the exact expression (3.134) at 500 eV. The ratio of the exact cross sections for excitation and those given by (3.137) are 2.69 at 4°, 1.23 at 5.4°, and 0.82 at 10° in the centre-of-mass system.

3.6.2 The eikonal approximation to the coupled channel equations

An equivalent approximation to the WKB solution of the radial equations can be obtained directly from the coupled channel equations before making an angular momentum decomposition. We start from (3.100), which we write in a condensed notation as

$$(\nabla_R^2 + \mathscr{K}^2)\mathbf{F}(\mathbf{R}) = 2\mu\mathbf{U}(\mathbf{R})\mathbf{F}(\mathbf{R}), \tag{3.138}$$

where

$$\mathbf{U} \equiv \mathbf{N}^{-1}[\mathbf{M} + \mathbf{P}\cdot\nabla_R], \tag{3.139}$$

and **N**, **M**, and **P** are defined by (3.67) and (3.68). The submatrices **S**, **V**, **K**, ... on which **N**, **M**, and **P** depend are in turn defined by (3.91) and (3.94). The general case will be considered in which $\mathbf{U}(\mathbf{R})$ depends on the orientation, as well as the magnitude, of **R**.

As before, we shall suppose that in each channel we can find a central potential $W_j(R)$ which determines the heavy particle motion, and also a *mean potential* $W(R)$ and *mean kinetic energy* $K^2/2\mu$ satisfying the condition (3.130). We can then express the coupled equations (3.138) as

$$[\nabla_R^2 + K^2 - 2\mu W(R)]F_j(\mathbf{R}) = 2\mu \sum_{n=1}^{N+M} [U_{jn}(\mathbf{R}) - W(R)\delta_{jn} - \Delta\bar{\varepsilon}_j\delta_{jn}]F_n(\mathbf{R}),$$

$$j = 1, 2, \ldots, N + M \tag{3.140a}$$

where

$$\Delta\bar{\varepsilon}_j \equiv \frac{K_j^2 - K^2}{2\mu}. \tag{3.140b}$$

The eikonal function

We now find the semi-classical solution of the uncoupled equations obtained by setting the right-hand side equal to zero:

$$[\nabla_R^2 + K^2 - 2\mu W(R)]\bar{F}(\mathbf{R}) = 0. \tag{3.141}$$

Setting

$$\bar{F}(\mathbf{R}) = \exp[\mathrm{i}S(\mathbf{R})], \tag{3.142}$$

where $S(\mathbf{R})$ is known as the eikonal function, it is found that

$$-\mathrm{i}\nabla_R^2 S(\mathbf{R}) + (\boldsymbol{\nabla}_R S(\mathbf{R}))^2 = p^2(\mathbf{R}), \tag{3.143a}$$

where

$$p^2(R) = K^2 - 2\mu W(R). \tag{3.143b}$$

In the region for which $K^2 \gg 2\mu W(R)$ and hence $(\boldsymbol{\nabla}_R S(\mathbf{R}))^2 \gg |\nabla_R^2 S(\mathbf{R})|$, the term in ∇_R^2 is small and can be dropped. In this approximation $S(\mathbf{R})$ satisfies the Hamilton–Jacobi equation:

$$(\boldsymbol{\nabla}_R S(\mathbf{R}))^2 = p^2(R). \tag{3.143c}$$

This equation defines all possible classical trajectories in the potential $W(R)$, at the energy $K^2/2\mu$. The trajectories are the normals to the surfaces $S(\mathbf{R}) =$ constant, and if we introduce a vector $\mathbf{p}(\mathbf{R})$, where $\mathbf{p}(\mathbf{R}) = \boldsymbol{\nabla}_R S(\mathbf{R})$, then $\mathbf{p}(\mathbf{R})$ represents the local classical momentum, $\mathbf{p}(\mathbf{R}) = \mu \mathbf{v}(t)$, where $\mathbf{v}(t) = \dot{\mathbf{R}}$. For a given energy, each classical trajectory depends on a two-dimensional impact parameter $\mathbf{b}$ and is a function of the time t

$$\mathbf{R} = \mathbf{R}(\mathbf{b}, t). \tag{3.144}$$

We shall take $\mathbf{b}$ to lie in the XY-plane (the Z-axis being the direction of incidence, as usual) and $t = 0$ is taken as the time of closest approach. The magnitude of $\mathbf{b}$ is related to the classical orbital angular momentum $J = (L + \frac{1}{2})$ by

$$J = Kb = (L + \tfrac{1}{2}), \tag{3.145}$$

and lies in the plane of scattering defined by $\Phi =$ constant, where (Θ, Φ) are the polar angles of $\mathbf{R}$.

If the distance along a trajectory is denoted by $\mathscr{s}$, then

$$S(\mathbf{R}) = \int_0^{\mathscr{s}} p(\mathbf{R})\, \mathrm{d}\mathscr{s}, \tag{3.146}$$

where $\mathbf{R}$ is to be expressed in terms of $\mathbf{b}$ and $\mathscr{s}$ rather than $\mathbf{b}$ and t, and $\mathscr{s} = 0$ is the point of closest approach. Since the trajectory is initially parallel to the Z-axis, we see that as $\mathscr{s} \to -\infty$, $\mathscr{s} \to KZ$ and

$$\bar{F}(\mathbf{R}) \underset{Z\to-\infty}{\simeq} \exp \mathrm{i}KZ. \tag{3.147}$$

Reduction of the coupled equations

We now express the channel functions $F_j(\mathbf{R})$ which satisfy equations (3.140a) as

$$F_j(\mathbf{R}) = \bar{F}(\mathbf{R}) a_j(\mathbf{b}, \mathcal{z}). \tag{3.148}$$

If the coupling between channels is small $a_j(\mathbf{b}, \mathcal{z})$ is expected to vary slowly with $\mathcal{z}$ compared with $S(\mathbf{R})$. Substituting (3.148) into (3.140) and neglecting both $\nabla_R^2 S$ and $\nabla_R^2 a_j$ compared with $|\mathbf{\nabla}_R S|^2$, we obtain

$$\mu^{-1} p(R) \mathrm{i} \frac{\mathrm{d} a_j(\mathbf{b}, \mathcal{z})}{\mathrm{d}\mathcal{z}} = \sum_{n=1}^{N+M} [U_{jn}(\mathbf{R}) - W(R)\delta_{jn} - \Delta\bar{\varepsilon}_j \delta_{jn}] a_n(\mathbf{b}, \mathcal{z}),$$

$$j = 1, 2, \ldots, N+M \tag{3.149a}$$

where, since $\mathbf{\nabla}_R S(\mathbf{R}) = \mathbf{p}(\mathbf{R})$, we can express $\mathbf{U}$ in the form

$$\mathbf{U} = \mathbf{N}^{-1}[\mathbf{M} + \mathrm{i}\mathbf{p}(\mathbf{R}) \cdot \mathbf{P}]. \tag{3.149b}$$

Changing the variable from $\mathcal{z}$ to t by noting that

$$\frac{\mathrm{d}\mathcal{z}}{\mathrm{d}t} = v(t) = \frac{p(R)}{\mu} = \left| \frac{\mathrm{d}\mathbf{R}}{\mathrm{d}t} \right|, \tag{3.150}$$

we obtain the time-dependent equations

$$\mathrm{i} \frac{\mathrm{d} a_j(\mathbf{b}, t)}{\mathrm{d}t} = \sum_{n=1}^{N+M} [U_{jn}(t) - W(t)\delta_{jn} - \Delta\bar{\varepsilon}_j \delta_{jn}] a_n(\mathbf{b}, t), \qquad j = 1, 2, \ldots, N+M. \tag{3.151}$$

Finally, by making the phase transformation

$$a_j(\mathbf{b}, t) = c_j(\mathbf{b}, t) \exp\left\{ -\mathrm{i} \int_0^t [W_j(t') - W(t') - \Delta\bar{\varepsilon}_j] \, \mathrm{d}t' \right\}, \tag{3.152}$$

we once again obtain equations of the form (3.131):

$$\frac{\mathrm{d} c_j(\mathbf{b}, t)}{\mathrm{d}t} = \sum_{n=1}^{N+M} [U_{jn}(t) - W_j(t)\delta_{jn}] c_n(\mathbf{b}, t)$$

$$\times \exp\left\{ \mathrm{i} \int_0^t [\Delta\varepsilon_{jn} + W_n(t') - W_j(t')] \, \mathrm{d}t' \right\}$$

$$j = 1, 2, \ldots, N+M, \tag{3.153}$$

where $\Delta\varepsilon_{jn} = \Delta\bar{\varepsilon}_j - \Delta\bar{\varepsilon}_n = E_j - E_n$. In view of (3.147) and (3.148), if the initial channel is labelled i, the equations are to be solved subject to the boundary condition

$$\lim_{t \to -\infty} c_j(\mathbf{b}, t) = \delta_{ji}. \tag{3.154}$$

The probability of finding the system in channel j after the collision is given by

$$P_{ji}(\mathbf{b}) = \lim_{t \to \infty} |c_j(\mathbf{b}, t)|^2 \tag{3.155}$$

and the total cross section for excitation or charge exchange from i to j is obtained by integrating over the two-dimensional impact parameter space

$$\sigma_{ji} = \int \mathrm{d}^2\mathbf{b} P_{ji}(\mathbf{b}). \tag{3.156}$$

To the extent that L can be treated as a continuous variable, we see that $\lim_{t \to \infty} c_j(\mathbf{b}, t) = c_j^L(\infty)$, with $b = (L + \frac{1}{2})/K$, and (3.156) agrees completely with the total cross section computed from $f_{ji}(\theta)$ given by (3.136).

3.7 THE IMPACT PARAMETER MODEL

The considerations of the previous paragraph suggest that, if the conditions for a common trajectory eikonal description hold, the resulting simplifications should be introduced immediately into the original Schrödinger equation (3.2), rather than at a later stage in the formalism. We start by expressing (3.2) as

$$H\Psi = \left[-\frac{1}{2\mu} \nabla_R^2 + W(R) + H_\mathrm{e} \right] \Psi = E\Psi, \tag{3.157}$$

where $W(R)$ is the potential defining the classical heavy particle motion and H_e is an electronic Hamiltonian defined by

$$H_\mathrm{e} = -\tfrac{1}{2}\nabla_s^2 + V - W, \tag{3.158}$$

where

$$V = V_{\mathrm{Ae}}(x) + V_{\mathrm{Be}}(r) + V_{\mathrm{AB}}(R), \tag{3.159}$$

Making the substitution

$$\Psi(\mathbf{s}, \mathbf{R}) = \bar{F}(\mathbf{R})\psi(\mathbf{s}, \mathbf{R}), \tag{3.160}$$

where $\bar{F}(\mathbf{R})$ satisfies the Schrödinger equation

$$\left[-\frac{1}{2\mu} \nabla_R^2 + W(R) - E \right] \bar{F}(\mathbf{R}) = 0, \tag{3.161}$$

we find

$$-\frac{1}{2\mu}\bar{F}(\mathbf{R})\nabla_R^2\psi(\mathbf{s},\mathbf{R}) - \frac{1}{\mu}\nabla_R\bar{F}(\mathbf{R})\cdot\nabla_R\psi(\mathbf{s},\mathbf{R}) + \bar{F}(\mathbf{R})H_e\psi(\mathbf{s},\mathbf{R}) = 0. \quad (3.162)$$

If semi-classical conditions apply, as discussed in the last section, $\bar{F}(\mathbf{R})$ can be approximated by (3.142) and (3.146), and the first term in (3.162) can be neglected compared with the second. Both ψ and H_e depend on $\mathbf{R}$ which can be expressed in terms of the impact parameter $\mathbf{b}$ and the time t by the trajectory equation (3.144) so that, in this semi-classical approximation, (3.162) can be written as a time-dependent Schrödinger equation

$$\mathrm{i}\left.\frac{\partial\psi(\mathbf{s},t)}{\partial t}\right|_{\mathbf{s}} = H_e(\mathbf{s},t)\psi(\mathbf{s},t). \quad (3.163a)$$

In obtaining (3.163a) we replaced ∇_R by a time derivative using the relations:

$$\frac{1}{\bar{F}(\mathbf{R})}\left[-\frac{1}{\mu}\nabla_R\bar{F}(\mathbf{R})\cdot\nabla_R\right] = -\frac{\mathrm{i}}{\mu}\mathbf{p}(\mathbf{R})\cdot\nabla_R = -\mathrm{i}\mathbf{v}(t)\cdot\nabla_R = -\mathrm{i}\frac{\partial}{\partial t}. \quad (3.163b)$$

In (3.163a) $\mathbf{b}$ enters as a parameter and the independent variables are t and $\mathbf{s}$. The vector $\mathbf{s}$ is defined with respect to space-fixed coordinates with the Z-axis parallel to the direction of incidence, which is the direction of $\mathbf{v}_0 = \mathbf{v}(-\infty)$. The operation $\partial/\partial t$ is performed keeping $\mathbf{s}$ fixed in the space-fixed coordinates.

It is interesting to note that the potential terms in H_e depending on $\mathbf{R}$ alone, $(V_{AB}(R) - W(R))$, can be removed from (3.163) by the phase transformation

$$\psi(\mathbf{s},t) = \tilde{\psi}(\mathbf{s},t)\exp\left[-\mathrm{i}\int_0^t (V_{AB}(R') - W(R'))\,\mathrm{d}t'\right], \quad (3.164)$$

where $\mathbf{R}' = \mathbf{R}(\mathbf{b}, t')$.

The transformed wave function $\tilde{\psi}(\mathbf{s},t)$ satisfies:

$$\mathrm{i}\left.\frac{\partial\tilde{\psi}(\mathbf{s},t)}{\partial t}\right|_{\mathbf{s}} = \tilde{H}_e(\mathbf{s},t)\tilde{\psi}(\mathbf{s},t), \quad (3.165)$$

with

$$\tilde{H}_e(\mathbf{s},t) = -\tfrac{1}{2}\nabla_s^2 + V_{Ae}(x) + V_{Be}(r). \quad (3.166)$$

This shows that the probability $P = |\langle\phi_f(t)|\psi(t)\rangle|^2$ of attaining a particular final state described by an unperturbed wave function ϕ_f at time t can be also expressed as $|\langle\phi_f(t)|\tilde{\psi}(t)\rangle|^2$. The potential $[V_{AB} - W]$ plays a role in determining the probability because $W(R)$ determines the trajectory $\mathbf{R} = \mathbf{R}(\mathbf{b}, t)$. If this trajectory is approximated by an undeviated straight line,

$$\mathbf{R} = \mathbf{b} + \mathbf{v}_0 t, \qquad \mathbf{b}\cdot\mathbf{v}_0 = 0, \quad (3.167)$$

then the potentials V_{AB} and W have no effect whatsoever on the calculation of probabilities. This is not the case for differential cross sections, the calculation of which depends on the phase of the wave function, which in turn depends on V_{AB}, even in the straight line approximation.

Asymptotic solutions

The solution of the time-dependent equations (3.163) or (3.165) parallels that of the time-independent equation discussed earlier. Asymptotic solutions are defined for large values of $|t|$, when A and B are well separated. If the electron is attached to the target ion B, and $r \ll R$, and if $[V_{Ae}(x) + V_{AB}(R) - W(R)] \to 0$ faster than $1/R$, the wave equation (3.165) becomes, in the limit of large $|t|$:

$$\left[-\tfrac{1}{2}\nabla_s^2 + V_{Be}(r) - \mathrm{i}\left.\frac{\partial}{\partial t}\right|_{\mathbf{s}}\right]\Phi_j^B(\mathbf{s}, t) = 0. \tag{3.168}$$

It should be noted that $\mathbf{r}$ and hence V_{Be} are functions of t, since

$$\mathbf{r} = \mathbf{s} + q_c\mathbf{R}(\mathbf{b}, t), \qquad q_c = \frac{M_A}{M_A + M_B}. \tag{3.169}$$

Solutions (3.168) are

$$\Phi_j^B(\mathbf{s}, t) = \phi_j(\mathbf{r}) \exp[\mathrm{i}\alpha_j(t)], \tag{3.170}$$

where $\phi_j(\mathbf{r})$ is an internal wave function of the atom $(B + e^-)$ satisfying (3.15) and where

$$\alpha_j = -\varepsilon_j t - q_c\mathbf{v}(t)\cdot\mathbf{s} - \tfrac{1}{2}q_c^2\int_0^t v^2(t')\,\mathrm{d}t'. \tag{3.171}$$

The relative velocity $\mathbf{v}$ of A and B becomes constant in the asymptotic region. For $t \to -\infty$, $\mathbf{v} \to \mathbf{v}_0$, where $\mathbf{v}_0$ is the incident velocity, which is parallel to the Z-axis. For $t \to +\infty$, $\mathbf{v} \to \mathbf{v}(\infty)$, where $\mathbf{v}(\infty)$ is parallel to the asymptote of the trajectory. That (3.170) is a solution of (3.168) is easily verified since for large $|t|$, $\dot{\mathbf{v}} = 0$ and

$$\mathrm{i}\left.\frac{\partial}{\partial t}\right|_{\mathbf{s}}\Phi_j^B(\mathbf{s}, t) = \{(\varepsilon_j + \tfrac{1}{2}q_c^2 v^2)\phi_j(\mathbf{r}) + \mathrm{i}q_c\mathbf{v}\cdot\nabla_r\phi_j(\mathbf{r})\}\exp(\mathrm{i}\alpha_j),$$

and we also have that

$$-\tfrac{1}{2}\nabla_s^2\Phi_j^B(\mathbf{s}, t) = -\tfrac{1}{2}\{[\nabla_r^2\phi_j(\mathbf{r})] - 2\mathrm{i}q_c\mathbf{v}\cdot\nabla_r\phi_j(\mathbf{r}) - q_c^2 v^2\phi_j(\mathbf{r})\}\exp(\mathrm{i}\alpha_j). \tag{3.172}$$

Using these results together with the wave equation (3.15) for $\phi_j(\mathbf{r})$, (3.168) is found to be satisfied. In the asymptotic region, $|t| \to \infty$, $\exp(\mathrm{i}\alpha_j) = \exp(-\mathrm{i}\varepsilon_j t)U$ where

$$U = \exp \mathrm{i}(-q_c \mathbf{v}\cdot\mathbf{s} - \tfrac{1}{2}q_c^2 v^2 t). \tag{3.173}$$

The physical significance of U is easily appreciated, since $-q_c\mathbf{v}$, is the velocity of the electron attached to B, relative to the chosen origin C, the centre of mass of A and B, while $\frac{1}{2}q_c^2 v^2$ is the corresponding kinetic energy. It follows that U is the operator which effects a Galilean transformation from a frame with origin at C to a frame with origin at B. In the asymptotic region, $|t| \to \infty$ (or in the straight line trajectory approximation for all values of t) $\dot{\mathbf{v}}(t) = 0$ and both the frames are inertial. On the other hand, in the interaction region for small t, if the trajectory is non-linear, $\dot{\mathbf{v}} \neq 0$, and only the centre-of-mass frame centred on C is a suitable inertial frame.

If the electron is attached to the projectile A in the asymptotic region $|t| \to \infty$, $x \ll r$, and if $[V_{\mathrm{Be}}(r) + V_{\mathrm{AB}}(R) - W(R)] \to 0$ faster than $1/R$, the asymptotic wave function $\mathrm{X}_n(\mathbf{s}, t)$ satisfies

$$\left[-\tfrac{1}{2}\nabla_s^2 + V_{\mathrm{Ae}}(x) - \mathrm{i}\left.\frac{\partial}{\partial t}\right|_{\mathbf{s}}\right]\mathrm{X}_n^{\mathrm{A}}(\mathbf{s}, t) = 0, \tag{3.174a}$$

where

$$\mathbf{x} = \mathbf{s} - p_c\mathbf{R}(\mathbf{b}, t), \qquad p_c = \frac{M_{\mathrm{B}}}{M_{\mathrm{A}} + M_{\mathrm{B}}}. \tag{3.174b}$$

The required solutions of (3.174) are

$$\mathrm{X}_n^{\mathrm{A}}(\mathbf{s}, t) = \chi_n(\mathbf{x}) \exp[\mathrm{i}\beta_n(t)], \tag{3.175}$$

where $\chi_n(\mathbf{x})$ satisfies (3.24) and

$$\beta_n(t) = -\eta_n t + p_c\mathbf{v}\cdot\mathbf{s} - \tfrac{1}{2}p_c^2 \int_0^t v^2(t')\,\mathrm{d}t'. \tag{3.176}$$

The momentum in the centre-of-mass frame of the electron, when attached to A, is $p_c\mathbf{v}$ and the corresponding kinetic energy is $\frac{1}{2}p_c^2 v^2 t$.

A modification to the asymptotic form arises if there is an overall Coulomb interaction at large separations in either the direct or the exchange channels. This will usually be the case since if Z_{A} and Z_{B} are the charges on the ion cores A and B, respectively, the overall interaction at large R in the direct channel is $-Z_{\mathrm{A}}(1 - Z_{\mathrm{B}})/R$ and in the exchange channel is $-Z_{\mathrm{B}}(1 - Z_{\mathrm{A}})/R$. Only in the case $Z_{\mathrm{A}} = Z_{\mathrm{B}} = 1$ is there no overall Coulomb interaction in either channel. For $Z_{\mathrm{B}} \neq 1$, asymptotic wave functions in the direct channel $\Phi_j^{\mathrm{B}}(\mathbf{s}, t)$ satisfies the equation (see (3.168))

$$\left[-\tfrac{1}{2}\nabla_s^2 + V_{\mathrm{Be}}(r) - \frac{Z_{\mathrm{A}}(1 - Z_{\mathrm{B}})}{R} - \mathrm{i}\left.\frac{\partial}{\partial t}\right|_{\mathbf{s}}\right]\Phi_j^{\mathrm{B}}(\mathbf{s}, t) = 0. \tag{3.177}$$

The asymptotic function Φ_j^{B} can be represented in the form (3.170), but the phase function $\alpha_j(t)$ is replaced by $\alpha_j^{\text{c}}(t)$ where

$$\alpha_j^{\text{c}}(t) = \alpha_j(t) + Z_{\text{A}}(1 - Z_{\text{B}}) \int^t \frac{1}{R(t')} \, \text{d}t' = \alpha_j(t) + \sigma(t),$$

with

$$\sigma(t) = -\frac{Z_{\text{A}}(1 - Z_{\text{B}})}{v_0} \log(R - v_0 t). \tag{3.178a}$$

The integration has been carried out along the straight line trajectory (3.167), which is a good approximation under most circumstances, and the irrelevant constant of integration is omitted. In the rearranged channel, a similar modification has to be employed when $Z_{\text{A}} \neq 1$, by introducing a phase β_n^{c} where

$$\beta_n^{\text{c}}(t) = \beta_n(t) + \bar{\sigma}(t),$$

and

$$\bar{\sigma}(t) = -\frac{Z_{\text{B}}(1 - Z_{\text{A}})}{v_0} \log(R - v_0 t). \tag{3.178b}$$

The variational method

To obtain an approximate solution of the time-dependent Schrödinger equation (3.163a), a trial wave function is now introduced of the form (cf. (3.37) and (3.38))

$$\psi(\mathbf{s}, t) = \sum_{j=1}^{N} \Phi_j(\mathbf{s}, t) A_j(\mathbf{b}, t) + \sum_{n=N+1}^{N+M} \text{X}_n(\mathbf{s}, t) A_n(\mathbf{b}, t), \tag{3.179}$$

where, in the asymptotic regions,

$$\begin{aligned} \Phi_j(\mathbf{s}, t) &\to \Phi_j^{\text{B}}(\mathbf{s}, t), \qquad |t| \to \infty \quad r \ll R, \\ \text{X}_n(\mathbf{s}, t) &\to \text{X}_n^{\text{A}}(\mathbf{s}, t), \qquad |t| \to \infty \quad x \ll R. \end{aligned} \tag{3.180}$$

We shall see later, in Chapters 7, 10, and 11, that in perturbative solutions of the impact parameter equation (3.163) it is very important to employ the phases α_j^{c}, β_n^{c} when there are overall Coulomb interactions in any channel. However, in the trial function (3.179) the Coulomb phase factor $\sigma(t)$, which is a function of **b** and t only, can be absorbed into the direct amplitudes $A_j(\mathbf{b}, t)$, $j \leq N$ and the corresponding Coulomb phase factor $\bar{\sigma}(t)$ can be absorbed into the exchange amplitudes $A_n(\mathbf{b}, t)$, $n > N + 1$. In what follows here, and in Chapter 5, it will be assumed that the Coulomb phases have been absorbed in this way, and correspondingly Φ_j^{B} is defined by (3.170) and (3.171), while X_n^{A} is defined by (3.175) and (3.176).

It will be recognized that $\Phi_j(\mathbf{s}, t)$ and $\text{X}_n(\mathbf{s}, t)$ are essentially the same as the functions $\Phi_j(\mathbf{s}, \mathbf{R})$ and $\text{X}_n(\mathbf{s}, \mathbf{R})$ defined in (3.89). The most important difference is that previously we had to take $\mathbf{v}_j^c$ and $\bar{\mathbf{v}}_j^c$ to be parallel to the Z-axis, while in the time-dependent formalism the direction of $\mathbf{v}(t)$ enters in a natural manner. For large negative values of t, $t \to -\infty$, $\mathbf{v}$ is parallel to the Z-axis, but elsewhere $\mathbf{v}(t)$ is tangential to the trajectory at each point.

Particular choices for the basis functions Φ_j and X_n will be the subject of subsequent chapters. These include atomic functions for which $\Phi_j = \Phi_j^B$ and $X_n = X_n^A$, and molecular functions for which Φ_j and X_n are taken to be the product of molecular orbitals multiplied by the asymptotic phases, in which case $\Phi_j = \psi_j^B(\mathbf{s}, \mathbf{R}) \exp i\alpha_j(t)$, $X_n = \psi^A(\mathbf{s}, \mathbf{R}) \exp i\beta_n(t)$ (see (3.72)). More flexible phase functions α_j and β_n, asymptotic to (3.171) and (3.176) can also be employed.

Equations to determine the unknown amplitude $a_j(\mathbf{b}, t)$ and $c_n(\mathbf{b}, t)$ can be found from the variational principle

$$\delta \int_{-\infty}^{\infty} \mathrm{d}t \int \psi^*(\mathbf{s}, t)\left[H_e - i\left.\frac{\partial}{\partial t}\right|_{\mathbf{s}}\right]\psi(\mathbf{s}, t)\, \mathrm{d}\mathbf{s} = 0, \tag{3.181}$$

under independent first-order variations in ψ and ψ^*. Varying the amplitudes a_j^* and c_n^*, we arrive at the coupled equations

$$\int \Phi_j^*(\mathbf{s}, t)\left[H_e - i\left.\frac{\partial}{\partial t}\right|_{\mathbf{s}}\right]\psi(\mathbf{s}, t)\, \mathrm{d}\mathbf{s} = 0; \qquad j = 1, 2, \ldots, N, \tag{3.182a}$$

$$\int X_n^*(\mathbf{s}, t)\left[H_e - i\left.\frac{\partial}{\partial t}\right|_{\mathbf{s}}\right]\psi(\mathbf{s}, t)\, \mathrm{d}\mathbf{s} = 0; \qquad n = N+1, N+2, \ldots, N+M. \tag{3.182b}$$

If the functions Φ_j and X_n satisfy the orthonormality conditions

$$\int \Phi_j^*(\mathbf{s}, t)\Phi_k(\mathbf{s}, t)\, \mathrm{d}\mathbf{s} = \delta_{jk}; \qquad \int X_n^*(\mathbf{s}, t)X_m(\mathbf{s}, t)\delta_{nm}\, \mathrm{d}\mathbf{s} \tag{3.183}$$

and the overlap integral $S_{jn}(t)$ is defined as

$$S_{jn}(t) = \int \Phi_j^*(\mathbf{s}, t)X_n(\mathbf{s}, t)\, \mathrm{d}\mathbf{s}, \tag{3.184}$$

the coupled equation (3.182) are easily reduced to the form

$$i\left[\dot{A}_j + \sum_{m=N+1}^{N+M} S_{jm}\dot{A}_m\right] = \sum_{k=1}^{N} H_{jk}A_k + \sum_{m=N+1}^{M+N} K_{jm}A_m; \qquad j = 1, 2, \ldots, N,$$

$$i\left[\sum_{k=1}^{N} \bar{S}_{nk}\dot{A}_k + \dot{A}_n\right] = \sum_{k=1}^{N} \bar{K}_{nk}A_k + \sum_{m=N+1}^{N+M} \bar{H}_{nm}A_m;$$

$$n = N+1, N+2, \ldots, N+M, \tag{3.185}$$

where a dot denotes differentiation with respect to t and where

$$
\begin{aligned}
\bar{S}_{nk} &= S^*_{kn} \\
H_{jk} &= \int \Phi^*_j \left[H_e - i \frac{\partial}{\partial t}\bigg|_s \right] \Phi_k \, ds, \qquad j \le N,\ k \le N \\
\bar{H}_{nm} &= \int X^*_n \left[H_e - i \frac{\partial}{\partial t}\bigg|_s \right] X_m \, ds, \qquad j > N,\ m > N \\
K_{jm} &= \int \Phi^*_j \left[H_e - i \frac{\partial}{\partial t}\bigg|_s \right] X_m \, ds, \qquad j \le N,\ m > N \\
\bar{K}_{nk} &= \int X^*_n \left[H_e - i \frac{\partial}{\partial t}\bigg|_s \right] \Phi_k \, ds, \qquad n > N,\ k \le N.
\end{aligned}
\tag{3.186}
$$

Boundary conditions

Boundary conditions similar to those of (3.154) and (3.155) apply. If the incident channel is channel i, we must have

$$\lim_{t \to -\infty} A_j(\mathbf{b}, t) = \delta_{ji}. \tag{3.187a}$$

The particular solution of the time-dependent Schrödinger equation (3.165) with this boundary condition will be denoted by $\psi_i(\mathbf{s}, t)$. For large positive values of t, the wave function becomes a linear combination of the asymptotic solutions Φ^B_j and X^A_n (see (3.179) and (3.180)) so that the probability amplitude for finding the system in the final state j is $\mathscr{A}_{ji}(\mathbf{b})$ where

$$\mathscr{A}_{ji}(\mathbf{b}) = \lim_{t \to \infty} A_j(\mathbf{b}, t). \tag{3.187b}$$

The corresponding total cross section is given by (3.156), where the probability $P_{ji}(b)$ is expressed as

$$P_{ji}(\mathbf{b}) = |\mathscr{A}_{ji}(\mathbf{b})|^2. \tag{3.188}$$

Conservation of probability

If the coupled equations (3.185), and hence (3.182), are satisfied it follows that

$$I = \int \psi^*(\mathbf{s}, t) \left[H_e - i \frac{\partial}{\partial t}\bigg|_s \right] \psi(\mathbf{s}, t) \, d\mathbf{s} = 0, \tag{3.189}$$

where ψ is the trial function (3.179). From this it can be shown that the normalization of ψ is independent of time, since

$$\begin{aligned}\frac{\mathrm{d}}{\mathrm{d}t}\int \psi^*(\mathbf{s},t)\psi(\mathbf{s},t)\,\mathrm{d}\mathbf{s} &= \int\left[\frac{\partial\psi^*}{\partial t}\psi + \psi^*\frac{\partial\psi}{\partial t}\right]\mathrm{d}\mathbf{s}\\ &= \mathrm{i}\int\{(H_\mathrm{e}\psi)^*\psi - \psi^*(H_\mathrm{e}\psi)\}\,\mathrm{d}\mathbf{s} \qquad (3.190)\\ &= 0,\end{aligned}$$

where the last line follows as H_e is Hermitian (Green 1965). The initial conditions (3.187) ensure that $\psi(\mathbf{s}, t)$ is normalized to unity at $t \to -\infty$, and it follows from (3.190) that $\psi(\mathbf{s}, t)$ is normalized to unity for all values of t. Since the overlap matrix S_{ij} vanishes as $|t| \to \infty$ for $i \neq j$, it follows that

$$\lim_{t\to\infty}\sum_{j=1}^{N+M}|A_j(\mathbf{b},t)|^2 = \sum_{j=1}^{N+M}|\mathscr{A}_{ji}(\mathbf{b})|^2 = 1, \qquad (3.191)$$

which is the statement of conservation of probability (unitarity). By inserting (3.179) into (3.190), and using the coupled equations (3.185), we find that the matrix elements defined in (3.186) satisfy

$$H_{jk} = H^*_{kj}; \qquad \bar{H}_{nm} = \bar{H}^*_{mn},$$

and

$$-\mathrm{i}\dot{S}_{jn} = K_{jn} - \bar{K}^*_{nj}. \qquad (3.192)$$

The last condition shows that the complete coupling matrix

$$\mathbf{Q} = \begin{pmatrix}\mathbf{H} & \mathbf{K}\\ \bar{\mathbf{K}} & \bar{\mathbf{H}}\end{pmatrix} \qquad (3.193)$$

is only Hermitian if the overlap matrix is independent of time.

Time reversal

Consider a particular solution $\psi_i(\mathbf{s}, t)$ of the time-dependent Schrödinger equation (3.165) corresponding to an incident channel i, so that the boundary condition is (3.187). From (3.179) and (3.180), we see that the probability amplitude $\mathscr{A}_{ji}(\mathbf{b})$ can be expressed as

$$\mathscr{A}_{ji}(\mathbf{b}) = \lim_{t\to\infty}\langle\eta_j(\mathbf{s},t)\,|\,\psi_i(\mathbf{s},t)\rangle \qquad (3.194)$$

where η_j is one of the asymptotic solutions of the Schrödinger equation, that is

$$\eta_j(\mathbf{s},t) = \begin{cases}\Phi^\mathrm{B}_j(\mathbf{s},t) & j = 1, 2, \ldots, N\\ \mathrm{X}^\mathrm{A}_j(\mathbf{s},t) & j = N+1, N+2, \ldots, N+M,\end{cases} \qquad (3.195)$$

and the notation $\langle A|B\rangle$ has been employed where

$$\langle A(\mathbf{s}, t)|B(\mathbf{s}, t)\rangle = \int A^*(\mathbf{s}, t)B(\mathbf{s}, t)\,d\mathbf{s}. \tag{3.196}$$

Since $\tilde{H}_e(-t) = \tilde{H}_e(t)$, the time-reversed wave function $\psi_j^T(\mathbf{s}, t)$ defined as

$$\psi_j^T(\mathbf{s}, t) = \psi_j^*(\mathbf{s}, -t) \tag{3.197}$$

satisfies the same equation as $\psi_j(\mathbf{s}, t)$, namely (3.165). The boundary condition satisfied by ψ_j^T is that as $t \to +\infty$, ψ_j^T coincides with the asymptotic state $\eta_j(\mathbf{s}, t)$:

$$\lim_{t\to\infty} \psi_j^T(\mathbf{s}, t) \to \eta_j(\mathbf{s}, t), \tag{3.198}$$

while as $t \to -\infty$, $\psi_j^T(\mathbf{s}, t)$ must approach a linear combination of the asymptotic states,

$$\lim_{t\to-\infty} \psi_j^T(\mathbf{s}, t) \to \sum_n \mathscr{A}_{nj}^T(\mathbf{b})\eta_n(\mathbf{s}, t). \tag{3.199}$$

In view of (3.198), we have from (3.194)

$$\mathscr{A}_{ji}(\mathbf{b}) = \lim_{t\to\infty} \langle\psi_j^T(\mathbf{s}, t)|\psi_i(\mathbf{s}, t)\rangle. \tag{3.200}$$

Proceeding as in (3.190), we see that if ψ_a, ψ_b are any two solutions of the Schrödinger equation (3.165), then

$$\frac{\partial}{\partial t}\langle\psi_a(\mathbf{s}, t)|\psi_b(\mathbf{s}, t)\rangle = 0. \tag{3.201}$$

Since $\psi_j^T(\mathbf{s}, t)$ and $\psi_i(\mathbf{s}, t)$ are such solutions, we see that

$$\mathscr{A}_{ji}(\mathbf{b}) \equiv \langle\psi_j^T(\mathbf{s}, t)|\psi_i(\mathbf{s}, t)\rangle, \tag{3.202}$$

is independent of t. Changing t into $-t$ we have

$$\begin{aligned}\mathscr{A}_{ji}(\mathbf{b}) &= \langle\psi_j^T(\mathbf{s}, -t)|\psi_i(\mathbf{s}, -t)\rangle\\ &= \langle\psi_i^T(\mathbf{s}, t)|\psi_j(\mathbf{s}, t)\rangle\\ &= \mathscr{A}_{ij}(\mathbf{b}).\end{aligned} \tag{3.203}$$

Similarly we find that the amplitudes $\mathscr{A}_{ij}^T$ satisfy

$$\mathscr{A}_{ij}^T = \mathscr{A}_{ji} = \mathscr{A}_{ij}. \tag{3.204}$$

The symmetry of the amplitude $\mathscr{A}_{ij}$ or $\mathscr{A}_{ij}^T$ is a consequence of the time-reversal invariance of the Schrödinger equation.

3.7.1 The connection between the impact parameter and eikonal models

Equations (3.185) for the impact parameter model can be written together in matrix form as

$$\mathrm{i}\dot{\mathbf{A}} = \mathbf{V}\mathbf{A} \tag{3.205}$$

where $\mathbf{A}$ is the column vector with components A_j and $\mathbf{V}$ is the $(N+M) \times (N+M)$-dimensional matrix

$$\mathbf{V} = \mathbf{N}^{-1}\mathbf{Q} \tag{3.206}$$

with

$$\mathbf{N} = \begin{pmatrix} 1 & \mathbf{S} \\ \bar{\mathbf{S}} & 1 \end{pmatrix}; \qquad \mathbf{Q} = \begin{pmatrix} \mathbf{H} & \mathbf{K} \\ \bar{\mathbf{K}} & \bar{\mathbf{H}} \end{pmatrix}. \tag{3.207}$$

The submatrices $\mathbf{S}, \bar{\mathbf{S}}, \mathbf{H}, \bar{\mathbf{H}}, \mathbf{K}, \bar{\mathbf{K}}$ have elements defined by (3.184) and (3.186). We shall now show that the solutions of (3.205) describe essentially the same approximation as do the solutions of (3.151) (or (3.153)), which were obtained from time-independent theory using the eikonal approximation. One advantage gained by reformulating the model in terms of the time-dependent impact parameter approach is that it is straightforward to devise schemes for optimizing the translational factors or other attributes of the trial functions by using the variational principle (3.181).

The potential matrix $\mathbf{U}$ which enters the eikonal equation (3.151) is of the form (using (3.149b) and (3.94))

$$\mathbf{U} = \mathbf{N}^{-1}\bar{\mathbf{Q}} \tag{3.208}$$

where $\mathbf{N}$ is the overlap matrix defined by (3.67) or (3.207), in terms of submatrices $\mathbf{S}$, determined by (3.184) or (3.91), and

$$\bar{\mathbf{Q}} = \begin{pmatrix} \mathbf{V} + \mathrm{i}\mathbf{p}\cdot\mathbf{W}, & \mathbf{K} + \mathrm{i}\mathbf{p}\cdot\mathbf{L} \\ \bar{\mathbf{K}} + \mathrm{i}\mathbf{p}\cdot\mathbf{L}, & \bar{\mathbf{V}} + \mathrm{i}\mathbf{p}\cdot\overline{\mathbf{W}} \end{pmatrix}. \tag{3.209}$$

The elements of the submatrices appearing in (3.209) are defined by (3.94), and $\mathbf{p}$ is the local classical momentum.

Consider the submatrix $\mathrm{i}\mathbf{p}\cdot\mathbf{W}$ which has elements

$$(\mathbf{p}\cdot\mathbf{W})_{jl} = -\frac{\mathrm{i}}{\mu}\int \Phi_j^*(\mathbf{s}, \mathbf{R})\mathbf{p}\cdot\nabla_R\Phi_l(\mathbf{s}, \mathbf{R})\,\mathrm{d}\mathbf{s}. \tag{3.210}$$

Since $\mathbf{p}/\mu = \dot{\mathbf{R}}$, we can write

$$(\mathbf{p}\cdot\mathbf{W})_{jl} = \int \Phi_j^*(\mathbf{s}, \mathbf{R})(-\mathrm{i})\left.\frac{\partial}{\partial t}\right|_{\mathbf{s}} \Phi_l(\mathbf{s}, \mathbf{R})\,\mathrm{d}\mathbf{s}. \tag{3.211}$$

Similarly, the submatrices $(\mathbf{p}\cdot\mathbf{L})$, $(\mathbf{p}\cdot\bar{\mathbf{L}})$, and $(\mathbf{p}\cdot\bar{\mathbf{W}})$ can be reduced to matrix elements of $-\mathrm{i}\partial/\partial t$. The submatrices $\mathbf{V}$, $\mathbf{K}$, $\bar{\mathbf{K}}$, and $\bar{\mathbf{V}}$ can be written as the sum of matrix elements of the total Hamiltonian and $\mathbf{N}$ times a diagonal matrix containing the internal energies ε_j, η_m and the kinetic energy terms $\frac{1}{2}v_j^{c^2}$, $\frac{1}{2}\bar{v}_m^{c^2}$. Since in the eikonal approximation the basis functions Φ_l and X_m are assumed to vary slowly with $\mathbf{R}$, the term $-(1/2\mu)\nabla_R^2$ contained in the total Hamiltonian H can be dropped in evaluating the potential matrix elements (3.94), so that H can be replaced by H_e (see (3.157)). It follows that

$$\mathbf{U} = \mathbf{N}^{-1}\bar{\mathbf{Q}} = \mathbf{N}^{-1}\mathbf{Q} + \mathbf{D}, \tag{3.212}$$

where $\bar{\mathbf{Q}}$ is defined by (3.209) and $\mathbf{Q}$ by (3.207), and where $\mathbf{D}$ is a diagonal matrix with elements $(-\varepsilon_j + \frac{1}{2}v_j^{c^2})$ in direct channels or $(-\eta_n + \frac{1}{2}\bar{v}_n^{c^2})$ in exchange channels. These diagonal terms can be combined with the remaining diagonal terms in (3.151). Thus

$$(-\varepsilon_j + \tfrac{1}{2}v_j^{c^2}) - W(t) - \Delta\bar{\varepsilon}_j = -W(t) + \frac{1}{2\mu}K^2 - E, \tag{3.213a}$$

and

$$(-\eta_n + \tfrac{1}{2}v_n^{c^2}) - W(t) - \Delta\bar{\varepsilon}_n = -W(t) + \frac{1}{2\mu}K^2 - E, \tag{3.213b}$$

where we have employed (3.86), (3.87), and (3.140b).

Using (3.213a and b), equation (3.151) takes the form

$$\begin{aligned} \mathrm{i}\dot{\mathbf{a}} &= \mathbf{N}^{-1}\mathbf{Q}\mathbf{a} + \left(\frac{1}{2\mu}K^2 - W(t) - E\right)\mathbf{a} \\ &= \mathbf{V}\mathbf{a} + \left(\frac{1}{2\mu}K^2 - W(t) - E\right)\mathbf{a}. \end{aligned} \tag{3.214}$$

Finally by introducing the phase transformation

$$\mathbf{a} = \mathbf{A}\exp\left[-\mathrm{i}\int_{-\infty}^{t}\left(\frac{1}{2\mu}K^2 - W(t') - E\right)\mathrm{d}t'\right] \tag{3.215}$$

we see that the amplitudes $\mathbf{A}$ satisfy (3.205), so that the eikonal and impact parameter models are *identical* in content, provided the conditions under which a common trajectory can be used are satisfied.

3.8 DIFFERENTIAL CROSS SECTIONS IN THE IMPACT PARAMETER MODEL

The differential cross sections for elastic scattering, excitation, or charge exchange can be calculated from the probability amplitudes $\mathscr{A}_{ji}(\mathbf{b})$ obtained from the impact parameter model. Various derivations of the semi-classical and impact parameter differential cross sections have been given in the literature. Particular mention should be made of the work of Bates and Holt (1966), McCarroll and Salin (1968), Schiff (1954), and Wilets and Wallace (1968). The desired expressions are most easily obtained by starting from the standard integral expression for the scattering amplitude $f_{ji}(\Theta, \Phi)$. The following expression applies to both direct and rearrangement scattering, so that the distinction made between f_{ji} and g_{ji} earlier in the chapter is not required.

$$f_{ji}(\Theta, \Phi) = -\frac{\mu}{2\pi}\int \mathrm{d}\mathbf{s} \int \mathrm{d}\mathbf{R}\Phi_j'^*(\mathbf{s}, \mathbf{R})V_j(\mathbf{s}, \mathbf{R})\Psi_i(\mathbf{s}, \mathbf{R}) \tag{3.216}$$

where Ψ_i is a wave function satisfying (3.2) or (3.157) corresponding to the boundary condition in which the incident particle is in channel i, Φ_j' is the unperturbed wave function in channel j, and V_j is the corresponding perturbation. The unperturbed functions Φ_j' satisfy the Schrödinger equation

$$(H - V_j)\Phi_j' = E\Phi_j' \tag{3.217a}$$

where, for excitation of the atom $(\mathrm{B} + \mathrm{e}^-)$,

$$V_j = V_{\mathrm{Ae}} + V_{\mathrm{AB}}, \tag{3.217b}$$

and for charge exchange leading to a level of the atom $(\mathrm{A} + \mathrm{e}^-)$

$$V_j = V_{\mathrm{Be}} + V_{\mathrm{AB}}. \tag{3.217c}$$

In the impact parameter model the wave function Ψ_i is expressed as (see (3.160))

$$\Psi_i(\mathbf{s}, \mathbf{R}) \approx \bar{F}_i(\mathbf{R})\psi_i(\mathbf{s}, t), \tag{3.218}$$

where $\bar{F}_i(\mathbf{R})$ is a semi-classical approximation to the solution of (3.141) and ψ_i satisfies (3.163). The dependence of ψ_i on t is through the predetermined classical trajectory $\mathbf{R} = R(\mathbf{b}, t)$. To start with, we shall consider the case of the straight line trajectory (3.167), so that $W(R) = 0$, and $\bar{F}_i(\mathbf{R})$ is the unperturbed plane wave:

$$\bar{F}_i(\mathbf{R}) = \exp[\mathrm{i}\mathbf{K}_i\cdot\mathbf{R}], \tag{3.219}$$

where $\mathbf{K}_i$ is along the direction of incidence and $|\mathbf{K}_i| = \mu v_0$. The unperturbed wave functions Φ_j' satisfying (3.217a) are the products of the plane wave $F_j'(\mathbf{R}) = \exp[\mathrm{i}\mathbf{K}_j\cdot\mathbf{R}]$ and an asymptotic solution $\eta_j(\mathbf{s}, t)$ of equation (3.163). The vector $\mathbf{K}_j$ is along the direction (Θ, Φ) and $|\mathbf{K}_j| = |\mathbf{K}_i|$. Thus we have

$$\Phi_j'(\mathbf{s}, \mathbf{R}) = \exp(\mathrm{i}\mathbf{K}_j\cdot\mathbf{R})\eta_j(\mathbf{s}, t), \tag{3.220}$$

where $\eta_j(\mathbf{s}, t)$ is defined through (3.195), (3.170), and (3.175). Using the trajectory equation (3.167) to express $d\mathbf{R}$ as $d\mathbf{R} = v_0\, d^2\mathbf{b}\, dt$ and using (3.218) and (3.220) in (3.216), we find that

$$f_{ji}(\Theta, \Phi) = \frac{\mu}{2\pi} v_0 \int_{-\infty}^{+\infty} dt \int d^2\mathbf{b} \exp[i(\mathbf{K}_i - \mathbf{K}_j)\cdot\mathbf{R}]\langle\eta_j|V_j|\psi_i\rangle. \quad (3.221)$$

At the velocities for which the impact parameter approximation is accurate, the probability of scattering is small except at small angles. Since $|\mathbf{K}_i - \mathbf{K}_j| = \mu v_0(1 - \cos\Theta)$, in the small-angle region the approximation can be made in which

$$(\mathbf{K}_i - \mathbf{K}_j)\cdot\mathbf{R} \approx \mathbf{K}\cdot\mathbf{b}, \quad (3.222)$$

where the momentum transfer $\mathbf{K}$ is of magnitude

$$K = |\mathbf{K}_i - \mathbf{K}_j| = 2\mu v_0 \sin\frac{\Theta}{2} \approx \mu v_0 \Theta. \quad (3.223)$$

Using (3.222) in (3.221), we have

$$f_{ji}(\Theta, \Phi) = -\frac{\mu v_0}{2\pi} \int_{-\infty}^{\infty} dt \int d^2\mathbf{b} \exp[i\mathbf{K}\cdot\mathbf{b}]\langle\eta_j|V_j|\psi_i\rangle. \quad (3.224)$$

This expression can be reduced by noting that

$$V_j\eta_j = \left(H_e - i\frac{\partial}{\partial t}\bigg|_s\right)\eta_j. \quad (3.225)$$

On integrating by parts with respect to t, and using the fact that H_e is Hermitian, together with (3.163a), we obtain

$$\begin{aligned} f_{ji}(\Theta, \Phi) &= -i\frac{\mu v_0}{2\pi} \int d^2\mathbf{b} \exp(i\mathbf{K}\cdot\mathbf{b})|\langle\eta_j(\mathbf{s}, t)\,|\,\psi_i(\mathbf{s}, t)\rangle|_{t=-\infty}^{\infty} \\ &= -i\frac{\mu v_0}{2\pi} \int d^2\mathbf{b} \exp(i\mathbf{K}\cdot\mathbf{b})\{\mathcal{A}_{ji}(\mathbf{b}) - \delta_{ji}\}. \end{aligned} \quad (3.226)$$

The two-dimensional impact parameter vector $\mathbf{b}$ lies in a plane which is normal to the direction of incidence. Its direction is defined by an azimuthal angle ϕ_b. By examining the interaction matrix elements it is straightforward to show that the amplitudes $\mathcal{A}_{ji}(\mathbf{b})$ for a reaction in which the target atom is in a state with magnetic number i and the final atomic state has a magnetic quantum number m_j are of the form

$$\mathcal{A}_{ji}(\mathbf{b}) = \mathcal{A}^{s}_{ji}(b) \exp[i(m_i - m_j)\phi_b] \quad (3.227)$$

where $\mathcal{A}^{s}_{ji}(b)$ is independent of the azimuthal angle ϕ_b. The amplitude $\mathcal{A}^{s}_{ji}(b)$ can be calculated by taking a space-fixed coordinate frame with the Z-axis in the direction of incidence, as usual, and the X-axis in the scattering plane of the heavy particles A and B. It should be noted that since

$$|\mathscr{A}_{ji}(\mathbf{b})| = |\mathscr{A}^{\text{s}}_{ji}(b)|, \tag{3.228}$$

total cross sections can be calculated directly from the amplitudes $\mathscr{A}^{\text{s}}_{ji}(b)$.

In the small-angle approximation (3.222), the momentum transfer vector **K** lies in the XY-plane in a direction specified by Φ, the azimuthal angle of $\mathbf{K}_j$. The angle between **K** and **b** is $(\Phi - \phi_b)$ so that,

$$\mathbf{K}\cdot\mathbf{b} = Kb\cos(\Phi - \phi_b). \tag{3.229}$$

From (3.226), and using the result

$$\frac{1}{2\pi}\int_0^{2\pi} \exp[-\text{i}n\phi + \text{i}x\sin\phi]\,\text{d}\phi = J_n(x), \tag{3.230}$$

we find with $\Delta m = m_j - m_i$

$$f_{ji}(\Theta, \Phi) = \text{i}\mu v_0 \exp[-\text{i}\Delta m\Phi](\text{i})^{\Delta m}\int_0^\infty b\,\text{d}b J_{\Delta m}(Kb)\{\mathscr{A}^{\text{s}}_{ji}(b) - \delta_{ji}\}. \tag{3.231}$$

The differential cross section for scattering from the state i to the state j is given in terms of the scattering amplitude $f_{ji}(\Theta, \Phi)$ by

$$\frac{\text{d}\sigma_{ji}}{\text{d}\Omega} = |f_{ji}(\Theta, \Phi)|^2. \tag{3.232}$$

The total cross section σ_{ji} is obtained by integrating the differential cross section over all solid angles:

$$\sigma_{ji} = \int_0^\pi \sin\Theta\,\text{d}\Theta\int_0^{2\pi} |f_{ji}(\Theta, \Phi)|^2\,\text{d}\Phi. \tag{3.233}$$

Since $f_{ji}(\Theta, \Phi)$ decreases rapidly with Θ, we can replace the integral over Θ by an integral over K

$$\int_0^\pi \sin\Theta\,\text{d}\Theta \approx \left(\frac{1}{\mu v_0}\right)^2\int_0^\infty K\,\text{d}K \tag{3.234}$$

where the upper limit has been extended to infinity. Using the orthogonality property of the Bessel functions

$$\int_0^\infty xJ_m(ax)J_m(bx)\,\text{d}x = \frac{1}{a}\,\delta(a-b), \tag{3.235}$$

σ_{ji} can be expressed as

$$\sigma_{ji} = 2\pi\int_0^\infty b\,\text{d}b|\mathscr{A}^{\text{s}}_{ji}(b) - \delta_{ji}|^2, \tag{3.236}$$

in complete agreement with (3.188) and (3.156).

Non-linear trajectories

In the common trajectory approximation with $W(R) \neq 0$, the high-energy form of $\bar{F}_i(\mathbf{R})$ can be obtained from (3.142) and (3.146) by expanding $p(R)$ defined by (3.143b). We have for large K,

$$p(R) \approx K - \frac{\mu}{K} W(R). \tag{3.237}$$

On changing the integration variable in (3.146) to t with $\mathrm{d}\jmath = v\,\mathrm{d}t \approx v_0\,\mathrm{d}t = (K/\mu)\,\mathrm{d}t$, $\bar{F}(\mathbf{R})$ can be written approximately as

$$\bar{F}_i(\mathbf{R}) \approx \exp\left[\mathrm{i}KZ - \mathrm{i}\int_{-\infty}^{t} W(R(t'))\right]\mathrm{d}t'. \tag{3.238}$$

Again carrying through the analysis starting from (3.216) and (3.218) the scattering amplitude, in the small-angle approximation, is found to be

$$f_{ji}(\Theta, \Phi) = -\mathrm{i}\mu v_0 \exp[-\mathrm{i}\Delta m\Phi](-\mathrm{i})^{\Delta m}$$
$$\times \int_0^{\infty} b\,\mathrm{d}bJ_{\Delta m}(Kb)\{\mathscr{A}^{\mathrm{s}}_{ji}(b)\exp[2\mathrm{i}\delta(b)] - \delta_{ji}\} \tag{3.239a}$$

where

$$2\delta(b) = -\int_{-\infty}^{\infty} W[R(t)]\,\mathrm{d}t. \tag{3.239b}$$

It is important to notice that any overall b-dependent phase of the wave function ψ_i affects the differential cross section, but, as we have seen, not the total cross section. For example, if A and B represent bare nuclei we may set $W(R) = V_{\mathrm{AB}}(R) = Z_{\mathrm{A}}Z_{\mathrm{B}}/R$, in which case $\mathscr{A}^{\mathrm{s}}_{ji}(b)$ is computed from the Hamiltonian $\tilde{H}_{\mathrm{e}}$ (see (3.166)). Using (3.239), we have

$$2\delta(b) = -Z_{\mathrm{A}}Z_{\mathrm{B}}\int_{-\infty}^{\infty}\frac{1}{R(t)}\,\mathrm{d}t = -2Z_{\mathrm{A}}Z_{\mathrm{B}}\int_0^{\infty}\frac{1}{R(t)}\,\mathrm{d}t, \tag{3.240}$$

and with $R = (b^2 + v_0^2t^2)^{\frac{1}{2}}$ and using (3.178),

$$\delta(b) = -\lim_{\tau\to\infty}\frac{Z_{\mathrm{A}}Z_{\mathrm{B}}}{v_0}|\log\{(b^2 + v_0^2t^2)^{\frac{1}{2}} - v_0t\}|_0^{\tau}$$
$$= \frac{Z_{\mathrm{A}}Z_{\mathrm{B}}}{v_0}\log(b) + \bar{\delta}, \tag{3.241}$$

where $\bar{\delta}$ is an infinite phase independent of b, which can be omitted since the differential cross section depends on the modulus of f_{ji}. With the omission of $\bar{\delta}$, we obtain (for $i \neq j$)

$$f_{ji}(\Theta, \Phi) = -i\mu v_0 \exp[-i\Delta m\Phi](i)^{\Delta m} \int_0^\infty b\,db J_{\Delta m}(Kb)\mathscr{A}^{s}_{ji}(b)b^{2iv} \quad (3.242)$$

where $v = Z_A Z_B/v_0$.

The partial-wave series It is interesting to note that the expressions we have obtained for the differential cross sections can be derived from the full quantum mechanical partial-wave series. The exact differential cross section for scattering from a state i to a state j is

$$f_{ji}(\Theta, \Phi) = -i \sum_{L, L'=0} (2L+1)^{\frac{1}{2}} \left(\frac{\pi}{K_i K_j}\right)^{\frac{1}{2}} i^{L-L'}[\mathscr{S}^{LL'}_{ji} - \delta_{ji}] Y_{L', m_i - m_j}(\Theta, \Phi). \quad (3.243)$$

This is a slight generalization of (3.134) allowing the initial and final atomic states to have non-zero magnetic quantum numbers m_i, m_j. In the impact parameter approximation, the heavy particle scattering does not depend on the electronic energies so that $K_i = K_j = K$ and in addition the total heavy particle angular momentum is quasi-classical and a constant of the motion. This implies that $\mathscr{S}^{LL'}_{ji}$ is diagonal in L (but not necessarily in m) and that the sum over L can be replaced by an integral as in (3.136). Thus we obtain

$$f_{ji}(\Theta, \Phi) = -\frac{i}{K}(2\pi)^{\frac{1}{2}} \int_0^\infty (L + \tfrac{1}{2})^{\frac{1}{2}}[\mathscr{S}^{L}_{ji} - \delta_{ji}] Y_{L, m_i - m_j}(\Theta, \Phi)\, dL. \quad (3.244)$$

The $\mathscr{S}$-matrix elements are probability amplitudes and, in the impact parameter approximation, must be identified with the amplitudes $\mathscr{A}^{s}_{ji}(b) \times \exp[2i\delta(b)](i)^{\Delta m}$ where b and L are related by (3.145), namely

$$L + \tfrac{1}{2} = Kb. \quad (3.245)$$

If the small-angle approximation is made in (3.244)

$$Y_{L, m_i - m_j}(\Theta, \Phi) \simeq \left(\frac{L + \frac{1}{2}}{2\pi}\right)^{\frac{1}{2}} \exp(-i\Delta m\Phi) J_{\Delta m}(L\Theta),$$

$$\text{with } \Delta m \equiv m_j - m_i; \quad \sin\Theta \leq 1/L \quad (3.246)$$

and recollecting that for small Θ, $L\Theta \approx Kb$, we regain the expression (3.242). A form of the differential cross section valid at large angles can also be obtained by using the asymptotic expression

$$Y_{L,m_i-m_j}(\Theta,\Phi) \simeq (\sin\Theta)^{-\frac{1}{2}}\exp(-\mathrm{i}\Delta m\Phi)\cos\left\{(L+\tfrac{1}{2})\Theta-(\Delta m+1)\frac{\pi}{2}\right\},$$

$$\sin\Theta > \frac{1}{L}. \quad (3.247)$$

Employing (3.247) in (3.244) and setting $\mathscr{S}^L_{ji} = \mathscr{A}^{\mathrm{s}}_{ji}(b)\exp(2\mathrm{i}\delta(b))(\mathrm{i})^{\Delta m}$ we obtain the large-angle approximation

$$f_{ji}(\Theta,\Phi) = -\frac{\mathrm{i}(2\pi)^{\frac{1}{2}}}{(\sin\Theta)^{\frac{1}{2}}}\int_0^\infty (Kb)^{\frac{1}{2}}\exp(-\mathrm{i}\Delta m\Phi)\mathscr{A}^{\mathrm{s}}_{ji}(b)\exp[2\mathrm{i}\delta(b)-\delta_{ji}]$$

$$\times\cos\left[Kb\Theta-(\Delta m+1)\frac{\pi}{2}\right]\mathrm{d}b. \quad (3.248)$$

The stationary phase approximation can be employed in (3.248) to obtain the result

$$\frac{\mathrm{d}\sigma_{ji}}{\mathrm{d}\Omega} \approx \left(\frac{\mathrm{d}\sigma}{\mathrm{d}\Omega}\right)_{\mathrm{class}}|\mathscr{A}_{ji}(b)|^2, \quad (3.249)$$

where $(\mathrm{d}\sigma/\mathrm{d}\Omega)_{\mathrm{class}}$ is the classical cross section for scattering by the effective potential $W(R)$; however, as we have already noted, this result is often not very accurate.

4

CHARGE EXCHANGE AT VERY LOW ENERGIES – THE PERTURBED STATIONARY STATE MODEL

In this chapter we shall discuss charge exchange at energies for which the relative velocity of the colliding ions is very small compared with typical orbital velocities of the pseudomolecule composed of the active electron moving in the field of the ionic cores A and B. A natural choice of basis functions in a coupled channel model for such a collision is a set of adiabatic molecular orbitals, modified by suitable translation factors as in equations (3.99). The coupled equations, determining the channel wave functions using the coordinate set (**s**, **R**) where **R** is the vector joining A to B and **s** is the position vector of the electron relative to C, the centre of mass of A and B, are then of the form (3.93), the overlap and potential matrix elements being defined by the expressions (3.91) and (3.94). In the *perturbed stationary state model*, which in the past was employed for the majority of calculations at very low energies, the translation factors which allow for the change in translational motion of the electron on capture are neglected completely. This model, originally developed by Massey and Smith (1933) is described in Section 4.1. At very low energies the number of significant partial waves is limited and a partial-wave expansion of the coupled equations is indicated. The general approach to a partial-wave expansion was outlined in Section 3.5, and in Section 4.3 we extend the method to the non-central potentials which are encountered. After discussing some examples of importance in astrophysical applications we shall also consider an improved model including translation factors. Finally we shall consider the process of radiative charge exchange which in some circumstances offers an alternative reaction path at low energies.

4.1 THE PERTURBED STATIONARY STATE MODEL

The adiabatic molecular orbitals from which the basis function of the perturbed stationary state model are constructed are defined, for a one-electron system (many-electron systems are discussed in Section 4.7 and Chapter 9) as solutions of the Schrödinger equation (see (3.71))

$$[-\tfrac{1}{2}\nabla_s^2 + V - E_n(R)]\psi_n(\mathbf{s}, \mathbf{R}) = 0, \tag{4.1}$$

where V is the total potential energy expressed by (3.13a). Equation (4.1) is solved for fixed values of $\mathbf{R}$, which enters as a parameter. The orbitals $\psi_n(\mathbf{s}, \mathbf{R})$ are mutually orthogonal and can be normalized at each value of $\mathbf{R}$, so that

$$\int \psi_n^*(\mathbf{s}, \mathbf{R})\psi_j(\mathbf{s}, \mathbf{R})\,\mathrm{d}\mathbf{s} = \delta_{nj}. \tag{4.2}$$

In the separated atom limit, $R \to \infty$, $\psi_j(\mathbf{s}, \mathbf{R})$ coincides with an atomic orbital centred on one of the colliding atoms A or B. As in the last chapter, this will be indicated, where necessary, by the addition of A or B as superscripts. If A and B are identical, a *homonuclear system*, combinations of the adiabatic orbitals separate to an atomic orbital attached to a given centre, but as we shall see in the next section this causes no difficulty.

In the formalism discussed in Section 3.4, the basis functions from which the coupled channel equations were derived were expressed in terms of the adiabatic orbitals $\psi_j^{\mathrm{A,B}}(\mathbf{s}, \mathbf{R})$ by

$$\Phi_j(\mathbf{s}, \mathbf{R}) = \psi_j^{\mathrm{B}}(\mathbf{s}, \mathbf{R}) \exp[\mathrm{i}\gamma_j^{\mathrm{B}}(\mathbf{s}, \mathbf{R})], \qquad j = 1, 2, \ldots, N, \tag{4.3a}$$

$$\mathrm{X}_n(\mathbf{s}, \mathbf{R}) = \psi_n^{\mathrm{A}}(\mathbf{s}, \mathbf{R}) \exp[\mathrm{i}\bar{\gamma}_n^{\mathrm{A}}(\mathbf{s}, \mathbf{R})], \qquad n = N+1, N+2, \ldots, N+M, \tag{4.3b}$$

where $\exp \mathrm{i}\gamma_j^{\mathrm{B}}$ and $\exp \mathrm{i}\bar{\gamma}_n^{\mathrm{A}}$ are translation factors. In the perturbed stationary state model these factors are replaced by unity. To examine the effect of this replacement, we shall first study the overlap and potential matrix elements in the Bates–McCarroll plane-wave approximation for which (see (3.96))

$$\gamma_j^{\mathrm{B}} = -\mathbf{v}^{\mathrm{c}}\cdot\mathbf{s}; \qquad \bar{\gamma}_n^{\mathrm{A}} = \bar{\mathbf{v}}^{\mathrm{c}}\cdot\mathbf{s}, \tag{4.4}$$

where $\mathbf{v}^{\mathrm{c}}$ and $\bar{\mathbf{v}}^{\mathrm{c}}$ are the velocities of A and B relative to the centre of mass, and the dependence of $\mathbf{v}^{\mathrm{c}}$ and $\bar{\mathbf{v}}^{\mathrm{c}}$ on the channel index, which is small, has been ignored. The simple plane-wave approximation is certainly not the best that can be devised for use at low energies and a better approach will be described in Section 4.6, but it will serve to illustrate the source of some difficulties which arise with the perturbed stationary state model.

Inserting (4.3), with γ_j^{B} and $\bar{\gamma}_n^{\mathrm{A}}$ given by (4.4), into the defining expressions for the potential matrix elements, (3.94), and using the electronic wave equation (4.1) together with the orthonormality relations (4.2), we obtain

$$\begin{aligned}
V_{jl}(\mathbf{R}) &= (E_j^{\mathrm{B}}(R) - \varepsilon_j)\delta_{jl} + \int \psi_j^{\mathrm{B}^*}(\mathbf{s}, \mathbf{R})\left[-\frac{1}{2\mu}\nabla_R^2 + \mathrm{i}\mathbf{v}^{\mathrm{c}}\cdot\nabla_s\right]\psi_l^{\mathrm{B}}(\mathbf{s}, \mathbf{R})\,\mathrm{d}\mathbf{s} \\
W_{jl}(\mathbf{R}) &= -\frac{1}{\mu}\int \psi_j^{\mathrm{B}^*}(\mathbf{s}, \mathbf{R})\nabla_R\psi_l^{\mathrm{B}}(\mathbf{s}, \mathbf{R})\,\mathrm{d}\mathbf{s} \\
K_{jm}(\mathbf{R}) &= S_{jm}(\mathbf{R})(E_m^{\mathrm{A}}(R) - \eta_m) + \int \exp(\mathrm{i}\mathbf{v}\cdot\mathbf{s})\psi_j^{\mathrm{B}^*}(\mathbf{s}, \mathbf{R}) \\
&\qquad \times \left[-\frac{1}{2\mu}\nabla_R^2 - \mathrm{i}\bar{\mathbf{v}}^{\mathrm{c}}\cdot\nabla_s\right]\psi_m^{\mathrm{A}}(\mathbf{s}, \mathbf{R})\,\mathrm{d}\mathbf{s} \\
\mathbf{L}_{jm}(\mathbf{R}) &= \int \exp(\mathrm{i}\mathbf{v}\cdot\mathbf{s})\psi_j^{\mathrm{B}^*}(\mathbf{s}, \mathbf{R})\nabla_R\psi_m^{\mathrm{A}}(\mathbf{s}, \mathbf{R})\,\mathrm{d}\mathbf{s}
\end{aligned} \tag{4.5}$$

where $\mathbf{v} = (\mathbf{v}^c + \bar{\mathbf{v}}^c)$ is the relative velocity of A and B which can be taken to be equal to $\mathbf{v}_0$, the relative velocity of A and B before the collision. Similar expressions are obtained for $\bar{V}_{nm}$, $\bar{\mathbf{W}}_{nm}$, $\bar{K}_{nl}$, and $\bar{\mathbf{L}}_{nl}$. The overlap matrix elements $S_{jm}(\mathbf{R})$ are given by

$$S_{jm}(\mathbf{R}) = \int \exp(\mathrm{i}\mathbf{v}\cdot\mathbf{s})\psi_j^{\mathrm{B}*}(\mathbf{s}, \mathbf{R})\psi_m^{\mathrm{A}}(\mathbf{s}, \mathbf{R})\, \mathrm{d}\mathbf{s},$$

and

$$\bar{S}_{jm}(\mathbf{R}) = S^*_{mj}(\mathbf{R}). \tag{4.6}$$

In (4.5), the molecular energies $E_j^{\mathrm{A,B}}(R)$ have been labelled according to which atomic limit is approached as $R \to \infty$, so that

$$E_j^{\mathrm{B}} - \varepsilon_j \to 0, \qquad R \to \infty, \tag{4.7a}$$

$$E_m^{\mathrm{A}}(R) - \eta_m \to 0, \qquad R \to \infty. \tag{4.7b}$$

We are now in a position to assess the effect of replacing the translational factors by unity, which is the same as setting $\mathbf{v}$, $\mathbf{v}^c$, and $\bar{\mathbf{v}}^c$ equal to zero in (4.5) and (4.6). The first consequence is that the oscillating factors $\exp(\pm\mathrm{i}\mathbf{v}\cdot\mathbf{s})$ which occur in the exchange and overlap integrals K_{jm}, $\mathbf{L}_{jm}$, S_{jm} and $\bar{K}_{mj}$, $\bar{\mathbf{L}}_{mj}$, $\bar{S}_{mj}$ are replaced by unity. This in itself will not cause a significant error provided that $(\mathbf{v}\cdot\mathbf{s}) \ll 1$ over the range of s for which the integrands are large. As an example if the maximum significant value of s is 20 a.u., v should be small compared with 0.05 a.u., which corresponds to an incident laboratory energy of $(60\ M_A)$ eV. A more serious source of error at these low energies is the omission of the terms in $(\mathbf{v}^c\cdot\boldsymbol{\nabla}_s)$ and $(\bar{\mathbf{v}}^c\cdot\boldsymbol{\nabla}_s)$ in the direct interaction matrices V_{jl} and $\bar{V}_{nm}$. We recollect that the individual terms $\Phi_j F_j$ and $X_n G_n$, where Φ_j and X_n are expressed by (4.3) and (4.4), satisfy the Schrödinger equation for the complete system in the separated atom limit, $R \to \infty$. This was shown in Section 3.4 (see (3.82)–(3.84)) and, as a consequence, the interactions between the direct channels vanish in this limit, for example

$$[V_{jl}(\mathbf{R}) + \mathbf{W}_{jl}(\mathbf{R})\cdot\boldsymbol{\nabla}_R]F_l(\mathbf{R}) \to 0, \qquad R \to \infty. \tag{4.8}$$

In obtaining this result, we used the approximation

$$\boldsymbol{\nabla}_R F_l(\mathbf{R}) \to \mathrm{i}\mathbf{k}_l F_l(\mathbf{R}), \qquad R \to \infty, \tag{4.9}$$

where $\mathbf{k}_l = M_{\mathrm{B}}\bar{\mathbf{v}}_l^c$ (see (3.79)), together with the fact that

$$\psi_l^{\mathrm{B}}(\mathbf{s}, \mathbf{R}) \to \phi_l(\mathbf{r}), \qquad R \to \infty. \tag{4.10}$$

As usual $\phi_l(\mathbf{r})$ denotes an atomic orbital centred on B and

$$\mathbf{r} = \mathbf{s} + \frac{\mu}{M_{\mathrm{B}}}\mathbf{R}. \tag{4.11}$$

It follows from (4.5) and (4.7a), with the use of (4.9) and (4.10), that

$$[V_{jl}(\mathbf{R}) + \mathbf{W}_{jl}(\mathbf{R})\cdot\nabla_{\mathrm{R}}]F_l(\mathbf{R}) \to$$
$$\left\{\int \phi_j^*(\mathbf{r})\left[-\frac{1}{2\mu}\nabla_R^2 + \mathrm{i}\mathbf{v}^{\mathrm{c}}\cdot\nabla_s - \frac{M_{\mathrm{B}}}{\mu}\mathbf{v}^{\mathrm{c}}\cdot\nabla_{\mathrm{R}}\right]\phi_l(\mathbf{r})\,\mathrm{d}\mathbf{s}\right\}F_l(\mathbf{R}). \quad (4.12)$$

Differentiation with respect to **R** must be carried out keeping **s** constant, and similarly differentiation with respect to **s** must be carried out with **R** held constant, so that

$$\nabla_s\phi_l(\mathbf{r}) = \nabla_r\phi_l(\mathbf{r}); \qquad \nabla_R\phi_l(\mathbf{r}) = \frac{\mu}{M_{\mathrm{B}}}\nabla_r\phi_l(\mathbf{r}). \quad (4.13)$$

The term $-\nabla_R^2/2\mu$ on the right-hand side of (4.12) is of the order m/M and can be neglected,† while the second and third terms within the brackets cancel. In the perturbed stationary state model the second term is absent and at large R we are left with a term coupling the channels j and l which is constant and independent of R:

$$-\frac{M_{\mathrm{B}}}{\mu}\int \phi_j^*(\mathbf{r})(\mathbf{v}^{\mathrm{c}}\cdot\nabla_r)\phi_l(\mathbf{r})\,\mathrm{d}\mathbf{s} = -\frac{M_{\mathrm{B}}}{\mu}\int \phi_j^*(\mathbf{r})(\mathbf{v}^{\mathrm{c}}\cdot\nabla_{\mathrm{r}})\phi_l(\mathbf{r})\,\mathrm{d}\mathbf{r}. \quad (4.14)$$

This coupling is non-zero whenever the atomic levels j and l are connected by a dipole transition. The existence of couplings at infinite separations makes it impossible, in principle, to calculate well-defined cross sections, without projecting the calculated wave function on to the asymptotically correct wave functions. Furthermore, the omission of the term in $(\mathbf{v}^{\mathrm{c}}\cdot\nabla_s)$ may lead to an overestimation of the effective coupling at finite values of R, where some degree of cancellation is to be expected. The exchange couplings automatically vanish as $R \to \infty$, because the overlap between the atomic orbitals on different centres becomes vanishingly small; however, the couplings at finite R may also be overestimated because of the corresponding lack of cancellation.

To overcome these difficulties, at least in part, several methods have been suggested. One possibility is to omit the terms $\mathbf{W}_{jl}$ and $\overline{\mathbf{W}}_{nm}$ beyond some cut-off point at $R = R_{\mathrm{c}}$, where R_{c} is large enough so that for $R > R_{\mathrm{c}}$ the wave functions are atomic in character. The sensitivity of the cross sections to alterations in the value of R_{c} can then be used to assess the consistency of the calculations. An alternative approach (Butler *et al.* 1980), useful when

† Since $-\nabla_R^2\phi_l(\mathbf{r})/2\mu = -(\mu/2M_{\mathrm{B}}^2)\nabla_r^2\phi_l(\mathbf{r})$, this term can be combined with the kinetic energy operator of the electron, producing a small change of order m/M in the reduced mass of the electron. In our development we have consistently neglected terms of order m/M in reduced masses. An interesting analysis in which these terms are retained has been presented by Thorsen and Delos (1978*a*).

$M_B \gg M_A$, in which case the centre-of-mass coordinate **s** nearly coincides with atomic coordinate **r**, is to replace the derivatives $\nabla_{R|\mathbf{s}}$ with **s** held constant by derivatives $\nabla_{R|\mathbf{r}}$ with **r** held constant. Since $\nabla_{R|\mathbf{r}}\,\psi_l^{\mathrm{B}}(\mathbf{s}, \mathbf{R})$ vanishes as $R \to \infty$, spurious couplings associated with the arrangement A + (B + e$^-$) are eliminated.

Despite these difficulties, the perturbed stationary state model has produced useful cross sections at very low energies and has the merit of considerable simplicity. A more consistent model based on the expansion in powers of the velocity of the translation operator (see (3.64)) with the consequential simplification of equations (3.46) is perfectly feasible (see Section 4.6) and may be expected to replace the more traditional model in the future.

When the approximation is made of neglecting the translational factors, the matrices **S** and $\bar{\mathbf{S}}$ vanish, which allows us to write the perturbed stationary state equations in the standard form

$$(\nabla_R^2 + \mathscr{K}^2(R))\mathbf{F}(R) = [\mathbf{M}(\mathbf{R}) + \mathbf{P}(\mathbf{R})\cdot\nabla_{\mathrm{R}}]\mathbf{F}(\mathbf{R}), \tag{4.15}$$

where, as usual, $\mathbf{F}(\mathbf{R})$ denotes the $(N + M)$-dimensional column vector of channel wave functions, $\mathscr{K}$ is the diagonal matrix with elements

$$K_{ij}(R) = -2\mu(E_j(R) - E)\delta_{ij}, \tag{4.16}$$

and **M** and **P** are the coupling matrices with elements

$$M_{ij}(R) = -\int \psi_i^*(\mathbf{s}, \mathbf{R})\nabla_R^2\psi_j(\mathbf{s}, \mathbf{R})\,d\mathbf{s}, \tag{4.17a}$$

$$P_{ij}(R) = -2\int \psi_i^*(\mathbf{s}, \mathbf{R})\nabla_R\psi_j(\mathbf{s}, \mathbf{R})\,d\mathbf{s}. \tag{4.17b}$$

The superscripts A and B have been omitted, since in the absence of translational factors, the trial function in the direct and exchange channels takes exactly the same form.

Before describing the partial-wave expansion of the channel functions $\mathbf{F}(\mathbf{R})$, it is necessary to discuss some properties of the adiabatic molecular orbitals from which **M** and **P** are constructed and we now turn our attention to this.

4.2 ADIABATIC MOLECULAR ORBITALS – ROTATING COORDINATES

Up to now we have been working in a frame of reference fixed in space, taking the direction of incidence as the Z-axis. To be consistent with this choice, the atomic and molecular basis functions must be quantized with respect to OZ, the axis of symmetry, and the atomic functions are eigenfunctions of $\mathbf{L}^2$ and L_z the component of the orbital angular momentum in the Z-direction. This presents no difficulty when an atomic basis is used. However, adiabatic molecular orbitals are usually calculated in a frame of reference fixed with respect to the interionic line AB, which is taken as the axis of quantization. In this section we shall first describe a few properties of the molecular orbitals in this 'body-fixed' frame of reference and then show how a transformation between this frame and the space-fixed frame may be effected.

In the body-fixed frame Cartesian axes $\overline{OX}$, $\overline{OY}$, and $\overline{OZ}$ will be taken, with $\overline{OZ}$ directed along the line from B to A, that is in the direction of the vector $\mathbf{R}$. The origin O will be, as before, the centre of mass of A and B although for the purely molecular problem any point along AB could be taken as origin.

In the body-fixed frame, the position vector of the electron will be denoted by $\bar{\mathbf{s}}$. Geometrically this is the same vector as $\mathbf{s}$, but $\bar{\mathbf{s}}$ is always expressed in terms of its components $(\bar{x}, \bar{y}, \bar{z})$ along the body-fixed axes, while $\mathbf{s}$ is always expressed in terms of its components (x, y, z) along the spaced-fixed axes. The molecular orbitals in the body-fixed frame, $\bar{\psi}_j(\bar{\mathbf{s}}, R)$ depend on the magnitude of $\mathbf{R}$ only and satisfy

$$H_e\bar{\psi}_j(\bar{\mathbf{s}}, R) = E_j(R)\bar{\psi}_j(\bar{\mathbf{s}}, R), \tag{4.18}$$

where

$$H_e \equiv -\tfrac{1}{2}\nabla_{\bar{s}}^2 + V_{Ae}(x) + V_{Be}(r) + V_{AB}(R). \tag{4.19}$$

In most cases of interest to us, spin–orbit coupling can be neglected and we shall assume the effective potentials V_{Ae}, V_{Be} describing the interaction of the electron with the ionic cores are central.

Since the line AB is an axis of symmetry, the orbitals $\bar{\psi}_j$ are eigenfunctions of the component $L_{\bar{z}}$ of the electronic orbital angular momentum in the direction of the $\bar{Z}$-axis:

$$\bar{L}_z\bar{\psi}_j(\bar{\mathbf{s}}, R) = \pm\lambda_j\bar{\psi}_j(\bar{\mathbf{s}}, R), \qquad \lambda_j = 0, 1, 2, \ldots. \tag{4.20}$$

It is conventional to refer to one-electron levels† with $\lambda_j = 0, 1, 2, \ldots$ as $\sigma, \pi, \delta, \phi \ldots$ levels. It is important to notice that the orbitals $\bar{\psi}_j$ cannot be eigenfunctions of $\mathbf{L}^2$, except in the united atom $(R \to 0)$, or separated atom $(R \to \infty)$, limits.

In the particular case in which A and B are bare nuclei and V_{Ae}, V_{Be} are Coulomb potentials, (4.18) is separable in spheroidal coordinates and an exact numerical solution can be obtained (see Appendix D). In other cases, the variational method can be used to obtain approximate orbitals, taking for example a trial function formed from a linear combination of Slater or

† For molecules with more than one electron the absolute value of the projection of the orbital angular momentum on the internuclear axis is denoted by Λ, with $\Lambda = 0, 1, 2, 3, \ldots$ The corresponding levels are designated by the letters $\Sigma, \Pi, \Delta, \Phi, \ldots$.

Gaussian functions centred at A and B (Pilar 1968; Maciás and Riera 1982; Burden and Wilson 1972).

Heteronuclear systems If V_{Ae} is not the same as V_{Be}, the molecule is heteronuclear. In this case the only additional symmetry is the symmetry of the Hamiltonian under a reflection in any plane containing the $\overline{OZ}$-axis (the line BA), which gives rise to a double degeneracy of each energy level with $\lambda_j \neq 0$. The different energy levels can be labelled by the index j and the eigenvalue λ_j. In the limit $R \to 0$, an orbital $\bar{\psi}_j(\bar{\mathbf{s}}, R)$ becomes equal to one of the atomic orbitals of the *united atom* (AB + e^-). In this limit the magnetic quantum number m_l of the united atom is equivalent to $\pm\lambda_j$ and each orbital can be designated by $nl\lambda_j$, where n and l are the principal and angular quantum numbers of the united atom limit. In the limit of large R, in the lowest approximation, each molecular orbital approaches an atomic orbital centred either on A or on B. A correlation diagram can be drawn which shows the connection between the levels of the united atom and those of the separated atoms for each molecular level. Such a schematic diagram is shown is Fig. 4.1 and, as an example of the corresponding potential energy curves, the case of $(HeH)^{2+}$ is shown in Fig. 4.2.

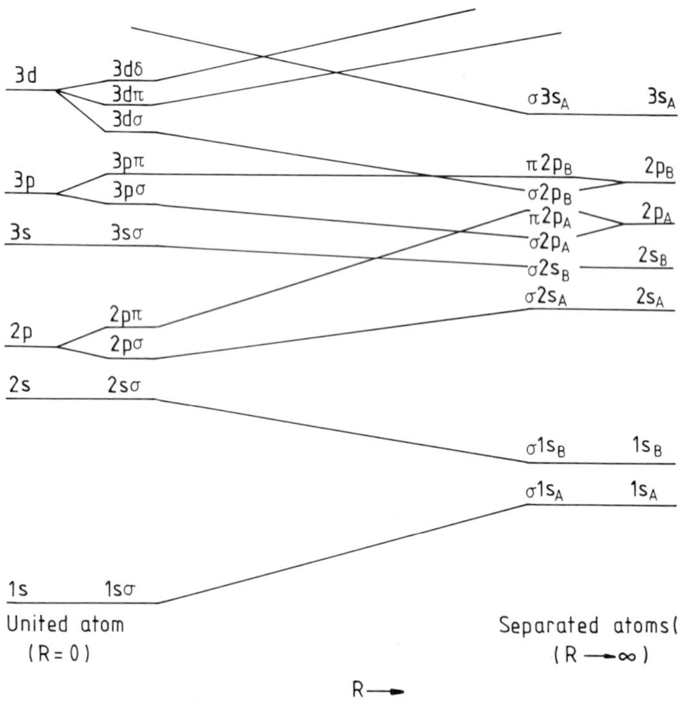

Fig. 4.1. A correlation diagram for a one-electron heteronuclear diatomic molecule. The diagram is not to scale and the actual energies vary from molecule to molecule.

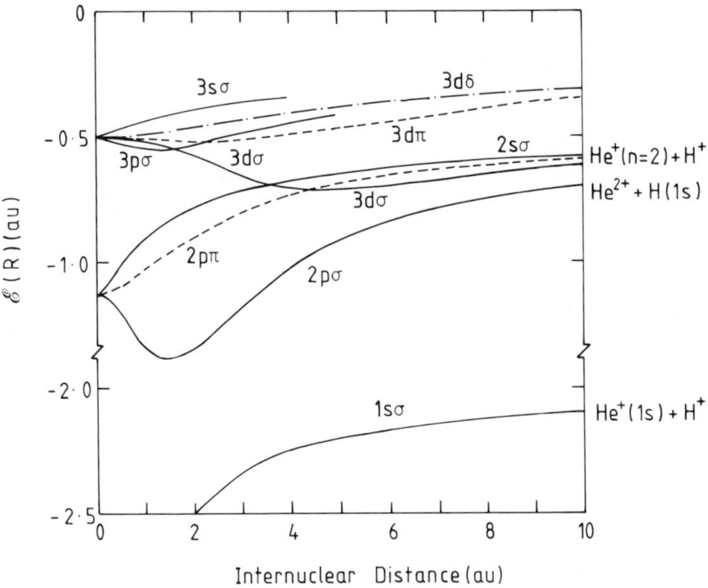

Fig. 4.2. Potential energies of (He H)$^{2+}$ as a function of the internuclear distance R. The internuclear potential is excluded so that $\mathscr{E}_j(R) = E_j(R) - V_{AB}(R)$. ——— σ states, ----- π states, —·—·— δ states (after Piacentini and Salin (1974)).

In drawing a correlation diagram, we note that λ_j becomes equal to the atomic magnetic quantum number m_l in both the united and separated atom limits. Also according to the *von Neumann–Wigner non-crossing rule* (see Bransden and Joachain 1983), energy levels corresponding to states of the *same symmetry* cannot cross as we let R vary from $R = 0$ to $R = \infty$. In the separated atom limit the levels can be labelled $(\lambda_j n'l')_{A,B}$ where the subscripts A or B refer to which atom the level separates. If either A or B represents a bare nucleus, the corresponding separated atom is hydrogenic and subject to Stark mixing of the degenerate levels with differing l' but this causes no fundamental difficulty.

Homonuclear systems If the potentials V_{Ae} and V_{Be} are the same, the electronic Hamiltonian is invariant under the inversion $\mathbf{s}_g \to -\mathbf{s}_g$ where $\mathbf{s}_g$ is the position vector of the electron with respect to the mid-point of the line AB. It should be noted that the heavy particles A and B need not be identical; for example the nuclei of A and B can be isotopes differing in mass but with the same charge. The operator P, which induces this inversion, commutes with L_z, so that orbitals which are simultaneous eigenfunctions of the electronic Hamiltonian, $\bar{L}_z$ and P can be constructed. Since applying P twice restores the original situation, $P^2 = 1$ and the eigenvalues of P are ± 1. States

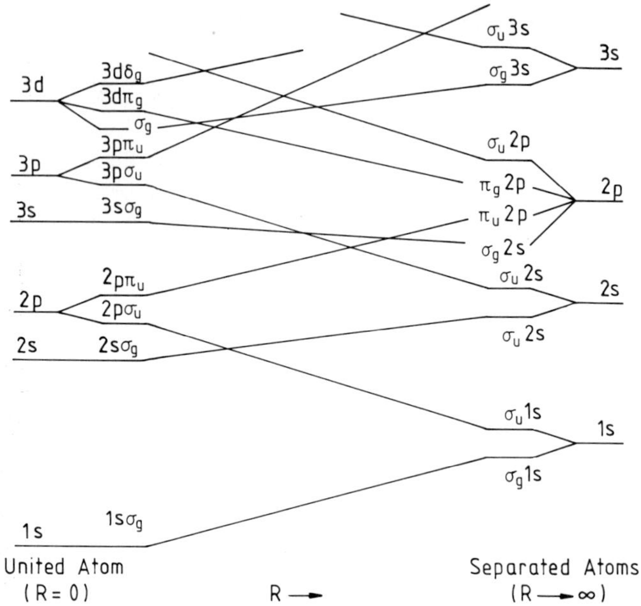

Fig. 4.3. A correlation diagram for a one-electron homonuclear diatomic molecule. The diagram is not to scale and the actual energies of the initial and separated atoms vary from molecule to molecule.

with even parity $(+1)$ are called gerade states and are labelled by the subscript g and those with odd parity (-1) are called ungerade states and are labelled by the script u

$$P\bar{\psi}_{j,\mathrm{g}}(\bar{\mathbf{s}}, R) = \bar{\psi}_{j,\mathrm{g}}(\bar{\mathbf{s}}, R); \qquad P\bar{\psi}_{j,\mathrm{u}}(\bar{\mathbf{s}}, R) = -\bar{\psi}_{j,\mathrm{u}}(\bar{\mathbf{s}}, R). \tag{4.21}$$

In the separated atom limit, the parity eigenfunctions $\bar{\psi}_{j,\mathrm{g}}$ and $\bar{\psi}_{j,\mathrm{u}}$ must approach a symmetrized combination of atomic orbitals, $\phi_j(\mathbf{r})$ centred on B and $\phi_j(\mathbf{x})$ centred on A:

$$\bar{\psi}_{j,\mathrm{g}} \approx \phi_j(\mathbf{x}) + \phi_j(\mathbf{r}); \qquad \bar{\psi}_{j,\mathrm{u}} \approx \phi_j(\mathbf{x}) - \phi_j(\mathbf{r}). \tag{4.22}$$

A correlation diagram for homonuclear molecules is shown in Fig. 4.3. If A and B are bare nuclei, then the degenerate separated atomic levels are subject to Stark mixing, which, as in the heteronuclear case, causes no particular difficulty. In the perturbed stationary state model, the parity eigenstates can be used as basis functions, but if translation factors are employed, it is necessary to identify combinations of orbitals which separate to an atomic orbital about a particular centre. The basis orbitals can be taken as

$$\mathrm{X}_j = \tfrac{1}{2}(\bar{\psi}_{j,\mathrm{g}} + \bar{\psi}_{j,\mathrm{u}}) \exp \mathrm{i}\bar{\gamma}_j, \tag{4.23a}$$

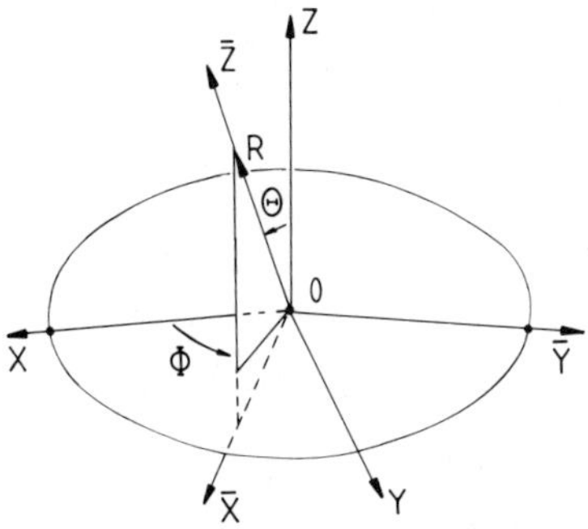

Fig. 4.4. The relationship between the space-fixed frame OX, OY, OZ and the molecular frame $\overline{\mathrm{O}X}$, $\overline{\mathrm{O}Y}$, $\overline{\mathrm{O}Z}$ with **R** directed along $\overline{\mathrm{O}Z}$.

which separates to the atomic orbital for the atom (A + e$^-$) modified by the correct translational factor and

$$\Phi_j = \tfrac{1}{2}(\bar{\psi}_{j,\mathrm{g}} - \bar{\psi}_{j,\mathrm{u}}) \exp \mathrm{i}\gamma_j, \tag{4.23b}$$

which separates to the atomic orbital for the atom (B + e$^-$) again with the appropriate translational factor.

Rotating coordinates

The clockwise rotations which bring the space-fixed axes OX, OY, and OZ into coincidence with body-fixed axes $\overline{\mathrm{O}X}$, $\overline{\mathrm{O}Y}$, $\overline{\mathrm{O}Z}$ (where $\overline{\mathrm{O}Z}$ is in the direction of the interionic line BA) can be specified in terms of Euler angles α, β, γ. We first rotate the axes through an angle α about OZ, producing new axes OX', OY', OZ; a second rotation through an angle β about OY' (the 'line of nodes') brings OZ into $\overline{\mathrm{O}Z}$. Finally a third rotation through an angle γ about $\overline{\mathrm{O}Z}$, brings the OX' and OY' axes into coincidence with $\overline{\mathrm{O}X}$ and $\overline{\mathrm{O}Y}$ (we follow the conventions and phases of Edmonds (1960)). If we set $\alpha = \Phi$, $\beta = \Theta$, and $\gamma = 0$, (Θ, Φ) are the polar angles of **R** in the space-fixed frame as shown in Fig. 4.4. With this choice the body-fixed Y-axis, $\overline{\mathrm{O}Y}$, lies in the space-fixed XY-plane. The components of the position vector of the electron in the body-fixed frame $(\bar{x}, \bar{y}, \bar{z})$ are related to those in the space-fixed frame (x, y, z) by

$$\begin{aligned} \bar{x} &= \cos\Theta \cos\Phi\, x + \cos\Theta \sin\Phi\, y - \sin\Theta\, z \\ \bar{y} &= -\sin\Phi\, x + \cos\Phi\, y \\ \bar{z} &= \sin\Theta \cos\Phi\, x + \sin\Theta \sin\Phi\, y + \cos\Theta\, z. \end{aligned} \tag{4.24}$$

The coupling matrix elements (4.17a, b) contain the operators ∇_R^2 and $\boldsymbol{\nabla}_R$ which can be expressed in terms of the derivatives $\partial/\partial R$, $\partial/\partial\Theta$, and $\partial/\partial\phi$. These derivatives act on the functions $\bar{\psi}_j(\bar{\mathbf{s}}, R)$ defined in the body-fixed frame, accordingly $\partial/\partial R$, $\partial/\partial\Theta$, and $\partial/\partial\Phi$ must be expressed in terms of $\partial'/\partial R$, $\partial'/\partial\Theta$, $\partial'/\partial\Phi$ where the primes denote that derivatives are to be taken with $\bar{\mathbf{s}}$ held constant in the body-fixed frame. Thus we have

$$\frac{\partial}{\partial R} = \frac{\partial'}{\partial R}, \tag{4.25a}$$

$$\frac{\partial}{\partial\Theta} = \frac{\partial'}{\partial\Theta} + \frac{\partial\bar{x}}{\partial\Theta}\frac{\partial}{\partial\bar{x}} + \frac{\partial\bar{y}}{\partial\Theta}\frac{\partial}{\partial\bar{y}} + \frac{\partial\bar{z}}{\partial\Theta}\frac{\partial}{\partial\bar{z}}, \tag{4.25b}$$

$$\frac{\partial}{\partial\Phi} = \frac{\partial'}{\partial\Phi} + \frac{\partial\bar{x}}{\partial\Phi}\frac{\partial}{\partial\bar{x}} + \frac{\partial\bar{y}}{\partial\Phi}\frac{\partial}{\partial\bar{y}} + \frac{\partial\bar{z}}{\partial\Phi}\frac{\partial}{\partial\bar{z}}. \tag{4.25c}$$

Using the relations (4.24), we find

$$\begin{aligned} &\frac{\partial\bar{x}}{\partial\Theta} = -\bar{z}; \quad &&\frac{\partial\bar{y}}{\partial\Theta} = 0; \quad &&\frac{\partial\bar{z}}{\partial\Theta} = \bar{x}; \\ &\frac{\partial\bar{x}}{\partial\Phi} = \cos\Theta\,\bar{y}; \quad &&\frac{\partial\bar{y}}{\partial\Phi} = -(\bar{x}\cos\Theta + \bar{z}\sin\Theta); \quad &&\frac{\partial\bar{z}}{\partial\Phi} = \sin\Theta\,\bar{y}, \end{aligned} \tag{4.26}$$

and combining these relations with (4.25b) and (4.25c), it is seen that

$$\begin{aligned} \frac{\partial}{\partial\Theta} &= \frac{\partial'}{\partial\Theta} - \mathrm{i}\bar{L}_y \\ \frac{\partial}{\partial\Phi} &= \frac{\partial'}{\partial\Phi} - \mathrm{i}\cos\Theta\,\bar{L}_z + \mathrm{i}\sin\Theta\,\bar{L}_x, \end{aligned} \tag{4.27}$$

where L_x, $\bar{L}_y$, and $\bar{L}_z$ are the components of the electronic angular momentum in the body-fixed frame. These results can now be used to evaluate the coupling matrices $\mathbf{M}(\mathbf{R})$ and $\mathbf{P}(\mathbf{R})$ defined by (4.17a, b). It is convenient to introduce the notation

$$\langle i|\mathbf{O}|j\rangle \equiv \int \bar{\psi}_i^*(\bar{\mathbf{s}}, R)\mathbf{O}\bar{\psi}_j(\bar{\mathbf{s}}, R)\,\mathrm{d}\bar{\mathbf{s}}, \tag{4.28a}$$

where $\mathbf{O}$ is an operator, so that

$$M_{ij}(R) = -\langle i|\nabla_R^2|j\rangle \equiv -\int \bar{\psi}_i^*(\bar{\mathbf{s}}, R)\nabla_R^2\bar{\psi}_j(\bar{\mathbf{s}}, R)\,\mathrm{d}\bar{\mathbf{s}}, \tag{4.28b}$$

$$\mathbf{P}_{ij}(R) = -2\langle i|\boldsymbol{\nabla}_R|j\rangle \equiv -2\int \bar{\psi}_i^*(\bar{\mathbf{s}}, R)\boldsymbol{\nabla}_R\bar{\psi}_j(\bar{\mathbf{s}}, R)\,\mathrm{d}\bar{\mathbf{s}}. \tag{4.28c}$$

Expressing ∇_R^2 in polar coordinates

$$\nabla_R^2 = \frac{\partial^2}{\partial R^2} + \frac{2}{R}\frac{\partial}{\partial R} + \frac{1}{R^2}\left[\frac{\partial^2}{\partial\Theta^2} + \cot\Theta\frac{\partial}{\partial\Theta} + \frac{1}{\sin^2\Theta}\frac{\partial^2}{\partial\Phi^2}\right], \tag{4.29}$$

and using (4.26), we obtain

$$M_{ij}(\mathbf{R}) = -\langle i|\frac{\partial^2}{\partial R^2} + \frac{\partial}{R}\frac{\partial}{\partial R}|j\rangle + \frac{1}{R^2}\langle i|\bar{L}_x^2 + \bar{L}_y^2|j\rangle$$
$$-\frac{2\cot\Theta}{R^2}\langle i|\bar{L}_x\bar{L}_z|j\rangle + \frac{\cot^2\Theta}{R^2}\langle i|\bar{L}_z^2|j\rangle. \qquad (4.30)$$

To evaluate $\mathbf{P}(\mathbf{R})\cdot\nabla_R$, the gradient operator can be written in polar coordinates as

$$\nabla_R = \hat{\mathbf{R}}\frac{\partial}{\partial R} + \frac{1}{R}\hat{\mathbf{\Theta}}\frac{\partial}{\partial\Theta} + \frac{1}{R\sin\Theta}\hat{\mathbf{\Phi}}\frac{\partial}{\partial\Phi}, \qquad (4.31)$$

where $\hat{\mathbf{R}}$ is a unit vector along $\mathbf{R}$, $\hat{\mathbf{\Theta}}$ is a unit vector normal to $\mathbf{R}$, in the plane of $\mathbf{R}$ and the OZ-axis, and $\hat{\mathbf{\Phi}} = \hat{\mathbf{R}}\times\hat{\mathbf{\Theta}}$ is a unit vector normal to the $(\mathbf{R}, OZ)$ plane. We find easily from (4.26):

$$\mathbf{P}_{ij}(\mathbf{R})\cdot\nabla_R = -2\langle i|\frac{\partial}{\partial R}|j\rangle\frac{\partial}{\partial R} + \frac{2\mathrm{i}}{R^2}\langle i|\bar{L}_y|j\rangle\frac{\partial}{\partial\Theta}$$
$$+\frac{2\mathrm{i}}{R^2\sin\Theta}[\langle i|\bar{L}_x|j\rangle + \cot\Theta\lambda_j\delta_{ij}]\frac{\partial}{\partial\Phi}. \qquad (4.32)$$

Since $\bar{\psi}_j$ is an eigenfunction of $\bar{L}_z$, we note that $\langle i|\bar{L}_x\bar{L}_z|j\rangle = \lambda_j\langle i|\bar{L}_x|j\rangle$ and $\langle i|\bar{L}_z^2|j\rangle = \lambda_j^2\delta_{ij}$. In addition since $\bar{L}_x$, $\bar{L}_y$, and $\bar{L}_z$ obey the usual commutation relations, $\langle i|\bar{L}_x^2 + \bar{L}_y^2|j\rangle$ connects states having the same value of λ, while $\langle i|\bar{L}_x|j\rangle$ and $\langle i|\bar{L}_y|j\rangle$ connect states differing in λ by one unit.

Inserting **M** and **P** as given by (4.30) and (4.32) into the coupled equations (4.15) and expressing ∇_R^2 in polar coordinates (see (4.29)), we obtain

$$\left(\frac{\partial^2}{\partial R^2} - \frac{\mathbf{J}^2 - \boldsymbol{\lambda}^2}{R^2} + \mathscr{K}^2(\mathbf{R})\right)(R\mathbf{F}(\mathbf{R})) = \{\mathbf{V}^{\mathrm{R}}(\mathbf{R}) + \mathbf{V}^{\mathrm{C}}(\mathbf{R})\}(R\mathbf{F}(\mathbf{R})), \quad (4.33)$$

where $\boldsymbol{\lambda}$ is a diagonal matrix with elements λ_j, and $\mathbf{J}^2$ is also a diagonal matrix defined as

$$\mathbf{J}^2 = -\left[\frac{\partial^2}{\partial\Theta^2} + \cot\Theta\frac{\partial}{\partial\Theta} + \frac{1}{\sin^2\Theta}\left(\frac{\partial}{\partial\Phi} - \mathrm{i}\boldsymbol{\lambda}\cos\Theta\right)^2 - \boldsymbol{\lambda}^2\right]. \quad (4.34)$$

We shall show below that $\mathbf{J}^2$ is the total orbital angular momentum of the system. The matrix $\mathbf{V}^{\mathrm{R}}$ is known as the *radial coupling* and contains only terms which do not connect levels with differing values of λ:

$$V_{ij}^{\mathrm{R}}(R) = \left[\langle i| - \frac{\partial^2}{\partial R^2} + \left(\frac{\bar{L}_x^2 + \bar{L}_y^2}{R^2}\right)|j\rangle - 2\langle i|\frac{\partial}{\partial R}|j\rangle\frac{\partial}{\partial R}\right]\delta_{\lambda_i\lambda_j}. \quad (4.35)$$

The matrix $\mathbf{V}^c$ represents the *rotational* or *Coriolis coupling* and contains terms which connect states differing in λ by one unit of angular momentum:

$$V_{ij}^{C}(R) = \frac{1}{R^2}\left[2i\langle i|\bar{L}_y|j\rangle \frac{\partial}{\partial\Theta} - 2i\langle i|\bar{L}_x|j\rangle \frac{1}{\sin\Theta}\frac{\partial}{\partial\Theta} - 2\lambda_j\langle i|\bar{L}_x|j\rangle \cot\Theta\right]. \tag{4.36}$$

4.3 PARTIAL-WAVE EXPANSIONS AND THE RADIAL EQUATIONS

To obtain coupled radial equations, suitable for numerical solution, the expansion of the channel function $\mathbf{F}(\mathbf{R})$ in partial waves must be inserted into equations (4.33). Various expansions are possible, for example in terms of the spherical harmonics $Y_{N,M_N}(\Theta, \Phi)$ where N is the quantum number of the orbital angular momentum $\mathbf{N}$ of the heavy particle motion, and M_N the corresponding magnetic quantum number. However, since neither $\mathbf{N}$ nor $\mathbf{L}$ (the electronic orbital angular momentum) are conserved, this choice would lead to large sets of coupled equations connecting radial functions for all the relevant values of N. In the absence of external fields, the total angular momentum is always conserved, and since we are neglecting spin–orbit interactions, the total *orbital* angular momentum $\mathbf{J} = \mathbf{L} + \mathbf{N}$ is also conserved, and $\mathbf{F}(\mathbf{R})$ can be usefully expanded in terms of simultaneous eigenfunctions of $\mathbf{J}^2$ and J_z where $\mathbf{J} = \mathbf{L} + \mathbf{N}$. Let us first show that the operator (4.34) represents $\mathbf{J}^2 = (\mathbf{L} + \mathbf{N})^2$. Using the relationships (4.24) and (4.25b, c), we easily find that

$$\begin{aligned} J_x + iJ_y &= (N_x + iN_y) + (L_x + iL_y) \\ &= e^{i\Phi}\left(\frac{\partial'}{\partial\Theta} + i\cot\Theta\frac{\partial'}{\partial\Phi}\right) + \frac{e^{i\Phi}}{\sin\Theta}\bar{L}_z, \end{aligned} \tag{4.37a}$$

where, in the second line, the derivatives with respect to Θ and Φ are performed with $\bar{\mathbf{s}}$ fixed and $\bar{L}_z$ is the component of the electronic angular momentum in the body-fixed frame. Likewise

$$(J_x - iJ_y) = e^{-i\Phi}\left(-\frac{\partial'}{\partial\Theta} + i\cot\Theta\frac{\partial'}{\partial\Phi}\right) + \frac{e^{-i\Phi}}{\sin\Theta}\bar{L}_z, \tag{4.37b}$$

$$J_z = -i\frac{\partial'}{\partial\Phi}. \tag{4.37c}$$

Since the individual terms in the trial function $\bar{\psi}_j(\bar{\mathbf{s}}, R)F_j(\mathbf{R})$ are eigenfunctions of $\bar{L}_z$ belonging to the eigenvalue λ_j, when $(J_x \pm iJ_y)$ operates on such a term, $\bar{L}_z$ can be replaced by λ_j. Forming $\mathbf{J}^2 = J_x^2 + J_y^2 + J_z^2$ we obtain the result

$$\mathbf{J}^2\{\bar{\psi}_j F_j\} = \bar{\psi}_j \mathbf{J}^2 F_j, \tag{4.38a}$$

where

$$\mathbf{J}^2 F_j = -\left[\frac{\partial'^2}{\partial\Theta^2} + \cot\Theta\,\frac{\partial'}{\partial\Theta} + \frac{1}{\sin^2\Theta}\left(\frac{\partial'}{\partial\Phi} - \mathrm{i}\lambda_j\cos\Theta\right)^2 - \lambda_j^2\right]F_j. \tag{4.38b}$$

Recollecting that when applied to a function of $\mathbf{R}$ alone, there is no distinction between $\partial'/\partial\Theta$ and $\partial/\partial\Theta$ or between $\partial'/\partial\Phi$ and $\partial/\partial\Phi$ we see that the result (4.38b) is identical with (4.34), which establishes that $\mathbf{J}^2$ in (4.33) indeed represents the square of the total orbital angular momentum.

The simultaneous eigenfunctions of $\mathbf{J}^2$ and J_z for a given value of λ are denoted by $\mathscr{H}^J_{M\lambda}(\Theta, \Phi)$

$$\mathbf{J}^2\mathscr{H}^J_{M\lambda} = J(J+1)\mathscr{H}^J_{M\lambda}; \qquad J_z\mathscr{H}^J_{M\lambda} = M\mathscr{H}^J_{M\lambda}. \tag{4.39}$$

In the body-fixed frame $\bar{N}_z = 0$, so that $\bar{J}_z = \bar{L}_z = \lambda$. It follows that, since $\bar{J}_z^2 \leq J^2$, then for a given λ, the quantum number J must satisfy $J \geq \lambda$. The Φ-dependence of the functions $\mathscr{H}^J_{M\lambda}$ is clearly of the form $\exp(\mathrm{i}M\Phi)$ and the Θ-dependence has been given by Landau and Lifshitz (1958) and by Edmonds (1960). In terms of the reduced rotation matrix of Edmonds, $d^J_{M\lambda}(\Theta)$, we have

$$\mathscr{H}^J_{M\lambda}(\Theta, \Phi) = (-1)^{\lambda+M} d^J_{M\lambda}(\Theta)\exp\mathrm{i}(M\Phi), \tag{4.40}$$

where $d^J_{M\lambda}$, the reduced rotation matrix, can be expressed as:

$$d^J_{M\lambda}(\Theta) = N^J_{M\lambda}\,\frac{(1-\mu)^{\frac{1}{2}(\lambda-M)}}{(1+\mu)^{\frac{1}{2}(\lambda+M)}}\left[\left(\frac{\partial}{\partial\mu}\right)^{J-M}(1-\mu)^{J-\lambda}(1+\mu)^{J+\lambda}\right]. \tag{4.41}$$

In (4.41) we have set $\mu \equiv \cos\Theta$, and $N^J_{M\lambda}$ is a normalization constant chosen so that

$$\int_{-1}^{+1} \mathrm{d}(\cos\Theta)\int_0^{2\pi}\mathrm{d}\Phi\,\mathscr{H}^{J^*}_{M\lambda}(\Theta,\Phi)\mathscr{H}^{J'}_{M'\lambda}(\Theta,\Phi) = \frac{4\pi}{2J+1}\delta_{JJ'}\delta_{MM'}. \tag{4.42}$$

In the particular case $M = 0$, the $\mathscr{H}$ functions reduce to spherical harmonics:

$$\mathscr{H}^J_{0\lambda}(\Theta,\Phi) = (-1)^{\lambda}\left(\frac{4\pi}{2J+1}\right)^{\frac{1}{2}} Y_{J\lambda}(\Theta, \Phi = 0). \tag{4.43}$$

Each channel function $F_j(\mathbf{R})$ can be expanded as (cf. (3.103))

$$F_j(\mathbf{R}) = R^{-1}\sum_{J=0}^{\infty} f^J_j(R)\mathscr{H}^J_{M\lambda_j}(\Theta,\Phi). \tag{4.44}$$

The radial equations, and hence the radial functions, are independent of M. However, the rotational part of the interaction couples channels with differing values of λ_j, from $\lambda_j = 0$ up to $\lambda_j = J$. Using the orthogonality properties of the $\mathscr{H}$ functions, together with the relations (Kronig 1930; Delos 1981a)

$$\left[\mp\frac{\partial}{\partial\Theta} + \frac{\mathrm{i}}{\sin\Theta}\frac{\partial}{\partial\Phi} + \lambda\cot\Theta\right]\mathscr{H}^{J}_{M\lambda}(\Theta,\Phi)$$
$$= [(J \pm \lambda + 1)(J \mp \lambda)]^{\frac{1}{2}}\mathscr{H}^{J}_{M\lambda\pm1}(\Theta,\Phi), \quad (4.45)$$

which are easily proved using (4.41), the radial equations can be reduced to the set

$$\left[\frac{\mathrm{d}^2}{\mathrm{d}R^2} + \frac{\lambda^2 - J(J+1)}{R^2} + \mathscr{K}^2(R)\right]\mathbf{f}^J(R) = \mathbf{V}^{\mathrm{R}}(R)\mathbf{f}^J(R) + \mathbf{V}^{\mathrm{C}}(R)\mathbf{f}^J(R) \quad (4.46)$$

where $\mathbf{V}^{\mathrm{R}}(R)$ has elements defined by (4.35), and

$$V^{\mathrm{C}}_{ij}(R) = -\delta_{\lambda_i,\lambda_j+1}[(J-\lambda_i)(J+\lambda_i+1)]^{\frac{1}{2}}\frac{\langle i|\bar{L}_x + \mathrm{i}\bar{L}_y|j\rangle}{R^2}$$
$$-\delta_{\lambda_i,\lambda_j-1}[(J+\lambda_i)(J-\lambda_i+1)]^{\frac{1}{2}}\frac{\langle i|\bar{L}_x - \mathrm{i}\bar{L}_y|j\rangle}{R^2}, \quad (4.47a)$$

which can also be written in the form

$$V^{\mathrm{C}}_{ij}(R) = -2\delta_{\lambda_i,\lambda_j+1}[(J-\lambda_i)(J+\lambda_i+1)]^{\frac{1}{2}}\frac{\langle i|\mathrm{i}\bar{L}_y|j\rangle}{R^2}$$
$$+2\delta_{\lambda_i,\lambda_j-1}[(J+\lambda_i)(J-\lambda_i+1)]^{\frac{1}{2}}\frac{\langle i|\mathrm{i}\bar{L}_y|j\rangle}{R^2}. \quad (4.47b)$$

The $\mathscr{S}$-matrix and cross sections

The radial equations (4.46) are in a form suitable for numerical solution. Solutions vanishing at the origin are required with the asymptotic form

$$f^J_j(R) \approx \frac{1}{k_j^{\frac{1}{2}}}\left\{\delta_{ji}\exp\left(-\mathrm{i}\left(k_jR - \frac{J\pi}{2}\right)\right) - \mathscr{S}^J_{ji}\exp\left(\mathrm{i}\left(k_jR - \frac{J\pi}{2}\right)\right)\right\}. \quad (4.48)$$

In (4.48), k_j and k_i are the channel momenta, such that $K_j(R) \to k_j$ and $K_i(R) \to k_i$ as $R \to \infty$. If there is an overall Coulomb interaction in the initial or final states or in both, the asymptotic form must be suitably modified, as indicated in Chapter 3:

$$\exp\left(\pm \mathrm{i}\left(k_j R - \frac{J\pi}{2}\right)\right) \rightarrow \exp(\pm \mathrm{i}(k_j R - \alpha_j(2k_j)^{-1} \log 2k_j R + \sigma_J - \tfrac{1}{2}J\pi)). \tag{4.49}$$

In (4.49), σ_J is the Coulomb phase shift, and $\alpha_j = 2\mu q_j^{\mathrm{A}} q_j^{\mathrm{B}}$ where $q_j^{\mathrm{A}} q_j^{\mathrm{B}}$ are the *overall* charges centred on A and B in the asymptotic region of channel j. The quantities $\mathscr{S}_{ji}^J$ form the matrix elements of the $\mathscr{S}$-matrix for scattering from an initial state i to a final state j which are eigenstates of the component of the angular momentum along the interionic line, with eigenvalues λ_i, λ_j. Provided no further approximations are made the $\mathscr{S}$-matrix obtained from the perturbed stationary state model is unitary and symmetrical.

If the initial state is spherically symmetrical with $\lambda_i = 0$, the scattering amplitude, for scattering into the state j, is

$$f_{ji}(\Theta) = \frac{\mathrm{i}}{2\sqrt{k_i k_j}} \sum_J [(4\pi)(2J+1)]^{\frac{1}{2}} \mathscr{H}_{0\lambda_j}^J(\Theta, \Phi)[\delta_{ij} - \mathscr{S}_{ji}^J], \tag{4.50}$$

which is the generalization of equations (3.106) and (3.107). The corresponding differential cross section is

$$\frac{\mathrm{d}\sigma_{ji}}{\mathrm{d}\Omega} = \frac{k_j}{k_i} |f_{ji}(\Theta)|^2, \tag{4.51}$$

and the corresponding total cross section is

$$\sigma_{ji} = \frac{\pi}{k_i^2} \sum_J (2J+1)|\mathscr{S}_{ji}^J|^2. \tag{4.52}$$

The S-matrix in other representations can be obtained straightforwardly from $\mathscr{S}_{ji}^J$ (Mott and Massey 1965, Chapter XIV). A complete discussion has been given by Nikitin and Umanskii (1984), and the paper by Gaussorgues *et al.* (1975*a*) can also be consulted with profit.

4.4 RADIAL AND ROTATIONAL COUPLINGS – DIABATIC REPRESENTATIONS

Radial coupling

The radial coupling matrix elements expressed in (4.35) only connect states i and j having the same symmetry, for which $\lambda_i = \lambda_j$. Some crude estimates of the size of each of the three terms can be given. The most important is the last term in (4.35):

$$-2\langle i|\frac{\partial}{\partial R}|j\rangle\frac{\partial}{\partial R}. \tag{4.53}$$

Introducing the electronic Hamiltonian defined by (4.19), we can write

$$\langle i|H_e\left(\frac{\partial}{\partial R}\right)|j\rangle-\langle i|\left(\frac{\partial}{\partial R}\right)H_e|j\rangle=-\langle i|\frac{\partial H_e}{\partial R}|j\rangle. \tag{4.54}$$

Using (4.18) the left-hand side of (4.54) becomes equal to $(E_i(R)-E_j(R))\langle i|\partial/\partial R|j\rangle$, so that the matrix element of the operator $\partial/\partial R$ denoted by A_{ij} can be expressed as

$$A_{ij}\equiv\langle i|\frac{\partial}{\partial R}|j\rangle=\begin{cases}-\dfrac{1}{E_i(R)-E_j(R)}\langle i|\dfrac{\partial H_e}{\partial R}|j\rangle & i\neq j\\ 0 & i=j.\end{cases} \tag{4.55}$$

where the second line for the case $i=j$ arises because the molecular orbitals are normalized to unity for every value of R. Since the electronic kinetic energy operator contained in H_e (see (4.19)) is independent of R, and since the term containing $V_{AB}(R)$ vanishes because of the orthogonality of the states i and j, $i\neq j$, the operator $\partial H_e/\partial R$ can be replaced in (4.55) by $\partial V/\partial R$ where V is the interaction energy

$$V(\bar{s},R)=V_{Ae}(x)+V_{Be}(r). \tag{4.56}$$

By the Wigner–von Neumann non-crossing rule, the difference of the adiabatic potential energies $(E_i(R)-E_j(R))$ can never vanish at any value of R, if the adiabatic states i and j have the same symmetry. However potential energy curves for states of the same symmetry *can approach* each other in a narrow region about some value of R and this is called an 'avoided crossing'. An example is shown in Fig. 4.5 for the system $(CH)^{4+}$. It is seen that for R close to 8.1 a.u., the σ1s level, which is asymptotic to $C^{+4}+H(1s)$ makes an avoided crossing with the σ3d level, which is asymptotic to $C^{+3}(3d)+H^+$. In the region of the avoided crossing we see from (4.55) that the matrix element $\langle i|\partial/\partial R|j\rangle$ is large. At low energies in many systems charge exchange only takes place with a significant probability in the region of one, or a few, avoided crossings of this type, and this fact in turn implies that accurate cross sections can be obtained from limited sets which contain these crossings.

Since $\partial f_j^J(R)/\partial R\approx k_j f_j^J(R)$ a rough estimate of the size of the interaction (4.53) is

$$\left|2\langle i|\frac{\partial}{\partial R}|j\rangle\right|\frac{\partial}{\partial R}\approx\frac{2k_j}{a|\Delta_{ij}|}|\langle i|V|j\rangle|, \tag{4.57}$$

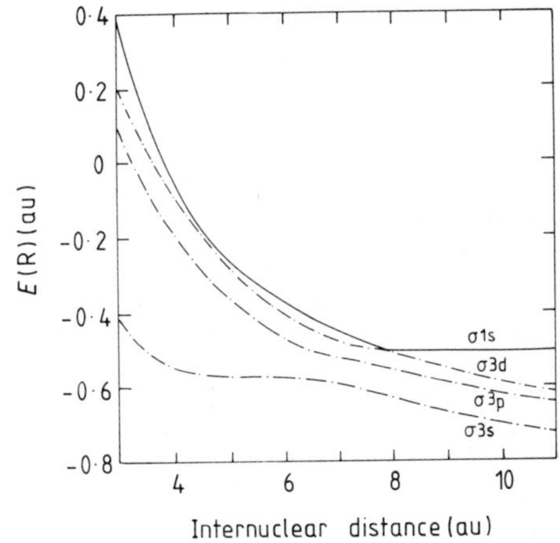

Fig. 4.5. Adiabatic potential energies $E(R)$ of $(CH)^{4+}$ as a function of internuclear distance R (after Gargaud and McCarroll (1985)).

where a is the extent of the region over which $(E_i - E_j)$ is small, and Δ_{ij} is the value of $(E_i - E_j)$ at the avoided crossing.

We now turn our attention to the first term in (4.35). This is generally small, but can be estimated as follows. We first notice that if **B** is the matrix with elements B_{ij} defined as

$$B_{ij} = \langle i| \frac{\partial^2}{\partial R^2} |j\rangle \tag{4.58a}$$

then

$$\mathbf{B} = \mathbf{A}^2 + \frac{\mathrm{d}}{\mathrm{d}R}\mathbf{A} \tag{4.58b}$$

where **A** is the matrix of elements A_{ij} defined by (4.55). If $\mathbf{A}^2$ is calculated approximately from a finite basis we have that $(A^2)_{ij} \simeq \sum_{k=1}^{N} A_{ik}A_{kj}$, and this relation provides an approximate means of calculating the elements B_{ij}. Although the diagonal elements of **A** vanish, it should be noticed that this is not the case for **B**. In estimating **B**, we notice the intermediate state k employed in calculating $\mathbf{A}^2$ can neither be equal to i nor j, since the diagonal elements of **A** vanish, and this implies that the term $\mathrm{d}\mathbf{A}/\mathrm{d}R$ is more important than the term $\mathbf{A}^2$. An estimate of the strength of the coupling is

$$\left|\langle i| \frac{\partial^2}{\partial R^2} |j\rangle\right| \approx \frac{1}{a^2|\Delta_{ij}|} |\langle i|V|j\rangle|. \tag{4.59}$$

This is small compared with (4.57) if $(k_j a) \gg 1$, a condition which is generally satisfied except for very low (thermal) energies.

The remaining term in (4.35), $R^{-2}\langle i|\bar{L}_x^2 + \bar{L}_y^2|j\rangle$ is of magnitude of order $1/R^2$ and again is unimportant compared with (4.57), if $k_j a$ is large.

Rotational coupling

Near an avoided crossing, the rotational coupling, between states differing in λ_j by one unit, is usually less important than the radial coupling between states having the same value of λ_j. In general, away from an avoided crossing, rotational coupling is not small compared with radial coupling and is often of comparable importance. A good example is where the $\sigma, \pi, \ldots$ states become degenerate in the united atom limit, as in a true one-electron system for which the united atom is hydrogenic. The coupling between the molecular states which approach the atomic states with the same n and different l is enhanced by the factor R^{-2} (see (4.47)) and is of major importance in determining the dynamics of such a system.

Diabatic representations

The radial coupling $A_{ij}(R) = \langle i|\partial/\partial R|j\rangle$ can be calculated by finite-difference methods (see for example Hirsch *et al.* 1980), but this is not always easy or convenient, since near an avoided crossing $A_{ij}(R)$ is nearly singular and changes vary rapidly. This is because the nature of the molecular orbital changes rapidly through the avoided crossing region, changing in nodal structure. Because of this it is often better to make a unitary transformation of the adiabatic basis functions to a new basis, called a *diabatic basis* in which the matrix A_{ij} is smoothly varying or zero (Garrett and Truhlar 1981). We shall consider the case in which the transformed matrix $\mathbf{A}$ is zero (Smith 1969; Heil *et al.* 1981). Let $\mathbf{C}(R)$ be the unitary matrix which transforms the (finite) basis of adiabatic molecular orbitals in the body-fixed frame, $\bar{\psi}_j(\bar{\mathbf{s}}, R)$ to a new set of diabatic orbitals $\bar{\psi}_j^{\mathrm{d}}(\bar{\mathbf{s}}, R)$

$$\bar{\psi}_j^{\mathrm{d}}(\bar{\mathbf{s}}, R) = \sum_k \bar{\psi}_k(\bar{\mathbf{s}}, R) C_{kj}(R). \tag{4.60}$$

The transformation will be chosen so that symmetries are preserved; a σ orbital transforms to a mixture of σ orbitals, a π orbital to a matrix of π orbitals, and so on. Under this transformation the column vector of radial channel functions $\mathbf{f}^J(R)$ transforms to a column vector $\mathbf{g}^J(R)$ where

$$\mathbf{f}^J(R) = \mathbf{C}\mathbf{g}^J(R). \tag{4.61}$$

Correspondingly the coupled equations (4.46) transform to

$$\left[\frac{\mathrm{d}^2}{\mathrm{d}R^2} + \frac{\lambda^2 - J(J+1)}{R^2} - 2\mu\mathbf{U}(R) + 2\mu E\right]\mathbf{g}^J(R) = [{}^{\mathrm{d}}\mathbf{V}^{\mathrm{R}}(R) + {}^{\mathrm{d}}\mathbf{V}^{\mathrm{C}}(R)]\mathbf{g}^J(R). \tag{4.62}$$

The radial and rotational interactions ${}^{d}\mathbf{V}^{R}$, ${}^{d}\mathbf{V}^{C}$ have exactly the same form as $\mathbf{V}^{R}$ and $\bar{\mathbf{V}}^{C}$; but the matrix elements are taken with respect to the diabatic basis states, rather than to the adiabatic basis states.

The matrix elements U_{ij} are the matrix elements of the electronic Hamiltonian with respect to the new basis,

$$U_{ij}(R) = \int \bar{\psi}_i^{d}(\bar{\mathbf{s}}, R) H_e \bar{\psi}_j^{d}(\bar{\mathbf{s}}, R)\, d\bar{\mathbf{s}}, \tag{4.63a}$$

and

$$-2\mu\mathbf{U} + 2\mu E = \mathbf{C}\dagger(R)\mathscr{K}^2(R)\mathbf{C}(R). \tag{4.63b}$$

In the new basis the electronic Hamiltonian is not diagonal, but we can ensure that $\mathbf{U}$ becomes diagonal in the limit $R \to \infty$, with elements equal to the energies of the asymptotic atomic levels, by requiring that the transformation matrix approaches the unit matrix in this limit

$$\mathbf{C}(R) \to \mathbf{I}; \qquad R \to \infty. \tag{4.64}$$

In the diabatic basis the radial coupling ${}^{d}\mathbf{V}^{R}$ contains contributions from matrix elements of $\partial/\partial R$, $\partial^2/\partial R^2$, and $(\bar{L}_x^2 + \bar{L}_y^2)/R^2$. The contribution from the matrix elements of $\partial/\partial R$ is obtained from the transform of the matrix $\mathbf{A}$ with elements given by (4.55). Denoting the transformed matrix by $\mathbf{A}^{d}$, we have that

$$\begin{aligned}(\mathbf{A}^{d})_{ij} &= \sum_{k,l} \langle \bar{\psi}_k | C_{ki}^{*}(R) \frac{\partial}{\partial R} C_{lj}(R) | \bar{\psi}_l \rangle \\ &= \sum_{k,l} C_{ki}^{*}(R) \langle k | \frac{\partial}{\partial R} | l \rangle C_{lj}(R),\end{aligned} \tag{4.65}$$

or, in matrix notation, since $\langle k | l \rangle = \delta_{kl}$

$$\mathbf{A}^{d} = \mathbf{C}\dagger\mathbf{A}\mathbf{C} + \mathbf{C}\dagger \frac{\partial}{\partial R}\mathbf{C}. \tag{4.66}$$

We can now ensure that in the new basis the derivative coupling vanishes by taking $\mathbf{C}$ to be the solution of the equation

$$\mathbf{A}\mathbf{C} + \frac{\partial}{\partial R}\mathbf{C} = 0, \tag{4.67}$$

using (4.64) as the boundary condition. In a finite diabatic basis, the second-derivative matrix $\mathbf{B}^{d}$ does not vanish. However, if equation (4.58b) is assumed, which is strictly correct only for a basis forming a complete set, it follows that $\mathbf{B}^{d}$ vanishes as well as $\mathbf{A}^{d}$. Under these conditions the only non-zero contribution to ${}^{d}\mathbf{V}^{R}$ is from the matrix elements, with respect to the diabatic basis, of $(\bar{L}_x^2 + \bar{L}_y^2)/R^2$. The rotational coupling $\mathbf{V}^c$ is unaltered in form, but the matrix elements are taken with respect to the diabatic, rather than the adiabatic states. The matrix $\mathbf{U}(R)$, which is a *smooth* function of R, takes the place of the interactions depending on derivatives with respect to R. It should be noted that if the basis forms a complete set, the matrix $\mathbf{A}^d$ must be identically zero, so that the diabatic basis functions must be independent of R (except for phase factors). The only such complete set satisfying (4.67) is the set of asymptotic atomic wave functions. This is not the case for *finite* basis sets, and diabatic representations offer a distinct method of calculation which is of utility.

For the particular case of a two-state approximation equations (4.67) can be solved. In this case $A_{11} = A_{22} = 0$ and the only non-zero element of $\mathbf{A}$ is $A_{12} = -A_{21}^*$. It is easy to verify that the solution is the rotation matrix:

$$\mathbf{C}(R) = \begin{pmatrix} \cos\omega(R) & \sin\omega(R) \\ -\sin\omega(R) & \cos\omega(R) \end{pmatrix}, \qquad (4.68)$$

where the angle of rotation, $\omega(R)$, is defined by

$$\omega(R) = \int_R^\infty A_{12}(R')\,\mathrm{d}R'. \qquad (4.69)$$

As $R \to \infty$, $\omega(R) \to 0$ so that $C(R)$ becomes equal to the unit matrix, and thus satisfies the boundary condition (4.64).

4.5 APPLICATIONS OF THE PERTURBED STATIONARY STATE MODEL

To illustrate some of the features of the perturbed stationary state model we shall first describe two applications to charge transfer in collisions between positive ions and atomic hydrogen at low energies, which are of importance for processes in stellar atmospheres, and then describe the symmetrical resonance reaction $H^+ + H(1s)$, which has a rather different character. References to some other typical calculations for collisions between charged ions and atomic hydrogen at very low energies are given in Table 4.1.

$N^{3+} + H(1s)$

This reaction has been studied by Gargaud *et al.* (1981) for centre-of-mass energies between 8.2×10^{-3} and 136 eV. The $N^{3+}(1s^2 2s^2)$ ionic core can be treated as inert, so that the system is an effective one-electron system. The interaction between the active electron, originally bound to the proton in the hydrogen atom, and the N^{3+} ionic core can be represented by the

Table 4.1. Partial-wave perturbed stationary state calculations selected for illustrative purposes.

System	Energy range eV/a.m.u.	Reference
$H^+ + H$	10^{-4}–10	Hunter and Kuriyan (1977*a*, *b*)
	0–250	Knudsen and Thorsen (1970)
$H + He^{2+}$	20–500	[1]van Hemert *et al.* (1985)
$H + C^{3+}$	0.27–8.1	Neil *et al.* (1981)
	0–5000	Bienstock *et al.* (1984)
	10–5000	[1]Opradolce *et al.* (1988)
$H + C^{4+}$	8×10^{-6}, 2.5–8.2	McCarroll and Valiron (1979)
	1–350	Gargaud and McCarroll (1985)
	1–1000	[1]Gargaud *et al.* (1987)
$H + N^{2+}$	0.27–8.1	Heil *et al.* (1981)
$H + N^{3+}$	8×10^{-6}, 8×10^{-3}–136	Gargaud *et al.* (1981)
	8×10^{-6}, 8×10^{-3}–10.0	Heil *et al.* (1983)
		Bienstock *et al.* (1984)
	1–200	Gargaud and McCarroll (1985)
$H + N^{5+}$	1–1000	Gargaud and McCarroll (1985)
$H + O^{4+}$, Si^{4+}	1–1000	[1]Gargaud and McCarroll (1988)
$H + Al^{3+}$, Ti^{4+}	1–1000	[1]Gargaud *et al.* (1988)
$He + O^{2+}$	0–5000	Bienstock *et al.* (1984)

[1] Modified perturbed stationary state calculation including translation factors.

potential (McCarroll and Valiron 1979)

$$V_{N^{3+}}(r) = -\frac{3 + (4 + 0.4608r)\exp(-2.422r)}{r}. \tag{4.70}$$

The parameters in this potential were determined to reproduce the energies of the lowest levels of the atom $N^{2+}(nl)$ as well as possible. The interaction between the proton and the N^{3+} core is sufficiently well represented by the Coulomb repulsion:

$$V_{AB} = \frac{3}{R}. \tag{4.71}$$

The molecular orbitals for the pseudomolecule $(NH)^{3+}$ were found by diagonalizing the electronic Hamiltonian in a basis of Slater orbitals using prolate spheroidal coordinates. The incident channel $N^{3+} + H(1s)$ is correlated to a σ level which we can designate as $\sigma 1s_H$. At zero impact velocity the corresponding molecular orbital is an eigenfunction of the total Hamiltonian. It is expected that at low velocities only a few other levels will be coupled significantly to the $\sigma 1s_H$ level, the number increasing with velocity. In fact this is a particularly simple case in that at the velocities considered

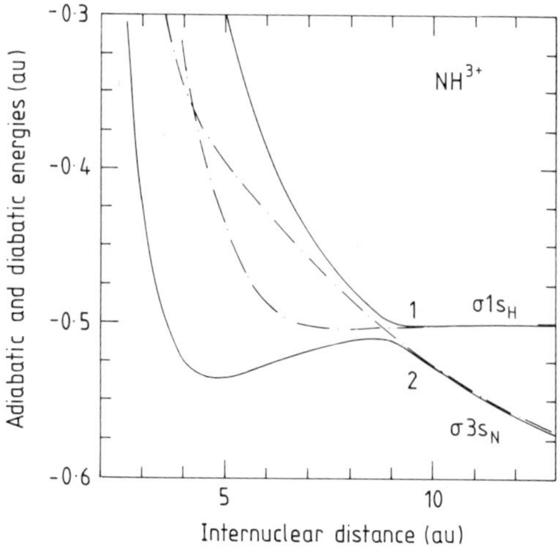

Fig. 4.6. Potential energies for NH^{3+}: ——— adiabatic energies showing an avoided crossing near $R = 9$ a.u.; —·—·— diabatic potentials showing crossings at 9 and 4.2 a.u. (after Gargaud *et al.* (1981)).

the only important coupling is to the $\sigma 3s_N$ level which is asymptotic to the configuration $H^+ + N^{2+}(3s)$. The adiabatic potential curves of the $\sigma 1s_H$ and $\sigma 3s_N$ levels make a *pseudocrossing* near $R = 9$ a.u., which is illustrated in Fig. 4.6. Also illustrated in Fig. 4.6 are the diagonal elements of the diabatic energy matrix U_{ij} (see (4.63a)) which make a *real crossing* in the same vicinity. The calculations of Gargaud *et al.* employed these $\sigma 1s_H$ and $\sigma 3s_N$ orbitals as a two-state basis. As only σ states are concerned there is no rotational coupling and only radial coupling need be considered. The matrix element of the radial coupling connecting the two levels is illustrated in Fig. 4.7, showing clearly the singular nature of the coupling in the region of the pseudocrossing.

A consistently formulated model, including translation factors in the trial function, should be invariant under a change of origin along the interionic line AB. This condition is not satisfied if the translation factors are omitted, as in the perturbed stationary state model. A measure of the importance of this violation of translational invariance can be obtained by evaluating the radial interaction elements using origins placed on each of the nuclei A and B in turn, rather than at the centre of mass. That is, the matrix elements $\langle i|\partial/\partial R|_r|j\rangle$ and $\langle i|\partial/\partial R|_x|j\rangle$ can be calculated, where the differentiation is performed holding r or x constant, respectively, in place of $\langle i|\partial/\partial R|_s|j\rangle$. In the present example, since the centre of mass is located near the N^{3+} ion, the matrix elements using the centre of mass and the N^{3+} ion as origins are nearly the same. However, except in the region of the crossing, placing the origin on the H^+ ion does change the matrix elements significantly, as shown

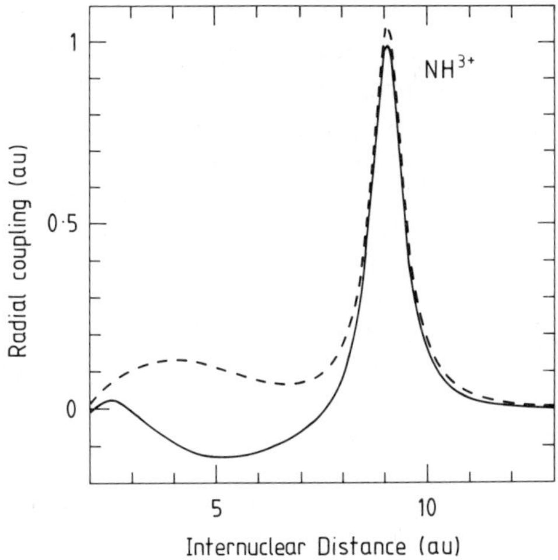

Fig. 4.7. Radial coupling between the $N^{3+} + H(1s)$ and $N^{2+}(3s) + H^+$ channels: ——— origin of coordinates on the N nucleus; ----- origin of coordinates on the H nucleus (after Gargaud *et al.* (1981)).

in Fig. 4.6. The connection between matrix elements of $\partial/\partial R$ with respect to different origins is easily obtained. If $\mathbf{s}' = \mathbf{s} + \Delta\mathbf{R}$ is the position vector of the electron with respect to an origin O taken so that $OA = pR$ and $OB = qR$, $p + q = 1$, and Δ is given by (3.9), then, in the body-fixed frame in which $\mathbf{R}$ is the Z-axis,

$$\langle i| \left.\frac{\partial}{\partial R}\right|_{\bar{s}} |j\rangle = \langle i| \left.\frac{\partial}{\partial R}\right|_{\bar{s}'} |j\rangle + \Delta\langle i| \frac{\partial}{\partial s'_z} |j\rangle$$

$$= \langle i| \left.\frac{\partial}{\partial R}\right|_{\bar{s}'} |j\rangle + \Delta(E_i - E_j)\langle i|\bar{s}'_z|j\rangle. \tag{4.72}$$

In Table 4.2 computed charge exchange total cross sections in the two-state approximation are shown taking different origins: (A) at the H^+ ion, (B) at the N^{2+} ion, and (M) at the mid-point of the line AB. As might be hoped, the lack of translational invariance exemplified by the spread of these results is very small below incident energies of about 100 eV. Further calculations with an enlarged basis set confirm that in this particular case the two-state approximation is accurate up to about 150 eV, at which capture into the 3p level of N^{2+} becomes significant at about 1 to 2 per cent of the total capture cross section.

The $N^{3+} + H(1s)$ collision has also been investigated by Heil *et al.* (1983) and Bienstock *et al.* (1984), who obtained the adiabatic potential energies from a multi-electron configuration interaction method, rather than by using

Table 4.2. Cross section for the reaction $N^{3+} + H(1s) \rightarrow N^{2+}(3s) + H^+$ (from Gargaud *et al.* 1981). A is origin at the H^+ ion; B is origin at the N^{3+} ion; M is origin at the mid-point of the interionic line AB.

E_L(eV/a.m.u.)	Cross sections (10^{-16} cm^2)		
	A	M	B
8.8×10^{-3}	19.74	19.75	19.70
8.8×10^{-2}	7.75	7.73	7.70
8.8×10^{-1}	18.39	18.34	18.39
8.8	35.96	35.84	35.73
29.1	36.01	36.01	36.18
146.0	26.04	27.16	28.98

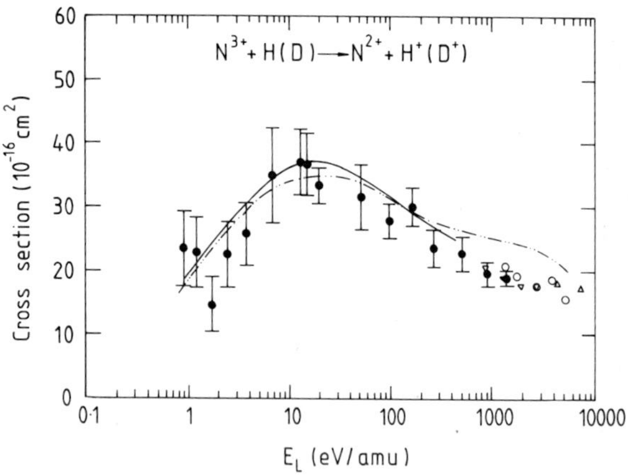

Fig. 4.8. Total capture cross section for electron capture by N^{3+} ions from H(1s) and D(1s). Calculated cross sections: ——— Gargaud and McCarroll (1985); —·—·—·— Bienstock *et al.* (1984). Experimental data: ⧫ Huq *et al.* (1989); ○ Crandall *et al.* (1979); △ Phaneuf *et al.* (1978); ▽ Seim *et al.* (1981).

a model potential. In addition to the cross section for capture in the $N^{2+}(2s^2 3s, {}^2S)$ state, which as we have noted is the dominant channel, Bienstock *et al.* calculated the cross section for capture into the $N^{2+}({}^2s2p^2, {}^2D)$ and $N^{2+}(2s^2 2p^2, {}^2S)$ states, in the energy range from 0.3 to 5,000 eV/a.m.u. The contribution of the $N^{2+}(2s2p^2, {}^2S)$ channel is less than 4 per cent even at 5 keV and that from the $N^{2+}(2s2p^2, {}^2S)$ channel is even smaller.

In Fig. 4.8, the calculated cross sections of Gargaud and McCarroll (1985) and of Bienstock *et al.* (1984) are compared with the experimental data. It is seen that at energies up to 500 eV/a.m.u., the agreement between the theoretical cross sections and the data is excellent, but at higher energies the

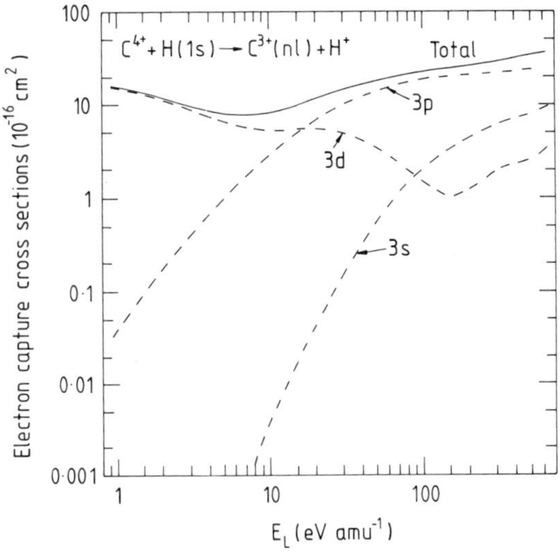

Fig. 4.9. Calculated cross sections for electron capture by C^{4+} ions from atomic hydrogen (from Gargaud and McCarroll (1985)).

theoretical cross section of Bienstock *et al.* is larger than the measured data by up to 25 per cent.

The behaviour of the cross section at low energies for this and similar systems depends on whether or not the avoided crossing is reached during the collision. If it is, the cross section is large; but, if the crossing region is inaccessible, the cross section is small. The interaction between a charged ion and a neutral target at large distances is attractive because of the dipole induced in the target atom (polarization). This controls the effective interaction in the entrance channel and ensures the crossing region is traversed. In the present example, the theoretical cross section shows a maximum at $E = 0.08$ eV, and at lower energies the cross section increases towards zero energy like $E^{-\frac{1}{2}}$.

C^{4+} + H(1s)

In this example, capture takes place primarily to the $(1s)^2 3d$ level of C^{3+} at low energies. However, above 20 eV/a.m.u. capture to the 3d level becomes less important in this region and the dominant final level is the $(1s)^2 3p$, as illustrated in Fig. 4.9. The cross section for capture into the $(1s)^2 3s$ level, which is very small at the lowest energies, rises sharply with increasing energy and makes an important contribution for energies greater than 100 eV/a.m.u. The calculations of Gargaud *et al.* (1981), and Gargaud and McCarroll (1985) are based on a four-state approximation which links the initial

$C^{4+} + H(1s)$ channel to the three important charge exchange channels:

$$\begin{aligned} C^{4+} + H(1s) &\rightarrow C^{3+}(3s) + H^+ \\ &\rightarrow C^{3+}(3p) + H^+ \\ &\rightarrow C^{3+}(3d) + H^+. \end{aligned} \tag{4.73}$$

As in the previous example, the interaction between the projectile and the active electron is represented by an effective potential. At low energies ($\lesssim 50$ eV/a.m.u.) only radial coupling is important and for this reason only σ states were retained in the basis. There are, in a separated atom designation, the σ1s, asymptotic to the incident channel $C^{4+} + H(1s)$, and the σ3s, σ3p, and σ3d states which are asymptotic to C^{3+} $(n = 3) + H^+$. Further work by Gargaud *et al.* (1987) employs a seven-state basis in which all orbitals asymptotic to the C^{3+} $(n = 3)$ are included. The three additional orbitals are π3p, π3d, and δ3d which are populated by rotational coupling. In agreement with earlier work of Hanssen *et al.* (1984) and Olson *et al.* (1978), it is found that rotational coupling is important above 50–100 eV/a.m.u. In the calculations of Gargaud *et al.* translation factors were included using the improved perturbed stationary state method outlined in Section 4.5 below. However, for this system the effect of allowing for translational invariance is small at the energies under discussion. Measurements of the total capture cross section have been made down to 15 eV/a.m.u. by Phaneuf *et al.* (1982). The agreement with the calculated cross section is good below ~ 50 eV/a.m.u., but at higher energies the results of the seven-state calculation exceed the data significantly as illustrated in Fig. 4.10. This disagreement is rather

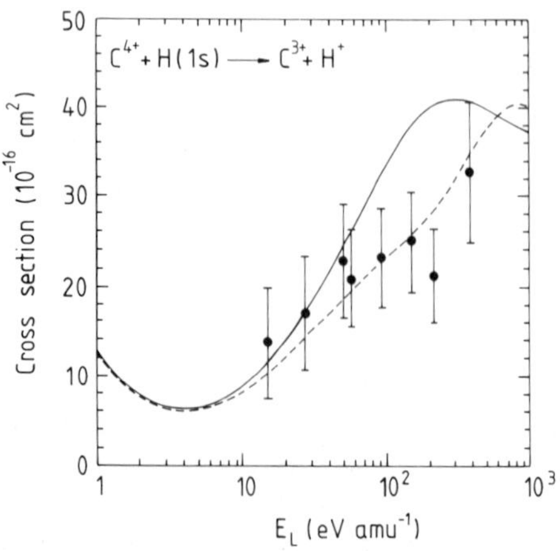

Fig. 4.10. Total cross section for electron capture by C^{4+} from atomic hydrogen. Theoretical cross section: seven-state ———, four-state – – – – –, Gargaud *et al.* (1987). ⧳ Experimental data, Phaneuf *et al.* (1982).

surprising since the calculated branching ratio in the seven-state approximation at $E_L = 1.18$ keV/a.m.u. of $P_{3s} = 0.35$, $P_{3p} = 0.41$, and $P_{3d} = 0.24$ is in excellent agreement with the measured values of Dijkkamp *et al.* (1985*b*). Above 100 eV/a.m.u. the calculated seven-state cross section is in harmony with other theoretical results by Hanssen *et al.* (1984) and somewhat exceeds the cross section computed by Fritsch and Lin (1984*a*) who employed an atomic two-centre basis.

$H^+ + H(1s)$

The simplest example in which the ionic cores are identical – the case of symmetrical resonance – is afforded by the system $H^+ + H(1s)$. Because of the identity of the two nuclei we shall have to discuss in what sense, if at all, a charge exchange cross section can be identified separately from elastic scattering. As the interactions V_{Ae}, V_{Be}, and V_{AB} are pure Coulomb potentials, the molecular orbitals and energy levels can be obtained as accurately as is desired without difficulty (see Appendix D). Because both nuclei have the same charge the orbitals divide into those of gerade and ungerade symmetry (defined in Section 4.2). The correlation diagram for homonuclear systems is shown in Fig. 4.3 and the corresponding potential energy curves are shown in Fig. 4.11 for the particular case of H_2^+. The lowest molecular levels, the $1s\sigma_g$ and $2p\sigma_u$ (using a united atom designation) both correlate to the channel $H^+ + H(1s)$. In view of the symmetry of the system the lowest molecular orbitals separate to the combinations

$$\begin{aligned}\psi_{1s\sigma_g}(\mathbf{s}, \mathbf{R}) &\underset{R\to\infty}{\approx} \frac{1}{\sqrt{2}}[\phi_{1s}(r) + \phi_{1s}(x)]\\ \psi_{2p\sigma_u}(\mathbf{s}, \mathbf{R}) &\underset{R\to\infty}{\approx} \frac{1}{\sqrt{2}}[\phi_{1s}(r) - \phi_{1s}(x)],\end{aligned} \tag{4.74}$$

where $\phi_{1s}(x)$ and $\phi_{1s}(r)$ are ground-state hydrogenic wave functions centred on A and B, respectively.

The total Hamiltonian, $H = -(1/2\mu)\nabla_R^2 + H_e$ commutes with the parity operator, so that equations (4.46) for the radial functions divide into separate sets for even and odd parity with solutions $\mathbf{f}_g^J(R)$ and $\mathbf{f}_u^J(R)$, respectively. In the perturbed stationary state model, the functions $\mathbf{f}_g^J(R)$ are associated with the even eigenfunctions of H_e, $\phi_{j,g}(\bar{\mathbf{s}}, R)$, and $\mathbf{f}_u^J(R)$ with the odd eigenfunctions, $\phi_{j,u}(\bar{\mathbf{s}}, R)$. In more elaborate models, for which the wave function contains translation factors, this is no longer the case and $\mathbf{f}_g^J(R)$ and $\mathbf{f}_u^J(R)$ are each associated with mixtures of the orbitals $\phi_{j,g}$ and $\phi_{j,u}$. In the perturbed stationary state model with no translation factors there is no coupling between the $1s\sigma_g$ and $2p\sigma_u$ levels. If the corresponding orbitals, and no others, are employed in the basis, the two radial functions $f_g^J(R)$ and

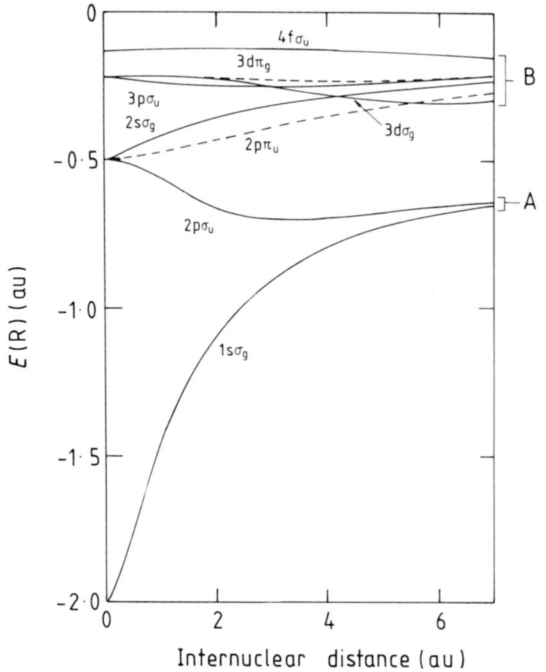

Fig. 4.11. Adiabatic potential energies of H_2^+ as a function of internuclear distance, excluding the internuclear potential: ——— σ states; - - - - - π states.

$f_u^J(R)$ satisfy the *uncoupled* equations

$$\left[\frac{d^2}{dR^2} - \frac{J(J+1)}{R^2} - 2\mu E_{g,u}(R) + 2\mu E\right] f_{g,u}^J(R) = 0, \tag{4.75}$$

where $E_g(R)$ and $E_u(R)$ are the potential energies of the $2s\sigma_g$ and $2p\sigma_u$ levels, respectively. Elastic scattering takes place in each of the uncoupled channels and (4.75) can be solved subject to the boundary conditions

$$f_{g,u}^J(\Theta) = 0; \qquad f_{g,u}^J(R) \underset{R\to\infty}{\sim} \sin(kR - \tfrac{1}{2}J\pi + \eta_{g,u}^J), \tag{4.76}$$

where $k^2 = 2\mu(E - \varepsilon_{1s})$ and η_g^J, η_u^J are phase shifts of order J.

Let us first consider the case in which the nuclei A and B are distinguishable, for example let A be a deuteron and B a proton. In the initial arrangement channel A is incident on the atom $(B + e^-)$, and this situation is represented by the radial wave functions $[f_g^J(R) + f_u^J(R)]/2$. The scattering amplitude in this channel, representing elastic scattering, is therefore

$$f(\Theta) = \tfrac{1}{2}[f_g(\Theta) + f_u(\Theta)], \tag{4.77}$$

where the scattering amplitudes $f_g(\Theta)$, $f_u(\Theta)$ are

$$f_{g,u}(\Theta) = \sum_{J=0}^{\infty} (2J+1)\frac{1}{2ik}(\exp(2i\eta^J_{g,u}) - 1)P_J(\cos\Theta). \tag{4.78}$$

The corresponding total elastic scattering cross section is

$$\sigma^{EL}(k) = \frac{2\pi}{k^2}\sum_{J=0}^{\infty}(2J+1)\{\sin^2\eta^J_g + \sin^2\eta^J_u - \tfrac{1}{2}\sin^2(\eta^J_g - \eta^J_u)\}. \tag{4.79}$$

Similarly the amplitude for finding the configuration $(A + e^-) + B$ is obtained from the radial function $[f^J_g(R) - f^J_u(R)]/2$, so that the amplitude for charge exchange $g(\Theta)$ is

$$g(\Theta) = \tfrac{1}{2}[f_g(\Theta) - f_u(\Theta)]. \tag{4.80}$$

Using the partial-wave series (4.78) for $f_g(\Theta)$ and $f_u(\Theta)$ in (4.80) the total charge exchange cross section is obtained:

$$\sigma^C(k) = \frac{\pi}{k^2}\sum_{J=0}^{\infty}(2J+1)\sin^2(\eta^J_g - \eta^J_u). \tag{4.81}$$

Unlike heteronuclear systems for which charge exchange requires either the radial or rotational coupling to be non-vanishing, in the homonuclear case charge exchange in the lowest (two-state) approximation is a result of the difference in phase of the scattering in the uncoupled gerade and ungerade levels.

To treat the case of the $H^+ + H(1s)$ system in which the nuclei are identical fermions of spin $\frac{1}{2}$, the total wave function must be antisymmetrical under the interchange of the (space + spin) coordinates of the two protons. In the singlet spin state ($S = 0$), the spatial wave function is symmetrical under the interchange $\mathbf{R} \to -\mathbf{R}$, and the elastic scattering amplitude is $f^+(\Theta)$ where

$$f^+(\Theta) = \tfrac{1}{2}\{f_g(\Theta) + f_g(\pi - \Theta) + f_u(\Theta) - f_u(\pi - \Theta)\}. \tag{4.82}$$

Likewise in the triplet spin state ($S = 1$), the spatial wave function is antisymmetrical under the interchange $\mathbf{R} \to -\mathbf{R}$, and the corresponding amplitude is

$$f^-(\Theta) = \tfrac{1}{2}\{f_g(\Theta) - f_g(\pi - \Theta) + f_u(\Theta) + f_u(\pi - \Theta)\}. \tag{4.83}$$

The elastic scattering differential cross section for an unpolarized beam of protons is

$$\frac{d\sigma^{EL}}{d\Omega} = \tfrac{3}{4}|f^-(\Theta)|^2 + \tfrac{1}{4}|f^+(\Theta)|^2. \tag{4.84}$$

In general, a separate charge exchange cross section cannot be identified for the case of indistinguishable ionic cores A and B. However, a spin exchange cross section can be defined in which an electron attached to a proton with Z-component of spin $S_z = \frac{1}{2}$ is transferred to a proton with $S_z = -\frac{1}{2}$. This cross section which is *identical to the cross section for charge exchange with distinguishable nuclei* (4.81), can in principle be measured using beams of spin-polarized protons.

Except at very low energies the amplitudes $f_g(\Theta)$ and $f_u(\Theta)$ are only large in a narrow cone about the forward direction $\Theta = 0$. There is therefore no interference between the terms in (4.82) and (4.83) $f_{g,u}(\Theta)$ and the terms $f_{g,u}(\pi - \Theta)$. Protons emerging close to $\Theta = 0$ can be ascribed to elastic scattering and those close to $\Theta = \pi$ to charge exchange, the corresponding cross sections being given by (4.79) and (4.81), respectively. For the $H^+ + H$ system, this behaviour is well established by $E_L = 4$ eV/a.m.u. (Smith 1966*a*, *b*) and it is only at smaller energies that the identity of the nuclei need be taken into account.

Accurate phase shifts for the two-state approximation have been calculated by Hunter and Kurijen (1977*a*) for all significant partial waves at a large number of energies between 10^{-4} and 10 eV. In the overlapping energy range 1 eV to 10 keV, similar results have been obtained by Dalgarno and Yadav (1953) from a semi-classical treatment (see also Smith 1966*a*, *b*). The theoretical cross section over this energy interval can be fitted closely to the form $A(B - \log v_0)^2$, where A and B are constants (Firsov 1951; Dalgarno and McDowell 1956). In Fig. 4.12 the theoretical spin exchange (capture)

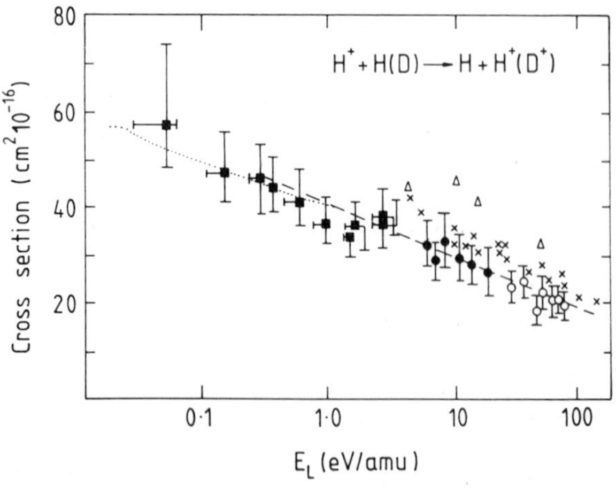

Fig. 4.12. Total cross sections for electron capture by H^+ from H(1s) or D(1s). Theoretical cross section: ——— Hunter and Kuriyan (1977*a*, *b*); ----- Dalgarno and Yadav (1953). Experimental data: ■, ● ($H^+ + D$), ○ ($H^+ + H$) Newman *et al.* (1982); × ($H^+ + H$) Fite *et al.* (1962); △ ($H^+ + H$) Belyaev *et al.* (1967).

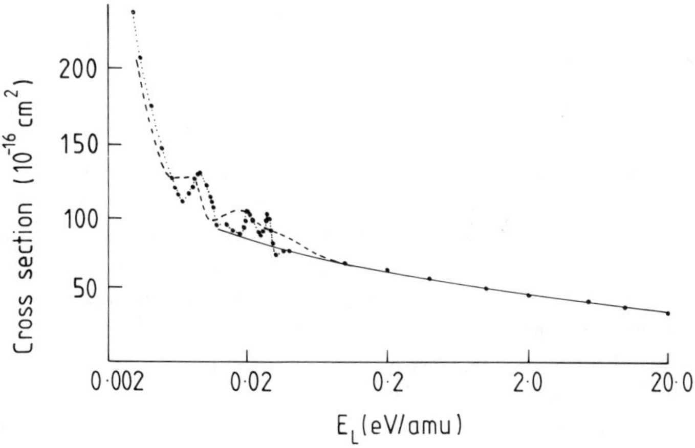

Fig. 4.13. Spin exchange (capture) cross section for $H^+ + H(1s)$ at low energies: •··•··• partial-wave calculation, Hunter and Kuriyan (1977*a*); ——— semi-classical calculation, Bates and Boyd (1962); ----- semi-classical calculation, Smith (1966*a*, *b*).

cross sections are compared with the experimental data in the energy range 0.2 eV $< E_L <$ 200 eV/a.m.u. The agreement is excellent. The behaviour of the spin exchange cross section at lower energies is illustrated in Fig. 4.13. Also shown in Fig. 4.13 are results obtained by calculating the phase shifts for scattering by the potentials $E_g(R)$ and $E_u(R)$ by the WKB approximation (see Chapter 2) and treating the angular momentum as a continuous variable (Bates and Boyd 1962; Smith 1966*b*). This is a very favourable case for such a semi-classical approximation since only elastic scattering by a potential is involved, and the semi-classical and partial-wave cross sections agree at energies above 0.04 eV/a.m.u.

Isotopic effects The charge exchange and elastic scattering cross sections for the interaction of H^+ with D(1s) and for the interaction of D^+ with H(1s) have been investigated by Hunter and Kuriyan (1977*b*) using the method outlined above, and also by Davies and Thorsen (1978) who used a method allowing for translational invariance. Except at very low energies there is no isotopic effect and the cross sections at a given relative velocity are the same for the $H^+ + H$ and $H^+ + D$ reactions. Below $E_L \approx 2$ eV/a.m.u., differences between the $H^+ + H$, $H^+ + D$, $D^+ + H$ cross sections begin to appear. The binding energy of D(1s) is greater than that of H(1s) by 0.0037 eV, so that the reaction $H^+ + D \rightarrow H + D^+$ has a finite threshold at $E_L = 0.0056$ eV/a.m.u. and the corresponding cross section becomes much smaller than that for the reaction $D^+ + H \rightarrow D + H^+$ as the incident energy decreases towards this value. The cross section for $D^+ + H \rightarrow D + H^+$ also differs from that for $H^+ + H \rightarrow H + H^+$ at energies less than $E_L \sim 0.04$ eV/a.m.u.

where the effects due to the identity of the two protons in the latter reaction become important.

Higher energies At energies above 10 eV, excitation of the target hydrogen atom to a 2p level begins to become important. This arises because of the strong rotational coupling at small R between the $2\mathrm{p}\sigma_\mathrm{u}$ level asymptotic to $\mathrm{H}^+ + \mathrm{H}(1\mathrm{s})$ and the $2\mathrm{p}\pi_\mathrm{u}$ level asymptotic to $\mathrm{H}^+ + \mathrm{H}(2\mathrm{p})$. Both levels are degenerate in the united atom limit, coinciding with the $2\mathrm{p}_0$ and $2\mathrm{p}_{\pm 1}$ orbitals of He^+. Knudsen and Thorson (1970) have shown that below 10 eV the coupling to the $2\mathrm{p}\pi_\mathrm{u}$ level is unimportant while in the interval $10 \leq E \leq 200$ eV a three-state model is adequate, giving rise to an uncoupled channel for the $1\mathrm{s}\sigma_\mathrm{g}$ level, as before, and two coupled channels describing the interaction between the $2\mathrm{p}\sigma_\mathrm{u}$ and $2\mathrm{p}\pi_\mathrm{u}$ levels. We shall defer further discussion of the $\mathrm{H}^+ + \mathrm{H}(1\mathrm{s})$ system until Chapter 6 where cross sections obtained in the impact parameter model will be discussed.

4.6 THE MODIFIED PERTURBED STATIONARY STATE MODEL

As we discussed in some detail in Section 4.1, the perturbed stationary state model, although in many cases reliable at thermal and very low energies has inherent defects. At moderate impact velocities, these defects can be corrected by introducing the Bates–McCarroll plane-wave translation factors expressed by (4.4), but at very low velocities it is better to employ more flexible translation factors, because for small distances of approach $R \to 0$, the electron should not be associated strongly with either of the centres A and B, and in addition the relative velocities $\mathbf{v}^\mathrm{c}$ and $\bar{\mathbf{v}}^\mathrm{c}$ cannot be taken to be always parallel to the Z-axis (the direction of incidence). Translation factors will be studied in more detail in the next chapter in connection with the impact parameter model. At the very low energies considered in this chapter, it is convenient to obtain the extra degree of flexibility by returning to trial functions of the form (3.75), which we shall write as

$$\Psi = \sum_{j=1}^{N+M} F_j(\boldsymbol{\zeta}_j)\psi_j(\mathbf{s}, \boldsymbol{\zeta}_j) \tag{4.85}$$

where, for $j \leq N$, $\psi_j \equiv \psi_j^\mathrm{B}$ and $\psi_j^\mathrm{B}(\mathbf{s}, \mathbf{R})$ is a molecular orbital asymptotic to an atomic wave function $\phi_j(\mathbf{r})$ centred on B, while for $j > N$, $\psi_j \equiv \psi_j^\mathrm{A}$ and $\psi_j^\mathrm{A}(\mathbf{s}, \mathbf{R})$ is a molecular orbital asymptotic to an atomic wave function $\chi_j(\mathbf{x})$ centred on A. The reaction coordinates $\boldsymbol{\zeta}_j$ are again expressed by (3.76a, b, c), (3.77),

$$\boldsymbol{\zeta}_j = \mathbf{R} - \frac{\mathbf{w}_j}{\mu} \tag{4.86}$$

with

$$\mathbf{w}_j = \tfrac{1}{2}\{f_j(\mathbf{s}, \mathbf{R}) + \lambda)\}\mathbf{s}_g + \tfrac{1}{8}\{1 - \lambda^2\}\mathbf{R} \qquad (4.87)$$

and

$$\mathbf{s}_g = \tfrac{1}{2}(\mathbf{x} + \mathbf{r})$$

with the consequence that for $j \leq N$, ζ_j approaches the centre-of-mass coordinate $\boldsymbol{\sigma}$ and for $j \geq N$, ζ_j approaches the centre-of-mass coordinate $\boldsymbol{\rho}$. This ensures that the correct boundary conditions are satisfied for large R. To express Ψ in terms of $\mathbf{s}$ and $\mathbf{R}$, we use the translation operator

$$T\left(-\frac{\mathrm{i}}{\mu}\mathbf{w}_j \cdot \nabla_R\right) = \exp\left[-\frac{\mathrm{i}}{\mu}\mathbf{w}_j \cdot \nabla_R\right],$$

(cf. 3.52), in (4.85) to obtain:

$$\Psi = \sum_{j=1}^{N+M} \exp\left[-\frac{\mathrm{i}}{\mu}\mathbf{w}_j \cdot \nabla_R\right] F_j(\mathbf{R})\psi_j(\mathbf{s}, \mathbf{R}). \qquad (4.88)$$

Coupled equations for the functions $F_j(\mathbf{R})$ are formed in the usual way by projecting $(H - E)\Psi$ from the left by the functions $\exp[(\mathrm{i}/\mu)\mathbf{w}_n \cdot \nabla_R]\psi_n^+(\mathbf{s}, \mathbf{R})$. We notice that at moderate and high velocities $(1/\mu)\nabla_R$ can be replaced by $\mathbf{v}_j$, and if the further approximation is made of setting $f_j = 1$ for $j \leq N$ and $f = -1$ for $j > N$, we recover the Bates–McCarroll factors discussed earlier. At low impact velocities it is evident that $(1/\mu)\nabla_R$ can be treated as a small quantity, and the translation operator can be expanded as (cf. 3.64)

$$\exp\left[-\frac{\mathrm{i}}{\mu}\mathbf{w}_j \cdot \nabla_R\right] \approx 1 - \frac{\mathrm{i}}{\mu}\mathbf{w}_j \cdot \nabla_R. \qquad (4.89)$$

In this approximation, the coupled equations are of the form (omitting terms in $1/\mu^2$)

$$\sum_j \left[\int \psi_n^*(\mathbf{s}, \mathbf{R})[H - E]\psi_j(\mathbf{s}, \mathbf{R})F_j(\mathbf{R})\,\mathrm{d}\mathbf{s} + \frac{\mathrm{i}}{\mu}\int \psi_n^*(\mathbf{s}, \mathbf{R})[\mathbf{w}_n \cdot \nabla_R H - H\mathbf{w}_j \cdot \nabla_R\}\psi_j(\mathbf{s}, \mathbf{R})F_j(\mathbf{R})\,\mathrm{d}\mathbf{s}\right] = 0. \qquad (4.90)$$

The first set of terms is identical with those terms arising in the perturbed stationary state model and the second set represents additional interactions. A detailed examination (Thorson and Delos 1978*a, b*) shows that to a good approximation these additional terms can be reduced to the form

$$\sum_j \frac{1}{\mu}\int [\psi_n^*(\mathbf{s}, \mathbf{R})\{\mathbf{w}_n H_e - H_e\mathbf{w}_j\}\psi_j(\mathbf{s}, \mathbf{R})] \cdot \nabla_R F_j(\mathbf{R})\,\mathrm{d}\mathbf{s}, \qquad (4.91)$$

where H_e is the electronic Hamiltonian (4.19). In this approximation the modified perturbed stationary state equations are (see (4.15))

$$(\nabla_R^2 + \mathscr{K}^2(\mathbf{R}))\mathbf{F}(\mathbf{R}) = [\mathbf{M}(\mathbf{R}) + \{\mathbf{P}(\mathbf{R}) + \mathbf{A}(\mathbf{R})\}\cdot\nabla_R]\mathbf{F}(\mathbf{R}) \qquad (4.92)$$

where $\mathscr{K}$, **M**, and **P** are defined as before in (4.16) and (4.17), while the additional interaction matrix **A** has elements A_{ij} where

$$\begin{aligned}\mathbf{A}_{ij} &= \frac{i}{\mu}\int \psi_i^*(\mathbf{s}, \mathbf{R})\{\mathbf{w}_j H_e - H_e \mathbf{w}_i\}\psi_j(\mathbf{s}, \mathbf{R})\, d\mathbf{s} \\ &= \frac{i}{\mu}\int \psi_i^*(\mathbf{s}, \mathbf{R})\{\mathbf{w}_j E_j(R) - E_i(R)\mathbf{w}_i\}\psi_j(\mathbf{s}, \mathbf{R})\, d\mathbf{s} \qquad (4.93) \\ &= \frac{i}{2}\int \psi_i^*(\mathbf{s}, \mathbf{R})\{(f_j(\mathbf{s}, \mathbf{R}) + \lambda)E_j(R) - (f_i(\mathbf{s}, \mathbf{R}) - \lambda)E_i(R)\}\mathbf{s}_g \psi(\mathbf{s}, \mathbf{R})\, d\mathbf{s}.\end{aligned}$$

Equations (4.90) and (4.92) are appropriate for basis functions ψ_j in the space-fixed frame. Equations based on the expansion in terms of the molecular orbitals $\bar{\psi}_j(\bar{\mathbf{s}}, R)$ quantized in the body-fixed frame, rotating with respect to the space-fixed system can be obtained following the analysis given earlier in this chapter. The final radial equations are of the form (4.33), but the radial and rotational interactions (4.35) and (4.36) are modified in two ways:

(1) by replacing the matrix element $\langle i|\partial/\partial R|j\rangle$ in the radial interaction given by (4.35), by

$$\left\{\langle i|\frac{\partial}{\partial R}|j\rangle + \tfrac{1}{2}\langle i|((f_j(\bar{\mathbf{s}}, R) + \lambda)E_j(R) - (f_i(\bar{\mathbf{s}}, R) + \lambda)E_i(R))\bar{z}_g|j\rangle\right\} \qquad (4.94)$$

(2) by replacing the matrix element $\langle i|i\bar{L}_y|j\rangle$ in the rotational interaction (4.36) by

$$\left\{\langle i|i\bar{L}_y|j\rangle + \frac{i}{2}\langle i|((f_j(\bar{\mathbf{s}}, R) + \lambda)E_j(R) - (f_i(\bar{\mathbf{s}}, R) + \lambda)E_i(R))\bar{y}_g|j\rangle\right\} \qquad (4.95)$$

where $\bar{z}_g$ and $\bar{y}_g$ are the Z- and Y-components of $\bar{\mathbf{s}}_g$. The choice of the switching factors, which must satisfy the conditions (3.77) is, to some extent, arbitrary and will be considered at length in Chapter 5.

He^{2+} – H *collisions*

An important application of the modified perturbed stationary state model to the scattering of He^{2+} ions by atom hydrogen has been carried out by van Hemert *et al.*† (1985). In this particular work *common switching factors* were used, independent of the channel index. Two forms were investigated. The first f^{VT} is due to Vaaben and Taulbjerg (1981) and is based on an analysis of the three-body electrostatic problem (see Chapter 5)

$$f_j(\bar{\mathbf{s}}, \mathbf{R}) = f^{\mathrm{VT}}(\bar{\mathbf{s}}, \mathbf{R}) = \frac{Z_{\mathrm{B}}x^3 - Z_{\mathrm{A}}r^3}{Z_{\mathrm{B}}x^3 + Z_{\mathrm{A}}r^3} + \frac{Z_{\mathrm{A}} - Z_{\mathrm{B}}}{Z_{\mathrm{A}} + Z_{\mathrm{B}}}, \tag{4.96}$$

where Z_{A}, Z_{B} are the charges on the nuclei A and B, while x and r are the distances of the electron from A and B, respectively. The second was introduced by Errea *et al.* (1982), partly because of its simplicity in calculating matrix elements:

$$f_j(\bar{\mathbf{s}}, \mathbf{R}) = f^{\mathrm{EWR}}(\bar{\mathbf{s}}, \mathbf{R}) = \frac{2R}{R^2 + p^2}\frac{1}{R + q}(\mathbf{s}_{\mathrm{g}} \cdot \mathbf{R}). \tag{4.97}$$

The parameters p and q were taken to be $p = 0.5$, $q = 1.0$. Since $f_q \to 0$ as $R \to 0$, the electron is not associated with either centre as the united atom limit is approached.

The ground state of the hydrogen atom has the same energy as the He^+ ion in the $n = 2$ level. This gives rise to strong coupling between the incident channel and the $H^+ + He^+$ $(n = 2)$ channels and at low energies the only significant processes are

$$\mathrm{He}^{2+} + \mathrm{H(1s)} \to \mathrm{He}^{2+} + \mathrm{H(1s)} \tag{4.98a}$$

$$\to \mathrm{He}^{+}(2\mathrm{s}) + \mathrm{H}^{+} \tag{4.98b}$$

$$\to \mathrm{He}^{+}(2\mathrm{p}_0) + \mathrm{H}^{+} \tag{4.98c}$$

$$\to \mathrm{He}^{+}(2\mathrm{p}_{\pm 1}) + \mathrm{H}^{+}. \tag{4.98d}$$

These four channels are correlated with four molecular orbitals, which in a united atom designation are $2\mathrm{p}\sigma$ (channel (a)) $2\mathrm{s}\sigma$ and $3\mathrm{d}\sigma$ which are asymptotic to Stark mixtures of $He^+(2s)$ and $He^+(2p_0)$ (channels (b) and (c)), and $2\mathrm{p}\pi$ (channel (d)). The lowest potential energy curves for the system $(\mathrm{H\ He})^{2+}$ have been shown in Fig. 4.2, where it can be seen that the $1\mathrm{s}\sigma_{\mathrm{g}}$ potential which is asymptotic to $He^+(1s) + H^+$ lies far below the entrance channel potential ($2\mathrm{p}\sigma$) in energy, with the result that the capture probability into the ground state of He^+ is negligible.

The calculations were performed at centre-of-mass energies between 20 eV and 500 eV. Other, semi-classical, cross sections were obtained up to 10 keV. These will be discussed in Chapter 5. It is interesting to note that by 500 eV

† It should be noted that the derivation followed here leading to equations (4.92) and (4.93) differs slightly from that employed by van Hemert *et al.*

Table 4.3. Cross sections for $He^{2+} + H(1s) \rightarrow He^{+}(n = 2) + H^{+}$ (units of 10^{-16} cm^2). T is calculated from the modified perturbed stationary state model; A is calculated from the unmodified perturbed stationary state model using the H^+ nucleus as origin; B is as for A, but using the He^{2+} nucleus as origin (after van Hemert *et al.* 1985).

E_L(eV/a.m.u.)	T	A	B
25	0.34×10^{-6}	0.32×10^{-6}	0.35×10^{-6}
62.5	0.46×10^{-4}	0.45×10^{-6}	0.56×10^{-6}
125	1.14×10^{-2}	1.09×10^{-2}	1.27×10^{-7}
250	1.31×10^{-1}	1.24×10^{-1}	1.73×10^{-1}
625	0.79	0.71	1.71

more than 2,500 partial waves were required for convergence. The computed total cross sections are shown in Table 4.3 for the switching function f^{VT}. The results are not sensitive to the form of the function and those obtained with f^{EWR} agree within 2 per cent with the results given in the table. Experimental data exist for energies above 500 eV (Shah and Gilbody 1978; Nutt *et al.* 1978). At 500 eV, the computed cross section is 20 per cent above the measured value, which however, is subject to a 15 per cent error. Results for the unmodified perturbed stationary state approximation are also included in Table 4.3. Since these are origin dependent two sets are shown, one with the origin on the H^+ nucleus and one with the origin on the He^{2+} nucleus. At the lowest energy the three sets of results are within 10 per cent of each other, but by 50 eV significant differences develop, indicating that translational effects must be taken into account at higher energies.

Further applications of the modified perturbed stationary state method

As further examples of the use of reaction coordinates, we may note the work of Gargaud et al. (1987) on the system $C^{4+} + H$; of Gargaud and McCarroll (1988) on $O^{4+} + H$ and $Si^{4+} + H$, and of Gargaud *et al.* (1988) on $Al^{3+} + H$; and $Ti^{4+} + H$; and of Opradolce *et al.* (1988) on $C^{3+} + H$. The method adopted in these calculations differed slightly from that outlined above in that rather than employing (4.88) to express Ψ in terms of **s** and **R**, equations similar to (4.92) were obtained for $F_j(\xi)$ expressed in terms of ξ the reaction coordinate. In addition a different form of switching factor was employed.

4.7 SYSTEMS WITH SEVERAL ELECTRONS

Although for convenience in notation we have discussed the theory of ion–atom interactions in terms of a one-active-electron model, there is no difficulty, in principle, in treating systems of several, or many, electrons. The basis of one-electron molecular orbitals $\psi_n(\mathbf{s}, \mathbf{R})$ is simply replaced by a basis of adiabatic molecular wave functions $\psi_n(\mathbf{s}_1, \mathbf{s}_2, \ldots, \mathbf{R})$ which are functions of the position vectors $\mathbf{s}_i$ of each electron. In light systems, in which magnetic interactions can be neglected, the total spin is a good quantum number so, for example, in a two-electron system two separate sets of singlet ($S = 0$) and triplet ($S = 1$) functions are formed, the singlet functions being symmetrical and the triplet functions antisymmetrical in the spatial coordinates in order to satisfy the Pauli principle. Exact energies and wave functions can only be found for one-electron systems, but for systems containing more than one electron approximate energies and wave functions can be obtained from the variational principle using standard methods of quantum chemistry (see, for example, Diercksen and Wilson (1983)) in which trial wave functions are expressed as linear combinations of products of one-electron orbitals $\phi_j^{\mathrm{A}}(\mathbf{x}_i)$, $\phi_n^{\mathrm{B}}(\mathbf{r}_i)$ centred on the two nuclei A and B. The orbitals ϕ_j^{A} or ϕ_n^{B} themselves are often expressed as linear combinations of Slater functions of the general form $r^q \exp(-\lambda r) Y_{lm}(\theta, \phi)$, or alternatively, of various forms of Gaussian functions, for example $r^q \exp(-\lambda r^2) Y_{lm}(\theta, \phi)$.

Historically it is interesting to note that the first perturbed stationary state calculation (Massey and Smith 1933) was for a two-electron system – electron capture by protons from helium – using explicitly two-electron wave functions. As a modern example of a specifically two-active-electron system, we may instance the calculation of Tan *et al.* (1987) of double electron capture from He by C^{4+}, using the quantal partial-wave method. Previous investigations in the semi-classical impact parameter model by Kimura and Olson (1984) had established that at low velocities ($v_0 < 0.2$ a.u.) the double electron capture reaction

$$C^{4+}(1s^2) + He(1s^2) \rightarrow C^{2+}(1s^2 2s^2) + He^{2+} \tag{4.99}$$

is the most important channel providing over 97 per cent of the total cross section at $v_0 = 0.12$ a.u. ($E_L = 360$ eV/a.m.u.). Tan *et al.* carried out a two-coupled state calculation, using the wave functions developed by Kimura and Olson and allowing for translation factors. The basis molecular orbital functions were constructed from Slater orbitals for the two active electrons, representing the effect of the $C^{4+}(1s^2)$ core by an effective potential. The differential cross sections for double charge exchange were computed at $E_L = 126$ and 42 eV/a.m.u., which required 1,500 and 900 partial waves, respectively, to obtain convergence. The results at $E_L = 126$ (eV/a.m.u.) ($v_0 = 0.071$ a.u.) are shown in Fig. 4.14, compared with the measurements

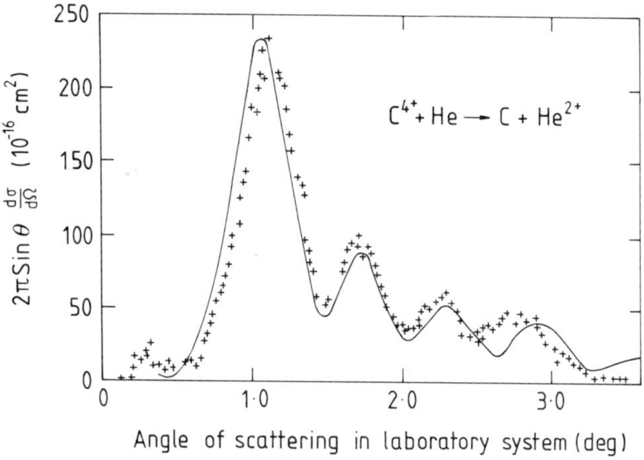

Fig. 4.14. Differential cross section for double charge exchange in collisions of C^{4+} ions with He at $E_L = 126$ eV/a.m.u. (after Tan *et al.* (1987)). Theoretical cross section from Tan *et al.* (1987); ——— Bárány *et al.* (1986).

of Bárány *et al.* (1986). The agreement is remarkable. Provided the number of important channels is small, which is generally the case in the low-energy region we are considering in this chapter, only a small set of basis functions is required and systems of several electrons can be treated using accurate molecular orbits allowing for correlation between the electrons. When large basis sets are necessary, for example at higher energies, methods based on simplified independent particle basis functions can be used and this topic will be treated briefly in Chapter 9.

4.8 RADIATIVE CHARGE TRANSFER

All collisions involving the acceleration of charged particles are accompanied by photon emission. In particular the acceleration undergone by the active electron in transferring from one ionic core to another during charge exchange can give rise to the emission of radiation:

$$A + (B + e^-) \rightarrow (A + e^-) + B + h\nu. \tag{4.100}$$

The probability of this radiative charge exchange is usually very small compared with non-radiative charge exchange, but should the non-radiative reaction have a particularly small cross section, the radiative reaction can be the most important exchange mechanism. There are two situations in which this occurs. At very high energies, the non-radiative charge exchange cross section decreases like E^{-6} (ultimately like $E^{-11/2}$), while the radiative cross section decreases as E^{-1} and for sufficiently high energies dominates.

At the opposite end of the energy spectrum, at thermal energies, the non-radiative charge exchange cross section can become very small. For example the probability of the $He^{2+} + H \rightarrow He^{+} + H^{+}$ reaction can be seen from Table 4.3 to be decreasing very rapidly with decreasing energy, and for this system radiative charge transfer is the main transfer process at energies below ~ 20 eV (West *et al.* 1982). From the viewpoint of the perturbed stationary state model, photon emission takes place at a particular value of R, the interionic separation, when the electron transfers from one level of the pseudomolecule formed during the collision to another. The process is entirely analogous to the radiation occurring between two bound electronic states of a molecule which gives rise to a band spectrum (Bransden and Joachain 1983); however as the transition is between continuum levels the frequency spectrum of the radiation is continuous.

In the electric dipole approximation, the transition rate for the radiative charge exchange reaction (4.100) is determined by the matrix element of the dipole operator $\mathbf{d} = -\mathbf{s}$ where $\mathbf{s}$ is the position vector of the active electron (for systems with more than one electron $\mathbf{d} = -\sum_n \mathbf{s}_n$). The heavy particle contribution to the dipole operator has been omitted since the corresponding transition matrix element is usually very small, and is zero in the lowest approximation (see Bransden and Joachain 1983, p. 443–4). The matrix element for a transition from an initial channel i to a final channel f is

$$M(\mathbf{k}_i, \mathbf{k}_f, \alpha) = \int \mathrm{d}\mathbf{R} \int \mathrm{d}\mathbf{s}\, \psi_f^{-*}(\mathbf{s}, \mathbf{R})(\mathbf{d}\cdot\boldsymbol{\varepsilon})\Psi_i^{+}(\mathbf{s}, \mathbf{R}), \tag{4.101}$$

where (α, β) is the direction of emission of the photon with polarization vector $\boldsymbol{\varepsilon}$.

In the perturbed stationary state model omitting translational factors, the wave functions of the initial and final states of the system are expressed as

$$\Psi_i^{+}(\mathbf{s}, \mathbf{R}) = \sum_{j=1}^{N+M} F_{i,j}^{+}(\mathbf{R})\psi_j(\mathbf{s}, \mathbf{R}) \tag{4.102a}$$

and

$$\Psi_f^{-}(\mathbf{s}, \mathbf{R}) = \sum_{j=1}^{N+M} F_{f,j}^{-}(\mathbf{R})\psi_j(\mathbf{s}, \mathbf{R}). \tag{4.102b}$$

The channel functions $F_{i,j}^{+}(\mathbf{R})$, $F_{f,j}^{-}(\mathbf{R})$ are determined from the coupled equations described earlier in this chapter, but with the following boundary conditions:

$$\begin{aligned} F_{i,j}^{+}(\mathbf{R}) &\approx \exp(\mathrm{i}\mathbf{k}_i\cdot\mathbf{R})\delta_{ij} + f_{i,j}^{+}(\omega, \Phi)R^{-1}\exp(\mathrm{i}k_jR), \\ F_{f,j}^{-}(\mathbf{R}) &\approx \exp(\mathrm{i}\mathbf{k}_f\cdot\mathbf{R})\delta_{fj} + f_{f,j}^{-}(\Theta, \Phi)R^{-1}\exp(-\mathrm{i}k_jR). \end{aligned} \tag{4.103}$$

The differential cross section in atomic units for the emission of a photon with angular frequency ω is given by

$$\frac{d\sigma}{d\omega} = \int d\Omega(\Theta, \Phi) \int d\Omega(\alpha, \beta) \frac{\mu^2\omega^3}{2\pi c^3} \frac{k_f}{k_i} |M|^2, \tag{4.104}$$

where the integrations are over all directions of scattering of the heavy particles, and over all directions of emission. For a given frequency ω, the kinetic energy of relative motion in the final state is determined by energy conservation

$$\frac{k_j^2}{2\mu} = \frac{k_i^2}{2\mu} + \Delta_{if} - \omega, \tag{4.105}$$

where Δ_{if} is the difference of the energies of the atomic levels in the initial and final states. Photons of all angular frequencies in the range $\omega \leq (k_i^2/2\mu + \Delta_{if})$ can be emitted for a given pair of initial and final states (Sobel'man 1972). If the only significant coupling between the levels of the pseudo-molecule formed during the collision is due to the radiative process, the problem is greatly simplified because Ψ_i^+ and Ψ_j^- can be approximated by the single channel functions

$$\begin{aligned} \Psi_i^+(\mathbf{s}, \mathbf{R}) &\approx F_i^+(\mathbf{R})\psi_i(\mathbf{s}, \mathbf{R}), \\ \psi_f^-(\mathbf{s}, \mathbf{R}) &\approx F_f^-(\mathbf{R})\psi_f(\mathbf{s}, \mathbf{R}). \end{aligned} \tag{4.106}$$

The channel functions $F_{i,f}^{\pm}(\mathbf{R})$ satisfy the potential scattering equations

$$(\nabla^2 + k_{i,f}^2 - E_{i,f}(R))F_{i,f}^{\pm}(\mathbf{R}) = 0, \tag{4.107}$$

where $E_i(R)$ and $E_f(R)$ are molecular potentials for the levels i and f. The calculation of the matrix elements (4.101) can then be reduced to the calculation of radial integrals if a partial-wave expansion of the functions $F_{i,f}^{\pm}$ is carried out.

An interesting example of such a calculation is provided by the work of Cooper *et al.* (1984) who have studied the reaction

$$He^+ + Ne \rightarrow He + Ne^+ + h\nu, \tag{4.108}$$

which competes with the radiative association process

$$He^+ + Ne \rightarrow (HeNe)^+ + h\nu. \tag{4.109}$$

Direct charge exchange at very low energies is not significant but the cross sections for the two reactions (4.106) and (4.107) are large enough for the emission spectrum to be detected by an experiment in which helium ions drift through a neon gas (Johnson 1983). The calculated cross sections given

Table 4.4. Cross sections from Cooper *et al.* (1984) in units of 10^{-21} cm^2 for the reactions: (a) $He^+ + Ne \rightarrow He + Ne^+ + h\nu$; (b) $He^+ + Ne \rightarrow (NeHe)^+ + h\nu$.

E(eV)	σ_a	σ_b
2.5×10^{-4}	5.7	4.0
5.0×10^{-4}	3.8	3.3
7.5×10^{-4}	3.7	3.1

in Table 4.4 are consistent with rate coefficients for each reaction of 0.5×10^{-15} cm^3 s^{-1} at 300 K, which is in fair agreement with the observed rate. Further examples are afforded by the work of Butler *et al.* (1977) on C^{2+}, C^{3+}, N^{2+} + H(1s), and of Zygelman *et al.* (1989) on He^+ + H.

5

THE IMPACT PARAMETER MODEL

In the last chapter the theory of ion–atom scattering at very low energies was discussed in terms of an angular momentum decomposition of the quantum mechanical coupled equations. As we saw in Chapters 2 and 3, the number of partial waves required for convergence becomes very large when the initial relative velocity v_0 is greater than a few tenths of an atomic unit of velocity (as shown in Table 3.1). Also, as v_0 is increased the kinetic energy of the heavy particle motion becomes large compared with changes in electronic energy. Under these conditions the impact parameter model is accurate and is more convenient to use than the partial-wave method. In the impact parameter model we have to solve the time-dependent Schrödinger equation (3.163a)

$$\mathrm{i}\left.\frac{\partial\psi(\mathbf{s}, t)}{\partial t}\right|_{\mathbf{s}} = H_{\mathrm{e}}(\mathbf{s}, t)\psi(\mathbf{s}, t), \tag{5.1}$$

where H_{e} is defined for a one-electron system by (3.158). In this equation the independent variables are the time t and the coordinate of the electron $\mathbf{s}$ with respect to the centre of mass of the heavy particles A and B as origin. The equation is integrated along a classical trajectory $\mathbf{R} = \mathbf{R}(\mathbf{b}, t)$ describing the relative motion of the heavy particles. At sufficiently high energies the classical trajectory can be taken to be a straight line traversed at the constant speed v_0. In this approximation the heavy particles are unaccelerated, so that any point on the internuclear line, for example the mid-point, can be employed as origin, since all such points define inertial frames.

Coupled channel equations of the form (3.185) are obtained from (5.1), by expressing the wave function $\psi(\mathbf{s}, t)$ as a linear combination of basis functions

$$\psi(\mathbf{s}, t) = \sum_{j=1}^{N} \Phi_j(\mathbf{s}, t)A_j(\mathbf{b}, t) + \sum_{n=N+1}^{N+M} \mathrm{X}_n(\mathbf{s}, t)A_n(\mathbf{b}, t), \tag{5.2}$$

and using the variational principle (3.181). Two sets of basis functions are employed, $\Phi_j(\mathbf{s}, t)$ and $\mathrm{X}_n(\mathbf{s}, t)$, which are asymptotic to the arrangements $\mathrm{A} + (\mathrm{B} + \mathrm{e}^-)$ and $(\mathrm{A} + \mathrm{e}^-) + \mathrm{B}$, respectively. The coupled equations (3.185) can be expressed as the matrix equations (as in (3.205)),

$$\mathrm{i}\dot{\mathbf{A}} = \mathbf{N}^{-1}\mathbf{Q}\mathbf{A} \equiv \mathbf{V}\mathbf{A}, \tag{5.3}$$

where the overlap and interactions matrices N and Q are defined through equations (3.207), (3.184), and (3.186). The $(N + M)$-dimensional column vector **A** has components $A_j(\mathbf{b}, t)$. When $1 \le j \le N$, $A_j(\mathbf{b}, t)$ represents the amplitude at time t of the wave function in one of the channels belonging to the arrangement $A + (B + e^-)$, while if $N + 1 \le j \le N + M$, $A_j(\mathbf{b}, t)$ represents the corresponding amplitude in one of the channels belonging to the arrangement $(A + e^-) + B$. The coupled equations (5.3) have to be solved with the boundary condition (3.187), that is, $\lim_{t\to-\infty} A_j(\mathbf{b}, t) = \delta_{ij}$, where i labels the incident channel. The probability for finding the system in the channel j after the collision is given by equations (3.188) in terms of $\mathscr{A}_j(\mathbf{b}) = \lim_{t\to\infty} A_j(\mathbf{b}, t)$, and the total cross section is given by (3.156).

In applications of the model, the key decisions are (a) the nature of the basis functions, for example, whether molecular or atomic, (b) the size of the basis, and (c) the type of translation factors to be employed. In this chapter, we first consider the use of a *molecular basis*, either adiabatic or diabatic. This is the most convenient form of basis to use at low velocities when the number of important coupled channels, which can be identified by curve crossing arguments, is small.

Later in the chapter, basis sets formed by placing *atomic orbitals* on each heavy particle centre are discussed. Such sets are generally employed as the velocity becomes greater ($v_0 \gtrsim 1$ a.u.), because the nearly adiabatic conditions under which only a few molecular orbitals are expected to be coupled together do not apply. The representation of the continuum states becomes particularly important at the higher energies and the atomic orbital sets must allow for this by the inclusion of *pseudostates* or by other means. Finally, expansions into other kinds of basis set and direct numerical solutions of the Schrödinger equation will be discussed briefly. A review of selected applications of the impact parameter coupled channel model will be left to the next chapter.

5.1 THE MOLECULAR ORBITAL BASIS

The adiabatic molecular orbitals forming the basis for an expansion of $\psi(\mathbf{s}, t)$ are defined as solutions of (4.1) and satisfy the orthonormality relations (4.2). Frequently the physical conditions of the system investigated make the use of diabatic molecular orbitals more convenient and these can be defined by the transformation (4.60). In either case, the basis functions of the molecular orbital impact parameter model can be written in a form similar to (4.3) as

$$\Phi_j(\mathbf{s}, t) = \psi_j^{\mathrm{B}}(\mathbf{s}, \mathbf{R}(t)) \exp[\mathrm{i}\gamma_j^{\mathrm{B}}(\mathbf{s}, t)], \qquad j = 1, 2, \ldots, N, \tag{5.4a}$$

$$\mathrm{X}_n(\mathbf{s}, t) = \psi_n^{\mathrm{A}}(\mathbf{s}, \mathbf{R}(t)) \exp[\mathrm{i}\gamma_n^{\mathrm{A}}(\mathbf{s}, t)], \qquad n = N + 1, N + 2, \ldots, N + M, \tag{5.4b}$$

where superscripts A, B are added to the molecular orbitals $\psi_j(\mathbf{s}, \mathbf{R}(t))$ and to the phases γ_j when we wish to denote to which arrangement the system separates. The phases $\gamma_j^{\rm B}$, $\gamma_n^{\rm A}$ are uniquely defined only in the asymptotic region, and for $|t| \to \infty$ satisfy the conditions:

$$\gamma_j^{\rm B}(\mathbf{s}, t) \to \alpha_j(\mathbf{s}, t); \qquad \gamma_n^{\rm A}(\mathbf{s}, t) \to \beta_n(\mathbf{s}, t), \tag{5.5}$$

where (see (3.171) and (3.176))

$$\alpha_j(\mathbf{s}, t) = -\varepsilon_j t - q_{\rm c}\mathbf{v}(t)\cdot\mathbf{s} - \tfrac{1}{2}q_{\rm c}^2 \int_0^t v^2(t')\,\mathrm{d}t', \tag{5.6a}$$

$$\beta_n(\mathbf{s}, t) = -\eta_n t + p_{\rm c}\mathbf{v}(t)\cdot\mathbf{s} - \tfrac{1}{2}p_{\rm c}^2 \int_0^t v^2(t')\,\mathrm{d}t'. \tag{5.6b}$$

As in Chapter 3 ε_j, η_n denote the binding energies of the levels j and n of the atoms $(\mathrm{B} + \mathrm{e}^-)$ and $(\mathrm{A} + \mathrm{e}^-)$, respectively, while $q_{\rm c}$, $p_{\rm c}$ are defined so that the origin at C, the centre of mass of A and B, divides the internuclear line in the ratio $q_{\rm c}/p_{\rm c}$ (see (3.169) and (3.174b)), with $\mathrm{CA} = p_{\rm c}R$, $\mathrm{BC} = q_{\rm c}R$, $p_{\rm c} + q_{\rm c} = 1$.

The important question of the choice of $\gamma_j^{\rm B}$ and $\gamma_n^{\rm A}$ outside the asymptotic region will be discussed later, in Section 5.3. First we turn to the question of the relation of the body-fixed and space-fixed coordinate system, which was discussed in Section 4.2 in the context of the partial-wave approach.

5.2 ROTATING COORDINATES IN THE IMPACT PARAMETER MODEL

The formulation of the scattering problem requires the use of a coordinate system fixed in space. As usual we shall take the Z-axis in the space-fixed system to be the direction of incidence. Provided that the effective potential employed to determine the trajectory $\mathbf{R} = \mathbf{R}(\mathbf{b}, t)$ is central, the heavy particle scattering is confined to one plane and this can be taken to be the XZ-plane. The operation $\partial/\partial t$ which occurs in the definition of the interaction matrix elements H_{jk}, $\bar{H}_{nm}$, K_{jm} and $\bar{K}_{nk}$ (see equations (3.186)) which make up the full matrix $\mathbf{Q}$, must be effected with the components (x, y, z) of the electronic position vector $\mathbf{s}$ held constant in the space-fixed coordinate system. However, as explained in Section 4.2, the molecular orbitals ψ_k, whether adiabatic or diabatic, are usually expressed in a body-fixed frame in which the interionic line is taken as the Z-axis. As in Chapter 4, we shall write the molecular orbitals expressed in the body-fixed frame as $\bar{\psi}_j(\bar{\mathbf{s}}, R(t))$, where $\bar{\mathbf{s}}$ is the same vector as $\mathbf{s}$ but expressed in terms of components x', y', z' along the body-fixed axes. Since the heavy particle scattering in this model is confined to the XZ-plane, the body-fixed and space-fixed Y-axes coincide, and the interionic line rotates about the Y-axis as shown in Fig. 5.1. If the vector $\mathbf{R}$ makes an angle $\Theta(t)$ with the space-fixed Z-axis at time t, then the relationship between the components of $\mathbf{s}$ in the rotating frame with those in

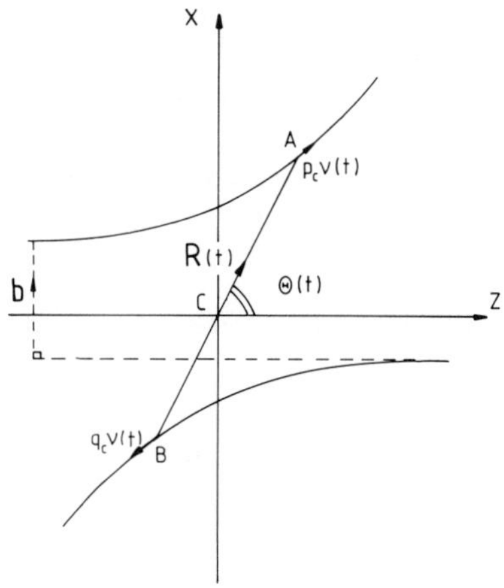

Fig. 5.1. The rotation of the interionic line about the Y-axis. A coordinate system is chosen with the centre of mass of A and B as origin. The Z-axis is in the direction of incidence and the XZ-plane is in the plane of scattering of A and B; CA $= p_c R$, BC $= q_c R$.

the space-fixed frame is

$$\begin{aligned} x' &= x \cos \Theta(t) - z \sin \theta(t) \\ y' &= y \\ z' &= x \sin \Theta(t) + z \cos \Theta(t). \end{aligned} \tag{5.7}$$

Using (5.7), we have

$$\begin{aligned} \left.\frac{\partial}{\partial t}\right|_{\mathbf{s}} \bar{\psi}_j(\bar{\mathbf{s}}, R(t)) &= \left[\dot{R}(t)\frac{\partial}{\partial R} - \dot{\Theta}(t)\left\{z'\frac{\partial}{\partial x'} - x'\frac{\partial}{\partial z'}\right\}\right]\bar{\psi}_j(\bar{\mathbf{s}}, R(t)) \\ &= \left[\dot{R}(t)\frac{\partial}{\partial R} - \mathrm{i}\dot{\Theta}(t)L_y\right]\bar{\psi}_j(\bar{\mathbf{s}}, R(t)), \end{aligned} \tag{5.8}$$

where L_y is the component of the orbital angular momentum operator along the Y-axis.

The choice of the effective interionic potential $W(R)$ (see 3.157) determines the trajectory $\mathbf{R} = \mathbf{R}(t)$ and hence $R(t)$ and $\Theta(t)$, following the procedures discussed in Chapter 2. The radial and angular velocities which appear in (5.8) can be expressed in the form

$$\dot{R}(t) = \pm v_0\left[1 - \frac{2W(R(t))}{\mu v_0^2} - \left(\frac{b}{R(t)}\right)^2\right]^{\frac{1}{2}}, \tag{5.9a}$$

$$\dot{\Theta}(t) = -\frac{v_0 b}{[R(t)]^2}, \tag{5.9b}$$

where v_0 is the incident velocity, b the impact parameter, and μ the reduced mass, $\mu = M_A M_B/(M_A + M_B)$. At the point of closest approach, $t = 0$, so that the negative sign is to be taken in (5.9a) for $t < 0$ and the positive sign for $t > 0$.

At the energies for which the straight line trajectory approximation is adequate, the following simplified expressions hold:

$$R(t) = (b^2 + v_0^2 t^2)^{\frac{1}{2}}, \qquad \Theta(t) = \sin^{-1}\left(\frac{b}{R(t)}\right), \tag{5.10a}$$

$$\dot{R}(t) = \frac{v_0^2 t}{R(t)}, \qquad \dot{\Theta}(t) = -\frac{v_0 b}{[R(t)]^2}. \tag{5.10b}$$

The matrix elements

Omitting, for the moment, the superscripts A, B on $\bar{\psi}_j$ and γ_j, all the matrix elements of **Q** (see (5.3)) can be expressed as

$$\begin{aligned} Q_{ij}(t) &= \int \bar{\psi}_i^*(\bar{\mathbf{s}}, R(t))\, e^{-i\gamma_i}\left(H_e - i\left.\frac{d}{dt}\right|_s\right)\bar{\psi}_j(\bar{\mathbf{s}}, R(t))\, e^{i\gamma_j}\, d\bar{\mathbf{s}} \\ &= \langle i|\, e^{-i\gamma_i}\left(H_e - i\left.\frac{d}{dt}\right|_s\right) e^{i\gamma_j}|j\rangle, \end{aligned} \tag{5.11}$$

where in the second line the notation introduced in Chapter 4 has been employed. The corresponding overlap matrix **N** has the elements

$$N_{ij} = \langle i|\, e^{i(\gamma_j - \gamma_i)}|j\rangle. \tag{5.12}$$

Using the expression (3.158) for H_e, Q_{ij} can be put in the form

$$Q_{ij}(t) = U_{ij}(t) + T_{ij}(t) + \Gamma_{ij}(t) \tag{5.13}$$

where

$$\begin{aligned} U_{ij}(t) &= \langle i|\exp[i(\gamma_j - \gamma_i)]H_e|j\rangle \\ T_{ij}(t) &= -i\langle i|\exp[i(\gamma_j - \gamma_i)]\left.\frac{d}{dt}\right|_s|j\rangle \\ \Gamma_{ij}(t) &= \langle i|\exp[i(\gamma_j - \gamma_i)]\{-\tfrac{1}{2}i\nabla_s^2\gamma_j + \tfrac{1}{2}(\nabla_s\gamma_j)^2 - i\nabla_s\gamma_j\cdot\nabla_s + \dot{\gamma}_j|j\rangle. \end{aligned} \tag{5.14}$$

In view of (5.8), the matrix elements T_{ij} can be expressed in terms of a radial coupling A_{ij} and a rotational coupling L_{ij}:

$$T_{ij}(t) = -k\dot{R}(t)A_{ij}(t) + i\dot{\Theta}(t)L_{ij}(t), \tag{5.15}$$

where

$$A_{ij}(t) = \langle i|\exp[\mathrm{i}(\gamma_j - \gamma_i)]\frac{\partial}{\partial R}|j\rangle, \tag{5.16}$$

and

$$L_{ij}(t) = \langle i|\exp[\mathrm{i}(\gamma_j - \gamma_i)]L_y|j\rangle. \tag{5.17}$$

It is important to notice that even if adiabatic orbitals are used, satisfying (4.18), the matrix U_{ij} is not, in general, diagonal because of the presence of the translation factors. The matrix Γ_{ij}, which depends directly on the phases γ_j has the effect of cancelling any finite couplings arising from the matrix T_{ij} as $|t| \to \infty$ (see Section 4.1).

5.3 CHOICE OF TRANSLATION FACTORS

To complete the evaluation of the matrix elemens Q_{ij} and N_{ij} a specific choice of the phase factors γ_i must be made. For notational simplicity, we shall discuss the case in which a straight line trajectory is employed with a constant velocity $\mathbf{v} = \mathbf{v}_0$. Any point of the interionic line can then be taken as the origin of coordinates and we shall take the mid-point with $p_c = q_c = \frac{1}{2}$. The position vector of the electron with respect to this origin is $\mathbf{s}_g$ (as illustrated in Fig. 3.1) and $\mathbf{s}_g$ replaces $\mathbf{s}$ in the appropriate formulae. The extension to the more general case of a non-linear trajectory, for which the inertial frame is the one with the centre of mass of A and B as origin, is straightforward.

It is convenient to separate the terms in the phase factors γ_i which depend on t alone. We can write for $j \leq N$ (the arrangement A + (B + e$^-$))

$$\gamma_j \equiv \gamma_i^{\mathrm{B}} = -\varepsilon_j t - \tfrac{1}{8}v_0^2 t + \pi_j^{\mathrm{B}}(\mathbf{s}_g, t) \tag{5.18a}$$

and for $j > N$ (the arrangement (A + e$^-$) + B)

$$\gamma_j \equiv \gamma_j^{\mathrm{A}} = -\eta_j t - \tfrac{1}{8}v_0^2 t + \pi_j^{\mathrm{A}}(\mathbf{s}_g, t). \tag{5.18b}$$

From (5.6), the redefined phase factors $\pi_j^{\mathrm{A,B}}(\mathbf{s}_g, t)$ have the asymptotic forms as $|t| \to \infty$

$$\pi_j^{\mathrm{B}}(\mathbf{s}_g, t) \to -\tfrac{1}{2}\mathbf{v}_0\cdot\mathbf{s}_g \quad \text{and} \quad \pi_j^{\mathrm{A}}(\mathbf{s}_g, t) \to \tfrac{1}{2}\mathbf{v}_0\cdot\mathbf{s}_g. \tag{5.19}$$

5.3.1 Plane-wave translation factors

When Bates and McCarroll first recognized that translation factors were required in order that the trial function satisfies the proper boundary conditions in each channel, they suggested that the phases $\pi_j^{\mathrm{A,B}}$ should be set equal to the asymptotic forms (5.19)

$$\pi_j^{\mathrm{B}} \equiv \pi^{\mathrm{B}} = -\tfrac{1}{2}\mathbf{v}_0 \cdot \mathbf{s}_{\mathrm{g}} \qquad \text{and} \qquad \pi_j^{\mathrm{A}} \equiv \pi^{\mathrm{A}} = \tfrac{1}{2}\mathbf{v}_0 \cdot \mathbf{s}_{\mathrm{g}}. \tag{5.20}$$

The direct matrix elements linking states belonging to the initial arrangement can be found from (5.14) and (5.20). If superscripts BB are added to denote the direct interaction in the channel $\mathrm{A} + (\mathrm{B} + \mathrm{e}^-)$ we have, using the Bates–McCarroll translational factors,

$$\begin{aligned}
U_{ij}^{\mathrm{BB}}(t) &= \langle i|H_{\mathrm{e}}|j\rangle \exp[\mathrm{i}(\varepsilon_i - \varepsilon_j)t] \\
T_{ij}^{\mathrm{BB}}(t) &= -\mathrm{i}\langle i| \left.\frac{\mathrm{d}}{\mathrm{d}t}\right|_{\mathbf{s}_{\mathrm{g}}} |j\rangle \exp[\mathrm{i}(\varepsilon_i - \varepsilon_j)t] \\
\Gamma_{ij}^{\mathrm{BB}}(t) &= -\varepsilon_j\delta_{ij} - \mathrm{i}\tfrac{1}{2}v_0\langle i| \frac{\partial}{\partial z_{\mathrm{g}}} |j\rangle \exp[\mathrm{i}(\varepsilon_i - \varepsilon_j)t], \qquad i, j \le N,
\end{aligned} \tag{5.21}$$

where z_{g} is the Z-component of the position vector $\mathbf{s}_{\mathrm{g}}$. In the rearranged channel the direct matrix elements U_{ij}^{AA}, $T_{ij}^{\mathrm{AA}}, \ldots$ have the same form as in (5.21) but η_j replaces ε_j and $-v_0$ replaces v_0. The exchange matrix elements, for $i > N$ and $j \le N$, denoted by U_{ij}^{AB}, $T_{ij}^{\mathrm{AB}}, \ldots$, are

$$\begin{aligned}
U_{ij}^{\mathrm{AB}}(t) &= \langle i| \exp(\mathrm{i}\mathbf{v}_0 \cdot \mathbf{s}_{\mathrm{g}})H_{\mathrm{e}}|j\rangle \exp[\mathrm{i}(\eta_i - \varepsilon_j)t] \\
T_{ij}^{\mathrm{AB}}(t) &= -\mathrm{i}\langle i| \exp(-\mathrm{i}\mathbf{v}_0 \cdot \mathbf{s}_{\mathrm{g}}) \left.\frac{\mathrm{d}}{\mathrm{d}t}\right|_{\mathbf{s}_{\mathrm{g}}} |j\rangle \exp[\mathrm{i}(\eta_i - \varepsilon_j)t] \\
\Gamma_{ij}^{\mathrm{AB}}(t) &= -\varepsilon_j N_{ij}(t) - \mathrm{i}\tfrac{1}{2}v_0\langle i| \exp(-\mathrm{i}\mathbf{v}_0 \cdot \mathbf{s}_{\mathrm{g}}) \frac{\partial}{\partial z_{\mathrm{g}}} |j\rangle \exp[\mathrm{i}(\eta_i - \varepsilon_j)t] \\
N_{ij}^{\mathrm{AB}}(t) &= \langle i| \exp(-\mathrm{i}\mathbf{v}_0 \cdot \mathbf{s}_{\mathrm{g}})|j\rangle \exp[\mathrm{i}(\eta_i - \varepsilon_j)t].
\end{aligned} \tag{5.22}$$

The corresponding exchange matrix elements for $i \le N$, $j > N$, $U_{ij}^{\mathrm{BA}}(t)$, $T_{ij}^{\mathrm{BA}}(t)$, etc. are obtained by replacing $\mathbf{v}_0$ by $-\mathbf{v}_0$ and replacing $(\eta_i - \varepsilon_j)$ by $(\varepsilon_i - \eta_j)$. Examining $Q_{ij}^{\mathrm{BB}}(t) = U_{ij}^{\mathrm{BB}}(t) + T_{ij}^{\mathrm{BB}}(t) + \Gamma_{ij}^{\mathrm{BB}}(t)$ we see that as $|t| \to \infty$, U_{ij}^{BB} cancels with the first term in $\Gamma_{ij}^{\mathrm{BB}}$, since the molecular orbital ψ_j^{B} is asymptotic to the atomic orbital $\phi_j(\mathbf{r})$.

The matrix T_{ij}^{BB} can be combined with the second term in $\Gamma_{ij}^{\mathrm{BB}}$ to obtain the matrix element $\bar{T}_{ij}^{\mathrm{BB}}$

$$\bar{T}_{ij}^{\mathrm{BB}} = -\mathrm{i}\langle i| \left.\frac{\mathrm{d}}{\mathrm{d}t}\right|_{\mathbf{s}_{\mathrm{g}}} + \tfrac{1}{2}v_0 \frac{\partial}{\partial z_{\mathrm{g}}} |j\rangle \exp[\mathrm{i}(\varepsilon_i - \varepsilon_j)t]. \tag{5.23}$$

Since $\mathbf{s}_{\mathrm{g}} = \mathbf{r} - \tfrac{1}{2}\mathbf{R}(t)$, where $\mathbf{r}$ is the position vector of the electron with respect to B we have the relation

$$\left.\frac{\mathrm{d}}{\mathrm{d}t}\right|_{\mathbf{s}_{\mathrm{g}}} + \tfrac{1}{2}v_0 \frac{\partial}{\partial z_{\mathrm{g}}} = \left.\frac{\mathrm{d}}{\mathrm{d}t}\right|_{\mathbf{r}}, \tag{5.24}$$

where $\mathrm{d}/\mathrm{d}t|_{\mathbf{r}}$ denotes a time derivative with $\mathbf{r}$ held constant. As in the limit $|t| \to \infty$, $\bar{\psi}_j^{\mathrm{B}} \to \phi_j$ we have that

$$\lim_{|t|\to\infty} \bar{T}_{ij}^{\mathrm{BB}} = -\mathrm{i} \int \phi_i^*(\mathbf{r}) \left.\frac{\mathrm{d}}{\mathrm{d}t}\right|_{\mathbf{r}} \phi_j(\mathbf{r})\,\mathrm{d}\mathbf{r} = 0. \tag{5.25}$$

Using the parallel argument for direct scattering in the channels $(\mathrm{A} + \mathrm{e}^-) + \mathrm{B}$, which is controlled by the interaction matrix Q_{ij}^{AA}, we find

$$\left.\frac{\mathrm{d}}{\mathrm{d}t}\right|_{\mathbf{s}_\mathrm{g}} - \tfrac{1}{2}v_0 \frac{\partial}{\partial z_\mathrm{g}} = \left.\frac{\mathrm{d}}{\mathrm{d}t}\right|_{\mathbf{x}}, \tag{5.26}$$

where $\mathbf{x}$ is the position vector of the electron with respect to A, from which it follows that all direct couplings vanish in the limit $|t| \to \infty$. In addition, the exchange matrix elements all vanish in the same limit since the overlap of the separated wave functions tends to zero. This shows directly that there are no spurious couplings between channels at infinite separations and that the boundary conditions are satisfied.

Plane-wave translation factors combined with molecular orbital bases have been employed rather infrequently. Some examples will be noted later in the next chapter. The reason for this is in part because of the complexity of the numerical evaluation of those matrix elements containing the exponential functions $\exp(\pm \mathrm{i}\mathbf{v}_0 \cdot \mathbf{s}_\mathrm{g})$. A more fundamental objection is that for low energies and small values of the interionic separation R, the electronic wave function in a given channel should not record explicitly to which centre the electron is attached in the asymptotic region ($R \to \infty$). However with plane-wave translational factors this is not the case since basis terms with the factor $\exp[-\mathrm{i}(\mathbf{v}_0 \cdot \mathbf{s}_\mathrm{g})/2]$ are associated with B, and those with $\exp[\mathrm{i}(\mathbf{v}_0 \cdot \mathbf{s}_\mathrm{g})/2]$ with A, for *all* values of R. This difficulty can be overcome by introducing more flexible forms of phase functions π_j^{A}, π_j^{B} in terms of switching functions (see (3.97), (3.98), and (3.77)),

$$\pi_j^{\mathrm{A}}(\mathbf{s}_\mathrm{g}, t) = \pi_j^{\mathrm{B}}(\mathbf{s}_\mathrm{g}, t) = -\tfrac{1}{2}(\mathbf{v}_0 \cdot \mathbf{s}_\mathrm{g}) f_j(\mathbf{s}_\mathrm{g}, \mathbf{R}) \tag{5.27}$$

where $f_j(\mathbf{s}_\mathrm{g}, \mathbf{R})$ satisfies (3.98)

$$f_j(\mathbf{s}_\mathrm{g}, \mathbf{R}) \to \begin{cases} 1 & r \ll x \text{ and } R \text{ large} \\ -1 & x \ll r \text{ and } R \text{ large.} \end{cases} \tag{5.28}$$

If functions satisfying (5.28) can be found with the property

$$f_j(\mathbf{s}_\mathrm{g}, \mathbf{R}) \to 0 \qquad \text{as } R \to 0, \tag{5.29}$$

then in the region of small R, the electronic wavefunction is not tagged by a factor depending on the asymptotic region. Two kinds of translational factor can be envisaged, *state-dependent factors* which vary with the index j, or *common translation factors*, which are independent of j, and which depend on a single switching factor $f(\mathbf{s}_\mathrm{g}, \mathbf{R})$.

5.3.2 Common translation factors

The use of a state-independent common translation factor depending on a single phase function $\pi(\mathbf{s}_g, t) = -\frac{1}{2}(\mathbf{v}_0 \cdot \mathbf{s}_g) f(\mathbf{s}_g, \mathbf{R})$ leads to a considerable simplification in the coupled equations (5.3) since the overlap matrix is then diagonal

$$N_{ij} = \delta_{ij}. \tag{5.30}$$

In this case the coupled equations (5.3) take the form

$$\mathrm{i}\dot{A}_i(\mathbf{b}, t) = \sum_j^{N+M} Q_{ij}(t) A_j(\mathbf{b}, t), \qquad i = 1, 2, \ldots, N + M. \tag{5.31}$$

Furthermore all the elements of the interaction matrix are simplified and from (5.14), we have

$$Q_{ij}(t) = [\tilde{U}_{ij}(t) + \tilde{T}_{ij}(t) + \tilde{\Gamma}_{ij}(t)] \exp[\mathrm{i}\{E_i(\infty) - E_j(\infty)\}t], \tag{5.32}$$

where

$$\begin{aligned} \tilde{U}_{ij}(t) &= \langle i | H_e - \delta_{ij} E_j(\infty) | j \rangle, \\ \tilde{T}_{ij}(t) &= -\mathrm{i}\langle i | \left.\frac{\mathrm{d}}{\mathrm{d}t}\right|_{\mathbf{s}_g} | j \rangle, \\ \tilde{\Gamma}_{ij}(t) &= \langle i | -\tfrac{1}{2}\mathrm{i}\nabla^2_{\mathbf{s}_g}\pi + \tfrac{1}{2}(\nabla_{\mathbf{s}_g}\pi)^2 - \mathrm{i}\nabla_{\mathbf{s}_g}\pi \cdot \nabla_{\mathbf{s}_g} - \tfrac{1}{8}v_0^2 + \left.\frac{\mathrm{d}\pi}{\mathrm{d}t}\right|_{\mathbf{s}_g} | j \rangle. \end{aligned} \tag{5.33}$$

In (5.33) we hae introduced the asymptotic energies $E_j(\infty)$ in each channel by setting $E_j(\infty) = \varepsilon_j, j \leq N$, and $E_j(\infty) = \eta_j, j > N$. It should be noted that the diagonal elements of $\tilde{T}_{ij}$ vanish, which can be proved by expressing $\mathrm{d}/\mathrm{d}t|_{\mathbf{s}_g}$ in terms of the radial and rotational coupling operators. The coupled equations can be simplified further if the basis functions $\bar{\psi}_j$ are adiabatic, rather than diabatic, in which case

$$H_e | j \rangle = E_j(R) | j \rangle, \tag{5.34}$$

from which

$$\tilde{U}_{ij}(t) = \delta_{ij}(E_j(R) - E_j(\infty)). \tag{5.35}$$

The secular terms, $\tilde{U}_{ii}$, can now be removed by the transformation

$$A_j(\mathbf{b}, t) = B_j(\mathbf{b}, t) \exp\left[-\mathrm{i}\int_0^t \{E_j(R') - E_j(\infty)\}\, \mathrm{d}t'\right] \tag{5.36}$$

where $R' \equiv R(t')$. The new amplitudes $B_j(\mathbf{b}, t)$ satisfy the equation

$$\mathrm{i}\dot{B}_i(\mathbf{b}, t) = \sum_{j=1}^{N+M} \tilde{Q}_{ij}(t) \exp\left[-\mathrm{i}\int_0^t \{E_j(R') - E_i(R')\}\,\mathrm{d}t'\right] B_j(\mathbf{b}, t),$$
$$i = 1, 2, \ldots, N+M \qquad (5.37)$$

with

$$\tilde{Q}_{ij} = \tilde{T}_{ij} + \tilde{\Gamma}_{ij}.$$

Some further simplification can be made in the matrix elements $\tilde{\Gamma}_{ij}$, using the identity

$$[H_e, F] = -\tfrac{1}{2}\nabla_{s_g}^2 F - \nabla_{s_g} F \cdot \nabla_{s_g}, \qquad (5.38)$$

where F is an arbitrary function of $\mathbf{s}_g$. In an adiabatic basis with F set equal to $\pi(\mathbf{s}_g, t)$

$$\tilde{\Gamma}_{ij} = (E_i(R) - E_j(R))\langle i|\pi(\mathbf{s}_g, t)|j\rangle + \langle i|\tfrac{1}{2}(\nabla_{s_g}\pi(\mathbf{s}_g, t))^2 - \tfrac{1}{8}v_0^2 + \left.\frac{\mathrm{d}}{\mathrm{d}t}\right|_{\mathbf{s}_g} \pi(\mathbf{s}_g, t)|j\rangle.$$
$$(5.39)$$

The second term is of order v_0^2, and if this is neglected we arrive at a low-energy approximation for $\tilde{Q}_{ij}$:

$$\tilde{Q}_{ij} \approx \langle i|\mathrm{i}\left.\frac{\mathrm{d}}{\mathrm{d}t}\right|_{\mathbf{s}_g} + \mathrm{i}[E_i(R) - E_j(R)]\pi(\mathbf{s}_g, t)|j\rangle. \qquad (5.40)$$

This is the impact parameter version of the modified perturbed stationary state equations (4.92).

Several specific forms of the common switching function $f(\mathbf{s}_g, \mathbf{R})$ have been suggested using physical arguments. A very simple form which satisfies the conditions (5.28) and (5.29) was suggested by Schneiderman and Russek (1969):

$$f(\mathbf{s}_g, \mathbf{R}) = \frac{R^2}{R^2 + \alpha^2}\cos\theta, \qquad (5.41)$$

where θ is the angle between the vectors $\mathbf{s}_g$ and $\mathbf{R}$. A cut-off parameter α is introduced, so that for $R < \alpha$, $f(\mathbf{s}_g, \mathbf{R})$ becomes proportional to R^2 and decreases for small R. Another form for $f(\mathbf{s}_g, \mathbf{R})$ has been suggested by Vaaben and Taulbjerg (1981) which is parameter-free. If the centres A and B are fully stripped the electron experiences electrostatic forces due to charges Z_A and Z_B, respectively. When the electron is closer to A than the interionic distance R, it translates with the velocity of A, and similarly it translates with the velocity of B when it is closer to B than the distance R. However, if R is smaller than x or r, the electron translates like the united atom, which is situated at the centre of charge. Expanding on these ideas, Vaaben and

Taulbjerg suggested that for any location of the electron, it should translate like the point of intersection of the line AB with the extension of the vector representing the electrostatic force on the electron. A switching factor satisfying these conditions is

$$f(s_g, R) = \frac{Z_B x^3 - Z_A r^3}{Z_B x^3 + Z_A r^3} + \frac{Z_A - Z_B}{Z_A + Z_B}, \tag{5.42}$$

and this expression has been applied with some success.

It should be noted that a standard computer program has been published (Salin 1991; Gaussorgues *et al.* 1975*b*) to solve the coupled impact parameter equations when a molecular basis with a common translation factor is employed.

5.3.3 Optimization

Numerical studies suggest that for the calculation of total cross sections at low velocities the precise form of switching factor is not critical (although differential cross sections are more sensitive) and a large number of *ad hoc* forms, similar in spirit to (5.41) and (5.42), have been employed. Clearly the use of a sufficiently large and flexible basis will tend to overcome any deficiencies in the choice of translation factor. Nevertheless a fundamental problem is to optimize the functional form of the translation factors with a basis of a given type and size. In the case of a common translation factor, this amounts to the determination of the optimal form of the switching function $f(\mathbf{s}_g, \mathbf{R})$.

To find this optimal form, perhaps the most satisfactory method is to appeal to the variational principal (3.181). For example, the parameter α in the Schneiderman–Russek expression (5.41) can be treated as a variational parameter and determined by solving the equation

$$\frac{\partial I}{\partial \alpha} = 0 \tag{5.43}$$

where

$$I = \int_{-\infty}^{\infty} dt \int d\mathbf{s}_g \psi^*(\mathbf{s}_g, t)\left[H_e - i\left.\frac{d}{dt}\right|_{\mathbf{s}_g}\right]\psi(\mathbf{s}_g, t). \tag{5.44}$$

This equation should be solved simultaneously with the coupled differential equations for the amplitudes $A_j(\mathbf{b}, t)$. More generally, equations can be obtained to determine the phases $\pi_j(\mathbf{s}_g, t)$ from the variational method (Riley and Green (1971)). Unfortunately these equations are *non-linear* and present a formidable numerical problem if no further approximations are made. For this reason there has been considerable interest in looking at simplified procedures.

Minimum norm optimization

An alternative to the variational method was originally suggested by some work of Rankin and Thorson (1978) on the calculation of ionization cross sections using a coupled channel approach. They observed that in their model the coupling between the incident bound state channel and the continuum was spuriously large, but this could be reduced by using certain forms of translation factors. This observation gave rise to the idea that the optimum translation factors are those which minimize the couplings between the states included in the expansion basis and all the remaining states. This principle has been formalized, and practical computational methods developed, in a series of papers by Riera and Errea and their co-workers (Riera 1984; Errea *et al.* 1985, 1987*a*, 1989*a*, *b*, *c*).

Let us split the *exact* wave function satisfying (5.1) into two terms

$$\psi(\mathbf{s}, t) = P\psi + Q\psi, \tag{5.45}$$

where P and Q are self-adjoint projection operators with the properties

$$P^2 = Q^2 = 1, \qquad PQ = QP = 0, \qquad P = P^\dagger, \qquad Q = Q^\dagger. \tag{5.46}$$

The operator P is chosen to project out the subspace spanned by the molecular orbitals ψ_j, $j \leq N + M$. The full Schrödinger equation (5.2) is equivalent to a pair of coupled equations for the wave functions $P\psi$ and $Q\psi$:

$$\begin{aligned} P\left(H_e - i\frac{d}{dt}\bigg|_s\right)P\psi &= -P\left(H_e - i\frac{d}{dt}\bigg|_s\right)Q\psi \\ Q\left(H_e - i\frac{d}{dt}\bigg|_s\right)Q\psi &= -Q\left(H_e - i\frac{d}{dt}\bigg|_s\right)P\psi. \end{aligned} \tag{5.47}$$

If the approximate wave function in the coupled channel approximation is denoted by $\tilde{\psi}$, we have

$$P\tilde{\psi} = \sum_{j=1}^{J} A_j(\mathbf{b}, t)\psi_j(\mathbf{s}, \mathbf{R}(t))\, e^{i\gamma_j}, \tag{5.48}$$

where $J = N + M$ and

$$\gamma_j = -E_j(\infty)t - \tfrac{1}{2}(\mathbf{v}_0 \cdot \mathbf{s})f(\mathbf{s}_g, \mathbf{R}). \tag{5.49}$$

The coupled equations for the amplitudes $A_j(\mathbf{b}, t)$ are equivalent to the P subspace equation

$$P\left(H_e - i\frac{d}{dt}\bigg|_s\right)P\tilde{\psi} = 0. \tag{5.50}$$

If the coupling between the P and Q subspaces were zero, then the coupled channel approximation in the P subspace would be exact. When this is not the case, a measure of the coupling between the P and Q subspaces is given by the norm† $M(t)$ of the state vector

$$\left[Q\left(H_e - i\frac{d}{dt}\bigg|_s\right)|P\tilde{\psi}\rangle\right],$$

where

$$M^2(t) = \left\|Q\left(H_e - i\frac{d}{dt}\bigg|_s\right)P\tilde{\psi}\right\|^2. \tag{5.51}$$

The idea of the *minimum norm method* is to adjust the function $f(\mathbf{s}_g, \mathbf{R})$ (or any parameter appearing in $f(\mathbf{s}_g, \mathbf{R})$) so that M is as small as possible. This can be done either at each of a number of specific values of t, or, on the average, by minimizing $\int_{-\infty}^{\infty} M^2(t)\, dt$. For practical calculations $M^2(t)$ can be written in either of two forms. To obtain the first form of M^2, we can start from the expression

$$M^2(t) = \langle P\tilde{\psi}|\left(H_e + i\frac{d}{dt}\bigg|_s\right)Q^2\left(H_e - i\frac{d}{dt}\bigg|_s\right)|P\tilde{\psi}\rangle. \tag{5.52}$$

Using the fact that the functions $\psi_j \exp(i\gamma_j)$, with $j > J$, span the Q subspace, so that

$$Q^2 = Q = \sum_{j>J} |\psi_j \exp(i\gamma_j)\rangle\langle\psi_j \exp(i\gamma_j)|, \tag{5.53}$$

$M^2(t)$ can be expressed as

$$M^2(t) = \sum_{k=1}^{J}\sum_{k'=1}^{J}\sum_{j>J} A_k^*(\mathbf{b}, t)A_{k'}(\mathbf{b}, t)\langle\psi_k \exp(i\gamma_k)|\left(H_e + i\frac{d}{dt}\bigg|_s\right)|\psi_j \exp(i\gamma_j)\rangle$$
$$\times \langle\psi_j \exp(i\gamma_j)|H_e - i\frac{d}{dt}\bigg|_s |\psi_{k'} \exp(i\gamma_{k'})\rangle. \tag{5.54}$$

In their work, Errea and his co-workers chose a simplified norm $N^2(t)$ consisting of those terms in (5.54) which are diagonal in k and k':

$$N^2(t) = \sum_{k=1}^{J}\sum_{j>J} |A_k(\mathbf{b}, t)|^2|\langle\psi_k \exp(i\gamma_k)|\left(H_e - i\frac{d}{dt}\bigg|_s\right)|\psi_j \exp(i\gamma_j)\rangle|^2. \tag{5.55}$$

The sum over j in (5.54) and (5.55) includes an integration over the continuum states of the molecule, and cannot easily be evaluated directly. A slightly simpler *Euclidean norm,* $N_E^2(t)$ is obtained from (5.55) if the

† The norm $\|g\|$ of a state vector $|g\rangle$ is defined so that $\|g\|^2 = \langle g|g\rangle$.

weighting by the collision amplitudes $|A_k(\mathbf{b}, t)|^2$ is omitted:

$$N_E^2(t) = \sum_{k=1}^{J} \sum_{j>J} |\langle \psi_j \exp(\mathrm{i}\gamma_j)| \left(H_e - \mathrm{i}\frac{\mathrm{d}}{\mathrm{d}t}\bigg|_s \right) |\psi_k \exp(\mathrm{i}\gamma_k)\rangle|^2. \qquad (5.56)$$

To obtain a second form for M^2, we again start from (5.52) and use the relation $Q^2 = Q = 1 - P$. We obtain

$$M^2(t) = \langle P\tilde{\psi}| \left(H_e + \mathrm{i}\frac{\mathrm{d}}{\mathrm{d}t}\bigg|_s \right) \left(H_e - \mathrm{i}\frac{\mathrm{d}}{\mathrm{d}t}\bigg|_s \right) |P\tilde{\psi}\rangle$$
$$- \sum_{j=1}^{J} |\langle \psi_j \exp(\mathrm{i}\gamma_j)| \left(H_e - \mathrm{i}\frac{\mathrm{d}}{\mathrm{d}t}\bigg|_s \right) |P\tilde{\psi}\rangle|^2, \qquad (5.57)$$

and this only depends explicitly on the P subspace basis set.

As with the variational method, the calculation of an optimized switching function by the minimum norm method is a complicated non-linear problem. Nevertheless, since this method leads to a minimization problem, it has a big advantage over the variational method, which only ensures that the integral I given by (5.44) is stationary and is not necessarily a minimum. In practice, the minimum norm method has not been used to determine switching functions f_j, or phase factors γ_j, but rather as a criterion in comparing different forms of switching functions, or for determining the best values of the parameters in a parametric form. As an example, in Fig. 5.2, some results obtained by Errea *et al.* (1985) are shown. The reaction studied is the collision between an α-particle and atomic hydrogen (see (4.98)) using a five-term molecular orbital basis of $1s\sigma$, $2s\sigma$, $2p\pi$, and $3d\delta$ orbitals. The results shown are for the particular choice of phase function $\pi(\mathbf{s}_g, t)$

$$\pi(\mathbf{s}_g, t) = -\tfrac{1}{2} f(\mathbf{s}_g, \mathbf{R})\mathbf{v}_0 \cdot \mathbf{s}_g - \tfrac{1}{8}(f^2(\mathbf{s}_g, \mathbf{R}) - 1)v_0^2 t, \qquad (5.58)$$

where the switching function f is

$$f(\mathbf{s}_g, \mathbf{R}) = \frac{-2}{R^2 + \beta^2}[\mathbf{s}_g \cdot \mathbf{R} + (\tfrac{1}{2} - p_0)R^2] + (1 - 2p_0). \qquad (5.59)$$

The parameters β and p_0 were adjusted so that the norm N_E was as small as possible over the important region of t. In the figure some results are shown for $v_0 = 0.447$ a.u. and $b = 5$ a.u. In this case, the curve labelled (a) with $\beta = 4$ and $p_0 = 0$ represents about the best that can be achieved with this type of function. In a later paper, Errea *et al.* (1987*a*) point out that the function $f(\mathbf{s}_g, \mathbf{R})$ approaches the limits ± 1 unrealistically slowly as $R \to \infty$ and correspondingly the norm N_E also decreases slowly, being for large t of order $1/R$. Further forms of phase function π are suggested which contain terms approaching the asymptotic values $\pm\frac{1}{2}(\mathbf{v}_0 \cdot \mathbf{s}_g)$ exponentially and these are shown to provide norms which are significantly smaller, and therefore

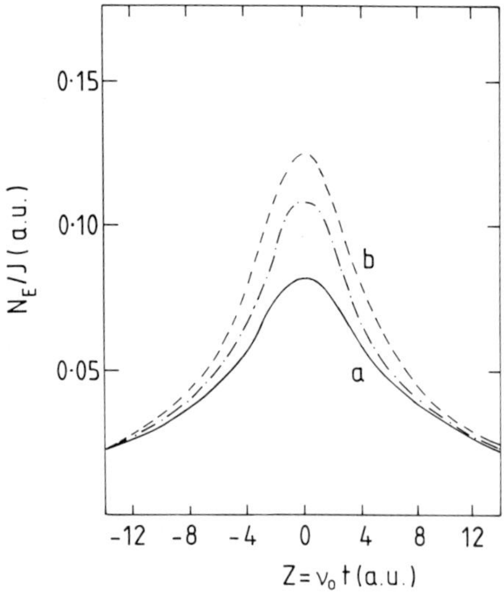

Fig. 5.2. Values of the Euclidean norm N_E/J with N_E given by (5.56) for the interaction of He^{2+} with H, in a five-state approximation at an energy of $E_L = 5$ keV/a.m.u. and an impact parameter $b = 5$ a.u. The curves are: ——— with $\beta = 4$, $p_0 = 0$ (see text); ----- with $\beta = 4$, $p_0 = 1$; —·—·— with $\beta = 0$, and any value of p_0 (after Errea *et al.* (1985)).

better. Some cross sections calculated by the minimum norm approach will be discussed in Chapter 6. In general good agreement with experimental data can be obtained at low energies ($E_L \lesssim 10$ keV/a.m.u.) with a common translational factor optimized by this method, provided a large enough basis is employed, but it is difficult to distinguish between results obtained using different forms of translational factor as the absolute values of charge exchange cross sections are, with some exceptions, not known with very great precision.

There is a very interesting connection between the norms of the type considered, which measure the residual coupling between the states included in a coupled channel calculation and those omitted, and the upper and lower bounds obtained by Spruch (1969) and Storm (1974) on the probability amplitude $\mathscr{A}_j(\mathbf{b}) = A_j(\mathbf{b}, \infty)$. If ψ is the approximate wave function corresponding to the (approximate) amplitude $\mathscr{A}_j(\mathbf{b})$, and $\mathscr{A}_j^E(\mathbf{b})$ is the exact value of $A_j(\mathbf{b}, \infty)$, then it is found that

$$\mathscr{A}_j(\mathbf{b}) - \Delta < A_j^E(\mathbf{b}, \infty) < \mathscr{A}_j(\mathbf{b}) + \Delta, \tag{5.60}$$

where the deviation Δ is defined as

$$\Delta = \int_{-\infty}^{\infty} D(t)\, \mathrm{d}t, \tag{5.61}$$

with

$$D^2(t) = \left\| \left(\mathrm{i} \frac{\mathrm{d}}{\mathrm{d}t}\bigg|_{\mathbf{s}} - H \right) \tilde{\psi} \right\|^2. \tag{5.62}$$

Since in the coupled channel approximation $\tilde{\psi}$ belongs to the P subspace, $P\tilde{\psi} = \tilde{\psi}$, and satisfies equation (5.50), in this approximation

$$D(t) = M(t)$$

and using (5.53) and (5.54) we see that

$$D^2(t) \leq JN^2(t). \tag{5.63}$$

As $D(t) > 0$ a knowledge of $N(t)$ provides through (5.61) an upper bound on the deviation Δ.

5.3.4 State-dependent phase factors

Although the use of a common translation factor with a phase function $\pi(\mathbf{s}_g, t)$ independent of the channel index is very appealing and simplifies the numerical problems encountered, there is evidence that such an approach becomes insufficient as the impact velocity increases. For instance, Hanssen *et al.* (1984) have employed a common translation factor in conjunction with a molecular orbital basis to calculate cross sections for electron capture by the ions C^{4+}, N^{5+}, and O^{6+} from atomic hydrogen and showed that while reasonable agreement with experiment could be obtained at low velocities ($v_0 \lesssim 1$ a.u.), for higher velocities the calculated cross sections failed to exhibit the steep decrease indicated by the data. The conclusion was reinforced by Newby (1985), who employed an atomic orbital basis, but with a number of different common translation factors. He showed, for the reaction $He^{2+} + H \rightarrow He^{+}$ $(n = 2) + H^{+}$, that for $v_0 \lesssim 0.5$ a.u., the results were rather insensitive to the type of translation factor. Above this velocity, not only did the calculated cross sections fail to fall rapidly as indicated by experiment but the differences between results obtained with different translation factors became significant. In this particular case total cross sections in good agreement with experiment are obtained when plane-wave translational factors are employed at the higher velocities. From these calculations and from extensive work by Errea *et al.* (1987*b*) it is clear that if a uniform model is required capable of providing accurate cross sections at both low and high energies, more elaborate translation factors must be employed, with phase factors depending on the channel index. Indeed, even if the low-energy region is of primary interest, the use of more flexible channel-dependent phase factors should lead to improvements in the calculated cross sections; particularly for differential cross sections, which require a more detailed knowledge of the probability amplitudes as a function of the impact parameter b.

Obviously the simplest state-dependent translation factor is the plane-wave factor we have already considered but, as we have seen, this is in principle inadequate in the adiabatic low-energy region. Several methods for determining state-dependent phases $\pi_j(\mathbf{s}_g, t)$, satisfying the boundary conditions (5.27) and (5.28), have been suggested. These include the *minimum norm method*, the *variational method*, and a *modified LCAO method* due to Thorsen *et al.* (1981). It is clear that from the derivation of the norms (5.54)–(5.57) that the phase factors γ_j contained in these expressions need not be expressed in terms of a single switching factor as in (5.49), but the more general expressions (5.18), depending on the functions π_j, can be used. In this case, the minimization of N_E or M can, in principle, be carried out with respect to each function π_j. The practicality of this procedure has yet to be demonstrated. A simplified method based on minimizing the couplings to the omitted discrete and continuum states on a channel by channel basis, has been used successfully by Kimura and Thorson (1981*b*) in studies of the $He^{2+} + H(1s)$ and $H^+ + He^+(1s)$ systems.

The modified LCAO approach

There is evidence from the variational method (Riley and Green 1971) that if a basis of atomic wave functions is employed, plane-wave translational factors are optimal even at small values of R. This fact suggests that if the molecular orbitals ψ_j are expressed in terms of a linear combination of atomic orbitals (LCAO), by associating plane-wave factors with the atomic orbitals, the appropriate molecular orbital translation factors can be derived (Thorsen *et al.* 1981). Let us express $\psi_j(\mathbf{s}_g, \mathbf{R})$ as

$$\psi_j(\mathbf{s}_g, \mathbf{R}) = \sum_n U_{nj}\phi_n^{A,B} \tag{5.64}$$

where $\phi_n^{A,B}$ is an atomic orbital centred on A or B. Switching factors f_j can now be defined so that

$$\psi_j(\mathbf{s}_g, \mathbf{R})\exp[-\tfrac{1}{2}i(\mathbf{v}_0\cdot\mathbf{s}_g)f_j(\mathbf{s}_g, \mathbf{R})] = \sum_n U_{nj}\phi_n^{A,B}\exp[\pm\tfrac{1}{2}i(\mathbf{v}_0\cdot\mathbf{s}_g)], \tag{5.65}$$

where the upper sign is taken with ϕ_n^A and the lower with ϕ_n^B. Expanding in powers of $\mathbf{v}_0$ and retaining the linear term we obtain

$$-\psi_j(\mathbf{s}_g, \mathbf{R})f_j(\mathbf{s}_g, \mathbf{R}) = \sum_n (\pm)U_{nj}\phi_n^{A,B}. \tag{5.66}$$

For example, the $1s\sigma_g$ level of H_2^+ is in the LCAO approximation

$$\psi_{1s\sigma_g} = N\{\exp(-\beta x) + \exp(-\beta r)\} \tag{5.67}$$

where $\beta = \beta(R)$ is a variational parameter, and N is a normalization factor. From (5.66), we immediately obtain

$$-f_{1s\sigma_g}(\mathbf{s}_g, \mathbf{R})\{\exp(-\beta x) + \exp(-\beta r)\} = \exp(-\beta x) - \exp(-\beta r), \qquad (5.68)$$

so that

$$f_{1s\sigma_g}(\mathbf{s}_g, \mathbf{R}) = \tanh\left[\frac{\beta}{2}(x - r)\right]. \qquad (5.69)$$

Although this expression is satisfactory, and in harmony with the conclusion of the minimum coupling approach, it appears the LCAO representation of the molecular orbitals is not sufficiently accurate in general for this approach to provide quantitative results for higher states. This difficulty can be overcome by examining the exact molecular orbitals (see Appendix D), which can be expressed in terms of the product of the functions $X_j(\xi)Y_j(\eta)$, where ξ and η are prolate spherical coordinates

$$\xi = \frac{x + r}{R}, \qquad \eta = \frac{x - r}{R}. \qquad (5.70)$$

In the asymptotic region in which either $x \to R$ or $r \to R$, we see that $\xi \to 1$ and $\eta \to \pm 1$ where the upper sign applies if the electron is attached to B ($r \ll R$) and the lower if the electron is attached to A ($x \ll R$). The function $Y_j(\eta)$ can be decomposed into terms which in the asymptotic region are associated either with the centre A or with the centre B, following which the switching factors can be introduced along the lines indicated by (5.66). The lowest-order result for the $1s\sigma_g$ level is the same as (5.69), and satisfactory switching factors can be obtained for other levels. Although this procedure is attractive, it can only be justified at small velocities and does not have the theoretical underpinning of the variational and minimum norm approaches to the problem.

The minimum coupling method

In the method of Kimura (1981), the parameters in an assumed form of switching factor are determined by the simplified form of the minimum coupling method taking the channels one by one. In the applications to the $He^{2+} + H(1s)$ and $H^+ + He^+(1s)$ collisions reported by Kimura and Thorsen (1981*b*), the switching factors were assumed to be of the form

$$f_j(\mathbf{s}_g, R) = \tanh[\beta_j(R)\{x - r - \alpha_j(R)\}], \qquad (5.71)$$

where $\beta_j(R)$ and $\alpha_j(R)$ are R-dependent parameters. The parameters for a given channel j, were adjusted at each value of R to minimize the couplings between this channel and channels omitted from the basis set. The results were found to be consistent with those based on the modified LCAO approach. Interestingly enough, for the $He^{2+} + H(1s)$ system, the charge exchange cross section in the energy interval 0.25 to 5 keV/a.m.u. is close both to the experimental data and to the results of Winter and Hatton (1978) and Hatton *et al.* (1979) who employed plane-wave translational factors. However, there are some significant differences between the cross sections for electron capture from $He^{+}(1s)$ by H^{+} calculated using the optimized switching factor and those using plane-wave translational factors (Winter *et al.* 1980), although both calculations agree with the experimental data within the errors. This will be discussed further in Chapter 6.

The variational method

As pointed our earlier, Euler–Lagrange equations can be obtained for the phases $\pi_j(\mathbf{s}_g, t)$ from the variational method, and this has been done by Riley and Green (1971). The equations are non-linear and should be solved simultaneously with the coupled equations for the amplitudes $A_j(\mathbf{b}, t)$. Some progress in this direction was made by Ponce (1979) who examined the $H^{+} + H$ system, but to obtain a tractable model, the approximation was made in which couplings between states were neglected in evaluating the phases π_j, leading to cross sections in disagreement with experiment. Rather more success has been obtained from a restricted variational scheme in which the phase $\pi_j(\mathbf{s}_g, t)$ is written in terms of a switching function $f_j(\mathbf{R})$ which is independent of the vector $\mathbf{s}_g$. Thus

$$\pi_j(\mathbf{s}_g, t) = -\tfrac{1}{2}(\mathbf{v}_0 \cdot \mathbf{s}_g) f_j(\mathbf{R}) \tag{5.72}$$

where, to satisfy the boundary condition (5.28),

$$f_j(\mathbf{R}) \to \begin{cases} 1 & |t| \to \infty \quad \text{and} \quad j \le N \\ -1 & |t| \to \infty \quad \text{and} \quad j > N. \end{cases} \tag{5.73}$$

Variationally determined switching factors of this type were introduced by Crothers and Hughes (1978*a, b*) for the $H^{+} + H$ system, and by Crothers and Todd (1981) for the heteronuclear case, and a similar approach has been taken by Green (1981*a, b*). Let us consider the heteronuclear case for which the trial wave function is

$$\psi(\mathbf{s}_g, t) = \sum_{j=1}^{N+M} \bar{\psi}_j(\bar{\mathbf{s}}_g, \mathbf{R}) \exp(i\gamma_j) A_j(\mathbf{b}, t). \tag{5.74}$$

The phases γ_j and π_j are connected through the relation

$$\gamma_j = -E_j(\infty) - \tfrac{1}{8} v_0^2 t + \pi_j(\mathbf{s}_g, t) \tag{5.75}$$

and $\bar{\psi}_j$ is an adiabatic molecular orbital calculated in the body-fixed frame. The variational functional I is given by (5.44). Imposing the condition $\delta I = 0$ under the independent variations $A_j \to A_j + \delta A_j$ and $f_j \to f_j + \delta f_j$ leads to the coupled equations (5.3) together with the equations

$$\operatorname{Re}\left\langle \frac{\partial \psi}{\partial f_j} \middle| H_e - \mathrm{i}\frac{\mathrm{d}}{\mathrm{d}t}\bigg|_{\mathbf{s}_g} \middle| \psi \right\rangle = 0. \tag{5.76}$$

From (5.72), (5.74), and (5.75)

$$\frac{\partial \psi}{\partial f_j} = -\mathrm{i}\tfrac{1}{2}(\mathbf{v}_0 \cdot \mathbf{s}_g)\bar{\psi}_j(\bar{\mathbf{s}}_g, \mathbf{R}) \exp(\mathrm{i}\gamma_j). \tag{5.77}$$

Combining this result with (5.76), we obtain

$$\operatorname{Im} \sum_{k=1}^{N+M} [-\mathrm{i}\mathscr{I}_{jk}\dot{A}_k(\mathbf{b}, t) + \mathscr{D}_{jk}A_k(\mathbf{b}, t)] = 0, \qquad j = 1, 2, \ldots, N + M \tag{5.78}$$

where

$$\mathscr{I}_{jk} = \langle \bar{\psi}_j \, \mathrm{e}^{\mathrm{i}\gamma_j} | z_g | \bar{\psi}_k \, \mathrm{e}^{\mathrm{i}\gamma_k} \rangle \tag{5.79}$$

and

$$\mathscr{D}_{jk} = \langle \bar{\psi}_j \, \mathrm{e}^{\mathrm{i}\gamma_j} | z_g \left(H_e - \mathrm{i}\frac{\mathrm{d}}{\mathrm{d}t}\bigg|_{\mathbf{s}_g} \right) | \bar{\psi}_k \, \mathrm{e}^{\mathrm{i}\gamma_k} \rangle. \tag{5.80}$$

Substituting for $\dot{A}_k$ from the equations (5.3), we find that (5.78) becomes

$$\sum_{k=1}^{N+M} \operatorname{Im}[(\mathscr{I} N^{-1} Q)_{jk} - \mathscr{D}_{jk}] A_k(\mathbf{b}, t) = 0, \qquad j = 1, 2, \ldots, N + M \tag{5.81}$$

Equations (5.81) are very complicated as they stand, but are greatly simplified if a weak coupling approximation (Crothers and Todd 1978) is made by setting $A_k(\mathbf{b}, t) = \delta_{kj}$. A linear equation for each factor f_j can then be found by expanding the terms $(\mathscr{I} N^{-1} Q)_{jj}$ and $\mathscr{D}_{jj}$ in powers of v_0 and retaining the first-order terms. The result is

$$\frac{f_j(\mathbf{R})}{2} = \frac{\langle \bar{\psi}_j | z_g \frac{\mathrm{d}}{\mathrm{d}t}\big|_{\mathbf{s}_g} | \bar{\psi}_j \rangle - \sum \langle \bar{\psi}_j | z_g | \bar{\psi}_k \rangle \langle \bar{\psi}_k | \frac{\mathrm{d}}{\mathrm{d}t}\big|_{\mathbf{s}_g} | \psi_j \rangle}{-\langle \bar{\psi}_j | z_g \frac{\partial}{\partial z_g} | \bar{\psi}_j \rangle + \sum \langle \bar{\psi}_j | z_g | \bar{\psi}_k \rangle \langle \bar{\psi}_k | \frac{\partial}{\partial z_g} | \bar{\psi}_g \rangle}. \tag{5.82}$$

The sums over k should be carried out over all states strongly coupled to the state j. Crothers and Todd presented numerical calculations of $f_j(\mathbf{R})$ for the $He^{2+} + H$ system including the five states 2pσ, 2pπ, 3dσ, 3dπ, and 2sσ. Of these, the 2pσ state separates to He^{2+} + H(1s) and belongs to the initial arrangement, while the remaining four states separate to the $He^{+} + H^{+}$ arrangement, and correspondingly as $R \to \infty$, $f_{2p\sigma}(\mathbf{R}) \to 1$, while for the remaining states $f_j(\mathbf{R}) \to -1$. In Fig. 5.3, $f_{2p\sigma}$ and $f_{2p\pi}$ are illustrated for a

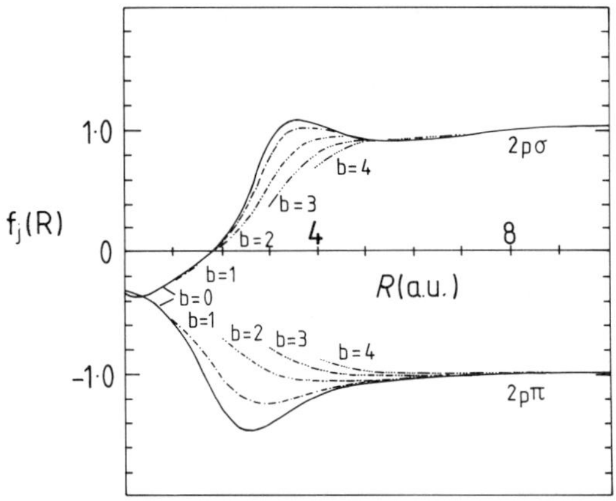

Fig. 5.3. Variationally determined switching functions $f_j(R)$ for $(\text{He H})^{2+}$ for different impact parameters. The upper curves relate to the 2pσ level and the lower to the 2pπ.

number of values of the impact parameter b. It is interesting to note that for each of the five states $f_j(0) = -\frac{1}{3}$. This shows that in the united atom limit the electron translates with the centre of charge which, for the $(\text{He H})^{2+}$ system, has a velocity $v_0/6$ with respect to the mid-point of the line AB, in conformity with the ideas of Vaaben and Taulbjerg (1981).

By restricting the switching factors to be functions of $\mathbf{R}$ alone the cross sections obtained with a given basis set are clearly not the most accurate that could be calculated by the variational method. Nevertheless the model has the important merit of computability and the switching factors reflect the physical features of the problem in a very satisfactory manner representing a great improvement over purely *ad hoc* approaches.

5.4 ATOMIC ORBITAL BASIS SETS

The simplest atomic orbital basis consists of a number of bound state atomic wave functions modified by plane-wave translational factors and centred on the two heavy particles A and B, chosen so that the incident channel and the most important charge exchange and excitation channels are represented. The trial wave function is of the form (3.179) where the basis functions Φ_j and X_n, centred on B and A respectively, are taken to be identical to the asymptotic forms Φ_j^{B}, X_n^{A}

$$\Phi_j(\mathbf{s}, t) = \Phi_j^{\mathrm{B}}(\mathbf{s}, t) = \phi_j(\mathbf{r}) \exp[\mathrm{i}\alpha_j(t)] \tag{5.83a}$$

$$X_n(\mathbf{s}, t) = X_n^{\mathrm{A}}(\mathbf{s}, t) = \chi_n(\mathbf{x}) \exp[\mathrm{i}\beta_n(t)]. \tag{5.83b}$$

The explicit form of the expansion becomes

$$\psi(\mathbf{s}, t) = \sum_{j=1}^{N} \phi_j(\mathbf{r}) \exp[\mathrm{i}\alpha_j(t)]A_j(\mathbf{b}, t) + \sum_{n=n+1}^{N+M} \chi_n(\mathbf{x}) \exp[\mathrm{i}\beta_n(t)]A_n(\mathbf{b}, t). \tag{5.84}$$

The phases α_j and β_n are defined by (3.171) and (3.176) and contain plane-wave translational factors. Although in principle one might consider more elaborate forms of translational factor, Riley and Green (1971) have shown, using the variational method, that the plane-wave form is optimal for an atomic basis.

The coupled equations (5.3) for the probability amplitudes depend on the interaction matrix **Q** and the overlap matrix **N**. These are in turn expressed in terms of the submatrices **J**, **K**, $\bar{\mathbf{H}}$, $\bar{\mathbf{K}}$, and **S** which are defined in (3.186). Provided $\phi_j(\mathbf{r})$ and $\chi_n(\mathbf{x})$ are exact eigenfunctions of the internal atomic Hamiltonians $H_{\mathrm{B}} = (-\frac{1}{2}\nabla_r^2 + V_{\mathrm{Be}}(r))$ and $H_{\mathrm{A}} = (-\frac{1}{2}\nabla_x^2 + A_{\mathrm{Ae}}(x))$, the expressions (3.186) can be simplified to become (using a straight line trajectory with $W(R) = 0$):

$$\begin{aligned}
H_{jk}(t) &= \exp[\mathrm{i}(\varepsilon_j - \varepsilon_k)t] \int \phi_j^*(\mathbf{r})[V_{\mathrm{AB}}(\mathbf{R}) + V_{\mathrm{Ae}}(\mathbf{x})]\phi_k(\mathbf{r})\,\mathrm{d}\mathbf{r} \\
\bar{H}_{jk}(t) &= \exp[\mathrm{i}(\eta_j - \eta^k)t] \int \chi_j^*(\mathbf{x})[V_{\mathrm{AB}}(\mathbf{R}) + V_{\mathrm{Be}}(\mathbf{r})]\chi_k(\mathbf{x})\,\mathrm{d}\mathbf{x} \\
K_{jk}(t) &= \exp[\mathrm{i}(\varepsilon_j - \eta_k)t] \int \exp(\mathrm{i}\mathbf{v}_0\cdot\mathbf{s}_{\mathrm{g}})\phi_j^*(\mathbf{r})[V_{AB}(\mathbf{R}) + V_{\mathrm{Ae}}(\mathbf{x})]\chi_k\,\mathrm{d}\mathbf{s}_{\mathrm{g}} \\
\bar{K}_{jk}(t) &= \exp[\mathrm{i}(\eta_k - \varepsilon_j)t] \int \exp(-\mathrm{i}\mathbf{v}_0\cdot\mathbf{s}_{\mathrm{g}})\chi_k^*(\mathbf{x})[V_{\mathrm{AB}}(\mathbf{R}) + V_{\mathrm{Be}}(\mathbf{r})]\phi_j(\mathbf{r})\,\mathrm{d}\mathbf{s}_{\mathrm{g}} \\
S_{jk}(t) &= \exp[\mathrm{i}(\varepsilon_j - \eta_k)t] \int \exp(\mathrm{i}\mathbf{v}_0\cdot\mathbf{s}_{\mathrm{g}})\phi_j^*(\mathbf{r})\chi_k(\mathbf{x})\,\mathrm{d}\mathbf{s}_{\mathrm{g}}.
\end{aligned} \tag{5.85}$$

5.4.1 Symmetries of the matrix elements

The matrix elements listed in (5.85) have some useful symmetry properties which can be used as an aid in numerical evaluations. We first notice that the Hamiltonian H_{e} is invariant under a reflection in the plane of heavy particle scattering. If the plane of scattering is the XZ-plane, then

$$H_{\mathrm{e}}(s_x, s_y, s_z, t) = H_{\mathrm{e}}(s_x, -s_y, s_z, t)$$

where **s** is the electronic position vector. To make use of this property, the atomic wave functions can be expressed in terms of the real spherical harmonics $Y^{+}_{l|m|}$, $Y^{-}_{l|m|}$, rather than the complex harmonics $Y_{l,m}$. The real harmonics are defined by (cf. Bransden and Joachain (1983) p. 89)

$$Y^{+}_{l|m|} = \begin{cases} Y_{l0}(\theta, \phi) & m = 0 \\ \dfrac{1}{\sqrt{2}}[Y_{l|m|} + Y^{*}_{l|m|}] & m \neq 0 \end{cases} \tag{5.86}$$

$$Y^{-}_{l|m|} = \frac{1}{\sqrt{2}}[Y_{l|m|} - Y^{*}_{l|m|}].$$

If the angles (θ, ϕ) are expressed in terms of the Cartesian variables x/r, y/r, and z/r, it is easy to see that the $Y^{\pm}_{l|m|}$ are eigenfunctions of the reflection $y \to -y$ with:

$$Y^{\pm}_{l|m|}(x, y, z) = \pm Y^{\pm}_{l,|m|}(x, -y, z). \tag{5.87}$$

By making use of this property the atomic wave functions can be divided into two sets $\phi^{\pm}_{k}$, $\chi^{\pm}_{j}$ of opposite parity under reflection in the XZ-plane so that

$$\begin{aligned} \phi^{\pm}_{k}(\mathbf{r}) &= R_{n_k l_k}(r) Y^{\pm}_{l_k|m_k|}(\theta_r, \phi_r) \\ \chi^{\pm}_{j}(\mathbf{x}) &= R_{n_k l_k}(x) Y^{\pm}_{l_j|m_j|}(\theta_x, \phi_x). \end{aligned} \tag{5.88}$$

From the symmetry of the Hamiltonian it follows that the states of different parity under the reflection are uncoupled. So, for example, if the target is in an s state ($l = 0$), all states coupled to it must be described by the eigenfunction of positive reflection parity, ϕ^{+}_{k} or χ^{+}_{j}.

To examine the relationship of the matrix elements (5.85) under the interchange $t \to -t$, we first notice that the atomic wave functions expressed in terms of Cartesian coordinates have the property

$$\phi_k(x, y, z) = (-1)^{l_k + |m_k|}\phi_k(x, y, -z) \tag{5.89}$$

and similarly for the functions χ_n. Recollecting that $\mathbf{r} = \mathbf{s}_g - \frac{1}{2}\mathbf{R}$, $\mathbf{x} = \mathbf{s}_g + \frac{1}{2}\mathbf{R}$ and that the Cartesian coordinates of $\mathbf{R}$ are $(b, 0, v_0 t)$, we find using (5.89) that

$$M^{*}_{jk}(-t) = (-1)^{l_j + l_k + |m_j| + |m_k|} M_{jk}(t) \tag{5.90}$$

where $\mathbf{M}$ is any one of the submatrices $\mathbf{H}$, $\bar{\mathbf{H}}$, $\mathbf{K}$, $\bar{\mathbf{K}}$, or $\mathbf{S}$.

In addition, it is clear from (5.85) that the submatrices $\mathbf{H}$ and $\bar{\mathbf{H}}$ satisfy

$$\begin{aligned} H_{jk}(t) &= H_{kj}(-t) = H^{*}_{kj}(t) \\ \bar{H}_{jk}(t) &= \bar{H}_{kj}(-t) = H^{*}_{kj}(t) \\ S^{*}_{jk}(t) &= S_{kj}(t), \end{aligned} \tag{5.91}$$

while $\mathbf{K}$ and $\bar{\mathbf{K}}$ are related by (3.192).

The symmetry properties (5.90) and (5.91) are easily generalized to the case of multi-electron systems.

Numerical evaluation

The direct interaction submatrices **H** and $\bar{\mathbf{H}}$ can be evaluated analytically without difficulty but the overlap and exchange matrix elements containing the factors $\exp(\pm i\mathbf{v}_0 \cdot \mathbf{s}_g)$ cannot be completely reduced analytically for atomic wave functions expressed in terms of Slater functions. Most of the computing time required for the evaluation of the coupled channel model goes into the evaluation of the overlap and exchange integrals, and some methods on which efficient programs can be based are considered in Appendix E.

5.4.2 Two-state approximation for charge exchange

A simple approximation in which only one direct and one rearranged state are retained in the expansion of the trial wave function was originally introduced by Bates (1958) and has since been employed to study many systems. As pointed out by Green *et al.* (1965), and also by Lin *et al.* (1978), there are particular circumstances in which the two-state approximation can produce cross sections of useful accuracy. These are that the coupling between the two states should be strong and energetically favoured, that the impact velocity should be near the velocity at which the cross section has its maximum value, and that the transition should derive from regions of large internuclear separation, so that the wave function should be of an atomic rather than a molecular character.

In the two-state approximation the coupled equations (3.185) or (5.3) reduce to

$$\begin{aligned} i[\dot{A}_i(t) + S_{if}\dot{A}_f(t)] &= H_{ii}A_i(t) + K_{if}A_f(t) \\ i[S_{fi}\dot{A}_i(t) + \dot{A}_f(t)] &= \bar{K}_{fi}A_i(t) + \bar{H}_{ff}A_f(t), \end{aligned} \tag{5.92}$$

where the dependence of the amplitudes $A_i(\mathbf{b}, t)$, $A_f(\mathbf{b}, t)$ on $\mathbf{b}$ has been suppressed. The index i labels the initial state of the atom $(B + e^-)$ and the index f the state of the atom $(A + e^-)$ into which the electron is captured. Solving the equations (5.92) for $\dot{A}_i$ and $\dot{A}_j$, we obtain

$$\begin{aligned} i\dot{A}_i(t) &= \frac{(H_{ii} - S_{if}\bar{K}_{fi})A_i(t) + (K_{if} - S_{if}\bar{H}_{ff})A_f(t)}{1 - |S_{if}|^2} \\ i\dot{A}_f(t) &= \frac{(\bar{K}_{fi} - S_{fi}H_{ii})A_i(t) - (\bar{H}_{ff} - S_{fi}K_{if})A_f(t)}{1 - |S_{if}|^2}, \end{aligned} \tag{5.93}$$

where we have used the fact that $S_{fi} = S^*_{if}$.

The secular terms in (5.93) can be removed by the phase transformation

$$A_i(t) = d_i(t)\exp\left\{-i\int_{-\infty}^{t}\delta_i(t')\,dt'\right\}$$
$$A_f(t) = d_f(t)\exp\left\{-i\int_{-\infty}^{t}\delta_f(t')\,dt'\right\}, \qquad (5.94)$$

where

$$\delta_i(t) = \frac{H_{ii} - S_{if}\bar{K}_{fi}}{1 - |S_{if}|^2}$$
$$\delta_f(t) = \frac{\bar{H}_{ff} - S_{fi}K_{if}}{1 - |S_{if}|^2}. \qquad (5.95)$$

Substituting expressions (5.94) into (5.93) we find

$$i\dot{d}_i(t) = Md_f(t)$$
$$i\dot{d}_f(t) = \bar{M}d_i(t) \qquad (5.96)$$

where

$$M = [1 - |S_{if}|^2]^{-1}[K_{if} - S_{if}\bar{H}_{ff}]\exp\left[i\int_{-\infty}^{t}\{\delta_i(t') - \delta_f(t')\}\,dt'\right]$$
$$(5.97)$$
$$\bar{M} = [1 - |S_{if}|^2]^{-1}[K_{fi} - S_{fi}\bar{H}_{ii}]\exp\left[i\int_{-\infty}^{t}\{\delta_i(t') - \delta_f(t')\}\,dt'\right]$$

Since $|d_i| = |A_i|$, equations (5.97) are to be solved subject to the boundary conditions

$$\lim_{t\to-\infty} d_i(\mathbf{b}, t) = 1, \qquad \lim_{t\to-\infty} d_f(\mathbf{b}, t) = 0 \qquad (5.98)$$

in which case the probability for charge exchange is

$$P_{fi}(v_0, \mathbf{b}) = \lim_{t\to\infty} |d_f(\mathbf{b}, t)|^2 = \lim_{t\to\infty} |A_f(\mathbf{b}, t)|^2. \qquad (5.99)$$

In (5.98) and (5.99) the dependence of d_f and A_f on both $\mathbf{b}$ and t has been displayed explicitly.

Symmetrical resonance

Despite their simplicity, equations (5.96) cannot be solved analytically and must be solved numerically, except in the special case of symmetrical resonance in which the initial and final atomic wave functions are identical.

Examples of symmetrical resonance are the reactions $H^+ + H(1s) \rightarrow H(1s) + H^+$ and $He^{++} + He^+(1s) \rightarrow He^+(1s) + He^{++}$. It is then easy to see that $\delta_i(t) = \delta_f(t)$ and $\bar{M}(t) = M(t)$, so that the coupled equations (5.96) reduce to

$$\begin{aligned} i\dot{d}_i(t) &= L d_f(t) \\ i\dot{d}_f(t) &= L d_i(t). \end{aligned} \tag{5.100}$$

The interaction L is real and defined by

$$L = \frac{K - SH}{1 - |S|^2}, \tag{5.101}$$

where we have set $S = S_{if} = S_{fi}$, $H = H_{ii} = \bar{H}_{ff}$, and $K = K_{if} = \bar{K}_{fi}$.

Uncoupled equations can be found by defining $C_\pm(t)$ as

$$C_\pm(t) = d_i(t) \pm d_f(t), \tag{5.102}$$

in which case the amplitudes $C_\pm(t)$ satisfy

$$i\dot{C}_\pm(t) = \pm L C_\pm(t). \tag{5.103}$$

The solution of equation (5.103) corresponding to the boundary condition (5.98) is

$$C_\pm(t) = \exp\left[\mp i \int_{-\infty}^{t} L(t')\, dt'\right], \tag{5.104}$$

in which case

$$\begin{aligned} \lim_{t\to\infty} d_f(\mathbf{b}, t) &= \tfrac{1}{2} \lim_{t\to\infty} [C_+(t) - C_-(t)] \\ &= -i \sin\left[\int_{-\infty}^{t} L(t')\, dt'\right]. \end{aligned} \tag{5.105}$$

The probability for charge exchange is then

$$P_{fi}(v_0, b) = \sin^2\left[\int_{-\infty}^{\infty} L(t)\, dt\right]. \tag{5.106}$$

The probability for elastic scattering $P_{ii}(v_0, b)$ can be calculated in the same way and is

$$P_{ii}(v_0, b) = \cos^2\left[\int_{-\infty}^{\infty} L(t)\, dt\right] \tag{5.107}$$

and we see explicitly that unitarity (probability conservation) is satisfied within the two-dimensional subspace spanned by the basis functions

$$P_{ii} + P_{if} = 1. \tag{5.108}$$

The probabilities P_{if} and P_{ii} are functions of both the impact parameter b and the incident velocity v_0. From the form of (5.106) it is clear that the probability for symmetrical resonance charge exchange $P_{fi}(v_0, b)$ at a fixed impact parameter oscillates as a function of the incident velocity and also at fixed impact velocity $P_{fi}(v_0, b)$ oscillates as a function of the impact parameter, between the limits zero and unity. How far this prediction is borne out in practice will be examined in Chapter 6.

The uncoupling of the equations for $C_+(t)$ and $C_-(t)$ is an immediate consequence of the invariance of the Hamiltonian for a symmetrical system under inversion with respect to the mid-point of the internuclear line $\mathbf{s}_g \rightarrow -\mathbf{s}_g$. If P is the inversion operator, $P^2 = 1$ and the eigenvalues of P are ± 1. The eigenfunctions of P, $\psi^{\pm}(\mathbf{s}_g, t)$, develop in time independently. The wavefunctions $\psi^{\pm}$ correspond to the gerade and ungerade states of the pseudomolecule formed by the system, and in the two-state approximation charge exchange does not result from an electronic transition of the pseudomolecule but arises from the phase difference between the amplitudes for elastic scattering in the lowest gerade and ungerade states. This has been discussed in molecular terms in Chapter 4.

5.5 THE ROLE OF THE CONTINUUM

A basis set formed from the bound states of the target $(B + e^-)$ and the rearranged system $(A + e^-)$ has a strictly limited range of applicability being confined to strongly coupled transitions between levels not differing greatly in energy and for which the interaction takes place at large values of the impact parameter. This is because the continuum part of the expansion in terms of the wavefunctions $\phi_j(\mathbf{r})$ or $\chi_n(\mathbf{x})$ often plays a major role in ion–atom collisions, both in direct scattering and in rearrangement. The physical reasons for this are various. For example, at small values of the interionic separation $\mathbf{R}$ and at low velocities, the electronic wave functions will resemble those of the united atom. Any particular state of the united atom will be described by a wave function which, of course, can be represented by the complete set of functions, including the continuum, ϕ_j or χ_n, but which may only be represented very poorly by the bound state part of those sets. In addition, as the incident energy increases, ionization becomes a dominant process and to allow for the flux entering the ionization channels, the continuum part of the basis wave functions must be represented. A good example of the importance, or otherwise, of the continuum is the system

$(\text{He H})^{2+}$. Charge exchange at low to intermediate energies ($v_0 \lesssim 1$ a.u.) in the reaction $He^{2+} + H(1s)$ is dominated by the resonant process

$$He^{2+} + H(1s) \to He^{+}(n = 2) + H^{+}. \tag{5.109}$$

The transition is strong and takes place at comparatively large separations, and a bound state basis set composed of the 1s wave function centred on B (the proton) and the 2s, $2p_0$, $2p_{\pm 1}$ wave functions centred on A (the α-particle) provides cross sections which agree with the data. On the other hand, the reaction

$$H^{+} + He^{+}(1s) \to H(nlm) + He^{2+} \tag{5.110}$$

is dominated by capture into the H(1s) level. However the difference in energy between $He^{+}(1s)$ and H(1s) is large and thc transition is weak. Accordingly, a two-state approximation containing the $He^{+}(1s)$ and H(1s) wave functions is very poor and the situation is not greatly improved by adding further bound states the computed cross section being roughly one-half that observed. However, when some allowance is made for the continuum by adding pseudostates (see below), calculated cross sections agreeing closely with experiment are achieved.

5.5.1 Eigendifferentials

The direct use of continuum atomic wave functions in coupled channel calculations is generally impracticable but various ways of representing the continuum approximately over the limited regions of configuration space which contribute to the interaction matrix elements have been devised. In general, all these methods involved the representation of the continuum atomic functions by square-integrable (L^2) functions of one kind or another. We shall discuss several approaches in turn.

Consider the complete set of wave functions for the atom ($B + e^{-}$). These can be divided into the bound state functions

$$\phi_{nlm}(r, \theta, \phi) = r^{-1} R_{nlm}(r) Y_{lm}(\theta, \phi)$$

and continuum functions $\phi_{\varepsilon lm}(r, \theta, \phi) = r^{-1} R_{\varepsilon lm}(r) Y_{lm}(\theta, \phi)$, where the (positive) energy of the continuum state is $\varepsilon(k) = k^2/2$ and k is the momentum. In the special case for which the interaction between B and e^{-}, $V_{Be^{-}}$, is a central potential the radial functions R_{nlm} and $R_{\varepsilon lm}$ are independent of m. For present purposes, we need only consider $l = 0$ states, since the theory for $l \neq 0$ follows the same pattern. With $R_n \equiv R_{n00}$ and $R_\varepsilon \equiv R_{\varepsilon 00}$, any function $f(r)$ of r can be expanded:

$$f(r) = \sum_n a_n R_n(r) + \int_0^\infty a(\varepsilon) R_\varepsilon(r)\, d\varepsilon, \tag{5.111}$$

and the problem is to replace the integral over ε by a discrete sum. The radial functions R_n and R_ε can be normalized according to the relations

$$\int_0^\infty R_n(r)R_{n'}(r')\,\mathrm{d}r = \delta_{nn'}, \qquad \int_0^\infty R_\varepsilon(r)R_{\varepsilon'}(r)\,\mathrm{d}r = \delta(\varepsilon - \varepsilon'),$$

and

$$\int_0^\infty R_n(r)R_\varepsilon(r)\,\mathrm{d}r = 0, \tag{5.112}$$

with the closure relation

$$\sum_n R_n(r)R_n(r') + \int_0^\infty R_\varepsilon(r)R_{\varepsilon'}(r')\,\mathrm{d}\varepsilon = \delta(r - r'). \tag{5.113}$$

Before the introduction of the Dirac delta function, the theory of the continuous spectrum of the Sturm–Liouville equation had been developed by Weyl (1910) (Kemble 1937). According to this theory integrals over the continuous eigenvalue ε, as in (5.111), can be interpreted by a limiting process. The integration interval is first divided into an infinite number of small strips, each of width 2Δ with the ith strip centred about the energy $\varepsilon = \varepsilon_i$. The integral is then approximated as

$$\int_0^\infty a(\varepsilon)R_\varepsilon(r)\,\mathrm{d}\varepsilon \to \sum_i \bar{a}_i \frac{1}{\sqrt{2\Delta}} \int_{\varepsilon_i - \Delta}^{\varepsilon_i + \Delta} R_\varepsilon(r)\,\mathrm{d}\varepsilon, \tag{5.114a}$$

where

$$\bar{a}_i = \frac{1}{\sqrt{2\Delta}}\,a(\varepsilon_i). \tag{5.114b}$$

The functions $\bar{R}_i(\Delta, r)$ are defined by the relation

$$\bar{R}_i(\Delta, r) = \frac{1}{\sqrt{2\Delta}} \int_{\varepsilon_i - \Delta}^{\varepsilon_i + \Delta} R_\varepsilon(r)\,\mathrm{d}\varepsilon, \tag{5.115a}$$

and are called eigendifferentials. The eigendifferentials can be shown to be square-integrable and form an orthonormal system so that

$$\int_0^\infty \bar{R}_i(r)\bar{R}_j(r)\,\mathrm{d}r = \delta_{ij}. \tag{5.115b}$$

The expansion of an arbitrary function $f(r)$ given by (5.111) can be replaced by the purely discrete expansion

$$f(r) = \sum_n a_n R_n(r) + \sum_i \bar{a}_i \bar{R}_i(\Delta, r), \tag{5.116}$$

and the limit $\Delta \to 0$ can be taken after the required matrix elements depending on $f(r)$ have been evaluated.

The eigendifferential theory suggests an *approximate* procedure to represent the continuum in the expansion of the total wave function (5.84) in the basis of travelling atomic orbitals. First an upper limit ε_{max} of the energy ε is determined on physical grounds, and then the range of integration is divided into a *finite* number I of strips of width 2Δ. The bound state terms in the expansion are augmented by I terms of the form (5.115a), obtaining a system of coupled channel equations in the usual way. This approximation procedure has been used successfully in a number of applications in nuclear physics (Austern *et al.* 1987), but has been somewhat neglected in atomic scattering theory.

As an illustration of the use of eigendifferentials in the expansion of the impact parameter wave function, let us take an origin on the target centre B and expand the $l = 0$ part of the wave function $\psi(\mathbf{r}, t)$ in the set of target eigenfunctions $\phi_j(\mathbf{r})$. The continuum part of the expansion will be of the form

$$r\psi_\varepsilon(\mathbf{r}, t) = \int_0^\infty A_\varepsilon(b, t)\,\mathrm{e}^{-\mathrm{i}\varepsilon t} R_\varepsilon(r)\,\mathrm{d}\varepsilon. \tag{5.117}$$

There is now a choice to be made. We can either proceed as in (5.114) and write

$$r\psi_\varepsilon(\mathbf{r}, t) \simeq \sum_{i=1}^{I} \bar{A}_{\varepsilon_i}(b, t)\,\mathrm{e}^{-\mathrm{i}\varepsilon_i t} \bar{R}_{\varepsilon_i}(\Delta, r), \tag{5.118a}$$

or we can attempt to include the oscillating factor $\exp(-\mathrm{i}\varepsilon t)$ in the integral over each strip of width Δ and write

$$r\psi_\varepsilon(\mathbf{r}, t) = \sum_{i=1}^{I} \bar{A}_{\varepsilon_i}(b, t)\bar{R}_{\varepsilon_i}(\Delta, r, t), \tag{5.118b}$$

where

$$\bar{R}_{\varepsilon_i}(\Delta, r, t) = \frac{1}{\sqrt{2\Delta}} \int_{\varepsilon_i - \Delta}^{\varepsilon_i + \Delta} R_\varepsilon(r)\,\mathrm{e}^{-\mathrm{i}\varepsilon t}\,\mathrm{d}\varepsilon. \tag{5.118c}$$

The time-dependent wave packets $\bar{R}_{\varepsilon_i}(\Delta, r, t)$ have been termed Weyl packets (Heinz *et al.* 1981) and for large t (or large ε_i) probably provide a better description of the continuum than the time-independent packets (5.117). In applications to excitation ionization in collisions of heavy ions at relativistic energies, the time-independent packet approach has been followed by Mehler *et al.* (1989) and the time-dependent approach by Momberger *et al.* (1990). The Weyl packets present considerable numerical complexity in practical calculations. However, as pointed out by Schiwietz (1990) a further approximation can be developed to overcome this. When $r\Delta \ll \pi$, the radial function $R_\varepsilon(r)$ is nearly independent of ε and can be taken to be constant in the integration; on the other hand for $\varepsilon t \ll \pi$ the function $\exp(-i\varepsilon t)$ is nearly independent of ε. This suggests the approximation

$$r\psi_\varepsilon(\mathbf{r}, t) \approx \sum_{i=1}^{I} \bar{A}_{\varepsilon_i}(b, t)\left[\frac{1}{2\Delta}\int_{\varepsilon_i-\Delta}^{\varepsilon_i+\Delta} \mathrm{e}^{-i\varepsilon t}\,\mathrm{d}\varepsilon\right]\bar{R}_{\varepsilon_i}(\Delta, r)$$

$$= \sum_{i=1}^{I} \{\Delta\bar{A}_{\varepsilon_i}(b, t)\}\left\{\frac{2\sin\Delta t}{\Delta t}\right\}\mathrm{e}^{-i\varepsilon_i t}\bar{R}_{\varepsilon_i}(\Delta, r), \qquad (5.119)$$

where $\bar{R}_{\varepsilon_i}(\Delta, r) \equiv \bar{R}_{\varepsilon_i}(\Delta, r, t = 0)$ is defined by (5.115a). Factors similar to $F = (2\sin\Delta t)/\Delta t$ were originally introduced by Reading *et al.* (1979) in connection with a pseudostate expansion and will be discussed in Chapter 8. The expansion (5.119) has been employed by Schiwietz in a coupled channel model of ionization at high ($20 \lesssim E_\mathrm{L} \lesssim 5{,}000$ keV/a.m.u.) but non-relativistic energies, but at the time of writing no applications to rearrangement collisions have been reported.

5.5.2 Pseudostates

A rather different approach (Burke and Webb 1970; Burke and Mitchell 1973; Callaway and Wooten 1973) attempts to represent the continuum in terms of a finite set of Q linearly dependent L^2 basis functions. To illustrate the procedure, we will again consider the continuum of the atom $(\mathrm{B} + \mathrm{e}^-)$ for which the exact wave functions are $\phi_j(\mathbf{r})$ and take the $l = 0$ case only. Suitable basis functions then might be Slater functions

$$F_i(r) = r^{p_i}\exp(-\lambda_i r), \qquad i = 1, 2, \ldots, Q \qquad (5.120a)$$

or Gaussian functions

$$F_i(r) = r^{p_i}\exp(-\lambda_i r^2), \qquad i = 1, 2, \ldots, Q \qquad (5.120b)$$

or the functions introduced by Zimmerman (1972) (Ford *et al.* 1972)

$$F_i = \exp\{-\lambda_i r(1 - p_i \exp i\phi_i)\} \qquad (5.120c)$$

where ϕ_i are points chosen at equally spaced intervals from 0 to 2π. In each case λ_i and p_i are pre-chosen parameters. The continuum functions are represented by a linear superposition of the functions $F_i(r)$

$$\bar{R}_i(r) = \sum_{j=1}^{Q} F_j(r)\, d^i_j, \qquad i = 1, 2, \ldots, Q. \qquad (5.121)$$

The constants d^i_j are determined so that the functions $\bar{R}_i(r)$, which are called *pseudostates*, diagonalize the internal Hamiltonian H_B of the target $(\mathrm{B} + \mathrm{e}^-)$.

For $l = 0$ this requires that

$$\int_0^\infty \bar{R}_i(r)\left[-\tfrac{1}{2}\frac{d^2}{dr^2} + V_{Be}(r)\right]\bar{R}_j(r)\,dr = \bar{\varepsilon}_i\delta_{ij}. \tag{5.122}$$

Each eigenvalue $\bar{\varepsilon}_i$ corresponds to the eigenvector $\mathbf{d}^i$ in this finite matrix representation of H_B. In general, as a result of this procedure some of the energies $\bar{\varepsilon}_i$ will be found to be negative and some positive. By the Hylleraas–Undheim theorem, if there are n negative energies $\bar{\varepsilon}_i$ these are upper bounds on the energies ε_i of the first n bound states of the atom $(B + e^-)$. The remaining positive energies $\bar{\varepsilon}_i$ and the corresponding radial functions $\bar{R}_i$ form a representation of the continuum. With a suitable basis of functions F_j, it can be arranged that all the bound states of importance in the scattering problem are well represented, together with an adequate representation of the continuum. For example, if the atom $(B + e^-)$ is hydrogenic, so that V_{Be} is a Coulomb potential, rather than some effective potential, and a Slater set of functions $\{F_i\}$ is employed, by including the function $F_1 = \exp(-Z_B r)$, $F_2 = \exp(-Z_B r/2)$, $F_3 = r\exp(-Z_B r/2)$, the first two radial functions R_{1s} and R_{2s} will be obtained exactly, together with an approximation to R_{3s}.

Provided the size of the basis $\{F_i\}$ is large enough it can be shown that the positive energy pseudostates $\bar{R}_i(r)$ corresponding to the energy $\bar{\varepsilon}_i$ are approximations to the exact radial functions $R_{\bar{\varepsilon}_i}(r)$ at the same energy over a finite interval of r. The normalization of the two functions $\bar{R}_i$ and R_{ε_i} is different since the L^2 pseudostate functions are normalized so that

$$\int_0^\infty \bar{R}_i(r)\bar{R}_j(r)\,dr = \delta_{ij}, \tag{5.123}$$

while, on the other hand, the exact continuum functions are normalized so that

$$\int_0^\infty R_\varepsilon(r)R_{\varepsilon'}(r)\,dr = \delta(\varepsilon - \varepsilon'). \tag{5.124}$$

It follows that over some finite region of r we have the relation (Bassichis *et al.* 1975, 1978; Hazi and Taylor 1970)

$$R_{\bar{\varepsilon}_i}(r) \approx N_i(\bar{\varepsilon}_i)\bar{R}_i(r), \tag{5.125}$$

where N_i is a normalization constant. The value of N_i can be obtained in a number of ways of which the following is an example. The $l = 0$ phase shift $\delta_0(\varepsilon)$ for scattering in the potential $V_{Be}(r)$ satisfies the integral equation

$$\sin\delta_0(\varepsilon) = -\sqrt{\frac{2\pi}{k}}\int_0^\infty \sin kr\, V_{Be}(r)R_\varepsilon(r)\,dr, \tag{5.126a}$$

where

$$R_\varepsilon(r) \sim \sqrt{\frac{2}{\pi k}}\sin(kr + \delta_0(\varepsilon)). \tag{5.126b}$$

By substituting $N_i\bar{R}_i$ for $R_\varepsilon(r)$ in (5.126a) and setting $\varepsilon = \bar{\varepsilon}_i$ an equation for N_i is obtained.

The closure relation for the exact wave function $R_\varepsilon(r)$ has been given in (5.113). If the integral in (5.113) is evaluated numerically, choosing the mesh points at the pseudostate energies $\bar{\varepsilon}_i$ with appropriate weights W_i, we have

$$\int_0^\infty R_\varepsilon(r)R_\varepsilon(r')\,\mathrm{d}\varepsilon \approx \sum_i W_i R_{\bar{\varepsilon}_i}(r)R_{\bar{\varepsilon}_i}(r'). \tag{5.127}$$

The closure relation will be satisfied approximately (and exactly in the limit of an infinite basis) by the pseudostate functions so that

$$\int_0^\infty R_\varepsilon(r)R_\varepsilon(r')\,\mathrm{d}\varepsilon \approx \sum_i \bar{R}_i(r)\bar{R}_i(r') \tag{5.128}$$

and in view of (5.127) and (5.125), we can write

$$W_i = N_i^{-2}. \tag{5.129}$$

Since the normalization factors are calculable, the weights W_i can also be calculated from (5.129) in principle. The idea that the weights and mesh points of an integration rule for integration over the continuum are generated automatically by the diagonalization of a single particle Hamiltonian has been established rigorously for certain specific potentials (Stelbovics 1989; Stelbovics and Slim 1986; Broad 1978; Heller *et al.* 1973*a*, *b*; and Heller 1973), including the Coulomb (Yamani and Reinhardt 1975) and oscillator potentials. Specific numerical applications (see an example given by Bransden and Stelbovics 1984) suggest that even where no rigorous proof exists the procedure leads to a reliable approximation.

Now let us return to the expansion of an arbitrary function $f(r)$ given exactly by (5.111). If the integral over ε is represented numerically with mesh points at the energies $\bar{\varepsilon}_i$, we have

$$\begin{aligned}\int_0^\infty a(\varepsilon)R_\varepsilon(r)\,\mathrm{d}\varepsilon &\approx \sum_i W_i a(\bar{\varepsilon}_i)R_{\bar{\varepsilon}_i}(r)\\ &\approx \sum_i W_i N_i a(\bar{\varepsilon}_i)\bar{R}_i(r)\\ &= \sum_i \bar{a}_i\bar{R}_i(r)\end{aligned} \tag{5.130}$$

with $\bar{a}_i = N_i a(\bar{\varepsilon}_i)$. Although we have only discussed the case $l = 0$, it is obvious that the approximation of an integral over the continuum by a sum over pseudostates in (5.130) can be made for states of any l. It follows that to take account of the continuum in the expansion of the full wave function, we may extend the sum over the internal wavefunctions $\phi_i(\mathbf{r})$ and $\chi_n(\mathbf{x})$ in (5.84) to include a set of positive energy pseudostates $\bar{\phi}_i$ and $\bar{\chi}_n$ which are determined by diagonalizing H_B or H_A on appropriate finite basis sets. Since the pseudostate function sets diagonalize H_B or H_A, the equations for the amplitude $A_j(\mathbf{b}, t)$ uncouple in the asymptotic region $|t| \to \infty$, and the probabilities for channels representing the excitation or capture into discrete states are given by $P_j(b) = |A_j(b, \infty)|^2$ as usual.

Ionization

If j labels one of the positive energy pseudostates, the probability flux into this state represents ionization. Two components contributing to ionization can be distinguished, that arising from the expansion in atomic orbitals centred on B, called *direct ionization*, and that arising from the expansion centred on A, called *charge exchange into the continuum*. If we were able to expand in the exact continuum functions ϕ_j with $j \equiv \varepsilon, l, m$, the ionization cross section would have been

$$\sigma_{\text{ion}} = \sum_{lm} \int_0^\infty \int |A_{\varepsilon lm}(\mathbf{b}, \infty)|^2 \, d^2\mathbf{b} \, d\varepsilon. \tag{5.131}$$

Replacing the integration over ε by a numerical integration, with the mesh points at the $\bar{\varepsilon}_i$, we can replace $A_{\bar{\varepsilon}_i lm}$ by $N_i \bar{A}_i$, where $\bar{A}_i$ is the probability amplitude for populating the pseudostate of energy $\bar{\varepsilon}_i$. Thus

$$\begin{aligned} \sigma^{\mathrm{D}}_{\text{ion}} &= \sum_i W_i \int |N_i \bar{A}_i(\mathbf{b}, \infty)|^2 \, d^2\mathbf{b} \\ &= \sum_i \int |\bar{A}_i(\mathbf{b}, \infty)|^2 \, d^2\mathbf{b} \end{aligned} \tag{5.132}$$

where the relationship $W_i N_i^2 \approx 1$ has been employed. Thus the direct ionization cross section can be computed approximately by summing the cross sections for populating the positive energy pseudostates centred on B, treated in the same way as bound states. Likewise the cross section for charge exchange into the continuum can be obtained by summing the cross sections for populating the positive energy pseudostates centred on A.

This procedure for calculating the ionization cross section can be modified in various ways. In practice, the diagonalization of the internal Hamiltonians on a finite basis of Slater functions will provide a certain number of pseudostates of negative energy, in addition to a finite number of exact bound states and a number of pseudostates of positive energy. Since the pseudostates are not eigenfunctions of the Hamiltonian the positive energy pseudostates overlap to some extent those bound states not represented exactly in the finite basis, and conversely the negative energy pseudostates overlap to some extent the exact continuum functions. A procedure, which has sometime been used to allow for this situation, is to calculate the overlap of the pseudostates with the bound state part of the spectrum and to correct the approximate ionization cross section accordingly. It is doubtful whether cross sections corrected in this way are, in fact, more accurate than those calculated from (5.132); the size of the basis is the critical factor.

The division of the cross section into a 'direct ionization' component and a 'charge exchange into the continuum' component corresponds to some extent to the fact, verified experimentally, that some ejected electrons are correlated with the motion of the heavy particle B and some with A (see Chapter 10). Nevertheless it should be kept in mind that the representation of the continuum by pseudostates centred on A and B leading to an apparent division of the ionization cross section is an artefact of the coupled channel model. It should be noted that since the target eigenfunctions $\phi_j(\mathbf{r})$ form a complete set, a formal expression for the *complete* ionization cross section is given by (5.131).

Tests for pseudostate sets

Each different primitive basis set of Slater (or Gaussian) functions generates a different set of pseudostate functions and energies. Fortunately, except for ionization, the cross sections calculated appear not to be very sensitive to the particular choice made, two or three positive energy pseudostates for each important value of the angular momentum often being sufficient to obtain accurate excitation and charge exchange cross sections. Nevertheless some criteria for the acceptability of a given basis set is desirable. For example, the dipole polarizability α of the level i is defined as

$$\alpha = 2 \sum_{n \neq i} \frac{\langle \phi_i | z | \phi_n \rangle \langle \phi_n | z | \phi_i \rangle}{E_n - E_i}, \tag{5.133}$$

where the sum includes an integration over the continuum. The exact value of α for the initial state of the target can be compared with the value computed by replacing the sum over n by a sum over the pseudostate basis. If i denotes an s state, the usual case, then only intermediate p states contribute to the sum, so the test is only relevant to pseudostates with $l = 1$. It is not a very severe test even then. For instance in the case of a H(1s) target a basis consisting of the exact 2p and 3p hydrogenic orbitals together with one or two p-pseudostates reproduces the value of α accurately.

A much better test directly related to the scattering problem is based on calculating the second-order term in the Born series for excitation of the number of low-lying states of the target. The second-order terms involve a sum over a complete set of target eigenfunctions in the intermediate state. This sum can be performed accurately by numerical methods and the results compared with those calculated by employing the target pseudostate basis in the intermediate states. This test was carried out by Bransden and Dewangan (1979) for the case of the excitation of atomic hydrogen in the ground state to the 2s, $2p_0$, and $2p_{\pm 1}$ levels, using a number of different pseudostate sets. The calculations were performed within the impact parameter formalism, in which the second-order term in the Born series can be written as

$$B_{ji}(b, v_0) = \frac{1}{v_0^2} \sum_n \int_{-\infty}^{\infty} dz \int_{-\infty}^{z} \exp[i(\omega_{jn} z + \omega_{ni} z')] \times \langle \phi_j | V(z) | \phi_n \rangle \langle \phi_n | V(z') \phi_i \rangle \, dz' \quad (5.134)$$

where $\omega_{ij} = (\varepsilon_i - \varepsilon_j)/v_0$ and the sum over n includes an integration over the continuum. The exact contributions to B_{ji} were obtained numerically for intermediate states with $l = 0$, 1, and 2. Some results are shown in Table 5.1 for the 1s → 2s transitions at an impact energy of 105 keV. The pseudostate

Table 5.1. The second-order term B_{ji} in the Born series for excitation of H(1s) by proton impact at 105 keV. Values of $10^2 \times B_{ji}$ are given for certain impact parameters b. E denotes the exact result; P1 and P2 are calculated from the pseudostate sets.

b	E	P1	P2
(a) Intermediate s states only			
0.606	20.14 + 0.15i	19.78 + 0.50i	19.91 + 0.38i
2.106	1.72 − 0.18i	1.67 − 0.17i	1.67 − 0.18i
5.986	0.004 − 0.001i	0.004 − 0.00i	0.004 − 0.001i
(b) Intermediate p states only			
0.606	−3.94 + 4.01i	−3.90 + 3.77i	−3.52 + 2.55i
2.106	0.30 + 2.68i	0.28 + 2.80i	0.30 + 2.76i
5.986	0.84 + 0.42i	0.77 + 0.41i	0.80 + 0.39i
9.472	0.264 + 0.088i	0.240 + 0.125i	0.238 + 0.113i
(c) Intermediate d states only			
0.606	0.23 − 0.11i	0.64 − 0.45i	0.65 − 0.47i
2.106	0.182 + 0.106i	0.260 + 0.056i	0.283 + 0.057i
5.986	0.086 + 0.011i	0.170 + 0.005i	0.167 − 0.004i
9.472	0.019 + 0.006i	0.031 − 0.006i	0.028 − 0.005i

set results shown in the table are:

P1: a set of five s, four p, and two d orbitals given by Callaway and Wooten (1974). This set contains the exact hydrogenic wave functions for $n = 1$ and $n = 2$ and the highest pseudostate energies are 0.8536 a.u. for the s levels, 0.6206 a.u. for the p levels, and 0.000317 a.u. for the d levels.

P2: this is an extension of the set P1 and was given by Callaway *et al.* (1975). It contains seven s, five p, and three d orbitals among which are the exact $n = 1$, 2, and 3 hydrogenic wave functions. The highest energies generated are 1.02 a.u. for s levels, 0.451 a.u. for p levels, and 0.021 a.u. for d levels.

Considering how small the pseudostate basis sets are, the agreement between the exact results and those obtained from the pseudostates is very good, the least accurate case being for the d intermediate state contribution to the 1s → 2s transition.

An important test of a different nature was introduced by Cheshire *et al.* (1970). As we have seen, one of the reasons for including continuum contributions in the expansion of the wave function in terms of the atomic orbitals $\phi_j(\mathbf{r})$ and $\chi_n(\mathbf{x})$, is so that at small interionic separations the wave function approximates to the united atom limit of the appropriate molecular orbital. Accordingly Cheshire *et al.* suggested that the choice of pseudostate sets should be made so that the maximum overlap between the basis functions and the united atom functions is obtained. Their work was concerned with the symmetrical reaction between protons and atomic hydrogen

$$\begin{aligned} \mathrm{H}^+ + \mathrm{H}(1\mathrm{s}) &\rightarrow \mathrm{H}^+ + \mathrm{H}(nl) \\ &\rightarrow \mathrm{H}(n'l') + \mathrm{H}'. \end{aligned} \tag{5.135}$$

In order to include the most important excitation and charge exchange channels the exact 1s, 2s, and 2p hydrogenic orbitals on each centre were included in the basis, together with two pseudostates, which can be designated $3\bar{\mathrm{s}}$ and $3\bar{\mathrm{p}}$, again on each centre. The radial pseudostate functions $R_{3\bar{\mathrm{s}}}$, $R_{3\bar{\mathrm{p}}}$ were chosen to have the simple forms

$$\begin{aligned} R_{3\bar{\mathrm{s}}}(r) &= N_{3\bar{\mathrm{s}}}(1 - \tfrac{16}{15}r + \tfrac{1}{5}r^2)\exp(-r) \\ R_{3\bar{\mathrm{p}}}(r) &= N_{3\bar{\mathrm{p}}}(1 - \tfrac{1}{2}r)\exp\left(-\frac{r}{2}\right) \end{aligned} \tag{5.136}$$

where $N_{3\bar{\mathrm{s}}}$, $N_{3\bar{\mathrm{p}}}$ are normalization factors. The exponents were arbitrarily chosen to be those occurring in the exact $n = 1$ and $n = 2$ bound states and the linear coefficients are determined so that these are the simplest functions orthogonal to those bound states. The overlap of the basis with the He^+

Table 5.2. Values of the overlap $S^N(n'l)$ between a hydrogenic pseudostate basis and the $He^+(n'l)$ wave functions.

$N/n'l$	1s	2s	2p
1	0.702	0.250	–
2	0.744	0.562	0.555
3	0.992	0.915	0.863

united atom wave functions $\psi_{n'l'}(He^+)$ is measured by the sums

$$S^N(n'l') = \sum_{n=1}^{N} |\langle \psi_{n'l}(He^+)|\phi_{nl}\rangle|^2. \tag{5.137}$$

As noted by Gallaher and Wilets (1968), for the He^+ ground state ($n'l = 1s$) the 1s hydrogen ground state ϕ_{1s} contributes $S^1(1s) = 0.7$. However if the sum is taken over all hydrogen bound states this is only increased to 0.76. It follows that the exact contribution of the continuum is 0.24, four times greater than the total from all the excited bound states. With the basis of Cheshire *et al.* the overlaps are shown in Table 5.2. It is seen that with the $n = 1$ and $n = 2$ bound states the overlap with the $\psi_{1s}(He^+)$ functions is only 0.744, but this is increased to 0.992 with the addition of a simple $3\bar{s}$ pseudostate, while large increases in overlap are also shown for the $n'l = 2s$ and 2p functions. This suggests that with pseudostate sets of modest size a good representation of the united atom wave functions can be obtained.

5.6 THE WAVE FUNCTION AT SMALL INTERNUCLEAR SEPARATIONS – THE UNITED ATOM LIMIT

Since the work of Cheshire *et al.* discussed in the last paragraph, several methods have been developed for ensuring that the wave function at small internuclear separations is well represented. These include the *AO^+ method* of Fritsch and Lin (1982*a*), the *triple centre expansion* of Anderson *et al.* (1974) and the *variable charge approximation* of Cheshire (1968). We shall consider each of these in turn.

The AO^+ pseudostate expansion

Fritsch and Lin (1982a) have taken the considerations of Cheshire *et al.* a stage further. A basis is formed from a number of bound state wave functions of the target atom ($B + e^-$) together with those wave functions of the united atom that correlation diagrams suggest should be important. Using this primitive basis, the internal Hamiltonian of the target is diagonalized as in (5.122), thus producing an orthonormal set of functions. The same procedure is carried out for the rearranged atomic system ($A + e^-$). The two-centre basis formed in this way has been termed by Fritsch and Lin and AO^+ basis, and it has proved very effective in practice (see Chapter 6). Although the AO^+ basis provides a representation of the continuum and of the united atom functions which does not differ greatly from using a primitive Slater (or Gaussian) basis of similar size, it has the great merit of providing a definite rule for the choice of the pseudostate parameters.

The triple centre expansions

Another method of introducing united atom wave functions explicitly into the basis is to expand the wave function about three centres – the heavy particle centres A and B and a third centre C. The centre C can be taken to be the centre of charge of the heavy particles, but a better choice, to be discussed shortly, may be the equiforce point so that $AC/CB = (Z_A/Z_B)^{\frac{1}{2}}$. The basis functions placed on the centre of charge can be taken to be the united atom wave functions, modified by the appropriate translational factor (Anderson *et al.* 1974, Lin *et al.* 1982). Thus the expansion of the wave function becomes

$$\psi(\mathbf{s}, t) = \sum_{j=1}^{N} \phi_j(\mathbf{r}) \exp[i\alpha_j(t)] A_j(\mathbf{b}, t) + \sum_{n=N+1}^{N+M} \chi_n(\mathbf{x}) \exp[i\beta_n(t)] A_n(\mathbf{b}, t)]$$

$$+ \sum_{m=N+M}^{I} \Phi_m(\mathbf{y}) \exp[i\delta_m(t)] A_m(\mathbf{b}, t) \tag{5.138}$$

where $\mathbf{y}$ is the position vector of the electron relative to the centre C, Φ_m are united atom orbitals, and $\delta_m(t)$ is defined analogously to α_j and β_n. The amplitudes of the basis functions placed on the centre of charge, $A_m(\mathbf{b}, t)$, $m > N + M$, do not vanish in the limit $t \to \infty$; however the probability flux entering these states can be attributed as a contribution to the ionization probability and indeed, the extra flexibility gained from these terms in addition to pseudostate terms placed on A and B allows accurate ionization cross sections to be calculated from basis sets of modest size. The choice of C as the equiforce point is suggested by the idea that, at low energies, ionization occurs in the vicinity of the equiforce point as well as in the vicinity of the two centres A and B (Winter and Lin 1984; Winter 1988) and some experimental evidence exists for this (Meckbach *et al.* 1986). The disadvantage of the triple centre expansion is that the interaction matrix elements become more complex and time consuming to compute, and it is doubtful if the method offers any real advantage over a two-centre expansion with a sufficiently large pseudostate or AO^+ basis.

The variable charge model

Yet another approach to improving the two-centre expansion in atomic orbitals was put forward by Russek (1966), in connection with the scattering of protons by atomic hydrogen. He suggested, in place of hydrogen radial wave functions centred on A and B of the form $R^{\mathrm{B}}_{nl}(Z, r)$ or $R^{\mathrm{A}}_{nl}(Z, x)$ where Z is the nuclear charge, in this case $Z = 1$, functions of the same form should be employed but with a variable charge, $Z_j(t)$, depending on the time and also on the channel index j. This idea was put into practice by Cheshire (1968) (see also Storm and Rapp (1976)) who used the variational principle to determine the optimum value of $Z_n(t)$. If each charge $Z_j(t)$ is varied, $Z_j(t) \to Z_j(t) + \delta Z_j(t)$ subject to the condition $\delta Z_j(t) \to 0$ as $|t| \to \infty$, the Euler–Lagrangian equations become

$$\frac{\partial I}{\partial Z_j} - \frac{\mathrm{d}}{\mathrm{d}t}\left(\frac{\partial I}{\partial \dot{Z}_j}\right) = 0, \qquad j = 1, 2, \ldots, \tag{5.139}$$

where I is the variational functional (5.44). These equations for $Z_j(t)$ have to be solved simultaneously with the coupled equations for the amplitudes, which now contain extra matrix elements because the time derivative of the basis functions produces additional terms of the form $\dot{Z}_j(t)\, \partial\phi^{\mathrm{A}}_j(\mathbf{r})/\partial Z_j$ and $\dot{Z}_j\, \partial\chi^{\mathrm{B}}_j(\mathbf{x})/\partial Z_j$. To solve the complete set of coupled equations (5.139) together with the modified equations for the amplitudes is too complex a procedure and Cheshire simplified the problem by determining each $Z_j(t)$ from a one-channel approximation. These calculated values are then used in the coupled equations for the amplitudes modified by the extra terms in $\dot{Z}_j$.

The effective charges depend both on the incident energy and the impact parameter. In the limit $E_{\mathrm{L}} = 0$ the corresponding wave function is an approximation to an adiabatic molecular orbital, but for finite values of the energy additional dynamical effects are represented. In Fig. 5.4 the effective charges for the two lowest levels of the $\mathrm{H}^+ + \mathrm{H}$ system are shown for particular values of the impact parameter corresponding to Coulomb scattering by the internuclear potential through a laboratory scattering angle of 3°. As we have seen the wave functions of the homonuclear system $\mathrm{H}^+ + \mathrm{H}$ divide into even and odd parity eigenstates, and with plane-wave translation factors the lowest basis functions of each parity are

$$\Phi^{\pm} = [\phi_{1\mathrm{s}}(Z^{\pm}, r)\, \mathrm{e}^{-\frac{1}{2}\mathrm{i}\mathbf{v}_0\cdot\mathbf{s}_g} \pm \phi_{1\mathrm{s}}(Z^{\pm}, x)\, \mathrm{e}^{\frac{1}{2}\mathrm{i}\mathbf{v}_0\cdot\mathbf{s}_g}]\, \mathrm{e}^{-\mathrm{i}(\varepsilon_{1\mathrm{s}} + v_0^2/8)t} \tag{5.140a}$$

where

$$\phi_{1\mathrm{s}}(Z^{\pm}, r) = \left(\frac{Z^{\pm}}{\pi}\right)^{\frac{1}{2}} \mathrm{e}^{-Z^{\pm}r}. \tag{5.140b}$$

In the asymptotic region $Z^{\pm} \to 1.0$ and $\phi_{1\mathrm{s}}(Z^{\pm}, r)$ becomes equal to the ground-state wave function of atomic hydrogen. In the case of the even parity

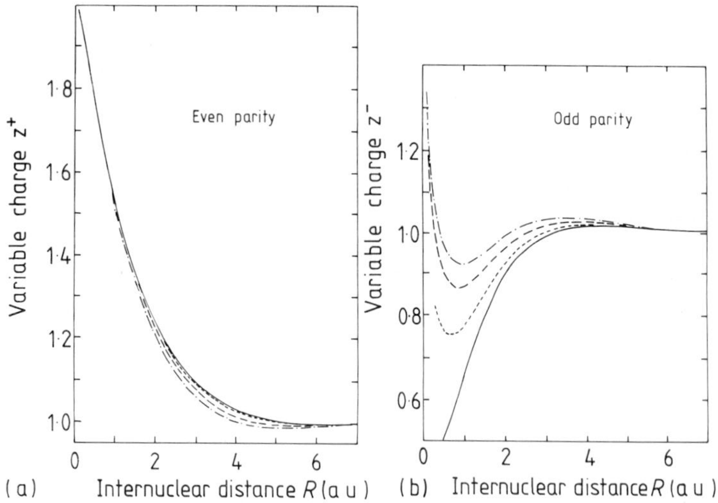

Fig. 5.4. The effective charge for the lowest even and odd parity state of $H^+ + H$: ——— $E_L = 0$; ----- $E_L = 2$ keV/a.m.u.; ——— $E_L = 7$ keV/a.m.u.; —·—·— $E_L = 13$ keV/a.m.u. The impact parameter is that corresponding to a scattering angle of 3° in the laboratory system (from McCarroll *et al.* (1970)).

(gerade) function the effective charge Z^+ at all the energies considered ($E_L < 13$ keV) is close to that for $E_L = 0$. However the energy effect is very large for the odd parity (ungerade) function. The low-energy capture differential and total cross sections for the $H^+ + H$ system are very well reproduced by the variable charge method if the 1s, 2s, and 2p hydrogenic functions are retained in the basis (see Chapter 6, particularly Fig. 6.1), but no extensive applications to other systems have been reported.

5.7 THE MIXED AO/MO EXPANSION

As we shall see in Chapter 6, the main methods we have described – the molecular orbital expansion and the atomic orbital expansion including pseudostates of various kinds – are all capable of providing accurate ion–atom cross sections, provided due allowance is made for the translational factors and the basis is adequate in size for the particular energy and system under consideration. As pointed out by Bates and Lynn (1959), if translational invariance is neglected, as in the perturbed stationary state model, the difference between calculations employing a molecular basis and those employing an atomic orbital basis can be analysed. Indeed, if a primitive basis of Slater functions is taken on each centre, we can either diagonalize

H_e to form molecular orbitals or diagonalize H_A and H_B, the internal Hamiltonians, to form atomic orbitals. Clearly the basis sets formed only differ by a unitary transformation and the computed cross sections from the coupled equations with the molecular orbital and atomic orbital bases will be identical. The difference between the molecular and atomic orbital approaches are therefore not very fundamental and reside in the use of different translation factors.

As we have seen when atomic orbital bases are used there is general agreement that plane-wave (Bates–McCarroll) factors are optimal; but no simple prescription exists for optimal factors for molecular orbitals, the optimization schemes based on the variational or minimum norm methods being in general rather complex. With this in mind the AO/MO mixed expansion has been introduced (Kimura and Lin 1985*a*, *b*, 1986; Winter and Lane 1985; Toepfer *et al.* 1987) in which, in an outer region $R > R_0$, an atomic orbital basis with plane-wave translational factors is employed, and in an inner region $R < R_0$ the expansion is in terms of molecular orbitals, either unmodified by translational factors, or, better, with translational factors depending on switching functions of suitable form. In this way the boundary conditions for large R are satisfied in each channel, while at small internuclear separations the molecular character of the wave function is represented. The calculation proceeds by integrating the coupled equations in the travelling atomic orbital basis from some initial time $t = -|t_0|$, where $|t_0|$ is large enough for the system to be represented by the initial unperturbed atomic state, to some time $-t_1$ at which $R(-t_1) = R_0$. At $t = -|t_1|$, the wave function has to be re-expanded (approximately) in terms of the molecular orbital basis. The coefficients of this expansion provide the boundary condition for the coupled equations in the molecular orbital basis, which are then integrated to the time $t = t_1$. At $t = t_1$, the wave function is re-expressed in the atomic orbital basis, and the equations integrated to $t = t_0$ to obtain the asymptotic amplitudes.

The success of the method can be judged by how far the calculated probabilities are independent of the position of the matching points at $R = R_0$. Since the basis sets are finite some loss of accuracy must occur at each of the two matching procedures, but nevertheless the method produces cross sections of useful accuracy, although it is not clear that the results are markedly superior to those obtained by other methods.

5.8 STURMIAN, GAUSSIAN, AND POLYNOMIAL EXPANSIONS

We now turn to methods in which the basic idea is to solve the impact parameter equation (5.1) by expansions into sets of functions which are mathematically convenient rather than functions dictated by the physical boundary conditions.

The first method we shall discuss has been developed specifically for the true one-electron problem in which A and B are bare nuclei and $V_{\mathrm{Ae}}(x)$, $V_{\mathrm{Be}}(r)$ are the Coulomb potentials

$$V_{\mathrm{Ae}}(x) = \frac{-Z_{\mathrm{A}}}{x}; \qquad V_{\mathrm{Be}}(r) = \frac{-Z_{\mathrm{B}}}{r}. \tag{5.141}$$

A two-centred expansion of the wavefunction ψ is employed of the form (5.2) where the basis functions $\Phi_j(\mathbf{s}, t)$ and $X_n(\mathbf{s}, t)$ are written in terms of travelling orbitals where (cf. (5.83))

$$\begin{aligned} \Phi_j(\mathbf{s}, t) &= \bar{\phi}_j(\mathbf{r}) \exp[\mathrm{i}\alpha_j(t)] \\ X_k(\mathbf{s}, t) &= \bar{\chi}_k(\mathbf{x}) \exp[\mathrm{i}\beta_k(t)]. \end{aligned} \tag{5.142}$$

The functions $\bar{\phi}_j(\mathbf{r})$ and $\bar{\chi}_k(\mathbf{x})$ are expressed as products of eigenfunctions of angular momentum, and radial functions $S_{nl}(r)$:

$$\begin{aligned} \bar{\phi}_j(\mathbf{r}) &= r^{-1} S_{n_j l_j}(r) Y_{l_j m_j}(\theta_r, \phi_r) \\ \bar{\chi}_k(\mathbf{x}) &= x^{-1} S_{n_k l_k}(x) Y_{l_k m_j}(\theta_x, \phi_x). \end{aligned} \tag{5.143}$$

At this point, the treatment differs from the usual expansion in eigenfunctions of H_{A} and H_{B}, in that the functions $S_{n_j l_j}(r)$ are chosen to be solutions, regular at the origin, of the equation

$$\left(-\frac{1}{2}\frac{\mathrm{d}^2}{\mathrm{d}r^2} + \frac{l(l+1)}{2r^2} - \frac{\alpha_{nl}}{r}\right) S_{nl}(r) = E_l S_{nl}(r). \tag{5.144a}$$

For a given value of the parameter E_l, regular solutions vanishing at $r \to \infty$ can only be obtained for discrete values of α_{nl}. The functions defined in this way are called Sturmian functions (Rotenberg 1962) and form an infinite discrete complete set. The Sturmian functions can be normalized and satisfy the orthogonality conditions

$$\int_0^\infty S_{nl}(r) S_{n'l}(r) r^{-1}\, \mathrm{d}r = \frac{\alpha_{nl}}{n^2}\, \delta_{nn'}. \tag{5.144b}$$

The analytic expressions for S_{nl} are scaled hydrogenic radial functions

$$S_{nl}(r) = \alpha_{nl}^{\frac{1}{2}} r^{-1} R_{nl}(\alpha_{nl} r). \tag{5.145}$$

If the target atom is hydrogen, $Z_B = 1$, the choice

$$E_l = -\frac{1}{2(l+1)} \tag{5.146}$$

ensures that the lowest Sturmian function for each l (that is the Sturmian for 1s, 2p, 3d, ...) coincides with the corresponding hydrogenic function, but all other Sturmians overlap both with the hydrogenic discrete levels and the continuum. Since the Sturmian set is complete and discrete, taking a sufficient number of terms in the basis should ensure that the hydrogenic continuum can be represented in a systematic way. With the choice (5.146) the boundary condition in the incident channel for scattering of a bare nucleus by a hydrogen atom in the ground state can be satisfied exactly. However since the Sturmian functions are not eigenfunctions of H_A and H_B, except for the lowest member of the set of each l, the solution has to be re-expanded in terms of hydrogenic eigenfunctions in order to calculate the physical amplitudes. A further disadvantage is that the representation of the long-range coupling between degenerate hydrogenic levels such as 2s and 2p require a rather large Sturmian basis, and in general the rate of convergence is slow. The Sturmian expansion was introduced to discuss the $H^+ + H$ system by Gallaher and Wilets (1968) and has also been employed by Shakeshaft (1975*a*, 1978*b*) and Winter (1982, 1986, 1987, 1991).

Gaussian expansions

The use of pseudostates, which are eigenfunctions of finite basis representations of H_A and H_B in terms of Slater orbitals or Sturmian functions does not, of course, exhaust the possibilities of a two-centred expansion of the general form expressed by (5.141) and (5.142). Another promising possibility is to replace the radial functions S_{nl} in the expressions for $\bar{\phi}_j(\mathbf{r})$ and $\bar{\chi}_k(\mathbf{x})$ by linear combinations of Gaussian functions. For example, $S_{nl}(r)$ could be replaced by $\bar{S}_{jl}(r)$ where

$$\bar{S}_{nl}(r) = r^{l+1} \sum_{\lambda} C_{n\lambda} \exp(-\eta_{n\lambda} r^2). \qquad (5.147)$$

The coefficients $C_{n\lambda}$ and parameter $\eta_{n\lambda}$ can be chosen in various ways, for example a pseudostate basis can be formed by diagonalizing H_A (and H_B) on this set of Gaussians rather than from Slater functions. The motivation for the use of a Gaussian basis is that, with plane-wave translational factors, all the elements of the overlap and interaction matrices can be evaluated analytically (Dose and Semini 1974; Errea *et al.* 1979). In contrast, the evaluation of matrix elements with a Slater basis requires one or two numerical integrations (Appendix E). Efficient methods developed in quantum chemistry for the calculation of molecular orbitals using Gaussian basis sets can be extended without difficulty to the Gaussian functions modified by plane-wave translational factors (see Appendix F). While the calculation of matrix elements, which is a major time consuming factor in all expansion methods, is faster with Gaussian functions, the disadvantage is that to represent the exact eigenfunctions of H_A and H_B, which is necessary in the asymptotic region to define the physical probability amplitude, the Gaussian basis must be fairly large. For example, to represent the H(1s) wave function, which is a single Slater orbital, at least three or four Gaussians are required. This disadvantage decreases when large basis sets are employed and the method has distinct advantages for application to systems of several electrons.

Orthogonal polynomials

One of the disadvantages of the various methods we have been considering is that, apart from the case of a Sturmian basis, it is difficult to devise a completely systematic method of increasing the size of the basis. This can be overcome by expanding the wave function $\psi(\mathbf{s}, t)$ at each value of the time t, in a complete set of orthogonal polynomials, chosen for convenience, and increasing the number of terms until convergence is obtained. A rather general method allowing for the use of different types of polynomial has been formulated and applied by Morrison and Öpik (1978, 1984) and a similar but somewhat simpler procedure has been described by Lüdde and Dreizler (1981, 1982*a*, *b*). Lüdde and Dreizler work in the body-fixed (rotating) frame of coordinates where the position vector of the electron $\bar{\mathbf{s}}_g(x, y, z)$ is with respect to the mid-point of the interionic line AB as origin. The wave function $\bar{\psi}(\bar{\mathbf{s}}_g, t)$ is expanded in a basis $\chi_\alpha(\bar{\mathbf{s}}_g, R(t))$:

$$\bar{\psi}(\bar{\mathbf{s}}_g, t) = \sum_\alpha C_\alpha(t)\chi_\alpha(\bar{\mathbf{s}}_g, R(t)). \tag{5.148}$$

The amplitudes $C_\alpha(t)$ satisfy a set of coupled first-order differential equations similar in form to (5.2):

$$\mathrm{i}\dot{\mathbf{C}}(t) = \tilde{N}^{-1}\tilde{\mathbf{Q}}\mathbf{C}(t) \tag{5.149}$$

which are solved by integration along the trajectory $\mathbf{R} = \mathbf{R}(t)$. Here, the overlap matrix $\tilde{\mathbf{N}}$ and the interaction matrix $\tilde{\mathbf{Q}}$ are defined by the equations

$$\tilde{N}_{\alpha\beta}(t) = \langle \chi_\alpha(\bar{\mathbf{s}}_g, t) \,|\, \chi_\beta(\bar{\mathbf{s}}_g, t)\rangle$$

and

$$\tilde{Q}_{\alpha\beta}(t) = \langle \chi_\alpha(\bar{\mathbf{s}}_g, t) \,|\, H_e - \mathrm{i}\dot{R}(t)\frac{\partial}{\partial R} - \dot{\Theta}(t)L_j \,|\, \chi_\beta(\bar{\mathbf{s}}_g, t)\rangle \tag{5.150}$$

where, as the rotating coordinate system is employed, the relationship (5.8) has been invoked.

The basis functions $\chi_\alpha(\bar{\mathbf{s}}, t)$ were chosen to be those introduced by Hylleraas (1931) for bound systems. These are expressed in prolate spheroidal coordinates ξ, η, ϕ where

$$\begin{aligned} x &= R(1 - \eta^2)(\xi^2 - 1) \cos\phi \\ y &= R(1 - \eta^2)(\xi^2 - 1)^{\frac{1}{2}} \sin\phi \\ z &= R\xi\eta \end{aligned} \tag{5.151}$$

and have the form

$$\chi_\alpha(\bar{\mathbf{s}}, t) = R^{-\frac{3}{2}} \exp\left(-\frac{K}{2}\right)(\xi^2 - 1)^{m/2} L_n^M(K) P_l^m(\eta) \cos m\phi \qquad (5.152)$$

with $K = (\xi - 1)/a$. The functions L_n^m and P_l^m are associated Laguerre and Legendre functions, respectively, and a is a parameter which is chosen to provide the most rapid convergence for given values of the interionic separation.

The expansion (5.148) is employed for values of t, $-|t_0| \leq t \leq t_0$ where $|t_0|$ is large enough, so that for $|t| > |t_0|$, the system is in the asymptotic region in which there is no coupling between the physical channels. The initial conditions at $t = -|t_0|$ are obtained by expanding the travelling atomic orbital representing the incident channel in terms of the Hylleraas basis to obtain the expansion coefficients $C_\alpha(-|t_0|)$. Correspondingly at $t = t_0$, the wave function is projected on a basis of travelling atomic orbitals, centred about A and B to obtain the physical probability amplitudes.

Lüdde and Dreizler investigated the scattering of a number of fully stripped ions (H^+, He^{2+}, Li^{3+}, Be, B^{5+}) by atomic hydrogen in the ground state at energies up to 50 keV. They found that rather large basis sets of up to $\sim$100 terms were required for convergence, but numerically the procedure is rather efficient and competitive with other methods.

5.9 SYSTEMS WITH SEVERAL ELECTRONS

As remarked in Section 4.7, there is no difficulty in principle in applying coupled channel models to several- (or many-) electron systems, although if a large basis set is required for convergence the computational effort required increases rapidly with increasing numbers of electrons. As in the one-electron case a basis of adiabatic or diabatic molecular functions can be employed; alternatively a set of many-electron atomic states and pseudostates can be constructed. In either, suitable translation factors must be introduced. As we have seen, with a molecular basis there is no unique way of choosing the translational factors. A generalization of state-dependent phase functions π_j^A, π_j^B given by (5.27) for a single electron has been given by Kimura *et al.* (1983) for a system of q electrons, in the form

$$\pi_j^A(\mathbf{s}_1, \mathbf{s}_2, \ldots, t) = \sum_{k=1}^{q} (\mathbf{v}_0 \cdot \mathbf{s}_k) f_j(\mathbf{s}_k; R) \qquad (5.153)$$

where the switching function f_j satisfies (5.28), and $\mathbf{s}_k$ is the position vector of the kth electron with respect to the mid-point of the internuclear line. The electronic wave function is then expanded in the form (assuming a straight line trajectory with constant velocity v_0)

$$\psi(\mathbf{s}_1, \mathbf{s}_2, \ldots, t) = \sum_{j=1}^{N+M} A_j(\mathbf{b}, t)\psi_j(\mathbf{s}_1, \mathbf{s}_2, \ldots, \mathbf{R}(t))F_j \tag{5.154}$$

with

$$F_j = \exp\left[-\mathrm{i}\sum_{k=1}^{q}\left\{(\mathbf{v}_0\cdot\mathbf{s}_k)\frac{f_j(\mathbf{s}_k;\mathbf{R})}{2} + \frac{v_0^2 t}{8}\right\} - \mathrm{i}\varepsilon_j t\right] \tag{5.155}$$

and where $\psi_i(\mathbf{s}_1, \mathbf{s}_2, \ldots, \mathbf{R}(t))$ denotes an adiabatic molecular wave function. The switching functions f_j can be chosen by the same procedures as for the one active electron case. In particular a common translation factor can be employed by using forms such as (5.41) or (5.42), or a minimum coupling method can be used to optimize a parameterized state-dependent switching factor. With an atomic basis plane-wave translational factors can be used as in the one-electron case. As an example let us consider electron capture by protons from helium initially in the ground state. The wave function is in the singlet spin state ($S = 0$) and is therefore symmetrical in the spatial coordinates of the two electrons. The complete wave function $\psi(\mathbf{s}_1, \mathbf{s}_2; t)$ can be expanded as

$$\begin{aligned}\psi(\mathbf{s}_1, \mathbf{s}_2, t) = {} & \sum_{n=1}^{N} A_n(\mathbf{b}, t)\Phi_n(\mathbf{r}_1, \mathbf{r}_2)\,\mathrm{e}^{\mathrm{i}\gamma_n^{\mathrm{B}}(\mathbf{s}_1,\mathbf{s}_2)} \\ & + \frac{1}{\sqrt{2}}\sum_{i=1}^{I}\sum_{j=1}^{J} A_{ij}(\mathbf{b}, t)[\phi_i(\mathbf{r}_1)\chi_j(\mathbf{x}_2)\,\mathrm{e}^{\mathrm{i}\gamma_{ij}^{\mathrm{B}}(\mathbf{s}_1,\mathbf{s}_2)} \\ & + \phi_i(\mathbf{r}_2)\chi_j(\mathbf{x}_1)\,\mathrm{e}^{\mathrm{i}\gamma_{ij}^{\mathrm{B}}(\mathbf{s}_2,\mathbf{s}_1)}].\end{aligned} \tag{5.156}$$

The first set of terms represents the arrangement in which both electrons are attached to the helium nucleus B. The functions $\Phi_n(\mathbf{r}_1, \mathbf{r}_2)$ are the symmetrical ground state and excited state wave functions of helium, both singly and doubly excited, and the set may include pseudostates. The phase factor γ_n^{B} is then

$$\gamma_n^{\mathrm{B}} = -\varepsilon_n t - \frac{v_0^2 t}{4} - \frac{\mathbf{v}_0\cdot(\mathbf{s}_1 + \mathbf{s}_2)}{2}. \tag{5.157}$$

The second set of terms represents one electron attached to the helium nucleus in a state with wave function ϕ_i and the other attached to the proton in a state with wave function χ_j. The phase factor $\gamma_{ij}(\mathbf{s}_1, \mathbf{s}_2)$ is then

$$\gamma_{ij}(\mathbf{s}_1, \mathbf{s}_2) = -(\varepsilon_i + \eta_j)t - \frac{v_0^2 t}{4} - \frac{\mathbf{v}_0\cdot(\mathbf{s}_1 - \mathbf{s}_2)}{2}. \tag{5.158}$$

A further set of terms representing double electron capture into the ground and doubly excited states of H^- could be added in an obvious way.

Computationally the matrix elements of the electron–electron interaction are rather time consuming, particularly electron exchange terms of the type

$$\left\langle \phi_i(\mathbf{r}_1)\chi_j(\mathbf{x}_2) \left| \frac{1}{|\mathbf{s}_1 - \mathbf{s}_2|} \exp\{\mathrm{i}[\gamma_{ij}^{\mathrm{B}}(\mathbf{s}_2, \mathbf{s}_1) - \gamma_{ij}^{\mathrm{B}}(\mathbf{s}_1, \mathbf{s}_2)]\} \right| \phi_i(\mathbf{r}_2)\chi_j(\mathbf{x}_1) \right\rangle \quad (5.159)$$

and this presents a limitation on the calculation of cross sections in several-electron systems. This can, in part, be overcome by representing the atomic wave functions by combinations of Gaussian rather than Slater functions; the matrix elements can then be reduced analytically (Dose and Semini 1974; Errea *et al.* 1979; Gramlich *et al.* 1989).

5.10 DIRECT SOLUTION OF THE TIME-DEPENDENT SCHRÖDINGER EQUATION

For a single-electron system it is possible to attempt a direct numerical solution of the time-dependent Schrödinger equation (5.1). If an exact solution is obtained it can serve as a bench-mark against which approximate solutions are tested. In addition, numerical solutions can illustrate new features of the collision mechanism. In particular the way in which the distribution of the electron density evolves during the collision can be studied. To obtain a numerical solution both finite-difference techniques (Maruhn-Rezwani *et al.* 1979; Grün *et al.* 1982; Grün and Scheid 1983; Horbatsch 1984) and finite-element methods (Bottcher 1982) have been employed.

In the approach of Grün *et al.*, equation (5.1) is written in the rotating body-fixed coordinate system. If the wave function is expressed in terms of $\bar{\mathbf{s}}$, the electronic position vector in the body-fixed frame with the centre of mass A and B as origin and the time t, we have, following the argument leading to (5.8),

$$\left.\frac{\partial}{\partial t}\right|_{\mathbf{s}} \bar{\psi}(\bar{\mathbf{s}}, t) = \left[\frac{\partial}{\partial t} - \mathrm{i}\dot{\Theta}(t)L_y\right]\bar{\psi}(\bar{\mathbf{s}}, t), \quad (5.160)$$

where, as before, $\Theta(t)$ is the angle that the internuclear line $\mathbf{R}$ makes with the space-fixed Z-axis, $\dot{\Theta}(t)$ being given in terms of $R(t)$ by (5.105). If a cylindrical system of coordinates (ρ, ϕ, z') is introduced with the Z'-axis along the internuclear line and

$$\rho = (x'^2 + y'^2), \qquad \phi = \tan^{-1}\left(\frac{y'}{x'}\right), \quad (5.161)$$

the Schrödinger equation becomes

$$i\frac{\partial}{\partial t}\bar{\psi}(\rho, \phi, z', t) = (-\tfrac{1}{2}\nabla^2 + V(\rho, z', t) - \dot{\Theta}L_y)\bar{\psi}(\rho, \phi, z', t). \quad (5.162)$$

In the cylindrical system the explicit forms of ∇^2, L_y, and V are:

$$\nabla^2 = \frac{\partial^2}{\partial\rho^2} + \frac{1}{\rho}\frac{\partial}{\partial\rho} + \frac{\partial^2}{\partial z'^2} + \frac{1}{\rho^2}\frac{\partial^2}{\partial\phi^2} \quad (5.163)$$

$$L_y = -i\left[z'\frac{\partial}{\partial\rho} - \rho\frac{\partial}{\partial z'}\right]\cos\phi + i\frac{z'}{\rho}\sin\phi\frac{\partial}{\partial\phi} \quad (5.164)$$

$$V(\rho, z', t) = -[\rho^2 + \{z' + \tfrac{1}{2}R(t)\}^2]^{-\frac{1}{2}} - [\rho^2 + \{z' - \tfrac{1}{2}R(t)\}^2]^{-\frac{1}{2}}. \quad (5.165)$$

The three-dimensional partial differential equation in ρ, ϕ, z' (for a given value of t) can be reduced to coupled two-dimensional equations in ρ and z', by expanding

$$\bar{\psi}(\rho, \phi, z', t) = \sum_m f_m(\rho, z', t)\exp(im\phi) \quad (5.166)$$

where, because of the symmetry under reflections in the plane of scattering,

$$f_m(\rho, z', t) = f_{-m}(\rho, z', t). \quad (5.167)$$

Using the orthogonality property of the function $\exp(im\phi)$,

$$\int_0^{2\pi}\exp[i(m - m')\phi] = 2\pi\delta_{mm'} \quad (5.168)$$

we obtain

$$\begin{aligned} i\frac{\partial}{\partial t}f_m(\rho, z', t) = {} & -\frac{1}{2}\left(\frac{\partial^2}{\partial\rho^2} + \frac{1}{\rho}\frac{\partial}{\partial\rho} + \frac{\partial^2}{\partial z'^2} - \frac{m^2}{\rho^2}\right)f_m(\rho, z', t) \\ & + V(\rho, z', t)f_m(\rho, z', t) \\ & + \tfrac{1}{2}i\dot{\Theta}(t)\left\{\left[z'\frac{\partial}{\partial\rho} - \rho\frac{\partial}{\partial z'} - z'\frac{(m-1)}{\rho}\right]f_{m-1}(\rho, z', t)\right. \\ & \left. + \left[z'\frac{\partial}{\partial\rho} - \rho\frac{\partial}{\partial z'} + z'\frac{(m+1)}{\rho}\right]f_{m+1}(\rho, z', t)\right\}. \end{aligned} \quad (5.169)$$

The calculation proceeds by replacing the derivatives with respect to ρ and z' by finite-difference expressions, obtaining finite-dimensional matrix equations. Finally, a numerical algorithm is employed to integrate the equations in small time steps Δt, starting from an initial condition in which the wave function coincides with the unperturbed wave function for the target atom with the correct plane-wave translational factor. The success of this method depends on the fact that the expansion in m converges rapidly at low energies, so that only two or three values of m need be considered.

In the initial configuration, the electronic probability density is concentrated about the target nucleus B, which is situated in the half-plane with $z' > 0$. During the collision the probability density becomes redistributed about both the centres A and B. The probability for charge exchange is then given by

$$P^c(\mathbf{b}) = 2\pi \sum_m \int \rho \, d\rho \int_{z' \leq 0} |f_m(\rho, z', T)|^2 \, dz' \tag{5.170}$$

where T is a time taken large enough for convergence. This particular method was tested for the H^+–H(1s) system at $E_L = 2$ keV, using a Coulomb trajectory and retaining terms with $m \leq 1$. Similar calculations have been reported (Grün and Scheid 1983) for the system Li^{3+} + H(1s) at $E_L = 1.5$ (keV/a.m.u.), again with a Coulomb trajectory and for $m \leq 3$. The calculated charge exchange probabilities for $m = 0$, 1, and 2 are shown as a function of impact parameters in Fig. 5.5. The contribution from $|m| = 3$ is negligible. Also shown in the figure are the results of coupled channel calculations by Fritsch (1983). The two methods agree rather well.

Although the direct numerical integration of the Schrödinger equation is

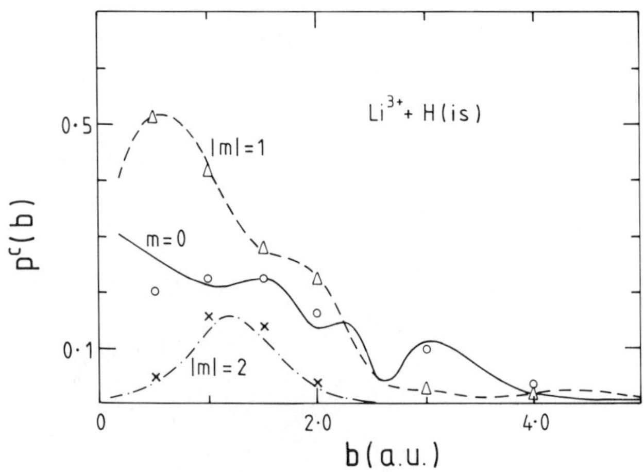

Fig. 5.5. The contribution to the charge exchange probability from the magnetic substates with $|m| \leq 2$ for the reaction Li^{3+} + H(1s) → Li^{2+} + H^+ at $E_L = 1.5$ keV/a.m.u., as a function of the impact parameter b. The corresponding contributions calculated from a coupled channel AO^+ model are shown by the symbols △, ○, × (after Grün and Scheid (1983)).

a successful and useful technique, there are some areas of difficulty. If the spatial extent of the ρ, z' box is too small, and ionization is appreciable, the probability density will be reflected from the boundary of the box instead of passing through the boundary. The boundary conditions can be modified to allow for the escape of ionization flux, but this is not straightforward. A further limitation is that it is difficult to achieve the accuracy that would allow the projection of the final wave function on to the states of the atom $(A + e^-)$, so that it is not easy to calculate the cross sections for transitions leading to particular final states without a very large expenditure of computing time.

6

APPLICATIONS OF THE COUPLED CHANNEL IMPACT PARAMETER MODEL

6.1 INTRODUCTION

In this chapter a number of illustrative examples of the application of the coupled channel impact paramcter model introduced in Chapter 3 and developed in Chapter 5 will be discussed. First let us recollect the conditions under which this model can be applied. A basic assumption is that the internuclear motion can be described by a single classical trajectory determined by an effective potential $W(R)$ at an energy in the centre-of-mass system of $E = \mu v_0^2/2$. The resulting coupled equations for the probability amplitudes $A_j(\mathbf{b}, t)$ are of the form (3.205). The conditions for the validity of these equations, or the equivalent set of equations (3.131), were analysed in Section 3.6. They can be summarized as follows.

1. The motion in each channel must be semi-classical as expressed by the condition (3.119)

$$\left| \frac{1}{p_j^2(R)} \frac{\mathrm{d}p_j(R)}{\mathrm{d}R} \right| \ll 1, \tag{6.1}$$

where $p_j(R)$ is the radial momentum in the jth channel defined by (3.115).

2. The classical trajectories for elastic scattering in each channel (that is in the absence of coupling between the channels) must all be close to a common trajectory for which the radial momentum is (see (3.129b))

$$p(R) = \left[K^2 - 2\mu W(R) - \frac{(L+\frac{1}{2})^2}{R^2} \right]^{\frac{1}{2}}, \tag{6.2}$$

so that

$$|p(R) - p_j(R)| \ll p(R). \tag{6.3}$$

3. The interaction matrix elements connecting different channels must all be small compared with the energy $p^2(R)/2\mu$,

$$|U_{jn}(R)| \ll \frac{p^2(R)}{2\mu}. \tag{6.4}$$

4. The coupling between channels in the vicinity of the common turning point should be small.

The first of these conditions, (6.1), is satisfied outside the turning point region down to very low energies of a fraction of an electron volt (see Chapter 2). Conditions (6.3) and (6.4) are more restrictive, but are satisfied provided changes in electronic energy are small compared with E, the energy of relative motion. This restricts the impact parameter method with a common trajectory to relative energies of $E = \mu v_0^2/2 \gtrsim 100$ eV. The last condition 4. would appear to be even more restrictive because in many cases the interchannel coupling is large at small internuclear separations. However, explicit calculations for the $H^+ + H(1s)$ system (Knudsen and Thorson 1970; Bates and Crothers 1970) have shown that accurate cross sections can be obtained from the impact parameter model even if the coupling in the turning point region is strong. This result has been confirmed by a study of a two-channel system using the momentum representation by Delos and Thorson (1972) who have shown that condition 4. can be relaxed in many cases.

From these considerations, it is expected that the impact parameter model should be applicable, provided the centre-of-mass energy is of the order of 100 eV or more, while at lower energies the partial-wave method should be employed.

A further question arises as to when the coupled equations should be integrated along the classical trajectory determined from the assumed potential $W(R)$, and when it is sufficient to employ a straight line trajectory at constant velocity. The answer to this question depends on the strength of the potential $W(R)$ and is system dependent. The Coulomb repulsion $Z_A Z_B/R$ between the ion cores dominates $W(R)$ at small R, and prevents the heavy particles A and B approaching as closely as would be the case with a straight line trajectory, for which the distance of closest approach is equal to the impact parameter b. It follows that trajectory effects will be most important for those systems dominated by short-range couplings and least important for those dominated by long-range coupling. At low energies the qualitative behaviour of cross sections can be completely changed by the action of the potential $W(R)$. For example, Bates and Boyd (1962) studied the low-energy behaviour of the charge exchange cross sections for the homonuclear systems $H^+ + H(1s)$ and $He^{2+} + He^+(1s)$. In the former case, allowing for the non-linear nature of the trajectory increases the cross section at low energies, the increase amounting to about 3 per cent at $E_L = 0.2$ eV/a.m.u., but does not qualitatively alter the very low-energy behaviour of the cross section. In contrast, for the more highly charged $(He\,He)^{3+}$ system, the effect is much larger becoming significant below $E_L = 2$ keV/a.m.u. The qualitative behaviour of the cross section is changed; instead of increasing logarithmically as the energy is decreased, the cross section exhibits a

Table 6.1. Comparison of partial wave and impact parameter cross sections for the reaction $He^{2+} + H(1s) \rightarrow He^{+}(nlm) + H^{+}$ for $n = 2$. The numbers in brackets represent the power of ten multiplying the preceeding figure. QM is the cross section computed by the partial-wave method; CT is the impact parameter cross section with a Coulomb trajectory; ST is the impact parameter cross section with a straight line trajectory. (Data from van Hemert *et al.* (1985).)

	Cross sections (10^{-16} cm^2)		
Final state	QM	CT	ST
$E_L = 0.125$ keV/a.m.u.			
2s	0.95(−3)	0.62(−3)	3.9(−3)
$2p_0$	1.18(−3)	0.64(−3)	4.8(−3)
$2p_{\pm 1}$	0.98(−2)	0.79(−2)	15.1(−2)
$E_L = 0.25$ keV/a.m.u.			
2s	1.93(−2)	1.76(−2)	2.66(−2)
$2p_0$	2.21(−2)	1.71(−2)	2.63(−2)
$2p_{\pm 1}$	9.9(−2)	14.5(−2)	20.3(−2)
$E_L = 0.625$ keV/a.m.u.			
2s	2.5(−1)	2.3(−1)	2.5(−1)
$2p_0$	2.6(−1)	2.6(−1)	2.7(−1)
$2p_{\pm 1}$	2.7(−1)	4.1(−1)	3.9(−1)

maximum near $E_L = 200$ eV/a.m.u., decreasing to zero as $E_L \rightarrow 0$. Another example is afforded by the heteronuclear system $C^{6+} + H(1s)$ for which the ratio of the capture cross section calculated with a Coulomb trajectory to that calculated with a straight line trajectory decreases from 0.98 at 1.0 keV/a.m.u. to 0.63 at 0.1 keV/a.m.u. (Fritsch 1982).

As an illustration both of the validity of the impact parameter model and of non-linear trajectory effects let us consider some results obtained by van Hemert *et al.* (1985) for collisions of He^{2+} ions with atomic hydrogen in the ground state. In this work coupled equations were obtained for a four-state molecular orbital (MO) basis both in the partial-wave formalism and in the impact parameter formalism. In each case, translational invariance was ensured by using a common translation factor and retaining terms of order $(1/\mu)$. Some of the results obtained in the partial-wave formalism have been discussed in Section 4.6. At energies below $E_L \simeq 25$ keV/a.m.u. the only channels of importance, other than the elastic scattering channel, are those leading to capture in the $n = 2$ level of He^{+}, which is degenerate in energy with the gound state of the target, H(1s). The four-state MO basis contained the $2s\sigma$ orbital asymptotic to the initial arrangement $He^{2+} + H(1s)$, the $2p\sigma$ and $3d\sigma$ orbitals which are asymptotic to Stark mixtures of the $He^+ + H(2s)$ and $He^+ + H(2p_0)$ channels, and the $2p\pi$ orbital which is asymptotic to the $He^+ + H(2p_{\pm 1})$ channel. In Table 6.1, cross sections are shown for capture into the 2s, $2p_0$, and $2p_{\pm 1}$ states of He^+ computed by the partial-wave method (QM) and by the impact parameter method either employing a Coulomb trajectory in which $W(R)$ is taken to be the internuclear potential (CT), or a straight line trajectory (ST). Examination of Table 6.1 shows at the lowest energy of 125 eV/a.m.u. the differences between the partial wave and impact parameter cross sections are large, but markedly smaller for the Coulomb than for a straight line trajectory. It is possible that the impact parameter results could be further improved with a better choice of the effective potential $W(R)$. The differences between the cross sections computed by the three methods decrease with increasing energy, and at $E_L = 625$ eV/a.m.u. reasonable agreement between all three is shown. While the trajectory effect is small for the calculation of total cross sections at $E_L \sim 625$ eV/a.m.u. and above, the effect on differential cross sections would be expected to be significant up to somewhat higher energies. These results support the general contention that with a suitable choice of trajectory, the impact parameter method provides cross sections of useful accuracy at incident energies of a few hundred electron volts and above.

6.2 HOMONUCLEAR ONE-ELECTRON SYSTEMS

6.2.1 $H^+ + H(1s)$

Low energies

In Chapter 4, the partial-wave theory of the $H^+ + H(1s)$ system was discussed using a molecular orbital (MO) basis. At very low energies only the elastic scattering channel

$$H_A^+ + H_B(1s) \rightarrow H_A^+ + H_B(1s), \tag{6.5}$$

and the *symmetrical resonance* capture channel

$$H_A^+ + H_B(1s) \rightarrow H_A(1s) + H_B^+, \tag{6.6}$$

are of importance. The electronic wave functions in these two channels are determined by combinations of the $1s\sigma_g$ and $2p\sigma_u$ molecular orbitals, for which the potential energy curves are shown in Fig. 4.11 (see (4.75)). In Chapter 4 partial-wave cross sections obtained from the simplest MO basis were discussed in which only the $1s\sigma_g$ and $2p\sigma_u$ orbitals were retained with no translation factors. It is of interest for comparative purposes to employ the same basis in the impact parameter formalism. The wave function can be expressed as (with no translation factor)

$$\psi(\mathbf{s}, t) = A_g(b, t)\psi_g(\mathbf{s}, R)\,e^{-i\varepsilon_{1s}t} + A_u(b, t)\psi_u(\mathbf{s}, R)\,e^{-i\varepsilon_{1s}t}, \tag{6.7}$$

where $\psi_g \equiv \psi_{1s\sigma_g}$ and $\psi_u \equiv \psi_{2p\sigma_u}$. Since asymptotically (see (4.74))

$$\begin{aligned} \psi_g(\mathbf{s}, R) &\underset{R\to\infty}{\sim} \frac{1}{\sqrt{2}}[\phi_{1s}(r) + \phi_{1s}(x)], \\ \psi_u(\mathbf{s}, R) &\underset{R\to\infty}{\sim} \frac{1}{\sqrt{2}}[\phi_{1s}(r) - \phi_{1s}(x)], \end{aligned} \tag{6.8}$$

the amplitudes for elastic scattering $a(b)$ and for charge exchange $c(b)$ are given by

$$a(b) = \lim_{t\to\infty} \frac{1}{\sqrt{2}}[A_g(b, t) + A_u(b, t)],$$

and (6.9)

$$c(b) = \lim_{t\to\infty} \frac{1}{\sqrt{2}}[A_g(b, t) - A_u(b, t)],$$

respectively. In order to satisfy the boundary condition that the electron should initially be in the hydrogen ground state centred on B,

$$\lim_{t\to-\infty} \psi(\mathbf{s}, t) = \phi_{1s}(r)\,e^{-i\varepsilon_{1s}t} \tag{6.10}$$

we must have that

$$A_g(t = -\infty) = A_u(t = -\infty) = \frac{1}{\sqrt{2}}. \tag{6.11}$$

By inserting the wave function (6.7) into the Schrödinger equation

$$i\frac{\partial\psi(\mathbf{s}, t)}{\partial t} = H_e\psi(\mathbf{s}, t) \tag{6.12}$$

and making use of the invariance of the Hamiltonian under the inversion $\mathbf{s}_g \to -\mathbf{s}_g$, uncoupled equations are found for A_+ and A_-:

$$i\dot{A}_{g,u}(b, t) = (E_{g,u}(R) - \varepsilon_{1s})A_{g,u}(b, t) \tag{6.13}$$

where $E_g(R)$ and $E_u(R)$ are the potentials for the $1s\sigma_g$ and $2p\sigma_u$ molecular levels, respectively

$$H_e\psi_{g,u}(\mathbf{s}, R) = E_{g,u}\psi_{g,u}(\mathbf{s}, R). \tag{6.14}$$

The solutions of (6.13) with boundary condition (6.11) are

$$A_{g,u}(b, t) = \frac{1}{\sqrt{2}} \exp\left\{-i \int_{-\infty}^{t} [E_{g,u}(R') - \varepsilon_{1s}]\, dt'\right\}, \tag{6.15}$$

with $R' = R(t')$, from which we find the probability for elastic scattering $P_{EL}(v_0, b)$ to be

$$P_{EL}(v_0, b) = |a(b)|^2 = \cos^2\left[\tfrac{1}{2}\int_{-\infty}^{\infty} \{E_g(R) - E_u(R)\}\, dt\right] \tag{6.16a}$$

and for capture to be

$$P_C(v_0, b) = |c(b)|^2 = \sin^2\left[\tfrac{1}{2}\int_{-\infty}^{\infty} \{E_g(R) - E_u(R)\}\, dt\right]. \tag{6.16b}$$

These expressions are of the same form as those obtained from the two-state *atomic* trial function ((5.106) and (5.107)), and show that in the two-state approximation P_{EL} and P_C oscillate with unit amplitude as a function of b, for fixed v_0, or of v_0 for fixed b.

If translation factors are included in the basis functions, the coupled equations still separate into two sets of parity eigenfunctions and the probabilities for elastic scattering and capture in the two-state approximation become

$$P_{EL}(v_0, b) = \cos^2\left[\xi + \tfrac{1}{2}\int_{-\infty}^{\infty} \{E_g(R) - E_u(R)\}\, dt\right], \tag{6.17a}$$

$$P_C(v_0, b) = \sin^2\left[\xi + \tfrac{1}{2}\int_{-\infty}^{\infty} \{E_g(R) - E_u(R)\}\, dt\right], \tag{6.17b}$$

where ξ is a function of v_0 and b. For plane-wave translation factors the explicit form of ξ has been given by Bates and McCarroll (1962). The presence of the function ξ has the effect of decreasing the capture cross section, as the energy increases, by as much as 45 per cent at 20 keV/a.m.u., but below $E_L \approx 2$ keV/a.m.u. it is unimportant (Ferguson 1961).

The oscillatory nature of the charge exchange and elastic scattering probabilities was demonstrated experimentally by Lockwood and Everhart (1962) and Helbig and Everhart (1965) who measured the cross sections at fixed scattering angles between 0.2° and 6.0° over the energy range 130 eV/a.m.u. to 150 keV/a.m.u. In Fig. 6.1, the capture probability P_0, defined as the ratio of the capture cross section $d\sigma^C/d\Omega$ to the sum of the capture and

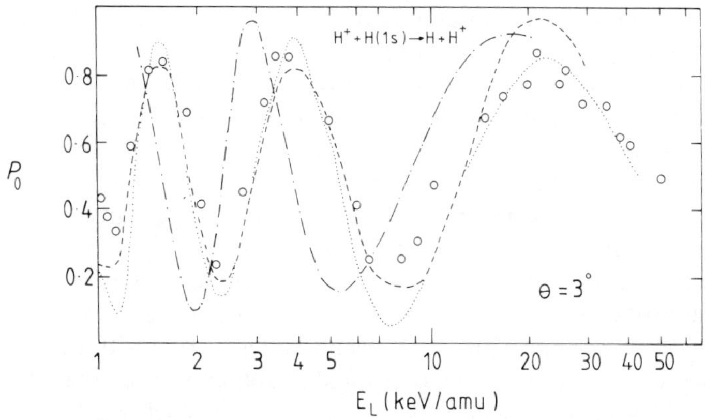

Fig. 6.1. The total charge exchange probability at a scattering angle of 3° in the laboratory system for electron capture by protons from atomic hydrogen in the ground state: ○ ○ ○ experimental data (Lockwood and Everhart 1962); —·—·— four-state atomic orbital basis (Wilets and Gallagher 1966); ——— variable charge model (Cheshire 1968); ······ pseudostate model (Cheshire *et al.* 1970).

elastic scattering cross sections $\{(d\sigma^{C}/d\Omega) + (d\sigma^{EL}/d\Omega)\}$,

$$P_0(v_0, \theta) = \frac{\dfrac{d\sigma^{C}}{d\Omega}}{\dfrac{d\sigma^{C}}{d\Omega} + \dfrac{d\sigma^{EL}}{d\Omega}}, \tag{6.18}$$

is shown as a function of energy at a scattering angle in the laboratory system of 3° for $1 \leq E_L \leq 50$ keV/a.m.u.

Let us take the case in which the trajectory in both the g and u states is determined by the Coulomb potential $1/R$ between the nuclei. Then (see (3.249)) in the stationary phase approximation

$$\frac{d\sigma^{C}}{d\Omega} = I_{\text{Coul}}(\theta)|c(b)|^2, \qquad \frac{d\sigma^{EL}}{d\Omega} = I_{\text{Coul}}(\theta)|a(b)|^2, \tag{6.19}$$

where $I_{\text{Coul}}(\theta)$ is the classical differential cross section for Coulomb scattering and b is the impact parameter corresponding to the angle θ. We then have

$$P_0(v_0, \theta) \approx \frac{|c(b)|^2}{|a(b)|^2 + |c(b)|^2}$$

$$\approx P_C(v_0, b). \tag{6.20}$$

In the present context 3° is a 'large angle' and P_0 is determined by close collisions at small values of the impact parameter. For $E_L\theta \gtrsim 8$ keV degrees, it can be shown (McCarroll and Salin 1968) that P_0 is nearly independent of angle and $P_0 \approx |c(b=0)|^2$. When $E_L\theta < 8$ keV degrees, P_0 depends on larger values of b and on θ. The simple two-state theory which leads to (6.16) predicts that P_0 should oscillate between 0 and 1, in contrast to the measurements, which show that this oscillation is 'damped', with P_0 ranging between $P_0 \approx 0.2$ and $P_0 \approx 0.8$. The energies predicted by (6.16) at which the maxima and minima occur are in fair, but not complete, agreement with the experimental data.

To obtain the observed damping of the oscillations in P_0 it is necessary to include further states in the expansion. A glance at Fig. 4.11 shows that in the united atom limit the $2p\sigma_u$ level becomes degenerate with the $2p\pi_u$ level, and in the region of small R the rotational coupling between these two levels is strong. At very low energies the system cannot reach small values of R because of the internuclear Coulomb repulsion, but for energies above $E_L \sim 40$ eV/a.m.u. the probability of excitation of the $2p\pi_u$ channel is appreciable. In the separated atom limit the $2p\pi_u$ orbital is asymptotic to $2p_{\pm 1}$ atom wave functions centred on A and B, with the consequence that above 40 eV/a.m.u. hydrogen atoms in the 2p level are produced both by excitation and by capture. Up to energies of ~ 1 keV only the $2p\pi_u$ orbital needs to be taken into account in the basis set in addition to the $1s\sigma_g$ and $2p\sigma_u$ orbitals (Bates and Williams 1964), and this provides most of the damping required to bring the theory into line with the experiments. There is, however, an additional source of damping which we shall now consider.

In deriving the two-state approximation leading to the expressions (6.16), it has been assumed that the same trajectory $\mathbf{R}(t)$ is used for both the gerade and ungerade states, which, for example, could be taken to be the Coulomb trajectory due to the internuclear potential. A better approximation (Smith 1964; Bates and Holt 1966) is to employ the potential $E_g(R)$ to determine $\mathbf{R}(t)$ in the $1s\sigma_g$ channel and $E_u(R)$ to determine $\mathbf{R}(t)$ in the $2p\sigma_u$ channel. Let $I_g(\theta)$ and $I_u(\theta)$ be the *classical* cross sections for scattering by E_g and E_u, respectively. Within the two-state model, and using the stationary phase approximation, the scattering amplitudes for scattering in the $1s\sigma_g$ and $2p\pi_u$ channels are $f_g(\theta)$ and $f_u(\theta)$, respectively, where

$$f_{g,u}(\theta) = [I_{g,u}(\theta)]^{\frac{1}{2}} A_{g,u}(\mathbf{b}, \infty), \tag{6.21}$$

the amplitudes $A_{g,u}(\mathbf{b}, \infty)$ being determined by (6.15). The scattering amplitudes for elastic scattering $f(\theta)$ and capture $g(\theta)$ can be expressed in terms of $f_g(\theta)$ and $f_u(\theta)$ by the relations

$$\begin{aligned} f(\theta) &= \frac{1}{\sqrt{2}}\{f_g(\theta) + f_u(\theta)\}, \\ g(\theta) &= \frac{1}{\sqrt{2}}\{f_g(\theta) - f_u(\theta)\}, \end{aligned} \tag{6.22}$$

from which we find

$$P_0 = \frac{1}{2}\left\{1 - g\cos\left[\int_{-\infty}^{\infty} \{E_g(R) - E_u(R)\}\, \mathrm{d}t\right]\right\}, \qquad (6.23)$$

where

$$g = \frac{2I_u^{\frac{1}{2}}(\theta)I_g^{\frac{1}{2}}(\theta)}{I_u(\theta) + I_g(\theta)}.$$

When $I_u(\theta)$ and $I_g(\theta)$ differ the oscillations in P_0 are damped to a greater or lesser extent, but if $I_u(\theta) \simeq I_g(\theta)$ then $g = 1$ and P_0 reduces to (6.20).

A very careful discussion of the angular distributions for elastic scattering, excitation, and charge exchange within the three state ($1s\sigma_g$, $2p\pi_u$, and $2p\sigma_u$) model has been given by Knudson and Thorson (1970) (see also Bates and Sprevak (1970)) for $E_L \leq 500$ keV/a.m.u. In the absence of translation factors (or with a common translation factor) the $1s\sigma_g$ channel uncouples from the $2p\pi_u$ and $2p\sigma_u$ channels. It follows that the trajectory $\mathbf{R} = \mathbf{R}(t)$ for the $1s\sigma_g$ channel should be calculated from the potential $E_g(R)$. For the two coupled ungerade channels it would be possible to use an average of the $2p\sigma_u$ and $2p\pi_u$ potentials to determine $\mathbf{R}(t)$, but in practice it was found sufficient to take the potential to be that associated with the initial $2p\sigma_u$ state. The work of Knudsen and Thorson was based on the stationary phase approximation, which is only of moderate accuracy; nevertheless reasonable agreement was found with the measurements of Helbig and Everhart (1965). The data at $E_L = 205$ eV/a.m.u. is shown in Fig. 6.2 for the large-angle region. At very

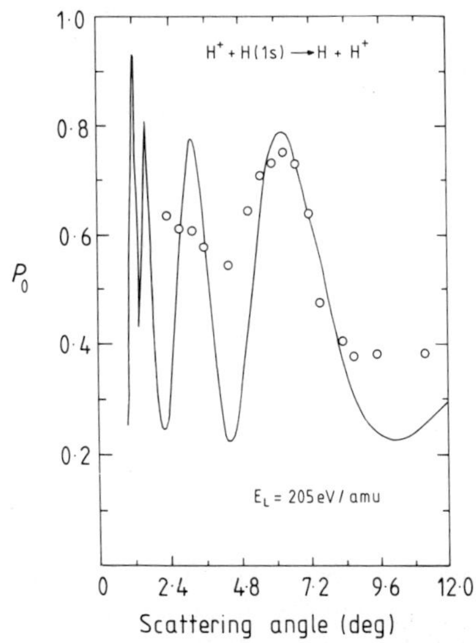

Fig. 6.2. The angular distribution for electron capture by protons from atomic hydrogen at $E_L = 205$ eV/a.m.u.: ○ ○ ○ experimental data of Helbig and Everhart (1965); ——— calculated cross section of Knudsen and Thorsen (1970).

small angles the measurements exhibit rapid oscillations due to rainbow scattering, which arises for the reasons discussed in Chapter 2.

The stationary phase approximation was avoided in similar three-state calculations by Gaussorgues *et al.* (1975*b*) who solved the semi-classical equations in the form (3.131) and then formed the scattering amplitudes by summing over the angular momentum using (3.133) and (3.134), thus avoiding either small-angle or large-angle approximations. Satisfactory agreement with the measurements of Houver *et al.* (1970) and Helbig and Everhart (1965) was found at energies up to 1 keV/a.m.u. The results at 500 eV/a.m.u. are illustrated in Fig. 6.3 for both linear and non-linear trajectories. The inadequacy of the straight line trajectory for small values of the impact parameter corresponding to large-angle scattering can be observed.

Higher energies

Below ~1 keV the only significant inelastic reactions are capture into the 1s level of atomic hydrogen and capture into or excitation of the $2p_{\pm 1}$ level.

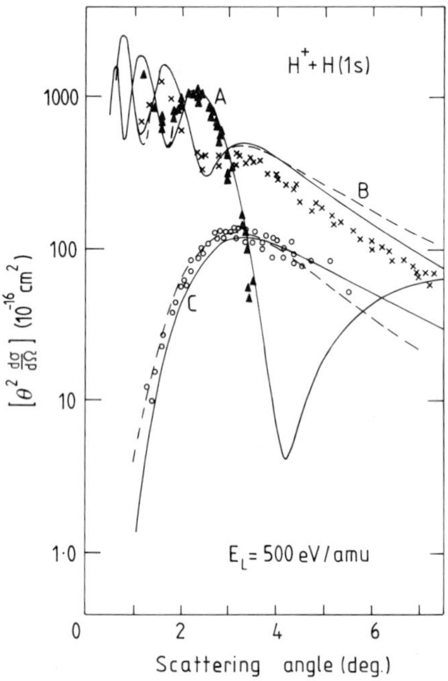

Fig. 6.3. Scaled differential cross section in the laboratory system for proton scattering by atomic hydrogen at 500 eV/a.m.u. (after Gaussorgues *et al.* (1975*b*)). Experimental data: ▲▲▲ elastic scattering; × × × charge exchange; ○○○ excitation from Houver *et al.* (1974). Theoretical curves from Gaussorgues *et al.* (1975*b*): A elastic scattering; B charge exchange; C excitation. Solid curves are with a common trajectory; dashed curves are for a straight line trajectory.

As the energy increases radial coupling between the $2p\sigma_u$, $3p\sigma_u$, and $4f\sigma_u$ levels and (to a lesser extent) that between the $1s\sigma_g$, $2s\sigma_g$, and $3d\sigma_g$ levels leads to capture into the 2s and $2p_0$ states of atomic hydrogen, and to the excitation of the target to these states. However it should be noted that capture into the 1s state remains the dominant mechanism below 40 keV/a.m.u. (see Fig. 1.1) above which ionization dominates. As a result the total capture cross section is close to that predicted by a two-state approximation (either MO or AO) even up to energies of 100 keV/a.m.u. and above. The increasing importance of coupling to the $2p_0$ state of hydrogen above 1 keV is illustrated by the measurements of Hippler *et al.* (1988), (see also Hippler (1988)) of the integral alignment parameter A_{20} which is defined as

$$A_{20} = \frac{\sigma_{2p_{\pm 1}} - \sigma_{2p_0}}{2\sigma_{2p_{\pm 1}} + \sigma_{2p_0}} \tag{6.24}$$

where σ_{nlm} is the partial cross section for excitation (or capture) leading to the state $|nlm\rangle$. The measurements shown in Fig. 6.4 show that at a few keV A_{20} is close to 50 per cent, the value that would be obtained if the only inelastic coupling of importance were to be to the $2p\pi_g$ molecular orbital which is asymptotic to the atomic $2p_{\pm 1}$ wave functions. The decrease of A_{20}

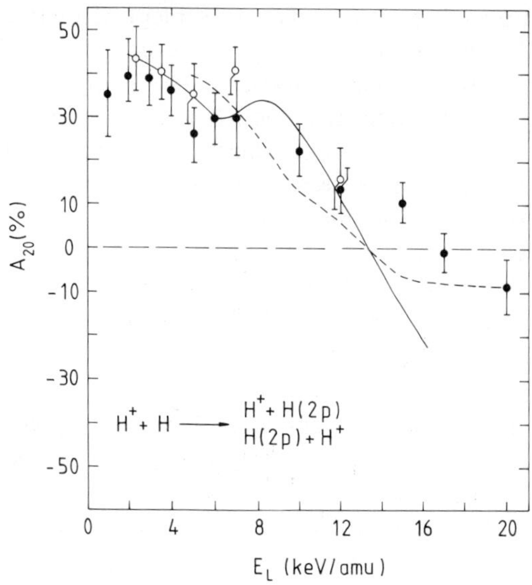

Fig. 6.4. The integral alignment A_{20} of the 2p level of atomic hydrogen formed in collisions of H^+ with H(1s). Experimental data: ●, ○ from Hippler *et al.* (1988). Theoretical data: ——— from Lüdde and Dreizler (1982*a*); – – – from Fritsch and Lin (1983*a*).

to values close to zero by 15 keV/a.m.u. shows the increasing importance of radial couplings to orbitals asymptotic to the $2p_0$ level.

To obtain a satisfactory description of the $H^+ + H$ system in the energy region extending from a few keV to a few hundred keV certain conditions must be satisfied. The basis set must allow for excitation of and capture into excited states, principally the $n = 2$ states of hydrogen; allowance should be made for ionization, at least for energies $\gtrsim 15$ keV (see Fig. 1.1); and proper account should be taken of translational (Galilean) invariance by the inclusion of translation factors in the basis. Capture by ions of low charge from neutral atoms takes place at small separations and this implies that if an MO expansion is not employed pseudostates or some other mechanism must be introduced to improve the wave function in the united atom limit.

Molecular orbital basis sets A molecular orbital basis containing up to ten states with variationally optimized switching functions defined by (5.72) was employed by Crothers and Hughes (1978*a*, *b*) to calculate the capture and excitation cross sections to the $n = 2$ levels of atomic hydrogen for an energy range extending to 25 keV/a.m.u. A possible defect in this work was an artificial symmetrization of the interaction matrix, and this was avoided in calculations by Kimura and Thorson (1981*a*) who used a similar basis with optimized state-dependent translational factors (Thorson *et al.* 1981) in the energy interval up to 7 keV/a.m.u. The agreement between the two calculations is generally satisfactory, although there are some differences in the shape of the $n = 2$ excitation cross sections. The agreement with experiment is excellent. This is illustrated in Table 6.2 where the turning points in the capture probability $P(E)$ at 3° computed by Crothers and Hughes (1978*a*) are compared with those measured by Lockwood and Everhart as shown in Fig. 6.1. Other work based on the molecular orbital expansion but omitting translational factors differs significantly from the calculations of Kimura and Thorson and of Crothers and Hughes and from the data (Rosenthal 1971; Chidichimo-Frank and Piacentini 1974; Schinke and Krüger 1976).

Table 6.2. The energies, in keV/a.m.u., of the turning points for the probability of capture at $\theta = 3°$ for $H^+ + H(1s) \rightarrow H + H^+$. Experimental results were obtained from Lockwood and Everhart (1962) (see Fig. 6.1); theoretical results are due to Crothers and Hughes (1978*a*).

	max	min	max	min	max	min	max
Experiment	0.78	1.11	1.57	2.39	3.92	7.69	20.1
Theory	0.81	1.11	1.59	2.40	3.92	7.40	19.8

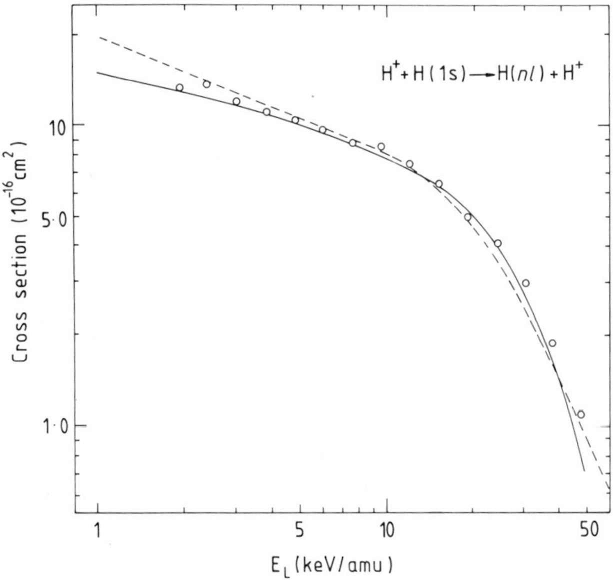

Fig. 6.5. The total cross section for electron capture by protons from atomic hydrogen in the ground state: ○ ○ ○ measurements of McClure (1966); ——— calculated cross section of Lüdde and Dreizler (1982*a*); – – – calculated cross section of Cheshire *et al.* (1970).

Atomic orbital basis sets Pioneer calculations based on a two-centre atomic orbital basis with plane-wave translational factors were performed by McCarroll (1961), who used a two-state approximation with 1s orbitals on each centre; by Lovell and McElroy (1965), using a four-state approximation with 1s, 2s orbitals on each centre; and by Wilets and Gallagher (1966) who used an eight-state approximation with 1s, 2s, $2p_0$, $2p_{\pm 1}$ orbitals on each centre. In later work by Rapp and Dinwiddie (1972) the 3s and 3p orbitals on each centre were added and it was demonstrated that the cross sections for capture into or excitation of these levels were very small compared with the $n = 2$ cross sections over the energy range up to 200 keV/a.m.u. None of these calculations give very satisfactory agreement with the 3° capture data of Lockwood and Everhart, and this can be attributed to the poor representation of the wave function at small internuclear separations when an expansion in *bound* atomic states is employed. This defect is removed in the variable charge model of Cheshire (1968) and in the pseudostate model of Cheshire *et al.* (1970), both of which were described in Chapter 5. Calculations based on these two models, which include both the ground state and $n = 2$ channels asymptotically, agree much better with the differential

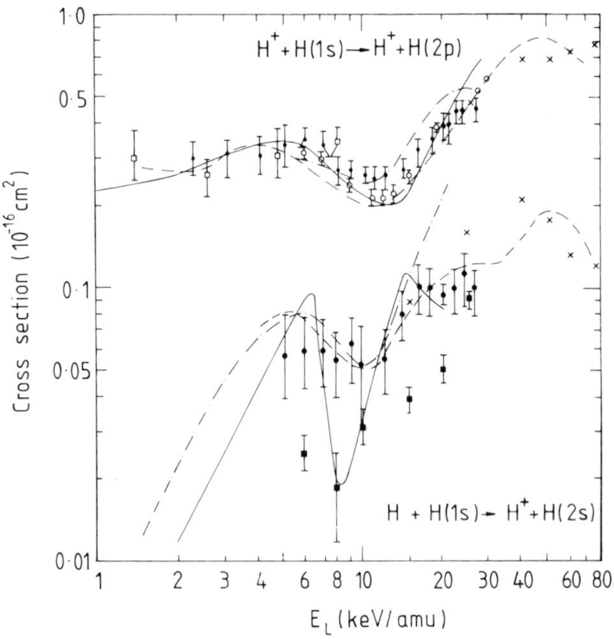

Fig. 6.6. Cross sections for excitation of the 2s and 2p levels of atomic hydrogen by proton impact. Experimental data: ● Morgan *et al.* (1973); □ Stebbings *et al.* (1965); ○ Kondow *et al.* (1974); ■ Chong and Fite (1977). Theoretical cross sections: —·—·— AO^+ basis with 22 states (Fritsch and Lin 1982*c*); ----- AO^+ basis with 40 states (Fritsch and Lin 1983*a*); ——— Hylleraas basis (Lüdde and Dreizler 1981, 1982*a*); × × × Basis of scaled hydrogenic functions (Shakeshaft 1978*b*).

cross section data (as shown in Fig. 6.1) than those based on bound state atomic orbital bases.

Although the adequate representation of the differential cross section data requires rather elaborate theoretical models, it is interesting to note that the *total* capture cross section given by the two-state atomic expansion is within 20 per cent of that given by more accurate calculations with large basis sets up to energies of $\sim E_L = 100$ keV/a.m.u. Total cross sections obtained using the pseudostate basis by Cheshire *et al.* are compared with total cross section measurements in Fig. 6.5 at energies up to 50 keV/a.m.u. However the data in this energy interval is not sufficiently precise to allow different theoretical predictions to be distinguished (Gilbody 1986).

In the earlier work based on two-centre expansions only basis sets of rather limited size could be employed. In more recent work the development of computing facilities has allowed the use of large two-centre basis sets, which enable the convergence of the computed cross sections to be assessed.

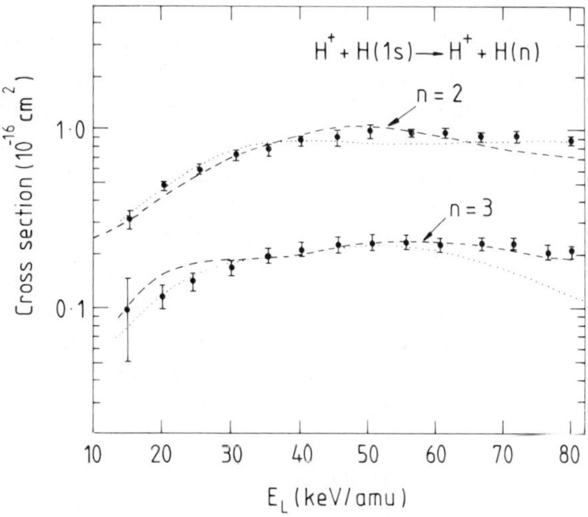

Fig. 6.7. Cross sections for excitation of atomic hydrogen to the $n = 2$ and $n = 3$ levels by proton impact. Experimental data from Park *et al.* (1975, 1976). Theoretical cross sections: ——— AO^+ basis with 40 states (Fritsch and Lin 1983*a*); ······ Basis of scaled hydrogenic functions (Shakeshaft 1978*b*).

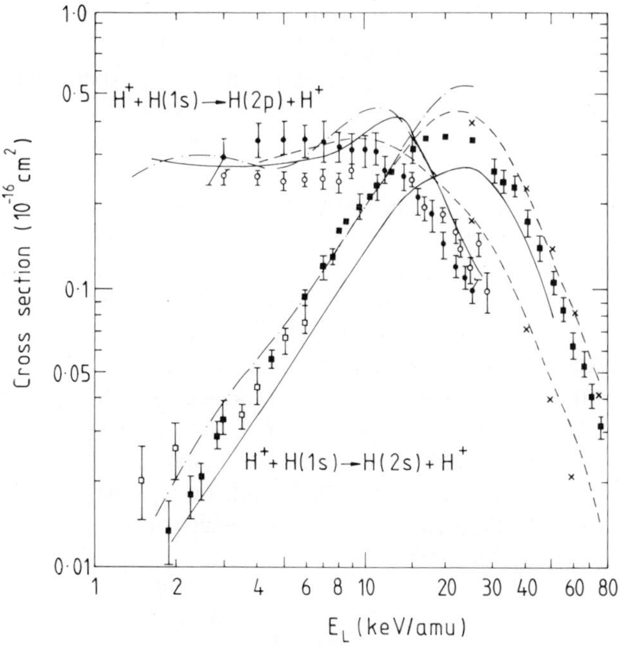

Fig. 6.8. Cross sections for capture into the 2s and 2p levels by protons incident on atomic hydrogen in the ground state. Experimental data: ● Morgan *et al.* (1973); □ Hill *et al.* (1979*a*, *b*); ■ Morgan *et al.* (1980); ○ Kondow *et al.* (1974). Theoretical cross sections are as in Fig. 6.6.

Shakeshaft (1975*a*, 1976, 1978*b*), following the earlier work of Gallagher and Wilets (1968), employed Sturmian basis sets with up to 38 terms on each heavy particle centre and similar basis sets have been employed by Winter (1986, 1987) to study collisions of protons with one-electron ions from He^+ to C^{5+} (see below). However the results suggest that this basis converges rather slowly, in part because several terms are required to represent the wave function in a given physical channel. More rapid convergence is shown by basis sets of the AO^+ type, in which the low-lying physical atomic orbitals are augmented by pseudostates chosen from the united atom orbitals. Large-scale calculations of this type have been made by Fritsch and Lin (1982*c*) using 22 states, and by Fritsch and Lin (1983*a*) using 40 states. Other calculations with a large basis include those of Winter and Lin (1984) who employed a 36-state three-centre basis and of Kimura and Lin (1985*b*) who employed the AO/MO method for energies below 15 keV/a.m.u., using a basis of 10 molecular orbitals in the inner region and 20 atomic orbitals in the outer region. In Fig. 6.6, the experimental data for excitation of the 2s and 2p levels of atomic hydrogen are compared with theoretical cross sections obtained from the AO^+ method (22 and 40 states) and a scaled hydrogenic (Sturmian) basis for energies below 80 keV. Corresponding data for excitation of the $n = 2$ and $n = 3$ levels is displayed in Fig. 6.7, and for capture into the 2s and 2p levels in Fig. 6.8. The theoretical cross sections obtained from three-centre basis calculations and from the AO/MO model agree equally well with the experimental data, but there are significant differences between the various theoretical results, not all of which may be fully converged.

Polynomial expansions Further results have been obtained for excitation and charge exchange by protons incident on atomic hydrogen by expanding the wave function in orthogonal polynomials (see Section 5.8) by Lüdde and Dreizler (1981, 1982*a*, *b*). The results obtained from a Hylleraas basis with up to 150 terms are shown in Figs. 6.5–6.8. In the interaction region no explicit translation factor was employed but to identify the physical probability amplitudes the wave function was projected onto hydrogenic wave functions with plane wave translation factors in the asymptotic region. In general, the method predicts cross sections for excitation of and capture into the $n = 2$ levels of hydrogen which, although reasonable, do not appear to fit the experimental data as closely as the results of other coupled channel models employing a large basis.

Ionization

The use of two-centre basis sets containing pseudostates, or of three-centre bases, allows an estimate of the ionization cross section. If orbitals are included in the basis which are asymptotic to atomic continuum states, the MO method can also be employed to calculate the probability of ionization.

In the case of the $H^+ + H(1s)$ system, ionization has been studied at low energies using the Sturmian expansion (Shakeshaft 1978*b*), the three-centre expansion (Winter and Lin 1984), the AO^+ expansion (Fritsch and Lin 1983*a*), and more recently the MO expansion (Kimura and Thorson 1989). At higher energies, for which capture is unimportant, ionization has been studied using a single-centred expansion based on the target atom (Fitchard *et al.* 1977; Reading *et al.* 1981) and the use of single-centred expansions will be discussed in Chapter 8. Below about 75 keV/a.m.u., capture cannot be neglected, but since capture into the ground state dominates, if the primary objective is to calculate accurate ionization and excitation cross sections, it is sufficient to place a single 1s orbital on the projectile and a very large atomic orbital basis set containing both real states and pseudostates on the target atom. Calculations of this type have been carried out by Ermolaev (1990*a*, *b*) using a basis set of 50 states on the target, containing 8s, 16p, 18d, and 8f states. In Fig. 6.9, the calculated cross sections for the ionization of H(1s) by proton impact provided by the model of Ermolaev are shown together with those of the AO^+ basis of Fritsch and Lin (1983*a*), the Sturmian basis of Shakeshaft (1978*b*), and the one-centred basis of Fitchard *et al.* (1977). The experimental data of Shah and Gilbody (1981) and Shah

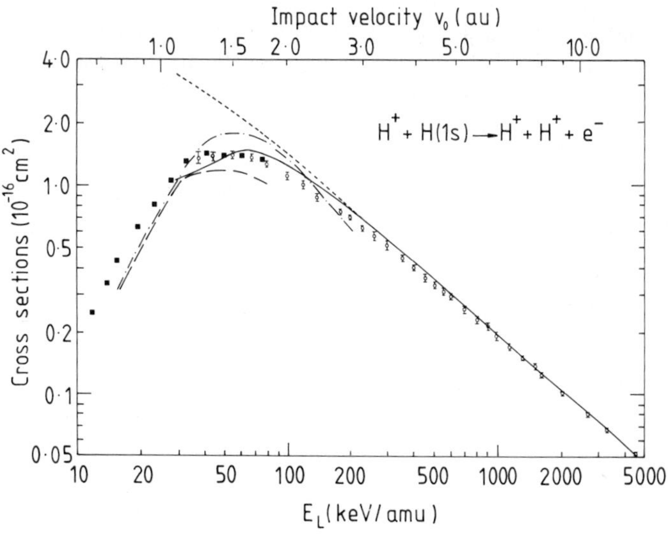

Fig. 6.9. The total ionization cross section for protons incident on atomic hydrogen. Experimental data: ○ from Shah and Gilbody (1981); ■ from Shah *et al.* (1987). Theoretical cross sections: ——— with pseudostate basis from Ermolaev (1990*a*); — — — with AO^+ basis from Fritsch and Lin (1983*a*); —·—·— with scaled hydrogenic (Sturmian) basis from Shakeshaft (1978*b*); ----- with one-centred pseudostate expansion (Fitchard *et al.* 1977). One-centred pseudostate expansions are discussed in Chapter 8.

et al. (1987) are also included for comparison. It can be seen that projectile-centred states must be included in the basis to obtain the correct shape of the cross section below ~100 keV/a.m.u. and that rather large basis sets are necessary to obtain sufficient overlap with the hydrogenic continuum.

Of particular interest in the study of ionization is the elucidation of the small differences between the high-energy cross sections for proton and antiproton ($\bar{p}$) impact at the same impact velocity. These differences result in a lower energy loss for antiprotons penetrating matter (the Barkas effect) compared with protons. To calculate antiproton cross sections, a single centred basis on the target may be employed, because the capture channels are entirely absent. Using antiprotons from the low-energy antiproton storage ring (LEAR) at CERN experiments can be made which determine ionization cross sections over an energy interval extending from 5 MeV down to ~30 keV. Calculations of the cross sections for the ionization of H(1s) by $\bar{p}$ impact over this energy range have been carried out by Martir *et al.* (1982) and with a larger basis set by Ermolaev (1987, 1988*a*, *b*, 1990*a*, *b*).

In Fig. 6.10, the calculated ratio of the total cross sections for ionization by proton and antiproton impact is shown. Above $E_L = 50$ keV/a.m.u. ($v_0 = 1.4$ a.u.), the difference between the ionization produced by proton and antiproton impact is small, the ratio of the cross sections not exceeding 12 per cent. For $E_L \lesssim 50$ keV/a.m.u. the ionization cross section by antiproton impact remains large, but that for proton impact decreases rapidly with decreasing energy. In this low-energy region, the proton–atomic hydrogen inelastic cross section is dominated by charge exchange, which of course is absent from the antiproton case. At high energies with $E_L > 3.5$ MeV/a.m.u. ($v_0 = 11.8$ a.u.), the ratio of the antiproton to proton cross sections is slightly less than unity, which implies that antiprotons have a longer range in hydrogen than protons.

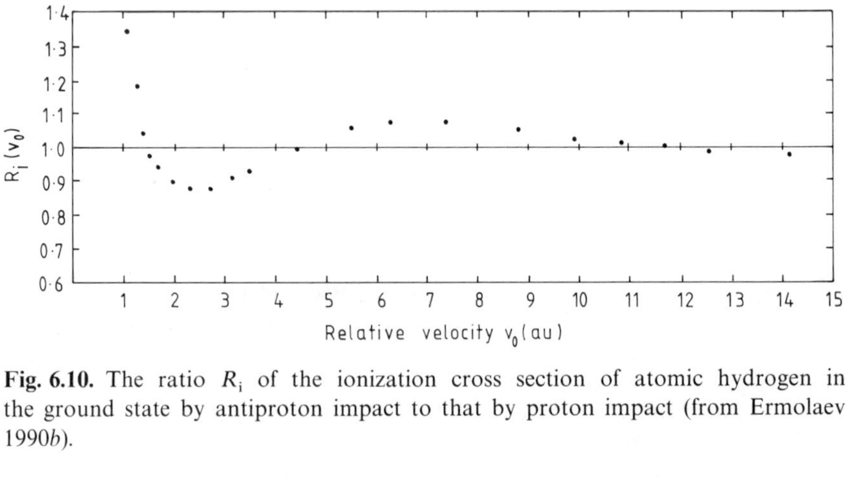

Fig. 6.10. The ratio R_i of the ionization cross section of atomic hydrogen in the ground state by antiproton impact to that by proton impact (from Ermolaev 1990*b*).

6.2.2 Systems with $Z_A = Z_B > 1$

All homonuclear one-electron systems, in which a fully stripped ion of charge Z_A is incident upon a hydrogenic ion in the ground state with nuclear charge $Z_B = Z_A$, behave like the $H^+ + H(1s)$ system. This can be seen from an argument due to Briggs and Macek (1973), who observed that the time-dependent Schrödinger equation

$$\left[-\tfrac{1}{2}\nabla^2 - \frac{Z_A}{|\mathbf{s} - \mathbf{R}/2|} - \frac{Z_A}{|\mathbf{s} + \mathbf{R}/2|} - \mathrm{i}\frac{\partial}{\partial t}\right]\psi(\mathbf{s}, t) = 0 \tag{6.25}$$

transforms to the equation:

$$\left[-\tfrac{1}{2}\nabla'^2 - \frac{1}{|\mathbf{s}' - R'/2|} - \frac{1}{|\mathbf{s}' + \mathbf{R}'/2|} - \mathrm{i}\frac{\partial}{\partial t'}\right]\psi(s', t') = 0, \tag{6.26}$$

where $\mathbf{s}' = Z_A\mathbf{s}$, $\mathbf{R}'(t') = Z_A\mathbf{R}(t)$, and $t' = Z_A^2 t$. Lengths, such as the impact parameter, transform by

$$b' = Z_A b, \tag{6.27}$$

and velocities transform by

$$v_0' = \frac{v_0}{Z_A}. \tag{6.28}$$

It follows that the probability $P(Z_A, b, v_0)$ for a process in a symmetrical system with nuclei of charge Z_A and the corresponding cross section $\sigma(Z_A, v_0)$, can be obtained from the probability and cross section, respectively, for the same process in the system $H^+ + H(1s)$ by the relations

$$P(Z_A, b, v_0) = P\left(1, Z_A b, \frac{v_0}{Z_A}\right), \tag{6.29}$$

$$\sigma(Z_A, v_0) = Z_A^{-2}\sigma\left(1, \frac{v_0}{Z_A}\right). \tag{6.30}$$

The scaling $\mathbf{R}'(t') = Z_A\mathbf{R}(t)$ is satisfied exactly for a straight line trajectory, and also for a Coulomb trajectory provided the ratio M_A/Z_A is kept constant, but is only approximate if the trajectory is determined from some other potential. The effect of the internuclear potential is to reduce the transition probabilities at small impact parameter and this effect increases with Z_A (at a given velocity) and increases with decreasing v_0 at a fixed value of Z_A. This is consistent with (6.30) where the cross section at v_0 for $Z_A > 1$ is seen to be related to the cross section in $H^+ + H$ systems for the smaller velocity v_0/Z_A.

The scaling laws (6.29), (6.30) can be used to analyse the production of K-shell vacancies in symmetrical collisions, for example $N^+ + N$ and $Ne^+ + Ne$. It was originally suggested by Lichten (1967) that if a vacancy occured in the $2p\pi_u$ molecular orbital before the collision, then the coupling between the $2p\sigma_u$ and $2p\pi_u$ orbitals may result in the excitation of one of the K-shell electrons to a 2p level, in just the same way that the $2p_{\pm 1}$ level is populated in $H^+ + H(1s)$ collisions. By employing an effective charge $Z'_A = Z_A - 1$ to allow for screening of the nuclear charge by the second K-electron, Briggs and Macek showed that the scaling law could reproduce approximately the data for the systems $C^+ + C$, $N^+ + N$, and $Ne^+ + Ne$. The extension of scaling laws to heteronuclear systems has been discussed by Briggs and Taulbjerg (1975) (see also Briggs 1976).

6.3 HETERONUCLEAR ONE-ELECTRON SYSTEMS

One-electron heteronuclear systems can be divided into two classes according to whether $Z_A > Z_B$ or $Z_A < Z_B$ where Z_A, Z_B are the charges on the projectile and target nuclei. As an illustrative example of the latter class, we shall take the much-studied system $H^+ + He^+(1s)$.

$H^+ + He^+(1s)$ Electron capture by protons from $He^+(1s)$ takes place almost entirely to the ground state of atomic hydrogen by the endothermic reaction

$$H^+ + He^+(1s) \rightarrow H(1s) + He^{++} - 1.5 \text{ a.u.}$$

The $1s\sigma$ and $2p\sigma$ molecular orbitals are asymptotic to the incident and exchange channels, respectively, and the corresponding adiabatic energies are well separated at all internuclear distances (see Fig. 4.12). The consequence of this large energy difference is that the capture cross section is small with a maximum of $\sim 3 \times 10^{-17}$ cm^2 near an energy of $E_L = 50$ keV/a.m.u. Another consequence of the large separation of the adiabatic potential for the entrance channel from that relating to the final state is that a two-state MO approximation is inadequate. This was shown clearly in the work of Winter *et al.* (1980) who employed MO bases containing 2, 3, 4, 5, and 10 states with plane-wave (Bates–McCarroll) translation factors to calculate the capture and excitation cross sections at energies in the range 2.0 to 20 keV/a.m.u. The composition of the bases employed was as follows

	Orbitals
2-state	$1s\sigma$, $2p\sigma$
3-state	$1s\sigma$, $2p\sigma$, $2p\pi$
4-state	$1s\sigma$, $2p\sigma$, $2p\pi$, $2s\sigma$
5-state	$1s\sigma$, $2p\sigma$, $2p\pi$, $2s\sigma$, $3d\sigma$
10-state	5-state + $3p\sigma$, $3p\pi$, $3p\delta$, $3d\pi$, $4s\sigma$

It was established that at least the 4 states $1s\sigma$, $2p\sigma$, $2p\pi$, and $2s\sigma$ were required to obtain converged capture cross sections. This is shown clearly in Fig. 6.11 where the probability for capture times impact parameter is shown for $E_L = 17.5$ keV/a.m.u. for the 2-, 3-, 4-, and 10-state bases. Winter *et al.* showed that for $E_L > 2$ keV/a.m.u. the use of a Coulomb trajectory did not change the cross sections appreciably from those obtained from a straight line trajectory. Further results using a 10-state MO basis have been computed by Kimura and Thorson (1981*b*) at 1–5 keV employing state-dependent translational factors, and by Errea *et al.* (1987*b*) at 5–20 keV, who employed an optimized common translational factor, using the minimum norm method. The three MO calculations agree fairly well at 5 keV/a.m.u.,

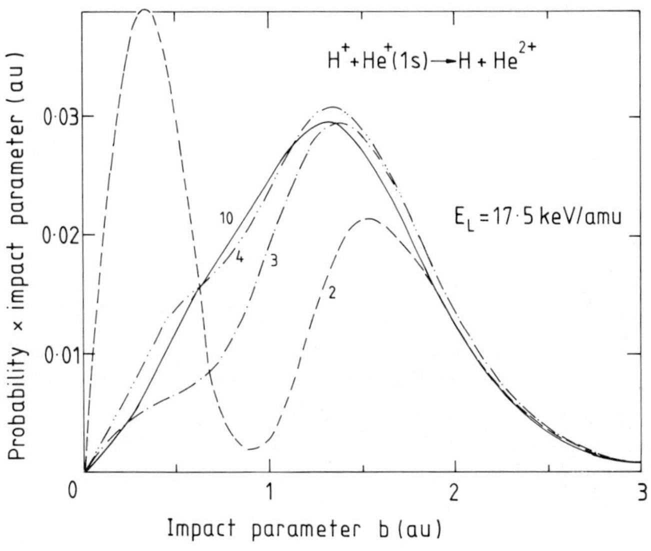

Fig. 6.11. The probability for charge exchange multiplied by the impact parameter, as a function of the impact parameter for protons incident on He^+ at $E_L = 17.5$ keV/a.m.u. The numbers against the curves indicate the size of the molecular orbital basis set employed (after Winter *et al.* 1980).

Table 6.3. Selected total electron capture cross sections for the reaction $H^+ + He^+(1s) \rightarrow H + He^{2+}$. The numbers in brackets represent the power of ten by which the preceding entry should be multiplied.

A 10-state MO basis (Kimura and Thorson 1981*b*); optimized translation factor;
B 10-state MO basis (Errea *et al.* 1987*b*); minimum norm translation factor;
C 23-state two-centre basis with pseudostates on the target (Bransden *et al.* 1983);
D 16-state two-centre AO^+ basis with united atom pseudostates (Fritsch and Lin 1982*b*); capture into H(1s) only;
E 35-state (21-state at 5 keV/a.m.u.) two-centre Sturmian basis (Winter 1986, 1987);
F up to 34 states, three-centre basis (Winter 1988);
G as for F but capture into H(1s) only.

Plane-wave translational factors are employed in C to G.

	Cross sections (10^{-16} cm^2)						
E_L (keV/a.m.u.)	2.0	5.0	17.5	31.25	50.0	125.0	187.5
A	2.84(−5)	1.99(−3)	–	–	–		
B	–	2.3(−3)	9.27(−2)	–	0.28		
C	–	–	–	0.21	0.25		
D	3.0(−5)	2.3(−3)	9.5(−2)	0.20	0.22		
E	–	2.1(−3)	9.64(−2)	0.236	0.272		
F	3.9(−5)	2.20(−3)	9.03(−2)	0.224	0.267	0.138	0.0583
G	3.9(−5)	2.17(−3)	8.85(−2)	0.219	0.235	0.108	0.0443

while at 20 keV/a.m.u. the results obtained with the plane-wave translational factor appear to be ~20 per cent too small when compared with the minimum norm calculations. This may well be due to the increasing importance of ionized states. Some calculated cross sections calculated with different basis sets are compared in Table 6.3.

As can be seen from Fig. 6.11, the capture probability is large for small values of the impact parameter $b \lesssim 2$. This suggests that in the trial wave function the united atom limit should be well represented, and this is borne out by calculations using two-centre basis sets. If the two-centre basis is composed of bound hydrogenic states about A and B, cross sections of about half the correct magnitude at the maximum are obtained. With the addition of a limited number of pseudostates overlapping the continuum and the united atom wave function, the capture cross section is increased to values which agree with experiment. Some results from two-centre calculations with plane-wave translational factors are shown in Table 6.3. These include a 23-state basis with hydrogenic states on each centre with $n \leq 2$ augmented by Callaway–Wooten pseudostates on one of the scattering centres (Bransden *et al.* 1983) and a 16-state AO$^+$ calculation containing united atom pseudostates (Fritsch and Lin 1982*a*). The cross sections for these calculations agree reasonably well between each other and with those obtained from 35-state Sturmian expansions (Winter 1982, 1986, 1987). The results of calculations with a *triple-centre* basis (Winter 1988) are also in harmony with those of the other calculations of the capture cross section. Bases of various size were investigated with up to 34 states containing all states with $n \leq 2$ centred on the proton, all with $n \leq 3$ centred on the α-particle, and all with $n \leq 4$ on the equiforce point. The results of the MO (minimum norm) calculation of Errea *et al.* (1987*b*), the two-centre pseudostate calculation of Bransden *et al.* (1983), the AO$^+$ two-centre calculation of Fritsch and Lin (1982*a*), and the three-centre calculation of Winter (1988) are compared with the experimental data in Fig. 6.12. Also included in the figure are the results of classical Monte Carlo calculations (Olson 1978) and of the OHCE model (Reading *et al.* 1982) to be discussed in Chapter 8. It can be seen that, up to energies ~150 keV/a.m.u., the general agreement between the impact

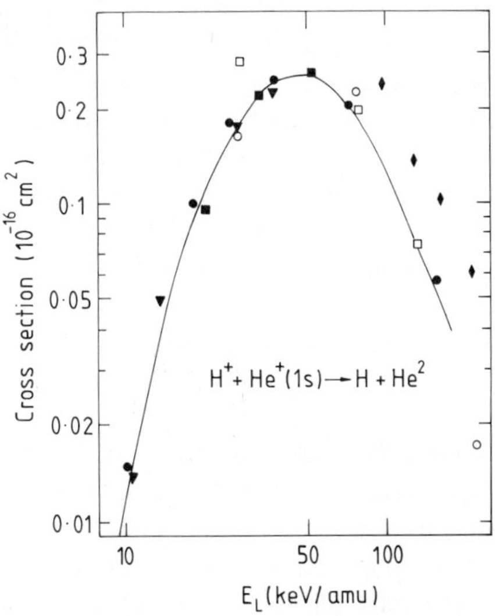

Fig. 6.12. The total cross section for electron capture by protons from He$^+$(1s). The solid line is representative of the experimental data of Peart *et al.* (1983), Rinn *et al.* (1985), and Watts *et al.* (1986). Theoretical cross sections: ○ pseudostate (Bransden *et al.* 1983); ● AO$^+$ (Fritsch and Lin 1982*a*); ▼ minimum norm MO (Errea *et al.* 1987*b*); ■ three-centre (Winter 1988); □ OHCE model (Reading *et al.* 1982); ◆ CTMC model (Olson 1978). The OHCE and CTMC models are discussed in Chapter 8.

parameter coupled channel calculations among themselves and with experiment is remarkably good.

Ionization

Theoretical models employing basis sets containing terms which overlap the continuum states of the target and/or the projectile allow an estimate of the ionization cross section to be made. In general to obtain an accurate ionization cross section it is necessary to represent the continuum states of the system with greater accuracy than is required for a calculation of excitation or charge transfer to low-lying bound states, and for this reason the calculation of the ionization cross section either requires the use of rather large basis sets, or of sets which can be shown to have a large overlap with the continuum. For the system under discussion, $H^+ + He^+$, ionization cross sections have been calculated from the two-centre Sturmian model (Winter 1986, 1987), the three-centre atomic orbital model (Winter 1988) and the two-centre pseudostate model of Fritsch and Lin (1983*b*). In addition Errea *et al.* (1989*b*, 1990) have introduced a method for forming molecular orbital pseudostates which belong to the continuum and have presented cross sections computed with the orbitals $1s\sigma$, $2s\sigma$, $2p\sigma$, $2p\pi$, $3p\sigma$, $3d\pi$, and $4\delta\sigma$, together with four pseudostate orbitals. The theoretical cross sections and the experimental data are illustrated in Fig. 6.13. It can be seen that the cross sections derived from the different variants of the coupled channel model differ quite widely, in contrast to the capture cross section for which the various methods provide essentially the same result. Further work to demonstrate which, if any, of the models have converged is required. On the positive side, the theoretical cross sections are in essential harmony with the experimental data, when the rather large errors are taken into account. Also included in Fig. 6.13 are results from the OHCE model of Reading *et al.* (1982) which will be described in Chapter 8. This is a high-energy method and as can be expected the computed cross sections agree well with the data at above $E_L \sim 100$ keV/a.m.u., but are clearly in error at lower energies

Targets with $Z_B > 2$

The calculations of Winter using the two-centre basis with Sturmian orbitals have been extended to determine cross sections for the capture and ionization of hydrogenic ions with nuclear charge $Z_B = 3, 4, 5$, and 6 by proton impact. In the case of a $Li^{2+}(1s)$ target ($Z_B = 3$) other calculations have been made in the two-centre pseudostate model (Ermolaev and McDowell 1987) and the OHCE model (Reading *et al.* 1982). It is expected that capture will take place most readily when the incident velocity v_0 is comparable in magnitude with the Bohr orbital velocity v_e of the target. This suggests that the capture cross sections for different Z_B should be considered as a function of the scaled energy $\tilde{E}_L = E_L/Z_B^2$. Winter's results show that the capture cross sections $\sigma^c(Z_B, \tilde{E}_L)$ all peak at the same value of the scaled energy, namely

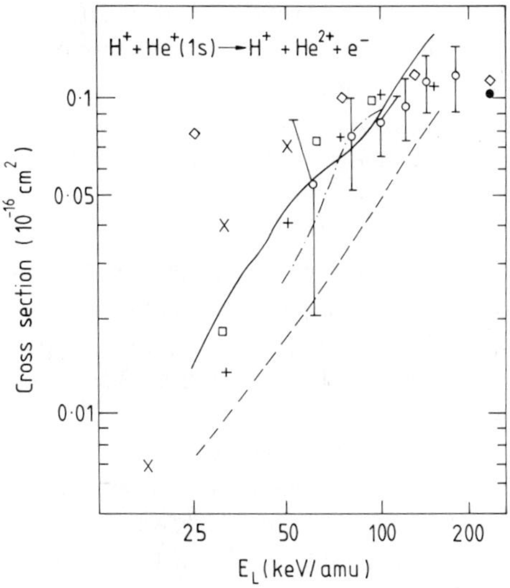

Fig. 6.13. The cross section for the ionization of $He^+(1s)$ by proton impact (after Errea *et al.* (1990)). Experimental data: ● Angel *et al.* (1978); —·—·— Peart *et al.* (1983); Watts *et al.* (1986). Theoretical cross sections: ——— upper and – – – lower estimates of cross sections by the molecular pseudostate model (Errea *et al.* 1990); □ AO^+ model (Fritsch and Lin 1983*b*); + Sturmian expansion (Winter 1986, 1987); × three-centre expansion (Winter 1988); ◇ OHCE model (Reading *et al.* 1982). The OHCE model is discussed in Chapter 8.

$\tilde{E}_L = 50$ (keV/a.m.u.) and the curves $Z_B^7\sigma^c(Z_B, \tilde{E}_L)$ all lie close to one another. The computed ionization cross sections multiplied by Z_B^4 (as suggested by the Born approximation) behave smoothly with Z_B, but the curves $Z_B^4\sigma^I(Z_B, \tilde{E}_L)$ lie less close to one another than those for capture.

Systems with $Z_A > Z_B$

The most-studied one-electron systems with $Z_A > Z_B$ are those in which the target is atomic hydrogen in the ground state ($Z_B = 1$). At low energies ($E_L \lesssim 25$ keV/a.m.u.) capture is the only important inelastic process; the cross sections are large and at the cross section maximum are approximately proportional to Z_A. States of the rearranged system ($A + e^-$) with principal quantum number $n = Z_A$ are degenerate with the initial atomic hydrogen ground state. This has the effect that for small values of Z_A, the capture cross section is dominated by transitions to states with n equal to or close to Z_A. We have already seen that for $He^{2+} + H(1s)$ collisions the only important final states at low energy are those with $n = 2$, and for $Li^{3+} + H(1s)$ the only important final states are those with $n = 3$. As Z_A increases the dominant final state occurs at values of $n < Z_A$ and the distribution in n broadens. For example, for $Z_A = 6$ the most important final states are those with $n = 4$, but states with $n = 5$ are also significant. Above the low-energy region capture becomes less state selective; for example, in the $He^{2+} + H(1s)$ system the cross section for capture into the $n = 3$ level is approximately 7 per cent of the total capture cross section at $E_L = 14$ keV/a.m.u., 15 per cent at $E_L = 25$ keV/a.m.u., and 24 per cent at $E_L = 75$ keV/a.m.u. (Bransden *et al.* 1983).

$He^{2+} + H(1s)$ The system $(He\,H)^{2+}$ was discussed from the partial-wave viewpoint in Chapter 4 and a comparison was made with the corresponding impact parameter model at the beginning of this chapter. At low energies ($\lesssim 1$ keV) a four-state MO basis ($2p\sigma$, $2p\pi$, $3d\sigma$, and $2s\sigma$) which correlates with the incident channel $He^{2+} + H(1s)$ and the exchange channels $He^{+}(n = 2) + H^{+}$ is sufficient (van Hemert *et al.* 1985). As the energy is increased a larger basis is required, and Winter and Hatton (1980) and Hatton *et al.* (1979) showed that the ten lowest states were necessary by $E_L \approx 20$ keV/a.m.u., the $3d\pi$ being particularly important. In the work of Winter and his colleagues plane-wave translation factors were employed, but further calculations employing MO basis sets with differing translation factors (Kimura and Thorsen 1981*b*; Errea *et al.* 1987*b*; Crothers and Todd 1981) are in close agreement with each other and with the plane-wave results, as shown in Table 6.4. The comparative unimportance of transitions at small

Table 6.4. Calculated total electron capture cross sections for the reaction $He^{2+} + H(1s) \rightarrow He^{+} + H^{+}$.

A four-state MO basis (van Hemert *et al.* 1985), common translation factor;
B 10-state MO basis (Hatton *et al.* 1979), plane-wave translation factor;
C 10-state MO basis (Errea *et al.* 1987*b*), minimum norm translational factor;
D 20-state two-centre basis (Bransden *et al.* 1983);
E Up to 24-state two-centre Sturmian basis (Winter 1982);
F Up to 35-state three-centre basis (Winter 1988).

	Cross sections (10^{-16} cm^2)				
E_L(keV/a.m.u.)	2.0	5.0	17.5	50.0	75.5
A	4.76	9.58	–	–	–
B	6.30	12.2	12.1	–	–
C	6.19	11.98	–	–	–
D	–	10.5	12.0	3.6	1.6
E	–	11.1	11.0	3.74	–
F	6.33	11.7	11.8	3.34	–

intermolecular separations and of coupling to the continuum can be illustrated from the calculations of Bransden and Noble (1981) and by Bransden *et al.* (1983) for the charge exchange cross section in the energy interval 2.5 to 228 keV/a.m.u. It was found that a two-centre expansion containing hydrogenic states with $n \leq 3$ on each centre provided cross sections which agreed well with experiment and, at the lower energies, with the results of the MO basis set calculations. The addition of pseudostates to the two-centre basis resulted in minor changes to the capture cross section, in contrast to those for the $H^+ + He^+(1s)$ system discussed above. More recently large-scale coupled channel calculations with two-centre Sturmian functions (up to 24 states) and with three-centre expansions (up to 36 states) (Winter 1986, 1988) have confirmed the comparative stability of the capture cross sections and allowed the ionization cross section to be predicted.

The cross sections calculated from a Hylleraas basis (Lüdde and Dreizler 1982*a*, *b*) are in fair agreement with the results of the other two-centre calculations. A comparison of selected theoretical cross sections with the experimental data for the total capture cross section and those for capture into the 2s and 2p states of He^+ are shown in Figs 6.14, 6.15, and 6.16. The ionization cross section has also been measured and is shown in Fig. 6.17

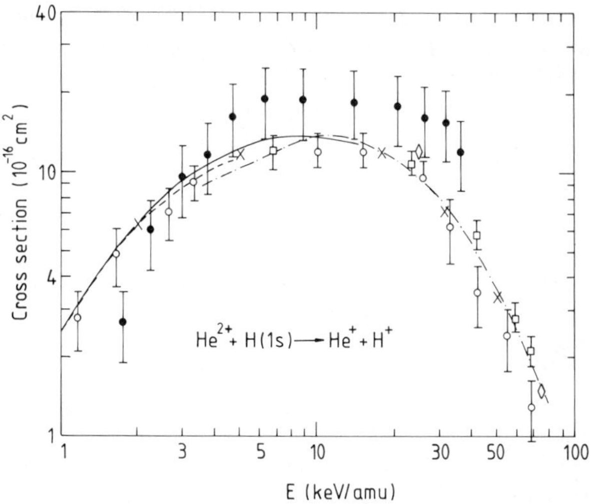

Fig. 6.14. The total cross section for electron capture by He^{2+} from H(1s). Experimental data: ○ Shah and Gilbody (1978) and Nutt *et al.* (1978); ● Bayfield and Khayrallah (1975*a*, *b*); □ Olson *et al.* (1977). Theoretical cross sections: × three-centre expansion (Winter 1988); ◇ OHCE model (Reading *et al.* 1982); ——— 10-state MO (plane-wave translation factor) (Hatton *et al.* 1979; Winter and Hatton 1980); ----- 10-state MO (optimized translation factor) (Kimura and Thorson 1981*b*); —·—·— 20-state AO (plane-wave translation factor) (Bransden *et al.* 1983). The OHCE model is discussed in Chapter 8.

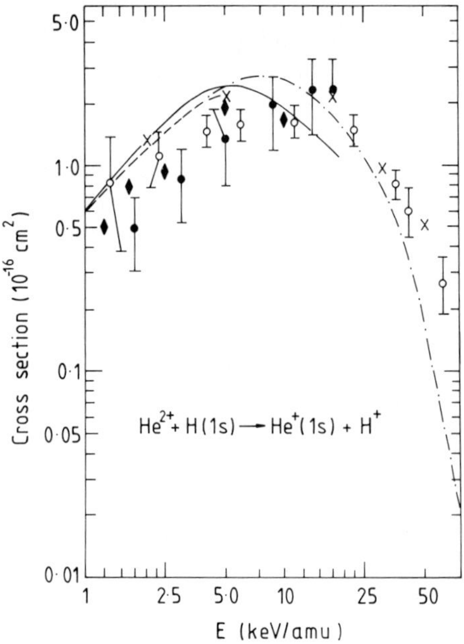

Fig. 6.15. The cross section for electron capture by He^{2+} from H(1s) into the 2s state of He^+. The notation is as in Fig. 6.14, with the addition of the experimental data of Ćirić *et al.* (1985*a*) ◆.

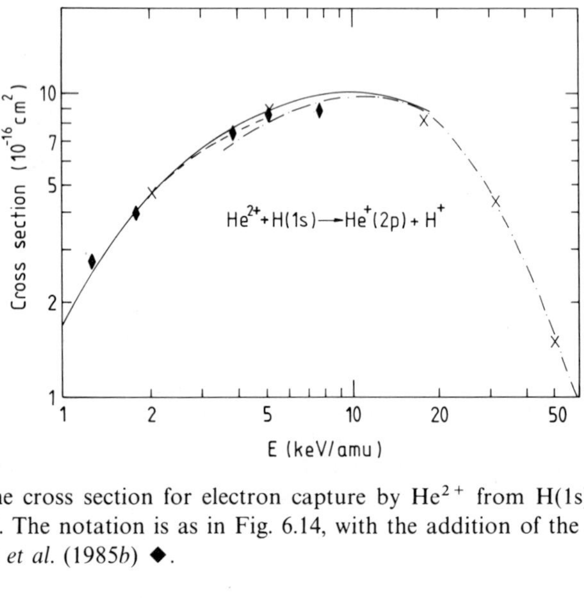

Fig. 6.16. The cross section for electron capture by He^{2+} from H(1s) into the 2p state of He^+. The notation is as in Fig. 6.14, with the addition of the experimental data of Ćirić *et al.* (1985*b*) ◆.

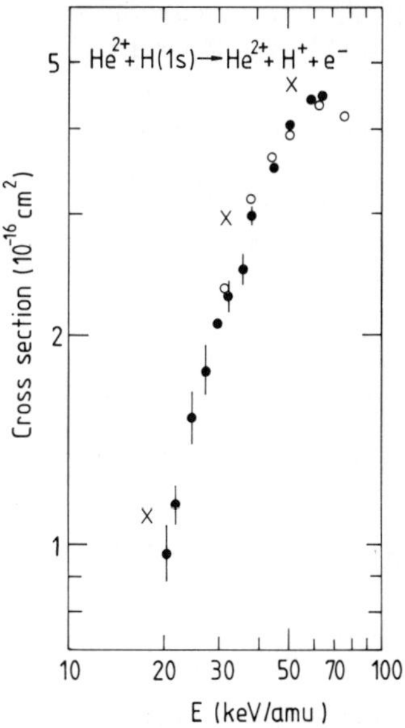

Fig. 6.17. Cross section for the ionization of H(1s) by He^{2+} impact (after Winter (1988)). Experimental data: ○ Shah and Gilbody (1981); ● Shah *et al.* (1988). Theoretical cross sections: × three-centre expansion (Winter 1988).

where it is seen that the calculated values of Winter (1988) have the correct energy variation but are larger in magnitude by 31 per cent at the lowest energy and by 10 per cent at the highest.

Li^{3+}, Be^{4+}, B^{5+}, C^{6+}, N^{7+} *and* O^{8+} + H(1s) Coupled channel impact parameter studies have been made for fully stripped ions with Z_A from 3 to 8 incident on H(1s) and there has been particular interest in predicting the distribution in *n*, *l*, and *m* of the final states of the one-electron ions formed by capture. In general, the total cross sections calculated with different basis sets (molecular orbital or two-centre) agree well with each other, and with the experimental data. (See, for example, the reviews of Gilbody (1986) and Janev and Winter (1985).) The characteristics of selected calculations are noted in Table 6.5. In contrast to the total capture cross section the cross sections for capture into the less important channels are less well determined. A particular example is the case of $Z_A = 6$, C^{6+} + H(1s). The 33-state MO basis set calculations of Green *et al.* (1982) agree well with the 35-state AO^+ calculation of Fritsch and Lin (1984*b*) and with the AO/MO calculations of Kimura and Lin (1985*b*) for capture into the dominant $n = 4$ level in an energy region extending from 1 to 10 keV/a.m.u. On the other hand, for capture into the $n = 5$ level, the partial cross section given by Green *et al.* (0.27×10^{-16} cm^2 at 5 keV/a.m.u.) differs considerably from those of Fritsch

Table 6.5. Selected coupled channel calculations for the capture reaction $A^{Z_A+} + H(1s) \rightarrow A^{(Z_A-1)} + H^+$, $3 \leq Z_A \leq 10$.

A^{Z_A+}	Characteristics of calculation[1]	Energy range (keV/a.m.u.)	Reference
Li^{3+}	six-state MO optimized switching factor	2–10	Kimura and Thorson (1981*b*)
	20-state AO all orbitals with $n \leq 3$ on each centre	1.4–200	Bransden and Noble (1982)
	Up to 34-state AO^+ including united atom pseudostate	0.2–20	Fritsch and Lin (1982*b*)
	Hylleraas basis with up to 150 terms[2]	2.25–49	Lüdde and Dreizler (1982*b*)
Be^{4+}	21-state, AO and AO^+	0.133–25	Fritsch and Lin (1984*b*)
	Hylleraas basis with up to 150 terms[2]	2.25–49	Lüdde and Dreizler (1982*b*)
B^{5+}	41-state, AO and AO^+	0.133–25	Fritsch and Lin (1984*b*)
	Hylleraas basis with up to 150 terms[2]	2.25–49	Lüdde and Dreizler (1982*b*)
C^{6+}	33-state MO; optimized state-dependent translation factor	0.013–27	Green *et al.* (1982)
	35-state AO	0.133–25	Fritsch and Lin (1984*b*)
	25/27 state MO/AO		Kimura and Lin (1985*b*)
N^{7+}	46-state AO	0.133–25	Fritsch and Lin (1984*b*)
O^{8+}	33-state MO; optimized state-dependent translation factor	0.013–27	Shipsey *et al.* (1983)
	46-state AO	0.1–30	Fritsch and Lin (1984*b*)

[1] Plane-wave translational factors used except where noted.
[2] No translation factor in the interaction region.

and Lin ($\sim 4 \times 10^{-16}$ cm^2) and Kimura and Lin ($\sim 2.5 \times 10^{-16}$ cm^2) at the same energy. These differences increase at lower energies, and while it is possible that the discrepancy might be due to an approximate treatment of the matrix elements employed by Green *et al.*, it points to the sensitivity of the smaller cross sections to the model employed and the difficulty of predicting the small cross sections with confidence.

6.4 QUASI-ONE-ELECTRON SYSTEMS

In many instances the inelastic reactions of a many-electron system can be described in terms of a one-electron model in which the interaction of the active electron with the ion cores centred on A and B can be described by effective potentials V_{Ae}, V_{Be}. Particularly favourable systems for a one-electron treatment are the alkalis, where the interaction of the valence electron with the core can be represented as a parameterized central potential. It is possible to determine the parameters of the potential so that the energy level spectrum of the alkali atom is reproduced accurately. (See, for example, Daniele (1979), Peach (1982), and Ermolaev and Hewitt (1985).) Quite successful one-electron models can be developed for other systems, for example those in which the target contains two electrons, H^-, He, $C^{4+}, \ldots$, but we shall confine our illustrative examples to some reactions evolving Li and Na atoms.

$He^{2+} + Li(2s)$ The interaction of He^{2+} ions with neutral Li atoms is of particular interest as it shows the very wide range of energies for which the coupled channel impact parameter model can provide accurate cross sections. Up to energies of the order $E_L \approx 50$ keV/a.m.u., the capture excitation and ionization reactions can be attributed solely to the 2s valence electron, the $Li^+(1s^2)$ core acting as a passive spectator

$$\begin{aligned} He^{2+} + Li(1s^2, 2s) &\to He^+(nlm) + Li^+(1s^2) \\ &\to He^{2+} + Li(1s^2, nlm) \\ &\to He^{2+} + Li^+(1s^2) + e^-. \end{aligned} \tag{6.31}$$

Above $E_L = 50$ keV/a.m.u., the situation changes in that capture then takes place predominantly from the K-shell. This again can be treated as a quasi-one-electron system using an effective potential which allows for screening of the nucleus by the second K-shell electron. The K-shell electrons also contribute to the ionization cross section, but not until much higher energies ~ 250 keV/a.m.u.

Using a molecular orbital basis with no translation factors Shipsey *et al.* (1978) studied the energy range from $E_L = 0.4$ to $E_L = 2.8$ keV/a.m.u. and showed that the capture reaction was extremely state selective leading to the formation of He^+ in the $n = 3$ shell. Further molecular orbital calculations allowing for translational invariance confirmed this result, but predicted cross sections in much better agreement with experiment (Sato and Kimura 1983). A large-scale two-centre AO^+ basis with united atom pseudostates and containing up to 40 terms extended the energy range to 16 keV/a.m.u. (Fritsch and Lin 1983*c*). Taking into account capture from the K-shell,

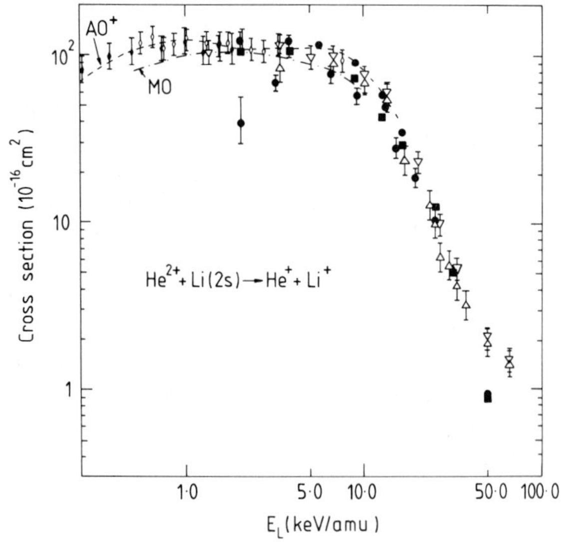

Fig. 6.18. Cross section for electron capture from the L-shell of Li by He^{2+} impact. Experimental data: ◇ Kadota *et al.* (1982); ● Murray *et al.* (1982); △ McCullough *et al.* (1982); ◆ Varghese *et al.* (1984); ▽ DuBois and Toburen (1985). Theoretical cross sections: —·—·— MO basis (Sato and Kimura 1983); ----- AO^+ basis (40 terms) (Fritsch and Lin 1983*c*), ■ AO pseudostate basis (32 terms) (Ermolaev *et al.* 1987*a*).

Ermolaev and Bransden (1984) computed capture and excitation cross sections up to 400 keV/a.m.u. using a two-centre expansion containing He^+ and Li bound states. The addition of pseudostates to the basis improved the convergence of the capture and excitation cross sections and allowed the computation of ionization cross sections over an extended interval 2–500 keV/a.m.u. (Ermolaev *et al.* 1987*a*, *b*).

The computed capture total cross sections are compared with experimental data in Fig. 6.18 for the MO, AO^+, and pseudostate basis sets for the energy interval in which capture from the L-shell dominates ($E_L < 50$ keV/a.m.u.). The results of the AO bound state basis (not shown) fall below the data at energies below 3 keV/a.m.u. because of the poor representation of the united atom limit. At the higher energies, the AO bound state basis provides accurate capture cross sections up to 16 keV/a.m.u. and for energies above 50 keV/a.m.u. Between 16 and 50 keV/a.m.u. the addition of pseudostates is important to obtain accurate capture cross sections and thus is directly related to the importance of the ionization of the valence electron over this interval. The cross sections for capture into some individual states of He^+ have been measured by Kadota *et al.* (1982) (2p, 3p, 3s + 3d, 4s + 4d) and by Boellard (1984) (2p, 3p, 4p). The agreement with the theoretical cross sections is in general good, with the smaller cross sections (2p, 4p, 4s + 4d) showing the greatest discrepancies, as expected. The largest capture cross sections (3p and 3s + 3d) are illustrated in Fig. 6.19. Rather similar studies have been carried out for the $H^+ + Li$, $H^+ + Na$, $He^{2+} + Na$, and $Be^{2+} + Li$ systems using both molecular orbital and two-centre expansions Ermolaev 1984, 1985; Fritsch 1984; Allan 1986; Shingal 1988*b*; Shingal *et al.* 1986*a*, 1987; Shingal and Bransden 1987; Allan and Hanssen 1985) and again rather good agreement is found both with the capture and, where it exists, the excitation data over a wide energy interval.

$Li^+ + Li$, $Na^+ + Na$, $Li^+ + Na$, $Na^+ + Li$ The symmetrical $Li^+ + Li$, $Na^+ + Na$ collisions are similar to those between H^+ and H(1s) in that the differential cross sections for elastic scattering, charge exchange, and excitation all show oscillations as a function of angle, due to interference between the lowest gerade and ungerade states. Similar oscillations are shown in the near symmetrical collisions $Li^+ + Na$, $Na^+ + Li$. Of particular interest is that, unlike the $H^+ + H$ system, these oscillations are also shown in the total cross sections as a function of energy. Coupled channel calculation using either an MO (Shimakura *et al.* 1981; Allan 1986; Allan and Hanssen 1985; Men *et al.* 1986) or a two-centre pseudostate basis (Shingal *et al.* 1986*b*; Shingal 1988*a*) can reproduce the main features of the experiments of Daley and Perel (1969), but it is clear that covergence is quite slow and rather large basis sets are required to obtain accurate cross sections. As an example, the total cross sections for electron capture by Li^+ from Li, computed with pseudostate basis sets containing 30 and 62 terms, are compared with the experimental data in Fig. 6.20.

6.5 TWO-ELECTRON SYSTEMS

Although in principle systems with two or more electrons can be treated by the coupled channel method in exactly the same way as a one- or quasi-one-electron system, the additional numerical complexity has limited the number of applications.

6.5.1 Homonuclear systems; H_2, He_2^+

For the fundamental system of two protons and two electrons certain reactions can be treated reasonably in a quasi-one-electron model. For example in the neutralization reaction:

$$H_A^+ + H_B^- \rightarrow H_A^* + H_B(1s) \tag{6.32}$$

the core can be described by a model in which an electron moves in the

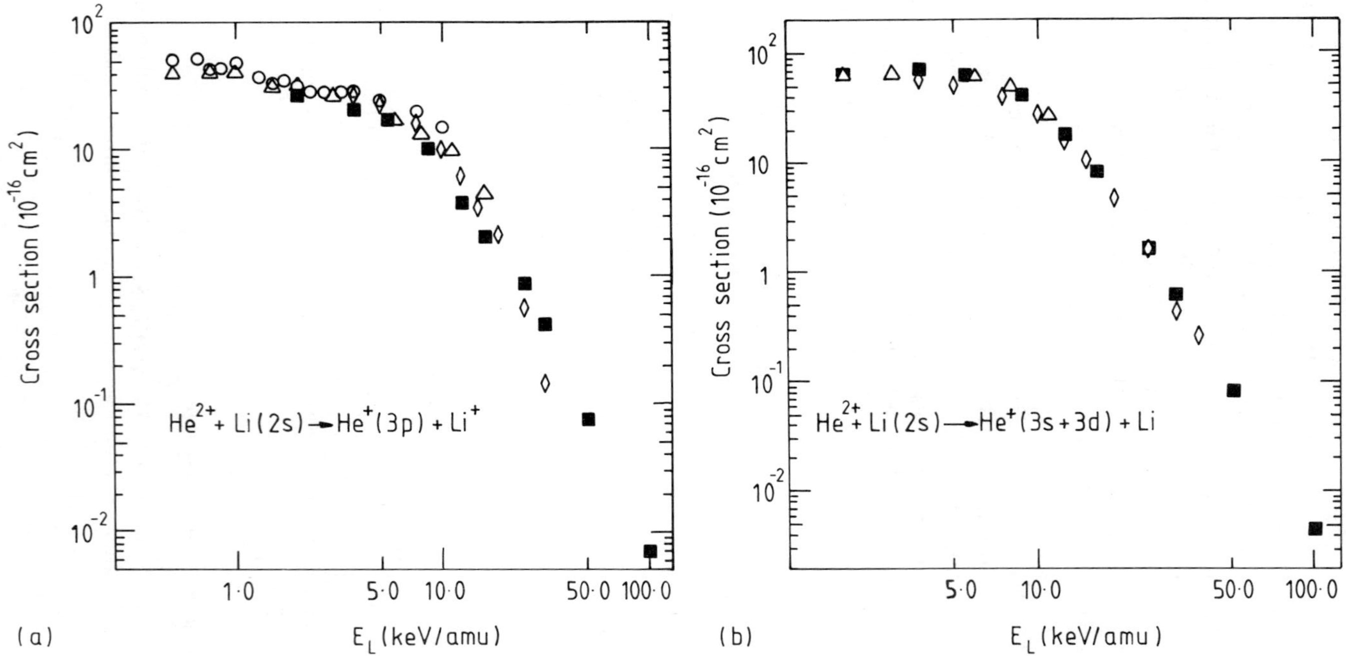

Fig. 6.19. Cross sections for capture into (a) the 2p level of He^+ and (b) the 3s + 3d levels of He^+ in the reaction $He^{2+} + Li \rightarrow He^+ + Li^+$. Experimental data: ○ Boellard (1984); ◇ Kadota *et al.* (1982). Theoretical cross sections: △ 40-term AO^+ basis (Fritsch and Lin 1983*c*); ■ 32-term AO pseudostate basis (Ermolaev *et al.* 1987*a*).

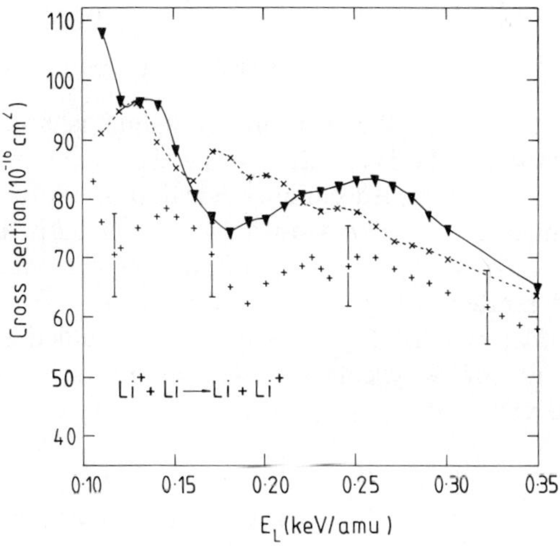

Fig. 6.20. Electron capture by Li^+ ions from Li (after Shingal 1988*a*). Experimental data: + Daley and Perel (1969). Theoretical cross sections: —▼—▼ 62-term pseudostate basis (Shingal 1988*a*); —∗—∗ 30-term pseudostate basis (Shingal 1988*a*).

Coulomb field of a proton and an effective potential due to an inert H(1s) core (Sidis *et al.* 1981, 1983; Ermolaev 1985). Obviously such a model cannot directly describe the double capture reaction

$$H_A^+ + H_B^- \to H_A^- + H_B^+, \tag{6.33}$$

nor can it describe neutralization in which the target is left in an excited state

$$H_A^+ + H_B^- \to H_A + H_B^*. \tag{6.34}$$

However, estimates of the probabilities for two-electron transitions, in terms of products of one-electron transition probabilities, can be made with the help of the independent particle model discussed in Chapter 9. The neutralization reaction (6.32) is well described over an extended energy region by an explicit two-electron coupled channel model (Shingal *et al.* 1985; Shingal and Bransden 1990) and qualitative agreement with experimental data has been obtained for the reactions

$$\begin{aligned} H(1s) + H(1s) &\to H_A^- + H_B^+ \\ &\to H_A^+ + H_B^-, \end{aligned} \tag{6.35}$$

and

$$H_A(1s) + H_B(2s) \to H_A^- + H_B^+. \tag{6.36}$$

Another reaction in which a two-electron treatment allowing for electron exchange is vital is excitation due to the collision of two ground-state hydrogen atoms:

$$\mathrm{H(1s) + H(1s) \rightarrow H(1s) + H(}nlm\mathrm{)}. \tag{6.37}$$

Early calculations, treating the electrons as distinguishable, produced cross section for excitation of the 2s and 2p states which are an order of magnitude too small. In contrast, coupled channel calculations (McLaughlin and Bell 1983, 1985; Shingal *et al.* 1989) with a two-centre basis allowing fully for exchange predict cross sections in qualitative agreement with the experimental data (illustrated in Fig. 6.21).

The symmetrical system He_2^{2+} has also been studied using either two-electron two-centre or two-electron molecular orbital basis sets. The reactions concerned are

$$\mathrm{He^+ + He^+ \rightarrow He + He^{2+}} \tag{6.38a}$$

$$\mathrm{\rightarrow He^+ + He^{2+} + e^-}, \tag{6.38b}$$

and

$$\mathrm{He^{2+} + He \rightarrow He^+ + He^+} \tag{6.38c}$$

$$\mathrm{\rightarrow He + He^{2+}} \tag{6.38d}$$

$$\mathrm{\rightarrow He^{2+} + He^+ + e^-}. \tag{6.38e}$$

All five reactions have been the subject of an extensive study over the energy range $4 \lesssim E_L \lesssim 100$ (keV/a.m.u.) by Gramlich *et al.* (1989) using a two-electron atomic orbital basis augmented by pseudostates for the calculation of ionization. The work is particularly interesting for the use of Gaussian orbitals rather than Slater functions to represent the atomic wave functions. As noted in Chapter 5, this technique, pioneered by Dose and Semini (1974), has the advantage that if plane-wave translational factors are employed all matrix elements can be evaluated analytically (see Appendix F). This is particularly important for multi-electron systems where the calculation of matrix elements using Slater functions is extremely time consuming. The calculated cross sections for reaction (a) are in good agreement with the two-centre AO^+ calculations of Fritsch and Lin (1987) and with those of Kimura (1988) who employed a molecular orbital basis. For reactions (c) and (d) comparison can also be made with the work of Kimura (1988) and with earlier calculations of Harel and Salin (1980) who omitted translation factors. In both cases the agreement is fair and there is reasonable agreement with the experimental data.

The most probable reactions between He^+ and He^+ and between He^{2+} and He are those leading to capture into singly excited helium and to single ionization. However, the reactions in which doubly excited autoionizing states are formed, and reactions leading to double ionization, are of particular interest because of correlation effects due to the electron–electron interaction. Similar formation of doubly excited states and multiple ionization

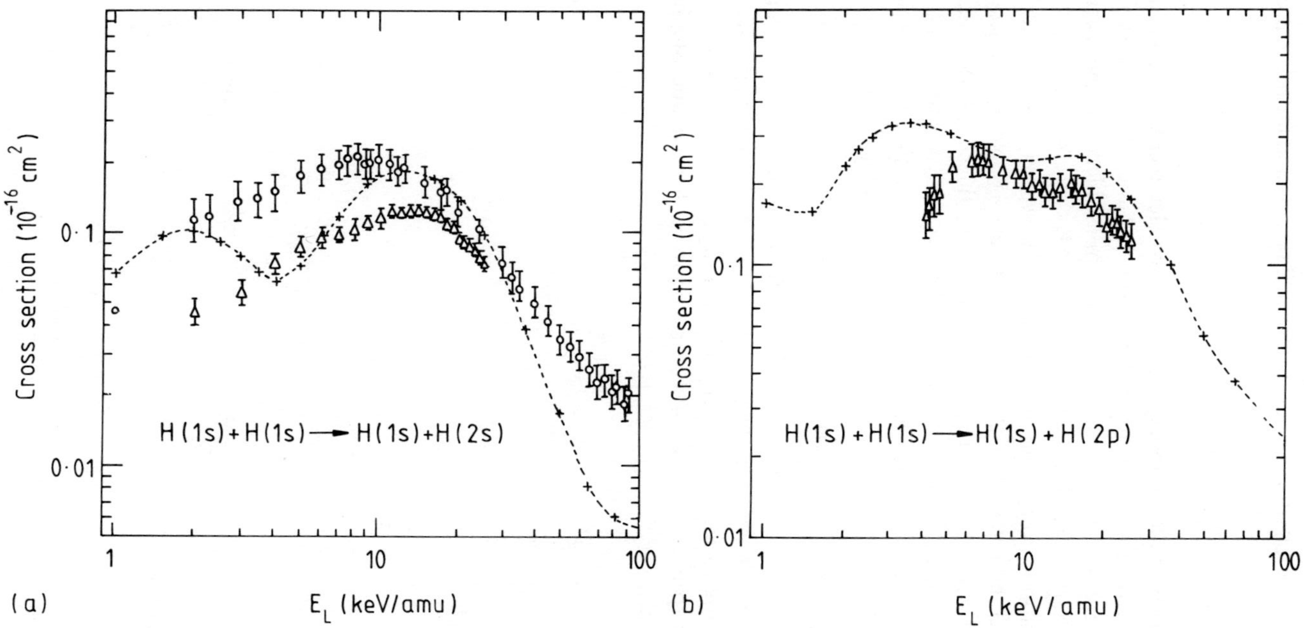

Fig. 6.21. Cross sections for (a) H(1s) + H(1s) → H(1s) + H(2s), and (b) H(1s) + H(1s) → H(1s) + H(2p). Experimental data: Fig. (a): △ Hill *et al.* (1979); ○ Morgan *et al.* (1980); ▽ Morgan *et al.* (1974). Theoretical cross sections: 22-term AO pseudostate basis +−+−+ (Shingal *et al.* 1989). Note: the experimental data are for the production of H(2s) and H(2p). Above 10 keV/a.m.u. contributions from double excitaton of the form H(1s) + H(1s) → H*(*nlm*) + H*(*n′l′m′*) become important.

can take place in all collisions between ions or atoms containing two or more electrons. The discussion of these topics will be deferred until Chapter 9.

6.5.2 Heteronuclear systems; $H^+ + He$

As an illustrative example of an asymmetrical two-electron system we shall discuss briefly electron capture by protons from helium. This process is of considerable historical interest since it was the subject of the earliest perturbed stationary state (PSS) calculation by Massey and Smith (1933). The PSS calculations were later improved by Haywood (1959), and the impact parameter model with a straight line trajectory using a molecular orbital basis, but with no translation factor, was introduced a little later by Colegrave and Stephens (1968). Capture takes place almost entirely into the ground state of atomic hydrogen leaving the helium ion also in the ground state:

$$H^+ + He(1s^2) \rightarrow H(1s) + He^+(1s). \tag{6.39}$$

From the molecular orbital viewpoint the reaction occurs via the coupling between the 1Σ state asymptotic to the $H^+ + He(1s^2)$ arrangement and the 2Σ state asymptotic to the $H(1s) + He^+(1s)$ arrangement (as shown in Fig. 6.22). The total capture cross section computed by Colegrave and Stephens

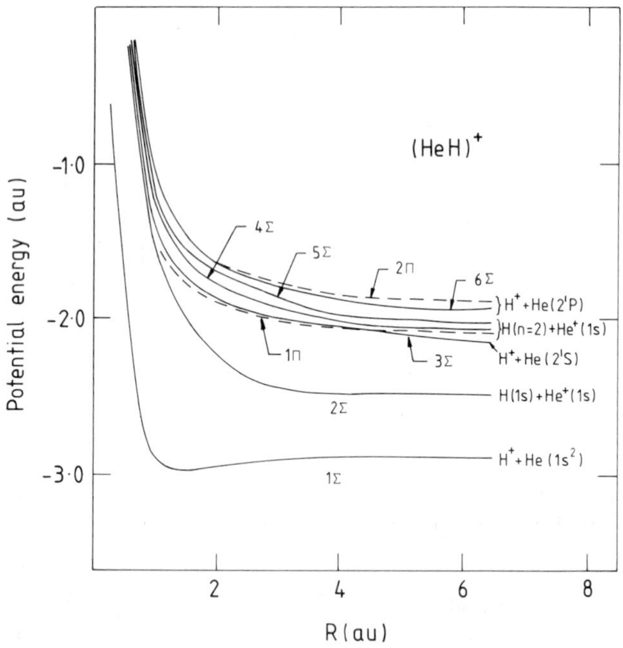

Fig. 6.22. Adiabatic potential energies for $(He\ H)^+$.

is in poor agreement with the experimental data near the cross section maximum at $E_L \approx 25$ (keV/a.m.u.) and this is certainly due to the unjustified omission of translational factors (see Crothers and Todd 1981). The first calculation by Green *et al.* (1965) consistent with translational invariance employed a two-state atomic orbital basis containing a ground-state helium wave function centred on the target nucleus in the incident channel and a symmetrical product of H(1s) and He^+(1s) wave functions in the rearranged channel; plane-wave translational factors were employed. If the calculated two-state capture cross sections are increased by a factor from 15 to 20 per cent to allow for capture into excited states reasonable agreement is found with the measured cross section over an energy range from $20 \lesssim E_L \lesssim 100$ (keV/a.m.u.) as illustrated in Fig. 6.23. The comparative success of the two-state atomic orbital approximation for ground-state capture can be attributed to the very compact nature of the helium atom, which ensures the wave function has an atomic character down to very small internuclear separations.

A comparison between the results of Green *et al.* and those of a two-state calculation of Bransden and Sin Fai Lam (1966), who treated the electrons in the final state as distinguishable, showed that electron exchange was unimportant for energies above $E_L \sim 10$ keV/a.m.u. as far as ground-state capture was concerned. For this reason, in the more elaborate multistate calculations of Sin Fai Lam (1967*a*, *b*), who used four states, and of Winter

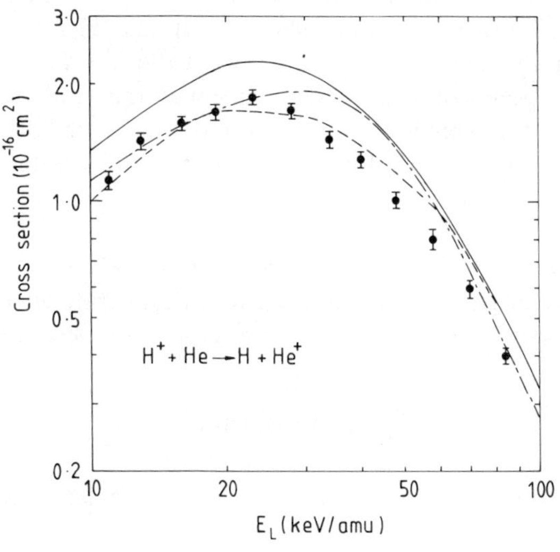

Fig. 6.23. The total cross section for electron capture by protons from helium. Experimental data: ● Shah *et al.* (1989). Theoretical cross sections: ——— 40-term pseudostate two-centre basis (Slim *et al.* 1991*a*); ----- 8/9-term MO/AO basis (Kimura and Lin 1986); —·—·— two-state AO approximation (Green *et al.* 1965).

and Lin (1974) who used 11 states, the distinguishable electron approximation was employed, but recent work (Slim *et al.* 1990) shows that this is a poor approximation for the smaller cross sections. In addition, to obtain accurate cross sections for the less important channels using a two-centre expansion, it is certainly necessary to add pseudostates to the basis.

Beyond the two-state approximation it is necessary, as a minimum, to include all the capture channels of the form $H(n = 2) + He^+(1s)$ and the excitation channels $H^+ + He(n = 2)$ all of which have closely spaced energies in the limit $R \to \infty$ (see Fig. 6.22). The characteristics of some calculations which include at least these channels are summarized in Table 6.6. The total capture cross sections calculated from molecular orbital and from MO/AO basis sets by Kimura and Lin (1986) agree closely and are also in fair agreement with the experimental data (as shown in Fig. 6.23). The total cross sections obtained from the two-centre pseudostate expansion of Slim *et al.* (1990) are slightly larger than those found from the MO/AO basis at energies below $E_L = 60$ keV/a.m.u., and this may be due to the small size of the latter basis. For capture into excited states, agreement between different calculations is more marked, as would be expected. In Fig. 6.24 the cross sections from the MO/AO basis of Kimura and Lin, and the pseudostate basis of Slim *et al.* for capture into the 2s and 2p levels of atomic hydrogen are shown, together with the experimental data. The theoretical cross sections show the qualitative features of the data, but close agreement is not obtained. Excitation cross sections for the 2^1S and 2^1P states of helium obtained from the MO/AO and the pseudostate calculations agree rather well with the experimental data (see Slim *et al.* 1990, 1991*a*); the differential cross sections for capture in all states and for excitation of the 2^1S and 2^1P helium state found from the pseudostate basis also agree with the available measurements (Slim *et al.* 1991*b*). This is illustrated in Chapter 7 where Fig. 7.6 displays the differential cross section for excitation of the 2^1S state of helium at an energy of 100 keV/a.m.u.

The density matrix for final states of the hydrogen atom

The excited hydrogen atoms produced in a degenerate level with principal quantum number n by capture are in a pure state described by the wave function

$$\Psi = \sum_{lm} \mathscr{A}_{nlm}(\mathbf{b})\phi_{nlm}(\mathbf{x}) \exp(-i\varepsilon_n t). \tag{6.40}$$

All observable quantities relating to this state can be described by the density matrix

$$\rho_{lm,l'm'}(\mathbf{b}) = \mathscr{A}_{nlm}(\mathbf{b})\mathscr{A}^*_{nl'm'}(\mathbf{b}). \tag{6.41}$$

The hydrogen atoms produced in actual experiments originate from all values of the impact parameter and the quantity of interest is the integrated

Table 6.6. Selected coupled channel calculations for $H^+ - He$.

Characteristics of calculation	Energy range (keV/a.m.u.)	Reference
Full two-electron models		
MO: eight states with channels asymptotic to H($n = 2$) and He($n = 2$); translation factor expanded to first order in v_0	1–40	Kimura (1985)
MO: 8 states with state-dependent translational factor (not expanded)		
MO/AO: 8 states as above in inner region 9 states in AO outer region including H(3p) inner MO region: no translational factor outer AO region: plane-wave translational factor Coulomb trajectory employed, also straight line	1–100	Kimura and Lin (1986)
Two-centre pseudostate He($n = 1$, $n = 2$) on target B together with $2s^2$, 2s2p, and $2p^2$ autoionizing states. H($n \leq 4$) on projectile A together with 8 pseudostates; plane-wave translational factors	20–100	Slim *et al.* (1991*a*, *b*)
Quasi-one electron model		
Two-centre pseudostate (AO^+) with plane-wave translation factors:		
Basis of He($n = 1$ and $n = 2$) states plus four pseudostates on centre B together with H($n \leq 3$) states plus two pseudostates on centre A	25–100	Jain *et al.* (1987*a*, *b*, 1988)
Large two-centre basis of 64 states and pseudostates	25–100	Shingal and Lin (1991*b*)

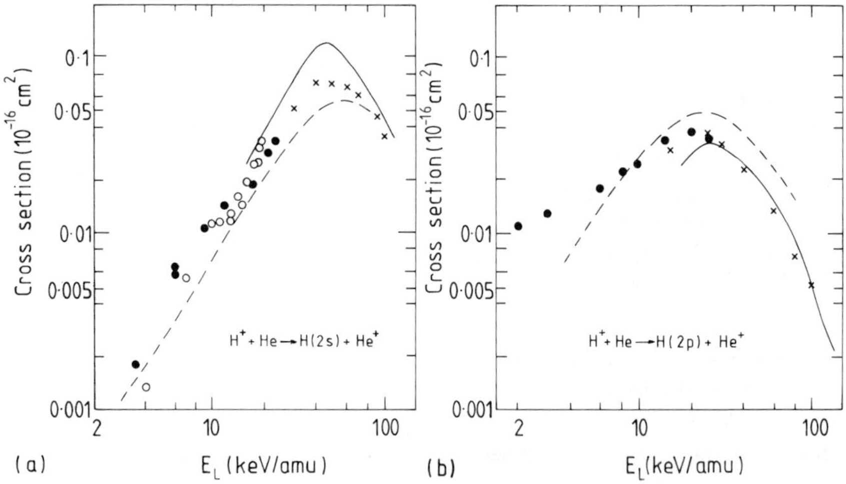

Fig. 6.24. Cross sections for electron capture by protons from helium (a) into the 2s level, and (b) into the 2p level of atomic hydrogen. Experimental data: ●● Jaecks *et al.* (1965); ○○ Crandall and Jaecks (1971); × × Hughes *et al.* (1971). Theoretical cross sections: ——— 40-term pseudostate two-centre basis (Slim *et al.* 1991*a*); ----- 8/9-term MO/AO basis (Kimura and Lin 1986).

density matrix $\bar{\rho}$ where

$$\bar{\rho}_{lm,l'm'}(\mathbf{b}) = \int d^2\mathbf{b}\mathscr{A}_{nlm}(\mathbf{b})\mathscr{A}^*_{nl'm'}(\mathbf{b}). \tag{6.42}$$

The density matrix must be Hermitian

$$\bar{\rho}^*_{lm,l'm'}(\mathbf{b}) = \bar{\rho}_{l'm',lm}(\mathbf{b}) \tag{6.43}$$

and because of the reflection symmetry in the plane of scattering of the heavy particles, the density matrix elements satisfy the relation

$$\bar{\rho}_{lm,l'm'}(\mathbf{b}) = (-1)^{m+m'}\bar{\rho}_{l-m,l'-m'}(\mathbf{b}). \tag{6.44}$$

In addition, the integrated density matrix (but not $\rho_{lm,l'm'}(\mathbf{b})$) must be diagonal in m, because of the cylindrical symmetry of the scattering system.

The symmetry requirements (6.43) and (6.44) reduce the number of independent elements. For example, for $n = 2$ there are three real diagonal and one complex off-diagonal elements, and for $n = 3$ there are six real diagonal and two complex off-diagonal elements. As is evident from (6.42), the diagonal elements of the integrated density matrix are equal to the partial cross sections for capture into the state with quantum numbers nlm. The off-diagonal elements are related to the phase differences (coherences) between the different states and can be interpreted in terms of the average

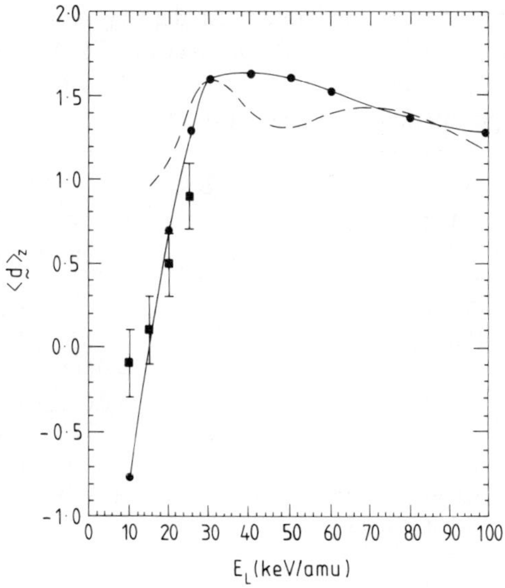

Fig. 6.25. The Z-component of the electric dipole moment for capture by protons from helium into the $n = 2$ level of atomic hydrogen. Experimental data: ● Hippler (1991*a*, *b*). Theoretical cross sections: —●— 40-term pseudostate two-centre basis (Slim *et al.* 1991*a*); ——— 72-term pseudostate two-centre basis employed in a one-electron model (Shingal and Lin 1991*b*).

electric dipole moment of the atoms produced in the level n and the first-order moment of the electron current distribution. As an illustration, in Fig. 6.25 the values of the electric dipole moment for capture to the $n = 2$ level calculated by Slim *et al.* (1990, 1991*a*) from a two-electron model and of Shingal and Lin (1991*b*) from a one-electron model are compared with the data of Hippler (1991*a*, *b*).

For the hydrogen atoms in the $n = 3$ level produced by capture by H^+ from helium, the diagonal elements have been measured by Brower and Pipkin (1989) using a microwave technique to induce fine structure transitions within the $n = 3$ level, observing the resulting effect on the Balmer α-radiation. By observing the polarization of the Balmer α-radiation when electric fields are applied parallel or antiparallel to the beam of hydrogen atoms Ashburn *et al.* (1989, 1990) have been able to calculate both the diagonal and off-diagonal elements of $\bar{\rho}$. The observed data are a function of the 10 real quantities that form the integrated density matrix, which can be determined by a statistical fitting procedure. Although the statistical uncertainties on some of the elements of $\bar{\rho}$ found in this way are large, the results provide a severe test for the theoretical models of the collision. The parameters of the integrated density matrix have been predicted by Jain *et al.* (1987*a*, *b*, 1988) and by Shingal and Lin (1991*b*) using one-electron

Table 6.7. The elements of the integrated density matrix for capture into $n = 3$ level of atomic hydrogen by the impact of 60 keV protons on helium. A represents the calculated density matrix elements (Slim *et al.* 1991*a*); B represents the experimental data (Ashburn *et al.* 1989, 1990).

	A	B
s_0	1.00	1.00 ± 0.008
p_0	0.203	0.213 ± 0.005
$p_{\pm 1}$	0.017	0.007 ± 0.003
d_0	0.016	0.030 ± 0.007
$d_{\pm 1}$	0.002	-0.005 ± 0.005
$d_{\pm 2}$	0.0004	0.011 ± 0.001
$(s_0 p_0)$	$0.391 - 0.013i$	$(0.324 \pm 0.006) + (-0.061 \pm 0.007)i$
$(s_0 d_0)$	$0.087 - 0.41i$	$(0.105 \pm 0.009) + (-0.154 \pm 0.024)i$
$(p_0 d_0)$	$0.046 + 0.030i$	$(0.057 \pm 0.002) + (0.026 \pm 0.002)i$
$(p_{\pm 1} d_{\pm 1})$	$0.004 + 0.003i$	$(0.004 \pm 0.002) + (0.002 \pm 0.001)i$

models, and by Slim *et al.* (1990, 1991*a*, *b*) using the explicit two-electron formulation. The agreement with the data, discussed in detail in the papers cited, is only partial, which is to be expected because the cross sections for capture to the $n = 3$ level are very small and are difficult to both calculate and to measure accurately. As an example, in Table 6.7 the computed values of the density matrix are compared with the measured data for $E_L = 60$ keV/a.m.u.

Further studies

Before leaving the $(\mathrm{H\,He})^+$ system we should note that similar studies can be carried out for the reaction channel

$$\mathrm{H}(1s) + \mathrm{He}^+(1s) \to \mathrm{H}^+ + \mathrm{He}, \tag{6.45}$$

and a recent discussion based on the molecular orbital expansion has been given by Errea *et al.* (1989*c*). In Chapter 9 we shall return to a discussion of two-electron transitions in the $(\mathrm{H\,He})^+$ system for which interesting studies have been made based on the atomic orbital expansion method (Fritsch and Lin 1988).

6.5.3 Quasi-two-electron systems

We have seen that it is often possible to make a model of a many-electron system in which one active electron moves in effective potentials representing the interaction with the nuclei and remaining electrons. This idea can be extended to a model in which two electrons move in effective potentials due to the remaining electrons and the nuclei. This extended model allows a detailed treatment to be given of many charge exchange reactions which could not be represented adequately by a one-active-electron model. Examples of such calculations are work on the He^{+}–Li^{0} collision in which the $Li(1s^{2})$ core is treated as giving rise to an effective potential acting on the remaining two electrons (Bransden *et al.* 1984) and the O^{6+} + He collision in which the O^{6+} core is represented by a potential (Fritsch and Lin 1986).

6.6 CONCLUSIONS

In this chapter we have examined briefly a number of illustrative examples of the application of the coupled channel impact parameter method. Although a detailed critical comparison with the experimental data has not been attempted, it is clear that, if the basis sets are carefully chosen, the impact parameter model can provide an accurate description of ion–atom scattering, although naturally for particular systems various points of detail remain to be resolved. In general, the large cross sections can be predicted with confidence, whether for capture, excitation, or ionization. The small cross sections present the most difficulty and are sensitive to the details of the basis employed. This is because direct coupling between the incident channel and the final channel of interest is often unimportant, the reaction proceeding via couplings to several or many intermediate channels. In turn this requires the basis set to be large, since if the intermediate channels of importance lie in the continuum a very good representation of the continuum is required, and pseudostate expansions of the wave function converge slowly. Of course the impact parameter model itself has its limitations. We have already seen that at very low energies the common trajectory model is not adequate and the partial-wave equations should be used, while at high energies, when v_0/c becomes close to unity, a relativistic version of the model must be introduced. This will be discussed in Chapter 11. A further difficulty is a practical, rather than a theoretical, one. For an adequate description of a one-electron system, the basis set can be rather large, often upwards of 50 terms being required. For a two-electron system, provided only single-electron excitation is important, basis sets of a similar size can be employed. This is the case, for instance, in the collision of two ground-state hydrogen atoms at low energies, for which the probability that one electron remains unexcited is large. On the other hand, in the same system at high energies both electrons can be excited, or ionized, and to obtain a basis of similar quality to a 50-term quasi-one-electron basis we would seem to require 50×50 terms. Some ideas for treating many-electron systems and simplifying this large basis problem will be considered in Chapter 9.

7

APPROXIMATE SOLUTIONS OF COUPLED EQUATIONS

Most of the theoretical descriptions of inelastic ion–atom collisions, discussed in the preceding chapters, have resulted in sets of coupled equations, either for radial functions in partial-wave models or for probability amplitudes in the case of the impact parameter model. These equations have to be solved numerically. However, under certain conditions, usually rather restrictive, approximate analytical solutions have been developed, which provide rapid means of calculating cross sections and which may also illuminate the general nature of the theory. In this chapter two rather different groups of approximation will be discussed. The first group is concerned with the approximate solutions of the two-state coupled equations, particularly in situations where a transition is dominated by a curve crossing between two molecular states, and the extension of these solutions to systems which exhibit multiple crossings. The second group is different in character and is based on perturbation and distorted wave methods, which apply in cases of weak coupling.

7.1 THE LANDAU–ZENER–STUECKELBERG APPROXIMATION

The two-state model

Let us consider a transition between a pair of adiabatic states 1 and 2 belonging to the same symmetry class and dominated by a single pseudo-crossing between the two potential energy curves. We shall assume that the molecular orbitals 1 and 2 correspond to σ states, and are asymptotic to s-state atomic orbitals, otherwise, as pointed out by Bates (1960), three or more molecular orbitals are required to take account of the strong coupling between magnetic substates due to the rotation of the internuclear line. If the asymptotic atomic states belong to the same arrangement, the transition between states 1 and 2 is an excitation, or if to different arrangements, the transition is charge exchange. In the simplest impact parameter model the wave function $\psi(\mathbf{s}, t)$ can be expressed in the rotating coordinate system as a linear combination of two adiabatic molecular orbitals $\bar{\psi}_i$ $(\bar{\mathbf{s}}, R)$, or of the two corresponding diabatic orbitals $\bar{\psi}_i^{\mathrm{d}}(\bar{\mathbf{s}}, R)$ (see Chapter 4). The orbitals

$\bar{\psi}_i$ and $\bar{\psi}_i^{\rm d}$ will be taken to be orthonormal

$$\langle\bar{\psi}_i|\bar{\psi}_j\rangle = \langle\bar{\psi}_i^{\rm d}|\bar{\psi}_j^{\rm d}\rangle = \delta_{ij}, \qquad i, j = 1, 2. \tag{7.1}$$

Using the diabatic basis, and ignoring translation factors, as in the perturbed stationary state model, the wave function $\psi(\mathbf{s}, t)$ can be written as

$$\psi(\mathbf{s}, t) = A_1(t)\bar{\psi}_1^{\rm d}(\bar{\mathbf{s}}, R) + A_2(t)\bar{\psi}_2^{\rm d}(\bar{\mathbf{s}}, R), \tag{7.2a}$$

where the interionic separation $\mathbf{R}$ is determined by a classical trajectory† $\mathbf{R} = \mathbf{R}(\mathbf{b}, t)$, and $\mathbf{b}$ is the impact parameter vector. The dependence of the amplitudes $A_i(t) \equiv A_i(\mathbf{b}, t)$ on the impact parameter $\mathbf{b}$ is not displayed in (7.2a) and in what follows, in order to simplify the notation. Following the usual variational method, it is found that the amplitudes $A_i(t)$ satisfy the coupled equations

$$\begin{aligned} \mathrm{i}\dot{A}_1(t) &= V_{11}(t)A_1(t) + V_{12}(t)A_2(t) \\ \mathrm{i}\dot{A}_2(t) &= V_{21}(t)A_1(t) + V_{22}(t)A_2(t). \end{aligned} \tag{7.2b}$$

The interactions $V_{ij}(t)$ form the elements of a Hermitian coupling matrix

$$V_{ij}(t) = \langle\bar{\psi}_i^{\rm d}|H_{\rm e}|\bar{\psi}_j^{\rm d}\rangle \tag{7.3a}$$

with

$$V_{ij}(t) = V_{ji}^*(t) \tag{7.3b}$$

where $H_{\rm e}$ is the electronic Hamiltonian (3.158). Since the states 1 and 2 have the same symmetry there is no rotational coupling. It should be noted that a Hermitian interaction matrix cannot be obtained if a non-orthogonal basis is employed, for example when plane-wave translation factors are used. However translational effects can be allowed for by using a common translation factor, in which case the basis remains orthogonal. Alternative procedures in which a non-orthogonal basis is employed, but the interaction matrix is artificially symmetrized to satisfy (7.3b) are not entirely satisfactory and may introduce significant errors (Green 1965).

Equations (7.2b) are integrated along the classical path subject to the boundary condition that in the limit $t \to -\infty$

$$A_1(t) \to 1, \qquad A_2(t) \to 0. \tag{7.4}$$

The case will be discussed in which the diabatic curves V_{11} and V_{22} expressed as functions of R possess a crossing at the point $R = R_{\rm c}$, as shown in Fig. 7.1. The corresponding adiabatic energies $E_i(R)$ which are defined as the eigenvalues of the equation

$$H_{\rm e}\bar{\psi}_i(\bar{\mathbf{s}}, R) = E_i(R)\bar{\psi}_i(\bar{\mathbf{s}}, R) \tag{7.5}$$

† We are assuming a common trajectory approximation in which the classical path is determined by an effective potential $W(R)$ with a single classical turning point at $R = R_0$, as described in Section 3.7.

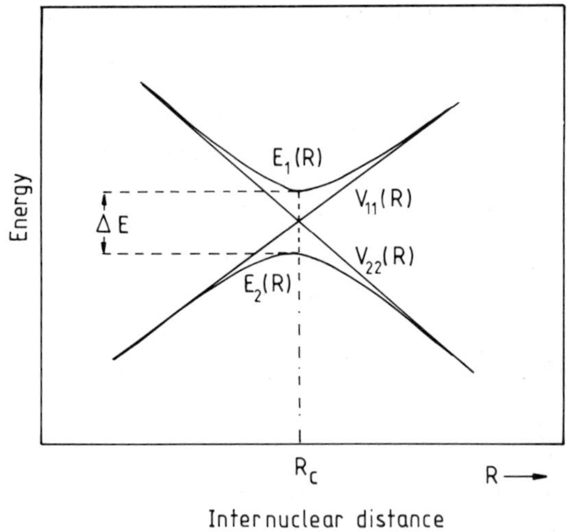

Fig. 7.1. Schematic diagram of adiabatic ($E_1(R)$, $E_2(R)$) and diabatic potentials ($V_{11}(R)$, $V_{22}(R)$) in the vicinity of a crossing at $R = R_c$.

do not cross, but there is a corresponding pseudocrossing in the vicinity of $R = R_c$, as illustrated. For future use it is convenient to relate the parameters of the adiabatic curves at the pseudocrossing with the parameters of the diabatic representation. Representing the adiabatic functions ψ_i, approximately, in terms of the diabatic functions ψ_i^d, we have

$$\bar{\psi}_i(\bar{\mathbf{s}}, R) = d_1^i \bar{\psi}_1^d(\bar{\mathbf{s}}, R) + d_2^i \bar{\psi}_2^d(\bar{\mathbf{s}}, R). \tag{7.6}$$

Inserting this expression into (7.5) and projecting the equation in turn with $\bar{\psi}_1^d$ and $\bar{\psi}_2^d$, we obtain

$$\begin{aligned} (V_{11} - E_i)d_1^i + V_{12}d_2^i &= 0 \\ V_{21}d_1^i + (V_{22} - E_i)d_2^i &= 0. \end{aligned} \tag{7.7}$$

These equations are compatible provided

$$(V_{11} - E_i)(V_{22} - E_i) - |V_{12}|^2 = 0. \tag{7.8}$$

The adiabatic energies are the roots of this quadratic equation, and we find

$$E_i(R) = \tfrac{1}{2}[(V_{11} + V_{22}) + (-1)^{i+1}\{(V_{11} - V_{22})^2 + 4|V_{12}|^2\}^{\frac{1}{2}}], \qquad i = 1, 2. \tag{7.9}$$

At the crossing point $R = R_c$, $V_{11} = V_{22}$ so that the energy difference at the adiabatic pseudocrossing is ΔE, where

$$\Delta E = E_1(R_c) - E_2(R_c) = 2|V_{21}(R_c)|. \tag{7.10}$$

If the system under consideration can be represented *exactly* in a two-state basis, the adiabatic and diabatic representations are related by a two-dimensional unitary transformation. In this case an exact calculation of the $\mathscr{S}$-matrix using either basis will provide the same results, although this will not be the case if approximations are made. In what follows, the two-state equations using the diabatic bases, (7.2b), will be employed, the connection to the parameters of the adiabatic potentials being effected through equation (7.9).

The $\mathscr{S}$-matrix

In the collision the crossing point at $R = R_c$ is traversed twice; once as the system passes along the classical path $\mathbf{R}(t)$ towards the distance of closest approach R_0, and once as the heavy particles A and B separate. For example, if a straight line trajectory is employed

$$R^2 = b^2 + v_0^2 t^2, \tag{7.11}$$

and the crossing point will be reached at a time $t = -t_c$, where $t_c = v_0^{-1}(R_c^2 - b^2)^{\frac{1}{2}}$, and again at the time $t = t_c$. In this case the distance of closest approach is $R_0 = b$, which is reached at $t = 0$.

The time development of the coupled equation (7.2b) is conveniently discussed in terms of the evolution matrix $\mathbf{U}(t, t')$, which is defined by the relation

$$\mathbf{A}(t) = \mathbf{U}(t, t')\mathbf{A}(t'), \tag{7.12}$$

where $\mathbf{A}(t)$ is the column vector with elements $A_i(t)$. The evolution matrix satisfies the differential equation

$$\mathrm{i}\frac{\mathrm{d}}{\mathrm{d}t}\mathbf{U}(t, t') = \mathbf{V}(t)\mathbf{U}(t, t') \tag{7.13}$$

where $\mathbf{V}(t)$ is the interaction matrix with elements $V_{ij}(t)$. It is easy to verify that $\mathbf{U}$ possesses the group properties

$$\begin{gathered}\mathbf{U}(t, t') = \mathbf{U}(t, t'')\mathbf{U}(t'', t)\\ \mathbf{U}(t, t) = \mathbf{I}, \qquad \mathbf{U}^{-1}(t, t') = \mathbf{U}(t', t),\end{gathered} \tag{7.14}$$

and that $\mathbf{U}$ is unitary

$$\mathbf{U}^{\dagger}(t, t')\mathbf{U}(t, t') = \mathbf{U}(t, t')\mathbf{U}^{\dagger}(t, t') = \mathbf{I}, \tag{7.15}$$

which follows from the Hermiticity of the coupling matrix $\mathbf{V}$ (see 7.36).

Let us suppose transitions between the states 1 and 2 only occur in a short interval of time of duration 2Δ about the times $\pm t_c$. Then using the group property of the evolution matrix $\mathbf{U}(t, t')$, the $\mathscr{S}$-matrix can be expressed as

$$\begin{aligned}\mathscr{S} = \mathbf{U}(\infty, -\infty) &= \mathbf{U}(\infty, t_c + \Delta)\mathbf{U}(t_c + \Delta, t_c - \Delta)\mathbf{U}(t_c - \Delta, -t_c + \Delta) \\ &\quad \times \mathbf{U}(-t_c + \Delta, -t_c - \Delta)\mathbf{U}(-t_c - \Delta, -\infty) \\ &\approx \mathbf{U}(\infty, t_c)\mathbf{U}(t_c)\mathbf{U}(t_c, -t_c)\mathbf{U}(-t_c)\mathbf{U}(-t_c, -\infty), \quad (7.16)\end{aligned}$$

where, in the second line, we have set

$$\mathbf{U}(-t_c) \equiv \mathbf{U}(-t_c + \Delta, -t_c - \Delta),$$

and

$$\mathbf{U}(t_c) \equiv \mathbf{U}(t_c + \Delta, t_c - \Delta). \quad (7.17)$$

It has been assumed that $|t_c| \gg \Delta$ and the approximation $\Delta = 0$ has been made in the remaining factors of (7.16).

Since, in this model, the coupling between the states 1 and 2 is taken to be zero except in the immediate vicinity of the crossing point, the matrices $\mathbf{U}(\infty, t_c)$, $\mathbf{U}(t_c, -t_c)$, and $\mathbf{U}(-t_c, \infty)$ are all diagonal with matrix elements given by

$$U_{ij}(t, t') = \delta_{ij} \exp\left\{-\mathrm{i}\int_{t'}^{t} V_{ii}(t'')\,\mathrm{d}t''\right\}. \quad (7.18)$$

In view of the boundary condition (7.4), for times $t < -t_c$, we have that

$$A_1(t) = \exp\left\{-\mathrm{i}\int_{-\infty}^{t} V_{11}(t')\,\mathrm{d}t'\right\}, \qquad A_2 = 0, \quad t < -t_c. \quad (7.19)$$

At the time $t = -t_c$, the action of the non-diagonal evolution matrix $\mathbf{U}(-t_c)$ will produce for $t > -t_c$, a new vector $\mathbf{A}(t)$, with $|A_1(t)| < 1$ and $|A_2(t)| > 0$. If P_{21} is the probability of a transition from the state 1 to the state 2 at the crossing, it follows that

$$P_{21} = |A_2(\tau)|^2 \quad \text{and} \quad 1 - P_{21} = |A_1(\tau)|^2, \quad (7.20)$$

where $\tau > -t_c$. In the interval $-t_c < \tau < t_c$, $\mathbf{U}(t, t')$ is again diagonal and the phases of $A_2(\tau)$ and $A_1(\tau)$ will continue to change, but $|A_2(\tau)|$ and $|A_1(\tau)|$ will be constant. Since the matrix $\mathbf{U}(-t_c)$ is unitary and $V_{11} = V_{22}$ at the crossing, the general form of $\mathbf{U}(-t_c)$ must be

$$\mathbf{U}(-t_c) = \begin{bmatrix} (1 - P_{21})^{\frac{1}{2}} \exp(-\mathrm{i}\alpha) & P_{21}^{\frac{1}{2}} \exp(\mathrm{i}\beta) \\ -P_{21}^{\frac{1}{2}} \exp(-\mathrm{i}\beta) & (1 - P_{21})^{\frac{1}{2}} \exp(\mathrm{i}\alpha) \end{bmatrix} \exp(\mathrm{i}\gamma) \quad (7.21)$$

where α, β, γ are real phases.

Time-reversal invariance requires that

$$\mathbf{U}(-t, -t_0) = \mathbf{U}^*(t, t_0) \quad (7.22a)$$

and using the fact that **U** is unitary it follows that

$$U_{ij}(-t, -t_0) = U_{ji}(t_0, t). \tag{7.22b}$$

Combining this result with the definitions of $\mathbf{U}(\pm t_c)$ given by (7.17), we have

$$U_{ij}(t_c) = U_{ji}(-t_c). \tag{7.23}$$

Using (7.23) in (7.16), together with (7.21), the overall probability of a transition from 1 to 2 during the collision is found to be $\mathscr{P}_{21}(v_0, b)$, where

$$\mathscr{P}_{21}(v_0, b) = |A_2(\infty)|^2 = 4P_{12}(1 - P_{12}) \sin^2\phi. \tag{7.24}$$

The phase ϕ is defined by the equation

$$\phi = \alpha + \beta - \tfrac{1}{2}\int_{-t_c}^{t_c} \{V_{11}(t') - V_{22}(t')\} \,\mathrm{d}t' \tag{7.25}$$

and depends both on the characteristics of the transition at the crossing point, through the phases α and β, and on the propagation of the system from $t = -t_c$ to $t = t_c$, through the integral term.

It should be noted that had we chosen to work in the adiabatic basis, the phase ϕ would be of the form $\phi = \phi_a$, where

$$\phi_a = \alpha + \beta - \tfrac{1}{2}\int_{-t_c}^{t_c} [\varepsilon_1(R') - \varepsilon_2(R')] \,\mathrm{d}t', \tag{7.26}$$

where $R' = R(t')$, and because the expression (7.24) is approximate, there will be some difference between results obtained in the two representations, although, of course, an exact solution of the coupled equations would lead to the same probability $\mathscr{P}$ in either case.

The total cross section $\sigma_{21}(v_0)$ is obtained from (7.24) by integration over the two-dimensional impact parameter space:

$$\sigma_{21}(v_0) = 2\pi \int_0^{R_c} b \,\mathrm{d}b \mathscr{P}_{21}(v_0, b), \tag{7.27}$$

where the upper limit of the integration has been taken to be R_c, since no transitions can occur if the crossing region is not reached during the collision. Because of the rapid oscillation of the factor $\sin^2\phi$ as b varies, it is often sufficient to replace $\sin^2\phi$ by its average of $\frac{1}{2}$ to obtain

$$\sigma_{21}(v_0) \approx 4\pi \int_0^{R_c} b \,\mathrm{d}b P_{21}(b)(1 - P_{21}(b)). \tag{7.28}$$

It should be noted that if the differential cross section is calculated from $A_2(b, \infty)$, by using (3.231), it is necessary to retain all dependence on phase, including the phase ϕ (Stueckelberg 1932).

The linear approximation

To calculate the transition probability P_{21} in the vicinity of the crossing at $R = R_c$, it is convenient to introduce new amplitudes $C_i(t)$ by the phase transformation:

$$C_i(t) = \exp\left[\mathrm{i}\int_0^t \mathrm{d}t' V_{ii}(t')\,\mathrm{d}t'\right] A_i(t) \tag{7.29}$$

which eliminates the secular terms in (7.2b). The amplitudes $C_i(t)$ satisfy the equations:

$$\mathrm{i}\dot{C}_1(t) = V_{12}(t)\exp(-\mathrm{i}\delta(t))C_2(t) \tag{7.30a}$$

$$\mathrm{i}\dot{C}_2(t) = V_{21}(t)\exp(\mathrm{i}\delta(t))C_1(t), \tag{7.30b}$$

where

$$\delta(t) = \int_0^t \{V_{22}(t') - V_{11}(t')\}\,\mathrm{d}t'. \tag{7.31}$$

Since V_{11} and V_{22} are real and $V_{21}(t) = V_{12}^*(t)$, it follows that the transition probability depends only on $V_{21}(t)$ and the real quantity $\delta(t)$. By substituting the solution of (7.30a) into (7.30b) a second-order equation for the amplitude $C_2(t)$ is obtained

$$\ddot{C}_2(t) - \{V_{21}\exp(\mathrm{i}\delta)\}^{-1}\frac{\mathrm{d}}{\mathrm{d}t}\{V_{21}\exp\mathrm{i}\delta\}\dot{C}_2(t) + |V_{12}(t)|^2 C_2(t) = 0 \tag{7.32}$$

which must be solved subject to the boundary conditions $C_2(-\infty) = \dot{C}_2(-\infty) = 0$.

To make further progress, a specific form for the matrix elements V_{ij} must be assumed in the region of $R = R_c$ $(t = \pm t_c)$. In the Landau–Zener–Stueckelberg model (Landau 1932; Zener 1932; Stueckelberg 1932), the linear term is retained in an expansion of $\{V_{22}(t) - V_{11}(t)\}$ about the crossing point at $t = -t_c$

$$V_{22}(t) = V_{11}(t) = a(t + t_c), \tag{7.33a}$$

where

$$a = \left\{\frac{\partial}{\partial t}[V_{22}(t) - V_{11}(t)]\right\}_{t=-t_c}. \tag{7.33b}$$

In addition $V_{12}(t)$ is assumed to be constant over the region in which transitions take place

$$V_{12}(t) = d. \tag{7.34}$$

By (7.10), $d = \Delta E/2$ so that d is determined by the energy difference of the adiabatic curves at $R = R_c$. With these approximations, equation (7.32) reduces to

$$\ddot{C}_2(t) - ia(t + t_c)\dot{C}_2(t) + d^2C_2(t) = 0. \tag{7.35}$$

Since taking the complex conjugate of this equation is equivalent to changing the sign of a, only the case $a > 0$ need be considered.

The first-order derivative in (7.35) can be eliminated by the changes of variable

$$z = a^{\frac{1}{2}} e^{i\pi/4}(t + t_c)$$

and

$$y(z) = C_2 \exp\left(-\frac{z^2}{4}\right). \tag{7.36}$$

It is found that the transformed function $y(z)$ satisfies Weber's equation (Whittaker and Watson 1927)

$$\frac{d^2y}{dz^2} + \left[\tfrac{1}{2} + \nu - \frac{z^2}{4}\right]y = 0 \tag{7.37}$$

where $\nu = -id^2/a$. Two independent solutions of this equation are the parabolic cylinder functions $D_{-\nu-1}(iz)$ and $D_{-\nu-1}(-iz)$. The asymptotic forms of the functions $D_\gamma(z)$ are, for $|z| \to \infty$, and fixed γ,

$$D_\gamma(z) \sim \begin{cases} \exp\left(-\frac{z^2}{4}\right)z^\nu & \text{for } |\arg z| < \tfrac{3}{4}\pi, \\ \exp\left(-\frac{z^2}{4}\right)z^\nu - \frac{(2\pi)^{\frac{1}{2}}}{\Gamma(-\nu)} e^{i\nu\pi} \exp\left(\frac{z^2}{4}\right)z^{-\nu-1} & \text{for } \tfrac{5}{4}\pi > \arg z > \tfrac{1}{4}\pi, \\ \exp\left(-\frac{z^2}{4}\right)z^\nu - \frac{(2\pi)^{\frac{1}{2}}}{\Gamma(-\nu)} e^{-i\nu\pi} \exp\left(\frac{z^2}{4}\right)z^{-\nu-1} & \text{for } -\tfrac{1}{4}\pi > \arg z > -\tfrac{5}{4}\pi. \end{cases} \tag{7.38}$$

Making use of these expressions it is found that the solution satisfying the boundary condition that $C_2(t) \to 0$ as $t \to -\infty$, corresponding to $iz \to |z| \exp(-i\pi/4)$, is obtained with the solution

$$y(z) = AD_{-\nu-1}(iz) \tag{7.39}$$

where A is a constant which can be determined from the conditions $|C_1(-\infty)| = |A_1(-\infty)| = 1$, using equation (7.29). By looking at the asymptotic form of $y(z)$ when t is large and positive, the value of $P_{21} = |A_2(t)|^2 = |C_2(t)|^2$ for $t > -t_c$, can be found and also the phases α and β. The results are, for P_{21},

$$P_{21} = 1 - \exp(-2\pi q), \tag{7.40a}$$

where

$$q = \frac{d^2}{|a|}, \tag{7.40b}$$

and for $\alpha + \beta$ (see (7.21)),

$$\alpha + \beta = \frac{\pi}{4} + q \log q - q - \arg \Gamma(1 + iq). \tag{7.41}$$

Expression (7.40a) was obtained originally by Landau (1932) from adiabatic perturbation theory, and by Zener (1932) from the diabatic equations, as here. The problem was also tackled by Stueckelberg (1932) who started from the coupled radial equations of the partial-wave formulation, using phase integral methods to obtain a semi-classical approximation. In the semi-classical approximation account has to be taken not only of the stationary phase region near the crossing, but also of the classical turning points where a connection formula must be devised. The principal advantage of the semi-classical treatment is that interference effects associated with different classical paths can be treated consistently. The Stueckelberg analysis has been extended by Crothers (1971) and reviewed by Crothers (1981), Nikitin and Umanskii (1984), and Eu (1984).

The potentials V_{ii} are functions of $R(t)$ only, and in the absence of translation factors, the dependence of V_{ii} on velocity is only through the trajectory equation $R = R(t)$. If a straight line trajectory is adopted $R^2 = b^2 + v_0^2 t^2$ and from (7.33b) we have

$$a = \frac{\partial}{\partial R}(V_{22}(R) - V_{11}(R))\bigg|_{R=R_c} \left(1 - \frac{b^2}{R_c^2}\right)^{\frac{1}{2}} v_0 \equiv \bar{a} v_0 \tag{7.42}$$

where $\bar{a}$ is independent of v_0.

Using (7.40a, b) it is seen that for large v_0, P_{12} is proportional to v_0^{-1}. This behaviour is incorrect. For large v_0, the first-order perturbation solution of equations (7.2b) should become exact, and the first-order probability $P_{21}^{(1)}$ is

$$P_{21}^{(1)} = \left| \int_{-\infty}^{\infty} V_{12}(t)\, \mathrm{d}t \right|^2. \tag{7.43}$$

For excitation, and also for charge exchange when translational factors are neglected, V_{12} is a function of R only, and on changing the integration variable from t to R we find

$$P_{21}^{(1)} \propto v_0^{-2}. \tag{7.44}$$

The reason for the incorrect variation of the Landau–Zener formula with v_0, for large v_0, can be seen if the size of the region over which transitions occur with a high probability is estimated. Let us assume that transitions are important in the interval in which the phase factors $\delta(t)$ in (7.31) change through an angle π. If this interval is of duration 2Δ and is centred about $t = -t_c$, then Δ is determined by the equation:

$$\int_{-t_c-\Delta}^{t_c+\Delta} |V_{22}(t') - V_{11}(t')|\, dt' = \pi. \tag{7.45}$$

Using the linear approximation (7.33), we find

$$\Delta = \left(\frac{\pi}{\alpha}\right)^{\frac{1}{2}} = \left[\frac{\pi}{(v_0\bar{a})}\right]^{\frac{1}{2}}. \tag{7.46}$$

The spatial extent of the region over which transitions take place is therefore of the order δR where

$$\delta R = 2\Delta v_0 = 2\left[\frac{\pi v_0}{\bar{a}}\right]^{\frac{1}{2}}. \tag{7.47}$$

The transition region is therefore unbounded as v_0 increases. This is clearly unphysical and accounts for the incorrect high-energy variation of the Landau–Zener formula. Since the validity of the Landau–Zener approximation depends on the transition region being narrow, we must have $\delta R \ll R_c$, or from (7.47), $v_0 \ll \bar{a}R_c^2/(4\pi)$, a condition which is often very restrictive.

The total cross section If the approximate formula (7.28) is employed, obtained by averaging over the phase difference ϕ between the two crossing points, the total cross section can be evaluated from the Landau–Zener expression (7.40) for P_{21}. Combining (7.28) and (7.40), we have

$$\sigma_{21}(v_0) = 4\pi \int_0^{R_c} b\, db(e^{-2\pi q} - e^{-4\pi q}). \tag{7.48}$$

The dependence of $2\pi q$ on the impact parameter is given by (7.40b) and (7.42),

$$2\pi q = 2\pi\, d^2\left[(\Delta V')v_0\left(1 - \frac{b^2}{R_c^2}\right)^{\frac{1}{2}}\right]^{-1} \tag{7.49}$$

where $\Delta V'$ is the difference in slope of the potentials V_{11} and V_{22} at the crossing

$$\Delta V' \equiv \frac{\partial}{\partial R}[V_{22}(R) - V_{11}(R)]_{R=R_c}. \tag{7.50}$$

Changing the variable of integration from b to $y = (1 - b^2/R_c^2)^{-\frac{1}{2}}$ in (7.48), we immediately find that

$$\sigma_{21}(v_0) = 4\pi R_c^2[E_3(z) - E_3(2z)] \tag{7.51}$$

where $z = 2\pi d^2/(\Delta V' v_0)$, and E_n is the exponential integral

$$E_n = \int_1^\infty \mathrm{d}y \frac{\exp(-zy)}{y^n}. \tag{7.52}$$

The accuracy of the formulae (7.40) and (7.51) can be tested by comparing the Landau–Zener values of P_{21} and σ_{21} with those obtained by integrating the two-state equations numerically. It is found (see, for example, Bates *et al.* 1964) that unless the conditions for the applicability of the Landau–Zener formula are very well satisfied, the computed Landau–Zener cross section can be in error by orders of magnitude.

Applications Despite the restrictive applicability of the Landau–Zener formula it has been used to discuss a variety of reactions. Particularly favourable cases arise for ion-pair formation in reactions such as

$$\mathrm{A + B^{2+} \rightarrow A^+ + B^+}, \tag{7.53a}$$

and

$$\mathrm{A + B \rightarrow A^+ + B^-}. \tag{7.53b}$$

In these cases, the crossing point is controlled by the long-range Coulomb and polarization forces and occurs at large atomic separations where the separated atom wave functions can be used as a diabatic basis. In reaction (7.53a), the interaction in the initial state is attractive because of the $1/R^4$ polarization potential between a charge and a neutral system (see Chapter 2), while in the final state the interaction is repulsive and proportional to $1/R$. For reaction (7.53b), the interaction is attractive and proportional to the van der Waals polarization potential which behaves like $1/R^6$, while in the final state the interaction is also attractive, but behaves like $1/R$.

Some examples for which numerical calculations have been made are (Bates and Moiseiwitsch 1954)

$$\begin{aligned} &\mathrm{H + Be^{2+} \rightarrow H^+ + Be^+} \\ &\mathrm{H + Si^{2+} \rightarrow H^+ + Si^+} \\ &\mathrm{H + Mg^{2+} \rightarrow H^+ + Mg^+}, \end{aligned} \tag{7.54}$$

and (Dalgarno 1954)

$$\begin{aligned} H + Al^{3+} &\to H^{+} + Al^{2+}(^{2}S) \\ &\to H^{+} + Al^{2+}(2p) \\ H + B^{2+} &\to H^{+} + B^{+}(^{1}P) \\ H + Li^{2+} &\to H^{+} + Li^{+}(^{3}S) \\ &\to H^{+} + Li^{+}(^{1}S) \\ H + Al^{2+} &\to H^{+} + Al^{+}. \end{aligned} \tag{7.55}$$

These processes involve the transfer of an electron from an orbital about one passive core to an orbital about another, and can be treated as one-electron reactions. The crossing points are determined by the charges and polarizabilities of the ions and occur at large distances of separation, for example in the group of reactions, (7.55), at 3.34, 6.58, 11.1, 8.80, 20.9, and 5.29 atomic units, respectively.

A particular example of astrophysical interest (Butler and Dalgarno, 1980) is the set of exothermic reactions between a positive ion A^{q+} and atomic hydrogen in the ground state,

$$A^{q+} + H(1s) \to A^{(q-1)+} + H^{+} + \Delta\varepsilon, \tag{7.56}$$

at collision energies up to ~10 eV/a.m.u., where A is C, N, O, or Ne and $q = 2$ or 3. To lowest order the diabatic curves are

$$V_{11} = \varepsilon_1 - 0.5 - \frac{\alpha q^2}{2R^4}, \tag{7.57}$$

$$V_{22} = \varepsilon_2 + \frac{q-1}{R}, \tag{7.58}$$

where α is the polarizability of atomic hydrogen, ε_1 is the energy of the ion A^{q+}, and ε_2 is the energy of the ion $A^{(q-1)+}$. The crossing points can be estimated by neglecting the term in α, so that

$$R_c = (q-1)[(\varepsilon_1 - \varepsilon_2) - 0.5]^{-1}. \tag{7.59}$$

The calculated matrix elements V_{12} are found to fit the formula

$$E_1(R_c) - E_2(R_c) = 2V_{12}(R_c) \approx R_c^2 \exp(-R_c) \approx 2d, \tag{7.60}$$

where $E_i(R)$ are the adiabatic energies. The remaining parameter required in the Landau–Zener formula is the difference in slopes $\Delta V'$ of the energies V_{11} and V_{22} at $R = R_c$. The short-range repulsion in the incident channel can be represented by a potential such as $A \exp(-BR)$ where it is found

$A \approx q$ and $B \approx 1.8$ and from (7.57)

$$V_{11}(R) = \varepsilon_1 - 0.5 - \frac{\alpha q^2}{2R^4} + A\exp(-BR). \qquad (7.61)$$

It is found that $V_{22}(R)$ is sufficiently well determined by the asymptotic expression (7.58).

Proceeding in this general way, the calculation of the charge exchange cross sections only requires a knowledge of the energy differences $(\varepsilon_2 - \varepsilon_1)$ for each ion. Some values obtained for charge transfer rate coefficients at a temperature of 10^4 K are shown in Table 7.1 compared with the results of coupled channel quantum calculations. It is seen that although no great accuracy is obtained from the Landau–Zener formula, in suitable cases it provides a useful estimate of the magnitude of the cross sections.

7.2 NON-LINEAR MODELS

The linear approximation discussed in the last section is not the only one that leads to an analytical solution of the two-state equations, and we shall now turn to consider models based on the assumption that the coupling varies exponentially over the transition region. The earliest exponential model was due to Rosen and Zener (1932) (see also Skinner (1961) and Bates (1962)) and extended models have been discussed by Demkov (1963) and by Nikitin (1962*a*, *b*, 1963, 1970).

7.2.1 The Rosen–Zener approximation

In the model suggested by Rosen and Zener a solution of the coupled equations (7.30a, b) is sought when the transitions between the two channels occur in a symmetrical region about a time $t = 0$. If $t = 0$ is the time of closest approach (as we have taken it to be in our previous discussions), the

Table 7.1. Rate coefficients for charge exchange at 10^4K (Butler and Dalgarno 1980).

Initial system	Landau–Zener	Quantum coupled channel
$C^{2+} + H$	0	1.0×10^{-12}
$C^{3+} + H$	1.8×10^{-9}	3.6×10^{-9}
$O^{2+} + H$	1.0×10^{-10}	7.7×10^{-10}
$O^{3+} + H$	2.8×10^{-9}	8.6×10^{-9}
$Ne^{3+} + H$	3.8×10^{-9}	5.7×10^{-9}

transition region is encountered once. If however the important transition region is centred about $R = R_p$, the time $t = 0$ can be taken to be the time at which $R = R_p$. In general the transition region is encountered twice, as in the Landau–Zener model. It is assumed that over the important transition region the potential difference $[V_{22}(t) - V_{11}(t)]$ can be taken to be a constant ω, while the coupling $V_{12}(t)$ is taken to be proportional to the bell-shaped function $[\cosh(Bt)]^{-1}$. In (7.30) we set

$$\delta(t) = \omega t, \qquad V_{12}(t) = V_{12}(t) = A[\cosh(Bt)]^{-1}, \tag{7.62}$$

where ω, A, and B are constants, and without loss of generality we take $B > 0$. In contrast to the Landau–Zener model the curves V_{22} and V_{11} do not cross, but are parallel over the transition region. An important example is afforded by the lowest two Σ states of $(\mathrm{H\ He})^+$ which are responsible for the charge exchange reaction $\mathrm{H}^+ + \mathrm{He} \rightarrow \mathrm{H}(1s) + \mathrm{He}(1s)$ for which both the adiabatic and diabatic energies are effectively parallel (see Fig. 6.22).

Substituting $\delta(t)$ and $V_{12}(t)$ given by (7.62) into (7.32), the equation for $C_2(t)$ becomes

$$\ddot{C}_2(t) + \left[-\mathrm{i}\omega + \frac{\mathrm{d}}{\mathrm{d}t}[\log\{\cosh(Bt)\}]\right]\dot{C}_2(t) + |A|^2\{\cosh(Bt)\}^{-2}C_2(t) = 0 \tag{7.63}$$

With the change of variable from t to y, where

$$y = \tfrac{1}{2}\{\tanh(Bt) + 1\}, \tag{7.64}$$

equation (7.63) becomes

$$y(1-y)\frac{\mathrm{d}^2C_2(y)}{\mathrm{d}y^2} + [\gamma - (\alpha + \beta + 1)y]\frac{\mathrm{d}C_2(y)}{\mathrm{d}y} - \alpha\beta C_2(y) = 0, \tag{7.65a}$$

where

$$\gamma = \tfrac{1}{2}\left(1 - \frac{\mathrm{i}\omega}{B}\right), \qquad \alpha = -\beta = \frac{A}{B}. \tag{7.65b}$$

The general solution of (7.65a) is in terms of hypergeometric functions ${}_2F_1$ (see Erdelyi *et al.* 1953) and it is found that

$$C_2(y) = \lambda\,{}_2F_1(\alpha, \beta; \gamma; y) + \mu y^{1-\gamma}\,{}_2F_1(\alpha + 1 - \gamma, \beta + 1 - \gamma; 2 - \gamma; y), \tag{7.66}$$

where λ and μ are constants which are to be determined from the boundary conditions

$$\lim_{t \to -\infty} |C_1(t)|^2 = 1, \qquad \lim_{t \to -\infty} |C_2(t)|^2 = 0. \tag{7.67}$$

As $t \to -\infty$, we see from (7.64) that $y \to 0$ and since ${}_2F_1(\alpha, \beta; \gamma; 0) = 1$ it follows that $\lambda = 0$. From (7.30b), we find

$$|C_1(t)| = |A||\dot{C}_2(t)\cosh(Bt)|. \tag{7.68}$$

Taking the limit $t \to -\infty$ determines the modulus of the constant μ:

$$|\mu| = \left|\frac{A}{B(1-\gamma)}\right|, \tag{7.69}$$

and finally, by taking the limit $t \to \infty$,

$$C_2(\infty) = \left|\frac{A}{B(1-\gamma)^2}\, {}_2F_1(\alpha + 1 - \gamma, \beta + 1 - \gamma; 2 - \gamma; 1)\right|. \tag{7.70}$$

The hypergeometric function ${}_2F_1(a, b, c, 1)$ can be expressed as

$${}_2F_1(a, b; c; 1) = \frac{\Gamma(c)\Gamma(c-a-b)}{\Gamma(c-a)\Gamma(c-b)} \tag{7.71}$$

and using the relation

$$\Gamma(c)\Gamma(1-c) = \frac{\pi}{\sin(\pi c)} \tag{7.72}$$

the probability for a transition from channel 1 to channel 2 is found to be

$$\mathscr{P}_{21}(v_0, b) = |C_2(t = \infty)|^2 = \frac{\sin^2(\pi|A|/B)}{\{\cosh(\pi\omega/2B)\}^2}. \tag{7.73}$$

The parameters $|A|$ and B depend on the impact velocity v_0 and the impact parameter b. By making some explicit assumptions about this dependence, Bates (1962) was able to obtain from (7.73) an approximate form for the cross section for a non-resonant ($\omega \neq 0$) collision at low velocities. He found

$$\sigma_{21}(v_0) = \frac{4v_0^2 a_1^2 a_3^2}{\pi\omega^2}\left(1 + \frac{\pi\omega}{v_0 a_3}\right)\exp(-\pi\omega/v_0 a_3) \tag{7.74}$$

where a_1 and a_3 are constants. This shows that, for $\omega \neq 0$, σ_{21} decreases exponentially as $v_0 \to 0$, this decrease occurring at very small values of v_0 if ω is small. However it should be recalled that the impact parameter formalism is not valid if v_0 is too small, so that this result, while indicative, must be used with discretion.

7.2.2 Nikitin–Demkov models

As the pioneer work of Rosen and Zener showed, exact solutions of the two-state coupled equations can be obtained in terms of hypergeometric functions by assuming that the coupling matrix elements V_{ij} in (7.30) vary exponentially with the interionic separation R. Models based on this fact have been developed by Ellison and Borowitz (1964), Demkov (1963), and Nikitin (1962*a*, *b*, 1963, 1970) to describe a situation in which transitions occur in a region centred about the internuclear separation $R = R_p$ (which may be, but is not necessarily, a crossing region). We shall define $t = 0$ to be the time at which $R = R_p$ on the first passage of the transition region. Since the two-state approximation can only be applied to s $\leftrightarrow$ s transitions, and since translation factors have been omitted, the matrix elements V_{ij} are functions of R only. In Nikitin's model the interaction matrix is assumed to satisfy the relations:

$$V_{22}(R) - V_{11}(R) = \omega[1 - \cos\theta \exp\{-\alpha(R - R_p)\}], \tag{7.75a}$$

$$V_{21}(R) = V_{12}(R) = \frac{\omega}{2}\sin\theta \exp\{-\alpha(R - R_p)\}, \tag{7.75b}$$

depending on the three parameters ω, θ, and α. The parameter ω is taken to be the energy difference in the separated atom limit, while the parameters α and θ can be determined by fitting to the energies obtained by diagonalizing V_{ij}. The model of Demkov (1963) is obtained as a special case of (7.75), in which θ is taken to be $\pi/2$. In Demkov's model, as in the Rosen–Zener model, there is a constant energy difference ω between the potentials V_{22} and V_{11}, which can be taken to be equal to $[V_{22}(R_p) - V_{11}(R_p)]$.

If R_p is far from the distance of closest approach, the local radial velocity is approximately constant so that

$$R(t) \approx R_p + v_p t. \tag{7.76}$$

For example, consider the case of a rectilinear trajectory for which $R(t) = [b^2 + v_0^2(t - t_0)^2]^{\frac{1}{2}}$, where t_0 is the time of closest approach. By expanding about $t = 0$, we obtain (7.76) with $v_p = -v_0^2 t_0/R_p$ and, eliminating t_0 we find

$$v_p = -v_0\left(1 - \frac{b^2}{R_p^2}\right)^{\frac{1}{2}}. \tag{7.77}$$

Using the relations (7.75) and (7.76), and setting $\tau = \alpha v_p t$ the coupled equations (7.30) take the form

$$\mathrm{i}\frac{\mathrm{d}C_1(\tau)}{\mathrm{d}\tau} = \tfrac{1}{2}\xi \sin\theta \exp(-\tau)\exp[-\mathrm{i}\delta(\tau)]C_2(\tau) \tag{7.78a}$$

$$\mathrm{i}\frac{\mathrm{d}C_2(\tau)}{\mathrm{d}\tau} = \tfrac{1}{2}\xi \sin\theta \exp(-\tau)\exp[\mathrm{i}\delta(\tau)]C_1(\tau), \tag{7.78b}$$

with

$$\delta(\tau) = \int_0^\tau [\xi - \xi\cos\theta\exp(-\tau')]\,\mathrm{d}\tau', \tag{7.79}$$

and

$$\xi = \frac{\omega}{\alpha v_p}. \tag{7.80}$$

By obtaining a single second-order equation for $C_2(\tau)$, following a procedure similar to that of the Landau–Zener and Rosen–Zener models, a solution can be obtained in terms of confluent hypergeometric functions. From this solution the probability of a transition from 1 to 2 during a double transit of the transition region is found to be

$$\mathscr{P}_{21}(v_0, \mathbf{b}) = \frac{4\sin^2\phi \exp(\pi\zeta\cos\theta)\sinh\left[\frac{\pi\xi}{2}(1-\cos\theta)\right]\sinh\left[\frac{\pi\xi}{2}(1+\cos\theta)\right]}{\sinh^2(\pi\xi)}. \tag{7.81}$$

It is generally sufficient to use the approximation $\alpha + \beta = 0$ in which case, assuming adiabatic propagation between the transition regions, ϕ is given by (see (7.25))

$$\phi = \int_0^{t_0} [(V_{22} - V_{11})^2 + 4V_{12}^2]^{\frac{1}{2}}\, dt' \tag{7.82}$$

where, as before, $t = 0$ is the time at which $R = R_p$ and $t = t_0$ is the time of closest approach. Explicit expressions for the dynamical phases α and β are given by Nikitin and Umanskii (1984). Provided ϕ is large, the factor $\sin^2\phi$ can be replaced by the average of $\frac{1}{2}$ when evaluating the total cross section by integrating over the impact parameter, the dependence on the impact parameter being determined for a straight line trajectory by (7.80) and (7.77).

By taking $\theta = \pi/2$, and modifying the phase factor ϕ, the Demkov approximation is obtained:

$$\mathscr{P}_{21}(v_0, b) = \frac{\sin^2\phi_D}{\cosh^2(\pi\xi/2)} \tag{7.83}$$

where

$$\phi_D = 2\int_0^{t_0} V_{12}\, dt. \tag{7.84}$$

We also note that (7.83) is identical with the Rosen–Zener expression (7.73) if the oscillating factors in both expressions are averaged, and with $\omega/B = \xi$.

Applications The exponential model has been applied to a variety of reactions, either in its original form or in slightly modified versions. Of particular interest is the work of Crothers and Todd (1978, 1980*a*) who investigated electron capture of H^+ from Be^+ and by the ions Zn^{2+}, Cd^{2+},

B^{2+}, Mg^{2+}, and C^{6+} from H(1s), mainly for low impact velocities $v_0 \lesssim 1$. An important conclusion reached by comparison with the numerical solution of the two-state equations for the reaction $C^{6+} + H(1s) \rightarrow C^{5+} + H^+$ is that given favourable conditions the model solutions can indeed represent the exact two-state solutions with useful accuracy. However it should again be emphasized that for the $(CH)^{6+}$ system and many others the basic two-state approximation is itself inaccurate (see Section 6.3). Also noteworthy are applications to ground-state capture by protons from helium by Burns and Crothers (1976) and Hughes and Crothers (1977).

The exponential model has been applied to inner shell processes (see Chapter 9) (Meyerhof *et al.* 1978; Stolterfoht 1979), but the analysis of Fritsch and Wille (1979) shows that although the model can be applied successfully to K-shell processes, it must be used with caution; for example K–L vacancy sharing cannot be properly described.

7.2.3 Further non-linear models

A number of different methods of representing the potentials $V_{22} - V_{11}$ and V_{12} have been developed which provide a more accurate description of certain systems, but which do not lead to an analytic solution of the two-state equations, for example the linear–exponential model of Bykhovskii and Nikitin (1965). A further case, which does have an analytical solution in terms of the confluent hypergeometric function, is that in which the potentials have the representation

$$V_{22}(t) - V_{11}(t) = \omega - \frac{1}{R(t)}$$
$$V_{12}(t) = \frac{\beta}{R(t)}. \tag{7.85}$$

For this model the transition probability is, for a head-on collision with $R(t) = V_0 t$, (Child 1971; Bandrauk 1972)

$$P_{21} = \frac{\frac{1}{2}[\exp(\pi/v_0) - \exp(-2\pi\delta)]}{\sinh(2\pi\delta)} \tag{7.86}$$

where

$$\delta = \left[\frac{4\beta^2 + 1}{4v_0^2}\right]^{\frac{1}{2}}. \tag{7.87}$$

This model has been applied to the ion–covalent coupling, such as that encountered in the ion-pair formation reaction

$$X + Y \rightarrow X^+ + Y^-. \tag{7.88}$$

For small v_0 and b, $\mathscr{P}_{21}$ reduces to the Landau–Zener formula.

7.3 THE MULTICHANNEL LANDAU–ZENER MODEL

Under certain conditions the Landau–Zener model can be generalized to treat a multichannel system. Consider, for example, electron capture by a fully stripped ion A of charge Z from atomic hydrogen in the ground state,

$$A^{Z+} + H(1s) \rightarrow A^{(Z-1)+}(n) + H^+. \tag{7.89}$$

The potential in the incident channel at large internuclear separations is

$$V_H(R) \simeq -\frac{1}{2} - \frac{Z^2\alpha}{R^4} + \cdots \tag{7.90}$$

where α is the polarizability of atomic hydrogen ($\alpha = 9/4$). The corresponding long-range potential in the rearranged system, for capture into a level with principal quantum number n, is

$$V_{A(n)}(R) \approx -\frac{1}{2}\frac{Z^2}{n^2} + \frac{Z-1}{R}. \tag{7.91}$$

Each potential $V_{A(n)}$ crosses the potential V_H at a certain value of R, $R = R_n$, as shown schematically in Fig. 7.2. If R_n is sufficiently large, the polarization term in (7.90) can be neglected and we find

$$R_n = (Z-1)\left(\frac{Z^2}{2n^2} - \frac{1}{2}\right)^{-1}, \qquad n < Z. \tag{7.92}$$

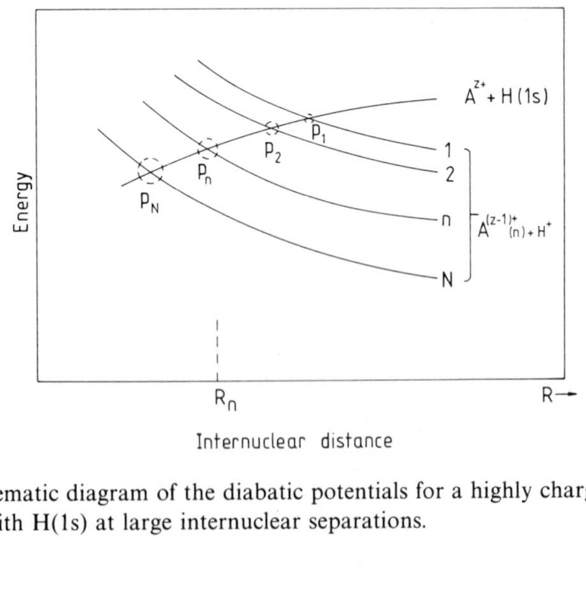

Fig. 7.2. Schematic diagram of the diabatic potentials for a highly charged ion A^{Z+} interacting with H(1s) at large internuclear separations.

Provided the transition zone about each crossing point is small compared with the distance between successive crossing points, the Landau–Zener formula can be applied at each crossing. For the reaction (7.89) this condition is satisfied for large values of Z since from (7.42) and (7.47) the width of the transition zone can be shown to be proportional to $Z^{-\frac{1}{2}}$ while from (7.92) the crossing spacing is proportional to $Z^{-\frac{1}{4}}$. By generalizing (7.16), the probability for capture into the nth level of the ion $P_n(b)$ can be obtained (Salop and Olson 1976) for a system with N crossing points. If the phases ϕ arising from propagation between each crossing point are averaged, P_n is found to be

$$\begin{aligned} P_n = p_1 p_2 \cdots p_n(1 - p_n)[1 &+ (p_{n+1} p_{n+2} \cdots p_n)^2 \\ &+ (p_{n+1} p_{n+2} \cdots p_{N-1})^2 (1 - p_N)^2 \\ &+ (p_{n+1} p_{n+2} \cdots p_{N-2})(1 - p_{N-1})^2 \\ &+ p_{n+1}^2 (1 - p_{n+2})^2 + (1 - p_{n+1})^2]. \end{aligned} \tag{7.93}$$

Here p_n is the probability calculated in the Landau–Zener approximation that at the nth crossing the system remains on the initial diabatic curve. From (7.40a)

$$p_n = \exp(-2\pi q_n), \tag{7.94}$$

where

$$q_n = \left.\frac{d^2}{a}\right|_{R=R_n}. \tag{7.95}$$

From (7.42), a has the value

$$a = \left|\frac{\partial}{\partial R}[V_n(R) - V_{\mathrm{H}}(R)]\right|_{R=R_n} \left(1 - \frac{b^2}{R_n^2}\right)^{\frac{1}{2}} v_0, \tag{7.96}$$

and using (7.90) and (7.91)

$$\left|\frac{\partial}{\partial R}[V_n(R) - V_{\mathrm{H}}(R)]\right|_{R=R_n} = \frac{Z-1}{R_n^2}. \tag{7.97}$$

The constant d in (7.95) is equal to the value of the coupling $V_{n\mathrm{H}}$ between the initial diabatic state and the nth rearranged state at $R = R_n$.

A numerical evaluation of $d|_{R=R_n}$ has been fitted to the following analytic expression by Olson and Salop (1976), and is accurate to ~ 17 per cent for $4 \leq Z \leq 54$:

$$d(R_n) = 9.13 Z^{-\frac{1}{2}} \exp(-1.324 R_n Z^{-\frac{1}{2}}). \tag{7.98}$$

The cross section for capture into the nth level of the ion is

$$\sigma_n = 2\pi \int_0^{R_n} bP_n(b)\,\mathrm{d}b, \tag{7.99}$$

and the total charge exchange cross section is

$$\sigma = \sum_{n=1}^{N} \sigma_N. \tag{7.100}$$

The multichannel Landau–Zener model naturally reflects the limitations of the two-state Landau–Zener approximation, particularly in the omission of translational effects and of rotational coupling. In addition, near-adiabatic conditions must hold and the impact velocity must be small, $v_0 \ll 1$. In the case of reaction (7.89), the degenerate hydrogenic levels of the ion $A^{Z+}(n)$ are mixed by rotational coupling. If account is taken of this (Abramov *et al.* 1979; Janev *et al.* 1983), the cross sections of the multichannel Landau–Zener model are increased in the low-energy region by up to a factor of two.

Applications

To illustrate the variation of the cross section for reactions of the type (7.89) with the charge on the incident ion, Fig. 7.3 shows the computed cross sections for values of Z, between 14 and 28, allowing for rotational coupling. To obtain some feeling for the accuracy of such cross sections, the multichannel Landau–Zener cross section (with allowance for rotational coupling) for the reaction $C^{6+} + H(1s) \rightarrow C^{5+} + H^+$ is compared in Fig. 7.4 with experimental data, and with cross sections computed from a coupled channel calculation employing an MO basis with 33 terms. Some data for the reaction $O^{6+} + H(1s) \rightarrow O^{5+} + H^+$ is also included in the figure. It can be seen that on the assumption that the 33-term MO basis cross sections are accurate, the Landau–Zener cross section is of about the correct order of magnitude near the cross section maximum, but does not reproduce the correct shape of the cross section curve as a function of energy. In applications to plasma physics where the general characteristics of the cross sections are required over a wide range of values of Z this limited degree of accuracy is useful, particularly for large Z for which coupled channel calculations are often impractical.

7.4 PERTURBATION SERIES AND THE BATES DISTORTED WAVE MODEL

The methods that have been studied in Sections 7.1 and 7.2 are useful at low impact velocities when the coupling between two levels is strong. If the coupling is weak, and the velocity large, perturbation methods are

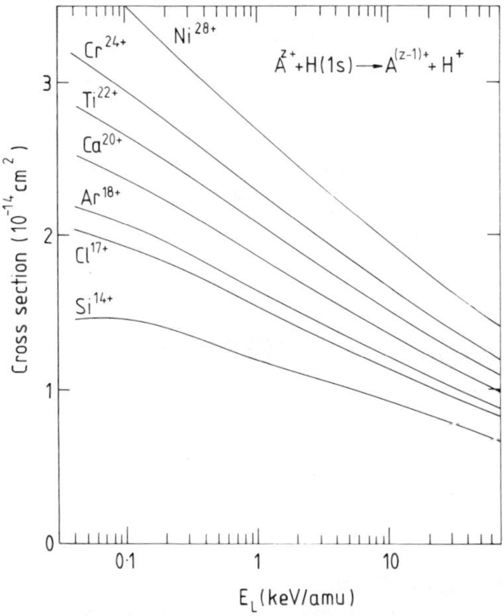

Fig. 7.3. Cross sections for the charge exchange reactions $A^{Z+} + H(1s) \rightarrow A^{(Z-1)+} + H^+$ calculated by the multichannel Landau–Zener model, with allowance for rotational coupling (after Janev *et al.* (1983)).

appropriate. In this chapter we shall consider the perturbation solution of the *coupled channel equations* obtained by an expansion on to a finite basis set, and leave the general question of the perturbation solution of the impact parameter equations (3.163a) to be discussed in Chapter 10.

Let us first consider the perturbation solution of the $N + M$ coupled equations (3.205)

$$\mathrm{i}\dot{\mathbf{A}} = \mathbf{N}^{-1}\mathbf{Q}\mathbf{A} = \mathbf{V}\mathbf{A} \tag{7.101}$$

where $\mathbf{N}$ is the overlap matrix and $\mathbf{Q}$ the interaction matrix. We recollect that, in general, with state-dependent translation factors and a non-orthogonal basis, $\mathbf{N}$ is not diagonal and $\mathbf{V}$ is not Hermitian, although unitarity (probability conservation) is satisfied provided the submatrices of $\mathbf{N}$ and $\mathbf{Q}$ satisfy (3.192).

If the system is initially in channel i, the boundary condition satisfied by the solution of (7.101) is

$$\lim_{t \to -\infty} \mathbf{A}(t) = \mathbf{A}^{(i)}, \tag{7.102}$$

where

$$A_j^{(i)} = \delta_{ij}. \tag{7.103}$$

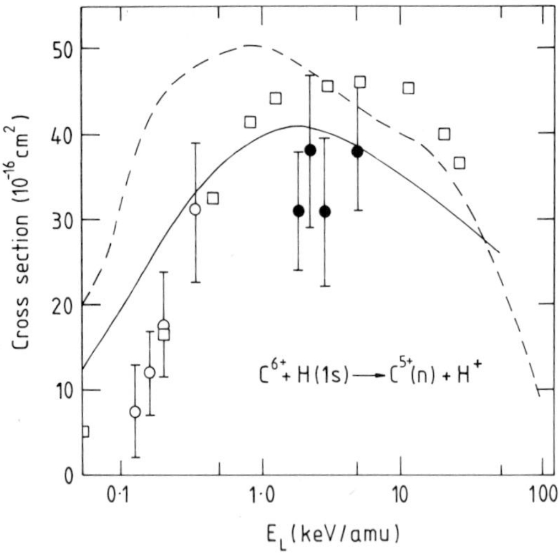

Fig. 7.4. Cross sections for electron capture by C^{6+} ions from H(1s). Experimental data: (a) for C^{6+} + H(1s) ⦶, Phaneuf *et al.* (1982); (b) for O^{6+} + H(1s) ●, Crandall *et al.* (1979). Theoretical cross sections: ——— Landau–Zener model with rotational coupling, Janev *et al.* (1983); □ □ □ 32-term MO coupled channel calculation, Green *et al.* (1982); – – – unitarized distorted wave (UDW) model, Ryufuku and Watanabe (1979).

A solution with this boundary condition satisfies the integral equation:

$$\mathbf{A}(t) = \mathbf{A}^{(i)} - \mathrm{i}\int_{-\infty}^{t} \mathbf{V}(t_1)\mathbf{A}(t_1)\,\mathrm{d}t_1 \tag{7.104}$$

or

$$\mathbf{A}(t) = \mathbf{A}^{(i)} - \mathrm{i}\int_{-\infty}^{\infty} \theta(t-t_1)\mathbf{V}(t_1)\mathbf{A}(t_1)\,\mathrm{d}t_1 \tag{7.105}$$

where $\theta(x)$ is the step function,

$$\theta(x) = \begin{cases} 1 & x > 0 \\ 0 & x < 0. \end{cases}$$

By iterating (7.105), we obtain the Born perturbation series for the coupled equations (7.101):

$$\mathbf{A}(t) = \left[\mathbf{1} - \mathrm{i}\int_{-\infty}^{t} \mathbf{V}(t_1)\,\mathrm{d}t_1 + \int_{-\infty}^{t} \mathrm{d}t_1 \int_{-\infty}^{t} \mathrm{d}t_2(-\mathrm{i})^2\theta(t_1-t_2)\mathbf{V}(t_1)\mathbf{V}(t_2) + \cdots\right]\mathbf{A}^{(i)}. \tag{7.106}$$

The second-order term I_2 of the series within the square brackets of (7.106) is

$$I_2 = (-\mathrm{i})^2 \int_{-\infty}^{t} \mathrm{d}t_1 \int_{-\infty}^{t} \mathrm{d}t_2 \theta(t_1 - t_2)\mathbf{V}(t_1)\mathbf{V}(t_2). \tag{7.107}$$

Following Dyson (1949) this can be written (by interchanging t_1 and t_2) in the form

$$\begin{aligned} I_2 &= (-\mathrm{i})^2 \tfrac{1}{2} \int_{-\infty}^{t} \mathrm{d}t_1 \int_{-\infty}^{t} \mathrm{d}t_2 [\theta(t_1 - t_2)\mathbf{V}(t_1)\mathbf{V}(t_2) + \theta(t_2 - t_1)\mathbf{V}(t_2)\mathbf{V}(t_1)] \\ &= (-\mathrm{i})^2 \tfrac{1}{2} \int_{-\infty}^{t} \mathrm{d}t_1 \int_{-\infty}^{t} \mathrm{d}t_2 T[\mathbf{V}(t_1)\mathbf{V}(t_2)]. \end{aligned} \tag{7.108}$$

In the second line the *time ordering operator* T has been introduced which is defined so that if T operates on a product of n time-dependent matrices (or operators) $\mathbf{A}_i(t_i)\mathbf{A}_j(t_j)\mathbf{A}_k(t_k)\ldots$, the factors are ordered so that the time arguments decrease from left to right

$$T[\mathbf{A}_i(t_i)\mathbf{A}_j(t_j)\mathbf{A}_k(t_k)\cdots] = \mathbf{A}_1(t_1)\mathbf{A}_2(t_2)\mathbf{A}_3(t_3)\cdots\mathbf{A}_n(t_n), \quad \text{where} \quad t_1 > t_2 > t_3 \cdots > t_n. \tag{7.109}$$

Thus in (7.108)

$$T[\mathbf{V}(t_1)\mathbf{V}(t_2)] = \begin{cases} \mathbf{V}(t_1)\mathbf{V}(t_2) & t_1 > t_2 \\ \mathbf{V}(t_2)\mathbf{V}(t_1) & t_2 > t_1. \end{cases} \tag{7.110}$$

In the nth term of the series (7.106), $n!$ interchanges between the times $t_1, t_2, \ldots, t_n$ can be made so that

$$I_n = (-\mathrm{i})^n \frac{1}{n!} \int_{-\infty}^{t} \mathrm{d}t_1 \int_{-\infty}^{t} \mathrm{d}t_2 \cdots \int_{-\infty}^{t} \mathrm{d}t_n T[\mathbf{V}(t_1)\mathbf{V}(t_2)\cdots\mathbf{V}(t_n)], \tag{7.111}$$

and we can re-express (7.106) as:†

$$\mathbf{A}(t) = \left[\mathbf{1} + \sum_{n=1}^{\infty} \frac{(-\mathrm{i})^n}{n!} \int_{-\infty}^{t} \mathrm{d}t_1 \int_{-\infty}^{t} \mathrm{d}t_2 \cdots \times \int_{-\infty}^{t} \mathrm{d}t_n T[\mathbf{V}(t_1)\mathbf{V}(t_2)\cdots\mathbf{V}(t_n)]\right]\mathbf{A}^{(i)} \tag{7.112}$$

$$= T\exp\left[-\mathrm{i}\int_{-\infty}^{t} \mathrm{d}t'\mathbf{V}(t')\right]\mathbf{A}^{(i)}. \tag{7.113}$$

† Following the same argument it is clear that the evolution matrix $\mathbf{U}(t, t')$ introduced in (7.12) can be expressed formally as

$$\mathbf{U}(t, t') = T\exp\left[-\int_{t'}^{t} \mathrm{d}t''\mathbf{V}(t'')\right].$$

The Born series for the probability amplitude $\mathscr{A}_j(b) = \lim_{t\to\infty} A_j(b, t)$ is correspondingly given by

$$\mathscr{A}_j(b) = \delta_{ji} + \left[\sum_{n=1}^{\infty} \frac{(-\mathrm{i})^n}{n!} \int_{-\infty}^{\infty} \mathrm{d}t_1 \int_{-\infty}^{\infty} \mathrm{d}t_2 \cdots \times \int_{-\infty}^{\infty} \mathrm{d}t_n\, T[\mathbf{V}(t_1)\mathbf{V}(t_2)\cdots\mathbf{V}(t_n)]\right]_{ji}. \tag{7.114}$$

Distorted wave series

On physical grounds it may be possible to find effective potentials W_j acting in each channel j which largely determine the amplitudes $A_j(\mathbf{b}, t)$ in that channel. If this is the case, the effect of these potentials can be allowed for to all orders (within the impact parameter formalism) by defining new amplitudes d_j by the transformation

$$A_j(t) = d_j(t) \exp\left\{-\mathrm{i} \int_{-\infty}^{t} \mathrm{d}t_1\, W_j(t_1)\right\} \tag{7.115}$$

in which case $\mathbf{d}(t)$ satisfies

$$\mathrm{i}\dot{\mathbf{d}} = \mathbf{M}\mathbf{d} \tag{7.116}$$

where

$$M_{kj}(t) = [V_{kj}(t) - W_j(t)\delta_{kj}] \exp\left\{-\mathrm{i} \int_{-\infty}^{t} \mathrm{d}t_1 [W_j(t_1) - W_k(t_1)]\right\}. \tag{7.117}$$

The distorted wave Born series for the amplitude $\mathscr{D}_j(\mathbf{b}) = \lim_{t\to\infty} d_j(\mathbf{b}, t)$ can be written in the form (7.114) as

$$\mathscr{D}_j(\mathbf{b}) = \delta_{ji} + \left[\sum_{n=1}^{\infty} \frac{(-\mathrm{i})^n}{n!} \int_{-\infty}^{\infty} \mathrm{d}t_1 \int_{-\infty}^{\infty} \mathrm{d}t_2 \cdots \times \int_{-\infty}^{\infty} \mathrm{d}t_n\, T[\mathbf{M}(t_1)\mathbf{M}(t_2)\cdots\mathbf{M}(t_n)]\right]_{ji}. \tag{7.118}$$

Since if W_j is real, $|\mathscr{D}_j(\mathbf{b})|^2 = |\mathscr{A}_j(\mathbf{b})|^2$, the distorted wave probability for a transition from channel i to channel j is

$$P_{ji}(\mathbf{b}) = |\mathscr{D}_j(\mathbf{b})|^2. \tag{7.119}$$

Overall Coulomb interactions If in a particular channel j both the interacting systems are charged, there is an overall Coulomb interaction at infinite separations between A and B. For example in the one-electron system $\mathrm{A} + (\mathrm{B} + \mathrm{e}^-)$ if the charges on the nuclei A and B are Z_A and Z_B, respectively, there is an overall Coulomb interaction at large distances of the form $Z_A(Z_B - 1)/R$. Correspondingly the diagonal elements $V_{jj}(t)$ for all channels in the arrangement behave like $V_{jj}(t) \sim Z_A(Z_B - 1)/R(t)$ for large $|t|$.

The second-order term in the Born series (see (7.107)) for the amplitude $\mathscr{A}_j$ is

$$(-i)^2 \sum_k \int_{-\infty}^{\infty} dt_1 \int_{-\infty}^{t_2} dt_2\, V_{jk}(t_1)V_{ki}(t_1) \tag{7.120}$$

and it follows that the term for which $k = j$, containing the diagonal element V_{jj}, is divergent since $\int^t [1/R(t')]\, dt'$ is logarithmically infinite as $t \to \infty$. Equally, all higher terms in the series are divergent. Under these circumstances the Born series does not exist, as pointed out by Mapleton (1972) and more recently by Dewangan and Eichler (1985). This difficulty is easily remedied by choosing distortion potentials W_j which cancel the long-range Coulomb interactions so that

$$W_j \sim -\frac{Z_A(Z_B - 1)}{R}. \tag{7.121}$$

Provided the long-range Coulomb interactions are cancelled in this way the distorted wave series exists and, when applied to a finite set of coupled equations, is convergent at sufficiently high energies. It should be noted that the divergence of the second-order Born approximation applies equally to excitation as to rearrangement if both partners in the collision are charged.

The Bates distorted wave model

A particular choice of distortion potentials satisfying the condition (7.121) was made by Bates (1958). He set W_j equal to the diagonal elements of $\mathbf{V}$ in all channels

$$W_j = V_{jj} \qquad j = 1, 2, \ldots, M + N. \tag{7.122}$$

In this model the diagonal elements of $\mathbf{M}$ vanish and the coupled equations (7.116) are in the 'interaction representation'.

The Bates distorted wave model was originally applied to the two-state equations (5.96) obtained using a two-centre atomic orbital basis, the channel i representing elastic scattering in the initial arrangement $A + (B + e^-)$ and the second channel f the rearranged system $(A + e^-) + B$. The first-order approximation to the amplitudes $d_f(b)$ is then, from (5.48):

$$\mathscr{D}_f(b) = -i \int_{-\infty}^{\infty} \frac{\bar{K}_{fi} - S_{fi}H_{ii}}{1 - |S_{if}|^2} \exp\left[-i \int_{-\infty}^{t} \{\delta_i(t') - \delta_f(t')\}\, dt'\right] dt \tag{7.123}$$

where the phases δ_i and δ_f are defined in terms of the potential and overlap matrix elements by (5.95):

$$\delta_i(t) = \frac{H_{ii} - S_{if}\bar{K}_{fi}}{1 - |S_{if}|^2} \tag{7.124a}$$

$$\delta_f(t) = \frac{\bar{H}_{ff} - S_{fi}K_{if}}{1 - |S_{if}|^2}. \tag{7.124b}$$

It should be noticed that, within the two-state approximation, since **M** has no diagonal terms there is no second-order term in the Bates distorted wave series, the first non-vanishing term beyond the first being the third-order term containing the interaction sequence $M_{12}(t_1)M_{21}(t_2)M_{12}(t_3)$.

In the particular case of symmetrical resonance, for example

$$\mathrm{H}^+ + \mathrm{H}(1s) \rightarrow \mathrm{H}(1s) + \mathrm{H}^+,$$

$\delta_i = \delta_f$, and

$$\mathscr{D}_f(b) = -\mathrm{i}\int_{-\infty}^{\infty} L(t)\,\mathrm{d}t, \tag{7.125}$$

where L is defined by (5.101). The probability P^B_{fi} of a transition, in the Bates approximation, is

$$P^B_{fi}(v_0, b) = \left|\int_{-\infty}^{\infty} L(t)\,\mathrm{d}t\right|^2, \tag{7.126}$$

which can be compared with the exact result (5.106)

$$P_{fi}(v_0, b) = \sin^2\left|\int_{-\infty}^{\infty} L(t)\,\mathrm{d}t\right|. \tag{7.127}$$

It follows that the distorted wave solution is only accurate if the approximation $\sin|\mathscr{D}_f(b)| \approx |\mathscr{D}_f(b)|$ is accurate.

A similar approximation to the Bates first-order expression (7.123) has been obtained by Bassel and Gerjuoy (1960). They applied conventional distorted wave theory (see, for example, Bransden (1983)) to the full Schrödinger equation (3.2), taking the distorting potentials to be H_{ii} in the initial and H_{ff} in the final state. They obtained an expression which, in the semi-classical approximation, reduces to

$$\mathscr{D}_f(b) = -\mathrm{i}\int_{-\infty}^{\infty} [\bar{K}_{fi} - S_{fi}H_{ii}]\exp\left[-\mathrm{i}\int_{-\infty}^{t} \{\delta_i(t') - \delta_f(t')\right]\mathrm{d}t. \tag{7.128}$$

Bassel and Gerjuoy made the further approximation of replacing the distorted waves by plane waves, which amounts in the impact parameter formulation to setting $\delta_i = \delta_f = 0$. Since $|S_{fi}|^2$ is generally small at the high energies at which perturbation theory can be applied, the Bates expression and the Bassel–Gerjuoy expressions are effectively the same for the case of symmetrical resonance. In general, however, the presence of the phases δ_i and δ_f has the effect of reducing the cross section considerably and the neglect of distortion is not justified.

Excitation

For sufficiently high energies the first Born approximation for the total cross section for excitation is accurate and has been widely applied. At the energies concerned, charge exchange is very small so that the submatrices **K** can be neglected and the overlap matrix **N** becomes equal to the unit matrix. Thus for excitation or direct ionization the matrix **V** in the Born series (7.106) can be replaced by the submatrix **H** defined by (3.186). For light systems such as $H^+ + H$, $H^+ + He$, comparison with the results of higher-order calculations suggests that for total cross sections the first Born approximation is accurate at energies above ~200 keV/a.m.u. for the excitation of strongly coupled states. This is not the case for differential cross sections at large angles which depend on small values of the impact parameter. For these nearly 'head on' collisions, allowance must be made for the nuclear deflection, either by solving the coupled equations or by using a distorted wave approximation. It should be recollected that whenever differential cross sections are calculated proper account must be taken of the phase generated by the internuclear interaction, as explained in Section 3.8.

An interesting case is the differential cross section for the excitation of the $n = 2$ level of atomic hydrogen by proton impact which is shown in Fig. 7.5. It is seen that although the cross section in the forward direction (and hence the total cross section) given by the Born approximation is fairly accurate, it fails completely at large angles compared with coupled channel calculations. As a further illustration, the differential cross section for the excitation of the 2^1S state of helium by protons is shown in Fig. 7.6 at $E_L = 100$ keV/a.u. The Born cross section agrees well with the data at angles up to 0.8×10^{-3} radians, except in the very small-angle region with $\theta < 0.2 \times 10^{-3}$ radians. The results of a coupled channel calculation with a two-centre basis on the other hand agree well with the data over the whole angular range covered by the measurements. The Born approximation for excitation has been discussed in detail by Bell and Kingston (1974, 1976) and coupled channel calculations for collisions involving H^+, H, He^+, and He have been reported by Flannery (1969*a*, *b*, 1970), Flannery and McCann (1973, 1974), and Bell and Kingston (1978).

The rate of convergence of the Born and Bates distorted wave series for a finite system of coupled channels has been examined in one particular case, the excitation of H(1s) by proton impact (Bransden *et al.* 1979). At the energies considered, $E \geq 75$ keV, charge exchange is small and a realistic

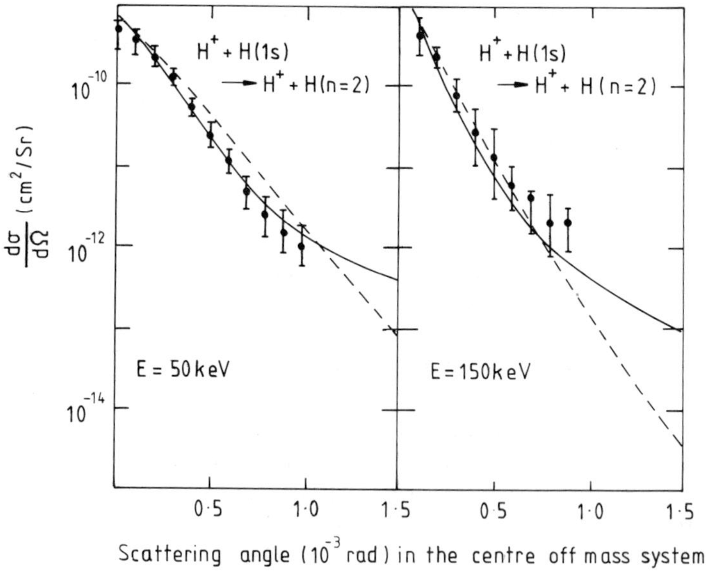

Fig. 7.5. Differential cross sections in the centre-of-mass system for the excitation of atomic hydrogen to the $n = 2$ level by proton impact at 50 and 100 keV/a.m.u. Experimental data: ⫯ Park *et al.* (1978). Theoretical cross sections: ------ Born approximation; ——— coupled channel calculations with a pseudostate basis, Bransden and Noble (1979).

coupled channel model is obtained by using a single-centre expansion about the target nucleus. A small basis set due to Callaway and Wooten (1974) consisting of the exact $n = 1$, $n = 2$, and 3d hydrogenic wave functions together with three s and three p pseudostates has been shown (Bransden and Dewangan 1979) to provide a good description of the $H^+ + H(1s) \to H^+ + H(n = 2)$ excitation, the total and differential cross sections computed agreeing well with the data of Park *et al.* (1976, 1978). The Born series to the coupled equations can be generated from the sequence of functions $A_j^{(n+1)}(t)$ satisfying (see (7.105))

$$A_j^{(n+1)}(t) = \delta_{ij} - \mathrm{i} \sum_{k=1}^{N} \int_{-\infty}^{t} H_{jk}(t_1) A_k^{(n)}(t_1)\, \mathrm{d}t_1, \tag{7.129}$$

where $A_k^{(1)}$ is defined as $A_k^{(1)} = \delta_{ik}$. In the same way the sequence of distorted wave amplitudes can be generated from

$$d_j^{(n+1)}(t) = \delta_{ij} - \mathrm{i} \sum_{k=1}^{N} \int_{-\infty}^{t} (1 - \delta_{kj}) H_{jk}(t_1) \exp[\mathrm{i}\delta_j(t_1) - \delta_k(t_1)]\, d_k^{(n)}(t_1)\, \mathrm{d}t_1, \tag{7.130}$$

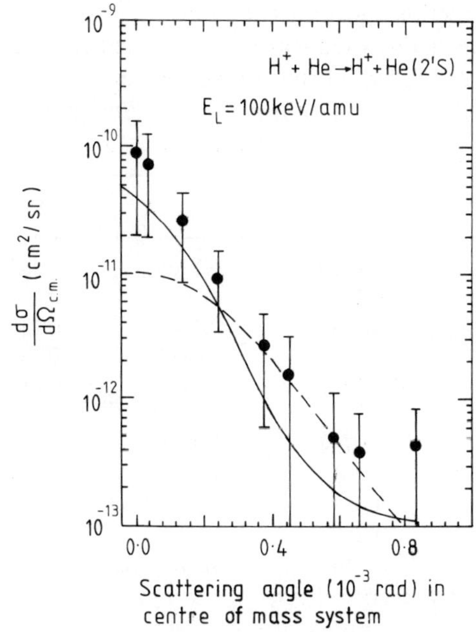

Fig. 7.6. Differential cross sections in the centre-of-mass system for the excitation of helium to the 2^1S state by proton impact at $E_L = 100$ keV/a.m.u. Experimental data: Kvale *et al.* (1985). Theoretical cross sections: ----- Born approximation; ——— coupled channel calculations with two-centre pseudostate basis, Slim *et al.* (1991*b*).

where

$$\delta_j(t) = \int_{-\infty}^{t} H_{jj}(t')\,\mathrm{d}t', \tag{7.131}$$

and $d_k^{(1)} = \delta_{ik}$. The matrix elements $H_{jk}(t)$ have the form, taking the target proton as the origin of coordinates,

$$H_{jk}(t) = \langle\phi_j(r)| -\frac{1}{x} + \frac{1}{R}|\phi_k(r)\rangle \exp[\mathrm{i}(\varepsilon_j - \varepsilon_k)t], \tag{7.132}$$

and

$$\delta_j(t) = \int_{-\infty}^{t} H_{jj}(t_1)\,\mathrm{d}t_1. \tag{7.133}$$

In Table 7.2, the cross sections for (2s) and (2p) excitation obtained by the numerical solution of the coupled equations are compared with those calculated in the first- and second-order distorted wave approximation and also with the first- and second-order Born approximations to the same set of equations. It is seen immediately that the second-order approximations

Table 7.2. Cross sections for the excitation of the 2s and 2p levels of hydrogen in the Born, distorted wave, and coupled channel models (units of πa_0^2). C is from the solution of the coupled equations; B1, B2 are first- and second-order Born approximations; D1, D2 are first- and second-order distorted wave approximations.

Energy		75 keV	105 keV	145 keV	200 keV
1s–2s	C	0.1886	0.1268	0.0871	0.0605
	B1	0.133	0.0981	0.0725	0.0533
	B2	0.374	0.220	0.136	0.0861
	D1	0.1024	0.0812	0.0632	0.0483
	D2	0.3283	0.1976	0.1227	0.0781
1s–2p	C	1.0134	0.9326	0.8058	0.6775
	B1	1.263	1.062	0.883	0.725
	B2	1.369	1.119	0.917	0.744
	D1	1.2671	1.0878	0.9058	0.7493
	D2	1.3641	1.1321	0.9251	0.7566

are in general worse than the first-order ones and that the distorted wave approximations are not more accurate for the total cross sections than the corresponding Born approximations.

In view of this, questions might be raised about the rate of convergence of both series. For the distorted wave series at the larger impact parameters ($b > 0.5$) excellent agreement was found with the numerical solutions of the coupled equations at fourth order, while for smaller impact parameters the rate of convergence was worse and agreement was not reached until sixth order. For the Born series, convergence was obtained at fourth order for $b > 1.0$, but for smaller b the convergence is less rapid. One can perhaps conclude that it is not profitable to employ perturbation methods until energies are reached where one can be confident that the first-order term is accurate, in this example for energies certainly in excess of 300 keV for the strong s $\rightarrow$ p transitions, and at still higher energies for weak transitions (see Bransden and Issa 1975). Although the first Born approximation is at least as accurate as the first-order distorted wave approximation to the total cross section, it must be emphasized that the distorted wave amplitude will provide much more accurate differential cross sections. This is because the large-angle scattering is essentially of the form of an elastic scattering followed by a transition, and the distorted wave amplitude contains the necessary information to describe such a process.

Similar first-order Born approximation calculations have been carried out for ionization both in the wave formulation (Bates and Griffing 1953) and in the impact parameter formulation (Bang and Hansteen 1959), and reviews of the results of calculations have been given by Bell and Kingston (1974, 1976). For collisions between light ions the Born approximation is accurate at velocities for which $v_0/v_e \gtrsim 2$, where v_e is the orbital velocity of the target electron; but when $Z_A \ll Z_B$, Born's approximation may be accurate at lower velocities because the perturbation $-Z_A/x$ is small compared with the binding potential. This situation, which occurs in the treatment of the ionization of the inner shells of heavy atoms, has been discussed by Briggs and Taulbjerg (1978). When the charge on the target nucleus is large then the integration must be carried out over a Coulomb rather than a straight line trajectory (see for example Kocbach (1976), Pauli and Trautmann (1978)).

Charge exchange

There have not been many direct applications of the first-order Bates distorted wave approximation (7.123) to charge exchange. The evidence suggests that, at the higher impact velocities, the first-order approximation agrees well with the results of a two-state approximation. For example, for the symmetrical resonance reaction $H^+ + H(1s) \rightarrow H(1s) + H^+$, McCarroll (1961) found that the cross section in the first-order approximation was 25 per cent smaller than the exact two-state result at 25 keV, but by 100 keV the first-order cross section was only 5 per cent less than the exact result, and by 200 keV there was excellent agreement. For symmetrical resonance there is almost no difference between the Bates approximation and that of Bassel and Gerjuoy, which neglects distortion. However, calculations by McCarroll and McElroy (1962) and McElroy (1963) for the reactions $He^{2+} + H(1s) \rightarrow He^+(n = 1, 2) + H^+$ and $H^+ + H(1s) \rightarrow H(n = 2) + H^+$ show that distortion can be important. For the non-symmetrical reaction $He^{2+} + H(1s) \rightarrow He^+(3s) + H^+$, it is interesting to compare the two-state calculations of Bransden *et al.* (1980) with the first-order Bates cross sections due to Saha *et al.* (1980). Below 100 keV, the first-order cross section exceeds those of the two-state approximation by about 50 per cent at 50 keV and a factor of five at 25 keV. However, at 100 keV and above there is no difference between the two cross sections as shown in Table 7.3.

Since at high energies the two-state approximation, and hence the distorted wave approximation, is a good approximation to the solution of the coupled equations using a basis of discrete *bound* states, it might be thought that the computed distorted wave cross sections for charge exchange represent the true high-energy limit. This is not the case since second-order transitions with continuum intermediate states become increasingly important at high energies. Nevertheless there is a velocity range extending from ~ 50 keV/a.m.u. to several hundred keV/a.m.u. in which the two-state and distorted wave approximations are useful. At velocities $v \approx v_e$, the two-state approximation may remain useful, but the first-order perturbation or distorted wave solutions may break down. This is often signalled by failure

Table 7.3. Cross sections, in units of cm^2, for the charge exchange reaction $He^{2+} + H(1s) \rightarrow He^{+}(3s) + H^{+}$. σ^{BDW} is the first-order Bates distorted wave approximation; σ^{N} is the numerical solution of the two-state equations.

E(keV)	σ^{BDW}	σ^{N}
25	2.08×10^{-18}	4×10^{-19} [1]
50	1.08×10^{-17}	7.1×10^{-18} [2]
100	2.14×10^{-17}	2.15×10^{-17}
200	1.24×10^{-17}	1.23×10^{-17}
400	3.31×10^{-18}	3.31×10^{-18}

[1] At $E = 20$ keV. [2] At $E = 40$ keV.

to preserve unitarity, that is, the computed value $P_{fi}(b)$ exceeds unity. Several attempts have been made to modify the distorted wave approximation to extend its range of applicability to lower energies. This has been largely in response to the need in thermonuclear plasma research for cross sections for electron capture by highly charged ions from atomic hydrogen. If the charge on the projectile Z_A is large the number of charge exchange channels also becomes large since capture is predominantly into levels of high n, and a straightforward coupled channel approach becomes increasingly difficult. We shall first outline the unitarized distorted wave model of Ryufuku and Watanabe (1978, 1979) and then discuss the Magnus approximation which also ensures unitarity.

7.5 THE UNITARIZED DISTORTED WAVE APPROXIMATION

The unitarized distorted wave approximation starts from the set of coupled equations (7.116), using a basis of atomic orbitals including pseudostates. The coupling matrix M_{kj} in (7.117) is completely defined by choosing the distortion potentials W_j to be equal to the diagonal elements of $\mathbf{V}$, $W_j = V_{jj}$, as in the Bates distorted wave model.

The probability amplitude for a transition from an initial state i to final state j is given by the series (7.118), where j can be either a state belonging to the original arrangement $A + (B + e^-)$ or to the rearranged system $(A + e^-) + B$. The basic approximations are now made that

(1) the time ordering is ignored in each term of the series;

(2) the only matrix elements retained are those coupling the initial state i to some other state in the basis k;

(3) the matrix **M** is treated as if it were Hermitian, that is, it is assumed that $M_{ik} = M^*_{ki}$.

Defining T_k as

$$T_k = \int_{-\infty}^{\infty} M_{ki}(t)\,\mathrm{d}t \tag{7.134}$$

and making approximations (1), (2), and (3) in (7.118) we find

$$\mathscr{D}_i(b) = 1 + \frac{(-\mathrm{i})^2}{2!} \sum_{k \neq i} |T_k|^2 + \frac{(-\mathrm{i})^4}{4!} \left\{ \sum_{k \neq i} |T_k|^2 \right\}^2 + \cdots \tag{7.135a}$$

and

$$\mathscr{D}_j(b) = (-\mathrm{i})T_j + \frac{(-\mathrm{i})^3}{3!} T_j \sum_{k \neq i} |T_k|^2 + \frac{(-\mathrm{i})^5}{5!} T_j \left\{ \sum_{k \neq i} |T_k|^2 \right\}^2 + \cdots, \qquad j \neq i \tag{7.135b}$$

where the sum over k is over all states in the basis, except the initial state i, since M_{kj} has diagonal elements which are zero.

The series (7.135a) and (7.135b) can be summed with the result

$$\mathscr{D}_i(b) = \cos(p^{\frac{1}{2}}), \tag{7.136}$$

$$\mathscr{D}_j(b) = -\mathrm{i}p^{-\frac{1}{2}} T_j \sin(p^{\frac{1}{2}}), \tag{7.137}$$

where

$$p = \sum_{k \neq i} |T_k|^2. \tag{7.138}$$

The probabilities for elastic scattering and inelastic scattering, P_{ii} and P_{ji}, are then

$$P_{ii}(b) = \cos^2(p^{\frac{1}{2}}); \qquad P_{ji} = |T_j|^2 \frac{1}{p} \sin^2(p^{\frac{1}{2}}). \tag{7.139}$$

Clearly, within the finite basis employed, unitarity is satisfied since

$$P_{ii}(b) + \sum_{j \neq i} P_{ji}(b) = 1. \tag{7.140}$$

In applications to charge exchange Ryufuku and Watanabe made some subsidiary approximations:

(4) the only levels contributing to the sum in (7.138) are discrete states of the rearranged system;

(5) the matrix elements M_{ki} are evaluated as in the two-state approximation, with $|S_{ki}|^2$ set equal to zero, so that (see (7.123))

$$M_{ki}(t) = (\bar{K}_{ki} - S_{ki}H_{ii})\exp\left[-\mathrm{i}\int_{-\infty}^{t} \mathrm{d}t'\, \{\delta_i(t') - \delta_k(t')\}\, \mathrm{d}t'\right]. \quad (7.141)$$

(6) The phases δ_i, δ_k given by (7.124a, b) (with k replacing f) are approximated by

$$\delta_i \approx H_{ii}, \qquad \delta_k \approx \bar{H}_{kk}. \quad (7.142)$$

The major approximations made in arriving at the expressions (7.139) are difficult to evaluate. Certainly, ignoring the time ordering operator can only be valid at high energies $v_0 \gtrsim v_e$, where v_e is the Bohr velocity of the target electron. Approximation (2) is also difficult to justify at low energies, because in the case of capture by a fully stripped ion the Stark mixing between the degenerate levels in the final state is ignored, and in addition, both coupling between the levels of the atom in the final state and excitation of the target followed by capture can be important processes. As the impact velocity increases intermediate continuum states become increasingly important, and no allowance is made for such transitions. It follows that the unitarized distorted wave (UDW) approximation can be expected to be useful over a range of velocities in the interval $v_e \lesssim v_0 \lesssim v_{max}$, where v_{max} is two or three times v_0. This is illustrated in Fig. 7.4 for the reaction $C^{6+} + H(1s) \rightarrow C^{5+} + H^+$. It is seen that near the cross section maximum, and for some way above, the UDW cross section is fairly close to the results of a coupled channel calculation with a large basis, which is expected to be accurate. Below the cross section maximum the UDW cross section does not display the correct energy variation and is too large compared with both the experimental data and the coupled channel cross sections.

Very extensive applications have been made for capture by fully stripped ions of charge Z from H(1s) and $1 \leq Z \leq 20$ (Ryufuku and Watanabe 1978, 1979; Ryufuku 1982). The levels included in the sum over k varied with the impact velocity and Z, but included all the important final discrete states. Both total capture cross sections and cross sections for capture into individual states have been reported.

A semi-empirical scaling law which fits the unitarized distorted wave cross sections is

$$\sigma(E_L) = Z^{1.07}\tilde{\sigma}(\tilde{E}); \qquad \tilde{E} = E_L/Z^{0.35}. \quad (7.143)$$

The reduced cross sections $\tilde{\sigma}$ is nearly the same for all Z for energies greater than 25 keV/a.m.u. ($v_0 > 1$). It is shown in Fig. 7.7, plotted against the

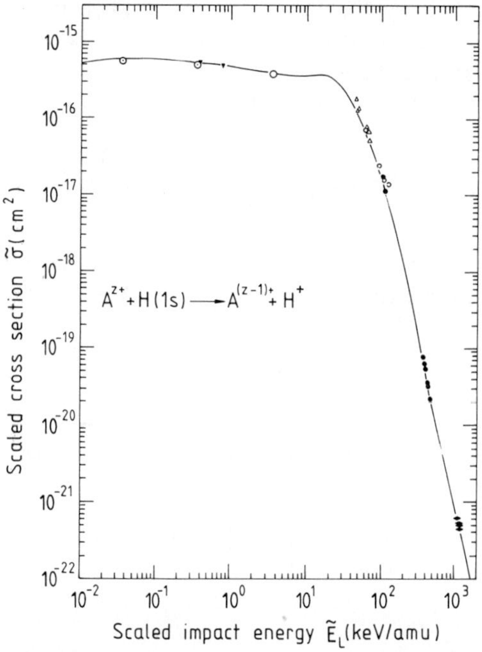

Fig. 7.7. Scaled cross sections as a function of scaled energy for electron capture for fully stripped ions of charge Z from H(1s) (after Ryufuku (1982)). The scaling relations are defined in (7.143). The solid curve shown is calculated for $Z = 14$ (silicon) but can be taken to be a universal curve applicable for ions with $Z > 14$. Specific results for $Z = 20$ are shown as ⊙ for comparison. Also included are data points, ○, △, ●, ◆, and ▼ for capture by various ions at various energies with $12 < Z < 25$. (For full references to the data see Ryufuku (1982).)

reduced energy $\tilde{E}$. Some experimental results are included for comparison.

The unitarized distorted wave model can equally well be applied to excitation or ionization, and extensive results for these cross sections have been reported by Ryufuku (1982) for fully stripped ions interacting with atomic hydrogen ($Z = 1, 3, 5, 6, 14$).

7.6 THE MAGNUS APPROXIMATION

The exact solution $\mathbf{A}(t)$ of the coupled equations (7.101) can be written in the form (7.113)

$$\mathbf{A}(t) = T \exp\left[-\mathrm{i} \int_{-\infty}^{t} \mathbf{V}(t')\,\mathrm{d}t'\right]\mathbf{A}^{(i)} \tag{7.144}$$

where T is the time ordering operator. In general this expression is purely formal and is defined by the infinite series (7.112). However, in the very special case that the matrices **V** commute at different times so that $\mathbf{V}(t_1)\mathbf{V}(t_2) = \mathbf{V}(t_2)\mathbf{V}(t_1)$, the order of the factors of each term of the series (7.112) is immaterial and we can write

$$\mathbf{A}(t) = \exp\left[-\mathrm{i}\int_{-\infty}^{t} \mathbf{V}(t')\,\mathrm{d}t'\right]\mathbf{A}^{(i)}. \tag{7.145}$$

An example for which (7.145) is an exact solution is the two-state equations (5.100) for symmetrized resonance. In this case $\mathbf{V}(t)$ has the form

$$\mathbf{V}(t) = \begin{bmatrix} 0 & L(t) \\ L(t) & 0 \end{bmatrix}, \tag{7.146}$$

where $L(t)$ is defined by (5.101). The exact $\mathscr{S}$-matrix is thus

$$\mathscr{S} = \exp\left[-\mathrm{i}\int_{-\infty}^{\infty} \mathbf{V}(t)\,\mathrm{d}t\right] \tag{7.147}$$

from which

$$\mathscr{S}_{12} = \mathscr{S}_{21} = -\mathrm{i}\sin\left[\left[\int_{-\infty}^{\infty} L(t)\,\mathrm{d}t\right]\right. \tag{7.148a}$$

$$\mathscr{S}_{11} = S_{22} = \cos\left[\left[\int_{-\infty}^{\infty} L(t)\,\mathrm{d}t\right]\right. \tag{7.148b}$$

so that the probability for charge exchange, $|S_{12}|^2$, is given by (5.106) as before.

In general, the interaction matrices at different times do not commute, in which case (7.145) is not exact and is known as the first-order Magnus approximation (Magnus 1954). It has the merit of providing a unitary $\mathscr{S}$-matrix (within the $(N + M)$-dimensional subspace spanned by the basis set) when the interaction matrix **V** is Hermitian, for example if a common translational factor is employed. To compute the matrix elements $[\exp\{-\mathrm{i}\int_{-\infty}^{\infty} \mathbf{V}(t)\,\mathrm{d}t\}]_{ij}$, it is necessary to find the transformation **T** that diagonalizes $\mathbf{U} = \int_{-\infty}^{\infty} \mathbf{V}\,\mathrm{d}t$ such that

$$(\mathbf{TUT}^{-1})_{ji} = v_i\,\delta_{ji} \tag{7.149}$$

in which case, from (7.144)

$$\mathscr{A}_j(\mathbf{b}) \equiv A_j(\infty) = \sum_{k=1}^{N+M} T_{jk}\,\mathrm{e}^{-\mathrm{i}v_k}T_{ki}. \tag{7.150}$$

The Magnus approximation has been used to study ion–atom collisions by Takayanagi (1963) and by Callaway and his co-workers (Callaway and Bauer 1965; Callaway and Bartling 1966; Callaway and Duggan 1966).

Clearly, if $t_1 - t_2$ is small enough, $\mathbf{V}(t_1)$ and $\mathbf{V}(t_2)$ will commute approximately. It follows that the Magnus approximation should become accurate at high velocities for which the collision time is short compared with characteristic atomic times.

The exact solution (7.144) can be developed as a product of exponential factors (Wilcox 1967). We may write

$$\mathbf{A}(t) = \exp[-\mathrm{i}\boldsymbol{\Omega}(t)]\mathbf{A}^{(i)} \tag{7.151}$$

where

$$\boldsymbol{\Omega}(t) = \sum_{n=1}^{\infty} \mathbf{U}_n(t) \tag{7.152}$$

with

$$\mathbf{U}_1(t) = \int_{-\infty}^{t} \mathbf{V}(t_1)\,\mathrm{d}t_1 \tag{7.153}$$

$$\mathbf{U}_2(t) = \frac{1}{2!}\int_{-\infty}^{t}\mathrm{d}t_1\int_{-\infty}^{t_1}\mathrm{d}t_2\,[\mathbf{V}(t_1), \mathbf{V}(t_2)]$$

$$\mathbf{U}_3(t) = \frac{1}{3!}\int_{-\infty}^{t}\mathrm{d}t_1\int_{-\infty}^{t_1}\mathrm{d}t_2\int_{-\infty}^{t_2}\mathrm{d}t_3\,[[\mathbf{V}(t_1), \mathbf{V}(t_2)], \mathbf{V}(t_3)] + [[\mathbf{V}(t_3), \mathbf{V}(t_2)], \mathbf{V}(t_1)]. \tag{7.154}$$

Greenland (1978, 1982) (see also Wille (1981, 1982)) has employed the third-order Magnus approximation in which $\mathbf{U}_1$, $\mathbf{U}_2$, and $\mathbf{U}_3$ are retained, to discuss the two-channel equations for a single pseudocrossing between two adiabatic potentials, neglecting translational factors, obtaining probabilities of a similar form to the Landau–Zener model.

The Magnus approximation has also been employed by Baye and Heenan (1973) (see also Baye (1970) and Baye *et al.* 1970) to obtain approximate solutions of the coupled equations obtained by a single-centre expansion about the target B. Excitation cross sections for proton impact on atomic hydrogen were calculated for the first-order approximation retaining $\boldsymbol{\Omega} = \mathbf{U}_1$ and the second-order approximation with $\boldsymbol{\Omega} = \mathbf{U}_1 + \mathbf{U}_2$. The procedure was to diagonalize the matrix $\boldsymbol{\Omega}(\infty)$ on the finite basis, by a unitary transformation matrix $\mathbf{S}$, so that

$$\langle k|\mathbf{S}^{\dagger}\boldsymbol{\Omega}(\infty)\mathbf{S}|j\rangle = \omega_j\,\delta_{kj}. \tag{7.155}$$

If the initial state is that labelled by i then from (7.151) the probability

Table 7.4. Cross sections for excitation of H(1s) by proton impact (in units of πa_0^2). A is the first-order Magnus approximation (4-state basis) (Baye and Heenan 1973); B is the second-order Magnus approximation (4-state basis) (Baye and Heenan 1973); C is the numerical solution of coupled equations (4-state basis) (Flannery 1969*a*).

E		1s–2s	1s–$2p_0$	1s–$2p_{\pm 1}$
100	A	0.109	0.402	0.631
	B	0.153	0.461	0.598
	C	0.179	0.467	0.573
200	A	0.0572	0.227	0.471
	B	0.0699	0.257	0.463
	C	0.0760	0.260	0.448
400	A	0.0287	0.118	0.315
	B	0.0321	0.130	0.313
	C	0.0344	0.135	0.310

amplitudes for the transition $i \to j$ are

$$A_{ji}(\mathbf{b}, t = \infty) = \sum_k S_{jk} \exp(\mathrm{i}\omega_k) S^{\dagger}_{ki}. \tag{7.156}$$

To assess the reliability of the approximation, Baye and Heenan (1973) compared their results for a four-state basis (1s, 2s, $2p_0$, $2p_{\pm 1}$) with those obtained by a numerical solution of the coupled equations with the same basis (Flannery 1969*a*). Some of their results are shown in Table 7.4. It is seen that the second-order approximation agrees well with the solution of the coupled equations at 400 keV, and the agreement is reasonable (at the 10 per cent level) at 200 keV. Of considerable significance is the comparison between the first- and second-order approximations, because this provides a measure of retardation effects in which the perturbation at one time is correlated with that at another. It is seen from Table 7.4 that the second-order term and hence the commutator in (7.154) is important even at the highest energy considered.

Baye and Heenan presented calculations with a large (20-term) bound state basis. These, while interesting, are limited by the omission of pseudo-states representing the continuum. Below 80 keV, where charge exchange is important, the single-centre expansion itself will not be adequate.

8
INTERMEDIATE ENERGIES

The full coupled channel model in which the direct and exchange channels receive equal representation in the wave function is the principal method for calculating ion–atom cross sections at low velocities, $v_0/v_e \leq 1$, where v_e is the Bohr velocity of the active electron in the target. This model can also be employed at arbitrarily high velocities, provided the continuum is sufficiently well represented in the eigenfunction expansions; but at high velocities, for example $v_0/v_e \geq 2$, it is natural to consider perturbation and distorted wave methods. Perturbation methods based on the coupled equations have been discussed in Chapter 7, and the topic will be taken up in a more general way in Chapter 10. Leaving aside such true high-energy methods, there are a number of approximations which are particularly suitable for the computation of cross sections outside the low-velocity region, up to moderate, but not extremely high, velocities. In this intermediate-energy interval possible approaches, described in this chapter, are to employ a simplified coupled channel model, based on single-centre expansions, to develop appropriate optical potentials, and, very different in spirit, to apply classical mechanics to the three-body problem presented by the two heavy particles and the active electron.

8.1 SINGLE-CENTRE EXPANSIONS

For systems in which excitation and ionization are the dominant processes and the flux entering the charge exchange channels is small, it is reasonable to assume the wave function can be approximated by a truncated single-centre expansion in terms of the target eigenfunctions. Although formally a single-centre untruncated expansion is complete, and must contain the charge exchange channels, the overlap matrix between any *finite* set of target eigenfunctions (or pseudostate functions) and a projectile-centred eigenfunction vanishes in the limit $|t| \to \infty$. It follows that a finite single-centre expansion can never represent rearrangement channels asymptotically. Charge exchange is a weak process at high energies and also when the projectile charge Z_A is small compared with that on the target Z_B. Accordingly it is expected that at high energies, and for $Z_A/Z_B \ll 1$, accurate excitation and ionization cross sections can be calculated from the coupled channel equations obtained by expanding in a finite set of target-centred states and pseudostates. For the single-electron system A + (B + e$^-$), the coupled equations are of the following form, obtained from (3.185) and (3.186), with **S** and **K** set equal to zero,

$$\mathrm{i}\dot{A}_j(\mathbf{b}, t) \sum_{k=1}^{N} H_{jk} A_k(\mathbf{b}, t) \tag{8.1}$$

with

$$H_{jk} = \int \mathrm{d}\mathbf{r} \Phi_j^{\mathrm{B}*} \left[H_{\mathrm{e}} - \mathrm{i} \frac{\mathrm{d}}{\mathrm{d}t}\bigg|_{\mathbf{r}} \right] \Phi_k^{\mathrm{B}} \tag{8.2a}$$

$$= \exp\{\mathrm{i}(\varepsilon_j - \varepsilon_k)t\} \int \mathrm{d}\mathbf{r} \phi_j^*(r) \{V_{\mathrm{Ae}}(\mathbf{x}) + V_{\mathrm{AB}}(R)\} \phi_k(\mathbf{r}) \tag{8.2b}$$

where a coordinate system (**r**, t) based on the heavy particle B as origin has been used, rather than the centre-of-mass system (**s**, t).

In Chapter 7, the Born approximation to equations (8.1), which are equivalent to the integral equations (7.129), has been discussed, and we have noted that satisfactory results can be obtained for excitation or ionization by first-order calculations at sufficiently high energies. For excitation, attempts to improve on the first-order Born approximation by 'close coupling' approximations in which a number of target bound states are included in the basis set and the corresponding coupled equations are solved numerically is not usually satisfactory because the continuum states of the target are important. This can be seen by examining the second-order Born approximation, which contains a sum over all target eigenstates both discrete and continuous. For instance, using a closure approximation to evaluate the sum over continuum intermediate states Holt and Moisewitsch (1968) and Holt (1969) showed that for excitation of H(1s) by proton impact to the 2s level, the continuum contributions increased the cross section at 100 keV from ~0.027 (πa_0^2) to ~0.036 (πa_0^2). This result has been confirmed by exact calculations, avoiding the closure approximation, of the second-order Born approximation for the same system (Bransden and Dewangan, 1979). As in the case of two-centre expansions (see Chapter 5), the continuum can be effectively taken into account by including a sufficient number of positive energy pseudostates in the expansion. These may be obtained by diagonalizing the target Hamiltonian on a finite basis of Slater or other suitable functions. Alternatively a subset of a complete, but discrete, set of functions such as Sturmians can be used as a basis set. Examples in which single-centred expansions including pseudostates have been employed are the calculation of the rate of production of inner shell vacancies by projectiles of various charges (Reading *et al.* 1976; Ford *et al.* 1977), and the analysis of the impact parameter dependence of ionization (Swafford *et al.* 1977). In the course of these calculations Reading and Ford (1979) have examined the general problem of replacing integrations over the continuous energy spectrum by sums over pseudostate energies and have developed improved practical procedures. We shall return to this point below, in connection with the calculation of charge exchange amplitudes.

The multipole expansion method

A method which is capable of providing a direct numerical solution of the time-dependent Schrödinger equation for a one-electron system, when charge exchange is negligible, was introduced by Cheshire and Sullivan (1967) following the related approach to electron scattering of Temkin (1962) (see also Temkin and Sullivan (1963) and Temkin and Bhatia (1964)). Taking the origin of coordinates on the target B, the Schrödinger equation (3.163) becomes

$$\left[-\tfrac{1}{2}\nabla_r^2 + V_{\mathrm{Ae}}(x) + V_{\mathrm{Be}}(r) + V_{\mathrm{AB}}(R) - \mathrm{i}\frac{\partial}{\partial t}\right]\psi(\mathbf{r}, t) = 0. \tag{8.3}$$

In the work of Cheshire and Sullivan, $\mathbf{R} = \mathbf{R}(\mathbf{b}, t)$ was taken to be a straight line trajectory, but a non-linear trajectory could be employed if necessary. The wave function is a function, for a given impact parameter $\mathbf{b}$, of $\mathbf{r}$ and t alone, so that an expansion can be made of the form

$$\psi(r, t) = r^{-1}\sum_{l=0}^{\infty}\sum_{m=-l}^{l}\mathscr{L}_{lm}(r, t)Y_{l,m}(\theta, \phi), \tag{8.4}$$

where the Y_{lm} are spherical harmonics and (θ, ϕ) are the polar angles of $\mathbf{r}$. By inserting (8.4) into the Schrödinger equation and projecting the equation with $Y^*_{l',m'}(\theta, \phi)$, a set of coupled partial differential equations is obtained for the functions $\mathscr{L}_{lm}$:

$$\left[-\frac{1}{2}\frac{\partial^2}{\partial r^2} + \frac{l(l+1)}{2r^2} + V_{\mathrm{Be}}(r) + V_{\mathrm{AB}}(R) - \mathrm{i}\frac{\partial}{\partial t}\right]\mathscr{L}_{l'm'}(r, t) = -\sum_{lm} A_{l'm'lm}(r, t)\mathscr{L}_{lm}(r, t), \tag{8.5}$$

where

$$A_{l'm'lm}(r, t) = \int Y^*_{l'm'}(\theta, \phi)V_{\mathrm{Ae}}(x)Y_{lm}(\theta, \phi)\,\mathrm{d}\Omega(\theta, \phi). \tag{8.6}$$

If the initial state of the target atom $(\mathrm{B} + \mathrm{e}^-)$ is a 1s state and the normalized radial wave function is $r^{-1}R_{1\mathrm{s}}(r)$, the boundary conditions satisfied by the functions $\mathscr{L}_{lm}(r, t)$ are

$$\mathscr{L}_{lm}(0, t) = 0, \qquad \mathscr{L}_{lm}(r, -\infty) = R_{1\mathrm{s}}(r)\exp(-\mathrm{i}\varepsilon_{1\mathrm{s}}t)\delta_{l0}\delta_{m0}. \tag{8.7}$$

The probability amplitude $\mathscr{A}_{nlm}(b)$ for excitation of a state of the target atom with quantum numbers nlm is then given by

$$\mathscr{A}_{nlm}(b) = \lim_{t\to\infty} \int_0^\infty \mathrm{d}r R_{nl}(r) \exp(\mathrm{i}\varepsilon_{nl} t) \mathscr{L}_{lm}(r, t). \tag{8.8}$$

Cheshire and Sullivan considered excitation of atomic hydrogen by proton impact using two approximations. In the first, only terms with $l = l' = 0$ were retained in (8.5) and, in the second, terms with $l = 0$ and 1 and $l' = 0$ and 1 were retained. The resulting equations were solved numerically. The first approximation is equivalent to a coupled channel calculation which retains all s states in the basis, both discrete and continuous, while the second is equivalent to employing a basis with all s and p states, both discrete and continuous. The results provide a bench-mark with which to test other approximations in which s and p target states are retained. For instance, Fitchard *et al.* (1977) have carried out coupled channel calculations for excitation of the $n = 2$ and $n = 3$ levels of H(1s) by proton impact between 15 and 200 keV, using a large basis with 10 s, 22 p, and 21 d states. The d states make a significant contribution, but if they are omitted from the basis, good agreement is obtained with the results of Cheshire and Sullivan showing that the pseudostates in the basis adequately represent the s and p states of the continuum.

Unfortunately the multipole expansion method cannot be used when charge exchange is significant. This is because the representation of the rearranged system in terms of the target eigenfunctions requires states of high angular momentum for which the coupled partial differential equations become difficult to solve. On the other hand, the multipole expansion method has shown itself to be useful in the calculation of ionization cross sections (Gayet 1989). In the work of Salin (1987) cross sections were computed for the ionization of helium (treated in an independent electron model) by fully stripped ions of charges with $1 \leq Z_{\mathrm{A}} \leq 44$ at an energy of 1.4 MeV/a.m.u.

The contribution to the cross section from states with $l = 0$ and 1 was calculated by the multipole method while the contribution from states with $l > 1$ was estimated by the first-order Born approximation. Some results for single ionization are shown in Table 8.1. It is seen that rather good agreement is obtained with experiment, although for $Z \geq 36$ charge exchange may be important and the results become less reliable.

While the Born approximation for $l \geq 2$ is expected to be accurate the first-order Born approximation for the lowest partial waves is poor. The failure of the Born approximation for $l = 0$ and $l = 1$ is due to the strength of the corresponding transitions, the calculated probabilities in the Born approximation exceeding unity for certain ranges of impact parameter.

Table 8.1. Single ionization cross sections of helium by ions of charge Z_A at 1.4 MeV/a.m.u. B is the first-order Born approximation; M is the multipole expansion method ($l = 0, 1$); MB is the multipole expansion method for $l \leq 1$, Born approximation for $l \geq 2$. Computed values are from Salin (1987); EXPT are experimental values from McGuire *et al.* (1987).

	Cross sections (10^{-16} cm^2)			
Z_A	B	M	MB	EXPT
1	16			
15	36	13.0	16.4	97.9
20	64	20.3	27.0	26.0
36	207	39.0	61.6	57.2
44	310	48.0	81.7	72.1

Charge exchange cross sections

Under circumstances in which the charge exchange flux is small, the wave function $\psi(\mathbf{r}, t)$, in a region close to the target, will be well represented by the single-centre expansion. The charge exchange amplitude can then be calculated approximately by substituting the single-centre expansion into the exact integral equation† (see Chapter 10)

$$\mathscr{A}_{ni}(\mathbf{b}) = -\mathrm{i}\int_{-\infty}^{\infty} \langle \mathrm{X}_n | V_{\mathrm{Be}}(r) + V_{\mathrm{AB}}(R) | \psi \rangle \, \mathrm{d}t, \tag{8.9}$$

where X_n is the unperturbed wave function belonging to the arrangement ($\mathrm{A} + \mathrm{e}^-$). In terms of the ($\mathbf{r}$, t) coordinates X_n has the form

$$\mathrm{X}_n(\mathbf{r}, t) = \chi_n(\mathbf{x}) \exp(-\mathrm{i}\eta_n t + \mathrm{i}\mathbf{v}_0 \cdot \mathbf{r} - \mathrm{i}\tfrac{1}{2}v_0^2 t), \tag{8.10}$$

with $\mathbf{x} = r - \mathbf{R}(t)$. The single-centre expansion of the wave function is

$$\psi(\mathbf{r}, t) = \sum_{j=1}^{N} A_j(\mathbf{b}, t)\bar{\phi}_j(\mathbf{r}) \exp(-\mathrm{i}\bar{\varepsilon}_j t), \tag{8.11}$$

where $\bar{\phi}_j(\mathbf{r})$ denotes a member of the basis set with energy $\bar{\varepsilon}_j$, whether representing a real state or a pseudostate. Inserting (8.11) into (8.9) we

† If the ion ($\mathrm{A} + \mathrm{e}^-$) formed by capture is charged overall, equation (8.9) must be corrected to satisfy the Coulomb boundary condition as discussed in Chapter 7.

obtain:

$$\mathscr{A}_{ni}(\mathbf{b}) = -\mathrm{i} \sum_{j=1}^{N} \int_{-\infty}^{\infty} \mathrm{d}t \exp[\mathrm{i}(\eta_n - \bar{\varepsilon}_j + \tfrac{1}{2}v_0^2)t]$$
$$\times \langle \chi_n(\mathbf{x})| \exp(-\mathrm{i}\mathbf{v}_0 \cdot \mathbf{r})(V_{\mathrm{Be}}(r) + V_{\mathrm{AB}}(R))|\bar{\phi}_j(\mathbf{r})\rangle A_j(\mathbf{b}, t). \quad (8.12)$$

As noted above, the approximation represented by (8.12) is expected to be accurate when charge exchange is a weak reaction, for example when $Z_{\mathrm{A}} \ll Z_{\mathrm{B}}$. However there is a possible difficulty as pointed out by Reading *et al.* (1979). If $Z_{\mathrm{A}} \ll Z_{\mathrm{B}}$, the most important final states of the rearranged system $(\mathrm{A} + \mathrm{e}^-)$ have small binding energies η_n, so that the total energy of the electron in the frame of reference centred on B is $\eta_n + \frac{1}{2}v_0^2 \approx \frac{1}{2}v_0^2$. This final state is a resonance with a continuum state of the target with corresponding energy ε, where $\varepsilon = \frac{1}{2}v_0^2$. It follows that if the pseudostate energies $\bar{\varepsilon}_j$ are all far from ε, the resonance region will be under-represented in (8.12), while if one of the $\bar{\varepsilon}_j$ coincides with $\frac{1}{2}v_0^2$, the resonance effect will be over-represented. To try and overcome this defect Reading and Ford (1979) have suggested an improved method of integrating over the continuum, which is similar in concept to the eigendifferential method discussed in Section 5.5.1. Let us consider the case of states with $l = 0$ as in Section 5.5.2, with pseudostate radial functions R_j defined by (5.121) and (5.122), the true continuum functions being denoed by $R_\varepsilon(r)$ normalized as in (5.124). The integral over the continuum of the product $R_\varepsilon(r)R_\varepsilon(r')$ (see (5.127)) can be expressed as the sum of integrals over a strip of width $2\Delta_j$ centred about each pseudostate energy $\bar{\varepsilon}_j$. Denoting the integral by I, we have

$$I = \int_0^{\infty} R_\varepsilon(r)R_\varepsilon(r')\,\mathrm{d}\varepsilon \approx \sum_j \int_{\bar{\varepsilon}_j - \Delta_j}^{\bar{\varepsilon}_j + \Delta_j} R_\varepsilon(r)R_\varepsilon(r')\,\mathrm{d}\varepsilon$$
$$\approx \sum_j 2\Delta_j R_{\bar{\varepsilon}_j}(r)R_{\bar{\varepsilon}_j}(r'). \quad (8.13)$$

In view of the relationship (5.125), the sum in the last line can be expressed in terms of the pseudostate functions $\bar{R}_j$:

$$I = \sum_j (2\Delta_j N_j^2)\bar{R}_j(r)\bar{R}_j(r'). \quad (8.14)$$

Since we require (5.128) to hold it follows that

$$2\Delta_j N_j^2 = 1, \quad (8.15)$$

and as the normalization function can be calculated, the weights $W_j = 2\Delta_j$ (see (5.127) and (5.129)) of the integration scheme can also be found. This particular choice of integration scheme is closely related to employing the eigendifferentials defined by (5.115a).

Now consider the spectral representation of the evolution operator

$U_{\rm B} = \exp(-{\rm i}H_{\rm B}t)$ where $H_{\rm B} = -\frac{1}{2}\nabla_r^2 + U_{\rm Be}$. Again taking only $l = 0$ states for illustrative purposes, we have

$$U_{\rm B} = \int_0^\infty {\rm e}^{-{\rm i}\varepsilon t} R_\varepsilon(r) R_\varepsilon(r')\,{\rm d}\varepsilon + \sum_j {\rm e}^{-{\rm i}\varepsilon_j t} R_j(r) R_j(r'), \qquad (8.16)$$

where the sum is over the discrete part of the spectrum.

Following the integration scheme we have outlined, the integral in (8.16) can be approximated as

$$\begin{aligned}\int_0^\infty {\rm e}^{-{\rm i}\varepsilon t} R_\varepsilon(r) R_\varepsilon(r')\,{\rm d}\varepsilon &= \sum_j \int_{\bar\varepsilon_j - \Delta_j}^{\bar\varepsilon_j + \Delta_j} {\rm d}\varepsilon\,{\rm e}^{-{\rm i}\varepsilon t} N_j^2 \bar R_j(r) \bar R_j(r') \\ &= \sum_j {\rm e}^{-{\rm i}\bar\varepsilon_j t} F_j \bar R_j(r) \bar R_j(r'), \qquad (8.17)\end{aligned}$$

where

$$F_j = \frac{\sin \Delta_j t}{\Delta_j t}, \qquad (8.18)$$

and the sum is over the possible energy pseudostates. The introduction of the factors F_j represents a correction to the usual pseudostate expansion of $U_{\rm B}$, which would again be of the form (8.17) but with all the F_j set equal to unity. The corresponding correction to the expression (8.12) for the amplitude $\mathscr{A}_{ni}(\mathbf{b})$ is

$$\begin{aligned}\mathscr{A}_{ni}(\mathbf{b}) = -{\rm i}\sum_j \int_{-\infty}^\infty {\rm d}t\,\exp[{\rm i}(\eta_n - \bar\varepsilon_j + \tfrac{1}{2}v_0^2)t] F_j \times \langle \chi_n(\mathbf{x})| \\ \times \exp(-{\rm i}\mathbf{v}_0\cdot\mathbf{r})(V_{\rm Be}(r) + V_{\rm AB}(R)|\bar\phi_j(\mathbf{r})\rangle A_j(\mathbf{b}, t) \qquad (8.19)\end{aligned}$$

where F_j is given by (8.18) for positive energy states and $F_j = 1$ for states of negative energy.

The correction factors F_j differ significantly from unity only when $\Delta_j t \gtrsim 1$. Let us see when this occurs. The energy difference between pseudostates is of the order $2\Delta_j \approx \gamma Z_{\rm B}^2$, where γ is a numerical factor with $\gamma \le 1$. To estimate the important range of t, we note that the size of the bound state $({\rm A} + {\rm e}^-)$ is of the order $1/Z_{\rm A}$, so that the interaction time is $\tau = (v_0 Z_{\rm A})^{-1}$. From the uncertainty principle, the width of the energy region which will be in resonance with the energy $v_0^2/2$ is of the order $\Delta\varepsilon \simeq 1/\tau = v_0 Z_{\rm A}$. It follows that to represent this resonance region, the relevant values of $\Delta_j t$ must be of the order $\Delta_j t = \gamma Z_{\rm B}^2/(2 v_0 Z_{\rm A})$. From this expression it is seen that the correction factors differ from unity only for large values of $Z_{\rm B}^2/Z_{\rm A}$, and for collisions in which $Z_{\rm A}$ and $Z_{\rm B}$ are of comparable magnitude the introduction of correction factors is not generally necessary. The integration scheme for the continuum in terms of the correcting factors F_j can be incorporated into the coupled channel equations themselves if these are written as integral equations.

8.2 CORRECTIONS TO THE SINGLE-CENTRE EXPANSION

As we have discussed, the single-centre expansion can only be employed for systems in which charge exchange is small, even for the calculation of excitation and ionization cross sections. Considerable improvement in the excitation and ionization cross sections can be obtained at the lower energies, even if a single term is retained in the expansion about the projectile A. For instance, Ermolaev has used such an approach to calculate excitation and ionization cross sections for a number of reactions, including the excitation and ionization of H(1s) by proton and antiproton impact in the energy range 30 keV to 5 MeV (Ermolaev 1990*a*, *b*).

Ermolaev calculated cross sections for the excitation of the $n = 2$ and $n = 3$ levels of H(1s) by H^+ over the energy range 30 to 500 keV. In this work a single 1s state was centred on the projectile A with a large basis on the target B consisting of 8s, 16p, 18d, and 8f states and pseudostates. At the lower energies Emolaev's cross sections are similar to those obtained from the OHCE model (see below) which are shown in Figs. 8.1 and 8.2. The calculated cross sections appear to have converged and good agreement is obtained with the experimental data and with the results of evaluations using basis sets with equal numbers of terms on each centre, for example those of Fritsch and Lin (1983*a*) and of Shakeshaft (1978*b*). The cross sections obtained from these symmetrical expansions were shown in Fig. 6.7 and agree well with the experimental data. Also shown in Fig. 8.1 are the results of the single-centre expansion presented by Fitchard *et al.* (1977). The failure of the single-centre expansion at energies below ~ 100 keV where charge exchange becomes significant is clearly seen.

Ermolaev's model was introduced with the objective of investigating differences between the stopping powers of protons and antiprotons in gases as discussed in Chapter 6. His calculated cross section for the ionization of atomic hydrogen by proton impact agrees very well with the experimental data which were shown in Fig. 6.9.

The OHCE model

The great advantage of a single centred expansion is that it is computationally rapid and cheap, since the matrix elements of the interaction depend only on R, the internuclear distance, and have only to be computed once, independently of the incident velocity. To retain this computational simplicity, and yet to allow for the flux entering the charge exchange channels, Reading *et al.* (1981) introduced what they termed the 'once-and-a-half-centred expansion' (OHCE). In this model the wave function is expanded in the form (using the coordinate system $\mathbf{r}$, t with B as origin)

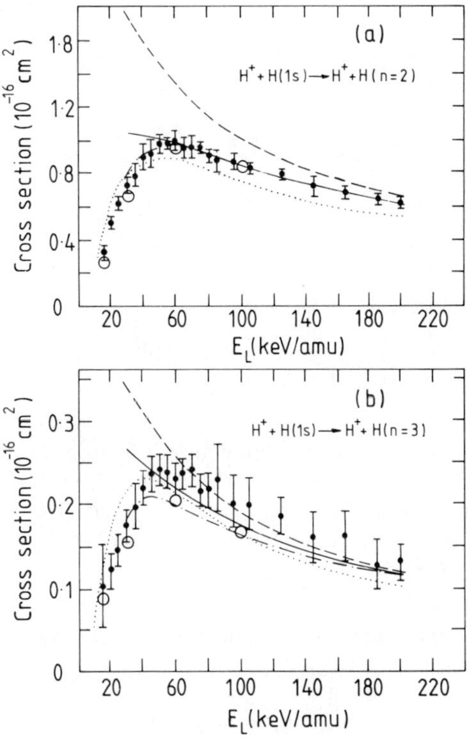

Fig. 8.1. Cross sections for the excitation of atomic hydrogen by proton impact: (a) to the $n = 2$ level, (b) to the $n = 3$ level (see also Fig. 6.7). Experimental data: $\bar{\underline{\bullet}}$, Park *et al.* (1975, 1976). Theoretical cross sections: ——— coupled channel model with single-centre basis, Fitchard *et al.* (1977); ○○○ OHCE model, Reading *et al.* (1981); — — — first-order Born approximation; · · · · · · single-centre optical potential, scaled theory, Ast *et al.* (1988).

$$\psi(\mathbf{r}, t) = \sum_{j=1}^{N} \bar{\phi}_j(\mathbf{r}) \exp[-i\bar{\varepsilon}_j t] A_j(\mathbf{b}, t) + \sum_{n=N+1}^{N+M} \chi_n(\mathbf{x}) \exp(i\beta_n)\gamma_n(t) A_n(\mathbf{b}, \infty),$$

with

$$\beta_n = -\eta_n t + \mathbf{v}_0 \cdot \mathbf{r} - \tfrac{1}{2} v_0^2 t. \tag{8.20}$$

The first set of terms contains a set of states (and pseudostates) centred on the target, with amplitudes $A_j(\mathbf{b}, t)$ of unknown functional form. The second set of terms contains wave functions $\chi_n(\mathbf{x})$ centred on the projectile A, together with the appropriate translation factors, and amplitudes $\gamma_n(t)A_n(\mathbf{b}, \infty)$, the functional form of the $\gamma_n(t)$ being *specified in advance.* To satisfy the boundary conditions the functions γ_n must be such that

$$\gamma_n(-\infty) = 0; \qquad \gamma_n(+\infty) = 1. \tag{8.21}$$

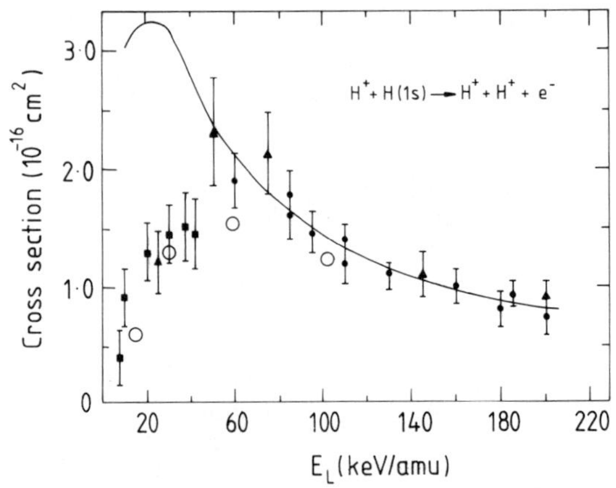

Fig. 8.2. Cross sections for the ionization of atomic hydrogen by proton impact (see also Fig. 6.9 for higher energies). Experimental data: ▲ Park *et al.* (1977); ■ Fite *et al.* (1960); ● Gilbody and Ireland (1963). Theoretical cross sections: ——— coupled channel model with single-centre basis, Fitchard *et al.* (1977); ○○○ OHCE mode, Reading *et al.* (1981).

A simple choice for the functions γ_n, followed by Reading and his co-workers, is to set

$$\gamma_n = \theta(t), \tag{8.22}$$

where

$$\theta(t) = \begin{cases} 0 & t < 0 \\ 1 & t \geq 0. \end{cases} \tag{8.22}$$

By inserting the expression (8.20) into the electronic Schrödinger equation and projecting with the functions $\bar{\phi}_j(\mathbf{r}) \exp(-i\bar{\varepsilon}_j t)$ and $\chi_n(\mathbf{x}) \exp(i\beta_n t)$ in turn, a set of first-order differential equations is found for the functions $A_j(\mathbf{b}, t)$, coupled to a set of algebraic equations for the constants $A_n(\mathbf{b}, \infty)$. The equations are:

$$\begin{aligned} \exp(i\bar{\varepsilon}_j t)\langle \bar{\phi}_j | H_e - i\frac{\partial}{\partial t} |\psi\rangle &= 0, \qquad j = 1, 2, \ldots, N, \\ \int_0^\infty dt \exp(-i\beta_n)\langle \chi_n | H_e - i\frac{\partial}{\partial t} |\psi\rangle &= 0, \qquad n = N+1, \ldots, N+M. \end{aligned} \tag{8.23}$$

where $\psi(\mathbf{r}, t)$ is expressed by (8.20). These equations are solved simultaneously to obtain amplitudes $A_j(\mathbf{b}, \infty)$, $j \leq N$, which describe excitation and ionization, and the amplitudes $A_n(\mathbf{b}, \infty)$ which describe charge exchange. In common with the usual set of two-centre coupled equations, obtained from the trial function (5.84), the model is consistent with the variational principle. The loss of flexibility in pre-choosing the functions $\gamma(t)$ is found not to be important for the determination of cross sections for excitation and ionization, but, with some exceptions, the computed charge exchange cross sections are only accurate at energies for which rearrangement is unlikely.

As a test of the OHCE model Reading *et al.* (1981) calculated cross sections for the symmetrical system $H^+ + H(1s)$. They took a similar target-centred basis to Fitchard *et al.* (1977) and added one ground-state term centred on A, with $\gamma_{1s}(t)$ chosen as described above. Their results for excitation and ionization, included in Figs 8.1 and 8.2, show reasonable agreement with the data down to 15 keV. Recollecting that charge exchange is more probable than direct excitation below ~ 80 keV, this agreement illustrates that the model is successful in predicting the effect of rearrangement on direct excitation and ionization. In Table 8.2 the calculated cross section for resonant charge exchange into the 1s state using the OHCE method is shown compared with those of the single-centre expansion (obtained with the use of (8.12)) and with the cross sections calculated from the elaborate two-centre expansion model of Shakeshaft (1978*a*, *b*). The general agreement between the OHCE, single-centred basis, and two-centred basis cross sections for capture into the 1s level down to 15 keV is reasonable. In particular it is surprising that the cross section obtained from the single-centre basis is fairly accurate for charge exchange in an energy region in which this approach fails for excitation and ionization. As we shall see, below this, success does not extend to the case of heteronuclear collisions. A slightly different version of the model in which the algebraic equations are obtained by projecting the Schrödinger equation with the functions $[\gamma_n(t)\chi_n(\mathbf{x})]\exp(i\beta_j(t))$, rather

Table 8.2. Electron capture cross sections for $H^+ + H(1s) \rightarrow H(1s) + H$.

A is single-centre expansion (Reading *et al.* 1981);
B is the OHCE model (Reading *et al.* 1981);
C is a two-centre expansion (Shakeshaft 1978*b*).

	Cross sections (10^{-16} cm^2)		
E(keV)	A	B	C
15	5.5	5.3	5.8
30	2.1	1.9	2.2
60	0.39	0.38	0.41
100	0.071	0.071	0.08

than with $[\chi_n(\mathbf{x})\exp(i\beta_j(t)]$, is less successful for charge exchange, and the resulting cross sections are not included in Table 8.2 or the figures.

The OHCE model has been applied to the more asymmetrical collisions $H^+ + He^+$, $He^{2+} + H$ (Reading *et al.* 1982), $H^+ + Li^+$, $H^+ + Li^{2+}$, and $Li^{2+} + H(1s)$, $Li^{3+} + H(1s)$ (Ford *et al.* 1982). In the case of ionization, the OHCE cross sections at low energies differ substantially from those of the single-centre expansion and are in better agreement with the experimental data. In contrast, the OHCE and single-centre cross sections for charge exchange differ rather little, and both show large differences from the measurements.

For capture by protons from $He^+(1s)$ and by He^{2+} ions from H(1s), the OHCE cross sections have been included in Figs 6.12 and 6.14, respectively, where it can be seen that agreement with the experimental data is good for energies above the cross section maximum. At lower energies the computed cross section becomes too large. In Fig. 6.13, the OHCE cross section for ionization of $He^+(1s)$ by proton impact is compared with experiment and with the results of other coupled channel calculations. The agreement is good in an energy interval extending down to ~ 50 keV/a.m.u.

The particular case of $Li^{3+} + H(1s)$ can be taken as a further example. A basis consisting of 9s, 9p, and 9d states was employed on the Li^{3+} projectile and a single 1s state on the target in the second group of terms in (8.20) with the pre-chosen function $\gamma(t)$. The results are illustrated in Fig. 8.3, where it can be seen that the calculated cross section for charge exchange is too large compared with the data. Above 100 keV/a.m.u. the OHCE cross sections agree reasonably with those obtained with a 20-state two-centre basis (Bransden and Noble 1982) which contained no pseudostates on either centre, and which agreed extremely well with the low-energy experimental data of Seim *et al.* (1980) covering the energy interval 1–6 keV/a.m.u. (not shown in Fig. 8.3). It seems safe to conclude that the failure of both the OHCE and two-centre calculations at high energies for this system is because of the omission of the target-centred continuum; the inclusion of the projectile-centred continuum through pseudostates in the OHCE calculation cannot compensate for this omission. It should be noted that the continuum distorted wave model (to be described in Chapter 10), which does allow completely for transitions via continuum intermediate states agrees very well with the data above 100 keV/a.m.u.; below 100 keV/a.m.u. it fails, as would be expected from what is essentially a perturbation model valid at high energies (Crothers and Todd 1980*b*).

8.3 THE OPTICAL POTENTIAL

The general form of the coupled equations for the probability amplitudes $A_j(\mathbf{b}, t)$ in the impact parameter model is given by (3.205), and is independent

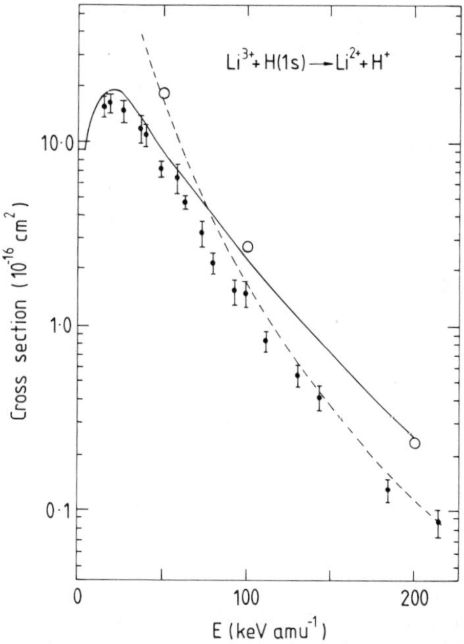

Fig. 8.3. Total cross section for electron capture by Li^{3+} ions from atomic hydrogen. Experimental data: ● Shah *et al.* (1978). Theoretical cross sections: ——— coupled channel model with a two-centre basis, Bransden and Noble (1982); ○○○ OHCE model, Ford *et al.* (1982); – – – Continuum distorted wave (CDW) model, Crothers and Todd (1980*b*).

of the type of basis set employed:

$$i\dot{\mathbf{A}}(\mathbf{b}, t) = \mathbf{V}(\mathbf{b}, t)\mathbf{A}(\mathbf{b}, t). \tag{8.24}$$

The $(N + M) \times (N + M)$ matrix $\mathbf{V}$ connects N direct and M rearranged channels and is defined in terms of the overlap matrix $\mathbf{N}$ and interaction matrix $\mathbf{Q}$ by (3.206), together with (3.207), (3.184), and (3.186). In the limit of an exact theory with a complete basis the number of channels becomes infinite. The idea leading to the introduction of an optical potential is to obtain an implicit representation of the infinite basis within a finite subspace spanned by a limited number of states, which include the initial and final states of interest and states strongly coupled to these. If the scattering channels are ordered so that the first J belongs to the finite subspace, a projection operator P singling out these channels can be introduced with the properties

$$PA_j(\mathbf{b}, t) = \begin{cases} A_j(\mathbf{b}, t) & j \leq J \\ 0 & j > J \end{cases} \tag{8.25}$$

and

$$PV_{jk} = \begin{cases} V_{jk} & j \leq J \text{ (any } k), \\ 0 & j > J \text{ (any } k). \end{cases} \tag{8.26}$$

The subspace complementary to the P-subspace is obtained by projecting with an operator Q defined as

$$Q = 1 - P. \tag{8.27}$$

Since P and Q are projection operators, they have the properties

$$P^2 = P, \qquad Q^2 = Q, \qquad QP = PQ = 0, \tag{8.28}$$

and both P and Q are self-adjoint.

Following Feshbach (1958, 1962), the coupled equations (8.24) can be expressed in the form:

$$\mathrm{i}P\dot{\mathbf{A}} = [P\mathbf{V}P + P\mathbf{V}Q]\mathbf{A} \tag{8.29a}$$

$$\mathrm{i}Q\dot{\mathbf{A}} = [Q\mathbf{V}Q + Q\mathbf{V}P]\mathbf{A} \tag{8.29b}$$

where, for notational convenience, the dependence of **A** and **V** on the impact parameter **b** and the time t has been suppressed.

The set of equations (8.29b) can be written in the form of an integral equation

$$Q\mathbf{A}(t) = \int_{-\infty}^{\infty} \mathbf{G}_Q(t, t')Q\mathbf{V}(t')P\mathbf{A}(t')\,\mathrm{d}t' \tag{8.30}$$

where the Green's function $\mathbf{G}_Q(t, t')$ is the inverse of the operator $Q[\mathrm{i}(\partial/\partial t) - \mathbf{V}]Q$:

$$\mathbf{G}_Q = \lim_{\varepsilon \to 0^+} \frac{1}{Q\left[\mathrm{i}\dfrac{\partial}{\partial t} + \mathrm{i}\varepsilon - \mathbf{V}\right]Q}. \tag{8.31}$$

The small positive quantity ε added to the denominator in (8.31) is to ensure that the boundary condition $\lim_{t\to-\infty} QA(t) = 0$ is satisfied, as will be explained in detail in Chapter 10 and Appendix G.

On substituting $Q\mathbf{A}(t)$ given by (8.30) into the P-subspace equations (8.29a), we obtain the coupled equations:

$$\mathrm{i}P\dot{\mathbf{A}}(t) = P\mathbf{V}(t)P\mathbf{A}(t) + \int_{-\infty}^{\infty} P\mathbf{W}(t, t')P\mathbf{A}(t')\,\mathrm{d}t', \tag{8.32}$$

where **W** is defined as

$$\mathbf{W} = P\mathbf{V}Q\mathbf{G}_Q Q\mathbf{V}P. \tag{8.33}$$

The matrix $\mathbf{W}$ represents an effective interaction, taking into account all states omitted from the P-subspace. It is known as a *matrix optical potential* and is both complex and non-local in time.

The set of equations (8.32) is exact, all the complexity of the problem being contained within the optical potential. In practice the optical potential must be approximated to perform numerical calculations and one possibility is to obtain a perturbation expansion of $\mathbf{W}$ in powers of the interaction $\mathbf{V}$, following the methods used to obtain the Born series that were described in Section 7.4.

The perturbation series for $Q\mathbf{A}$ obtained from equations (8.29b) is (see (7.106), (7.112), and (7.113)):

$$Q\mathbf{A}(t) = \sum_{n=1}^{\infty} \frac{(-\mathrm{i})^n}{n!} \int_{-\infty}^{t} \mathrm{d}t_1 \int_{-\infty}^{t} \mathrm{d}t_2 \cdots \int_{-\infty}^{t} \mathrm{d}t_n \times T[Q\mathbf{V}(t_1)Q\mathbf{V}(t_2)Q \cdots Q\mathbf{V}(t_n)]P\mathbf{A}(t_n), \tag{8.34}$$

where T is the time ordering operator. By substituting this expression into (8.29a), the corresponding expansion of the optical potential $\mathbf{W}$ is obtained. It is found that

$$\mathbf{W}P\mathbf{A} = P\mathbf{V}(t) \sum_{n=1}^{\infty} \frac{(-\mathrm{i})^n}{n!} \int_{-\infty}^{t} \mathrm{d}t_1 \int_{-\infty}^{t} \mathrm{d}t_2 \cdots \int_{-\infty}^{t} \mathrm{d}t_n \times T[Q\mathbf{V}(t_1)Q\mathbf{V}(t_2)Q \cdots Q\mathbf{V}(t_n)]P\mathbf{A}(t_n) \tag{8.35}$$

The second-order potential method

In the second-order potential method of Bransden and Coleman (1972) the basis set employed was the single-centre set of target eigenfunctions and only the first term in (8.35) was retained, so that

$$\mathbf{W}P\mathbf{A} \approx -\mathrm{i}P\mathbf{V}(t) \int_{-\infty}^{t} Q\mathbf{V}(t_1)P\mathbf{A}(t_1)\, \mathrm{d}t_1. \tag{8.36}$$

If a single-centre target basis is employed and if the interionic potential V_{AB} has been removed by the phase transformation† (3.164), the matrix elements $V_{ij}(t)$ have the explicit form

$$V_{ij}(t) = \exp\{\mathrm{i}(\varepsilon_i - \varepsilon_j)t\}h_{ij}(t), \tag{8.37}$$

where

$$h_{ij}(t) \equiv \langle i|V_{\mathrm{Ae}}(t)|j\rangle. \tag{8.38}$$

The explicit form of the coupled equations in P-space using the second-order

† In applications, a straight line trajectory with $W(R) = 0$, as in (3.167), has been used, but a non-linear trajectory could be employed without difficulty.

potential model is, from (8.32), (8.36), and (8.37),

$$\begin{aligned} \mathrm{i}\dot{A}_i(t) = \sum_{j=1}^{J} \exp\{\mathrm{i}(\varepsilon_i - \varepsilon_j)t\}h_{ij}(t)A_j(t) \\ + \sum_{k=J+1}^{\infty} \int_{-\infty}^{t} \exp\{\mathrm{i}(\varepsilon_i - \varepsilon_k)t\}h_{ik}(t)\exp\{\mathrm{i}(\varepsilon_k - \varepsilon_j)t_1\} \\ \times h_{kj}(t_1)A_j(t_1)\,\mathrm{d}t_1, \qquad i = 1, 2, 3, \ldots, J. \end{aligned} \tag{8.39}$$

The second term on the right-hand side of (8.39) contains an infinite sum over the states with $k \geq J + 1$. One way of obtaining an approximation to this sum is to replace ε_k, $k \geq J + 1$, by an average energy $\bar{\varepsilon}$. In this case the closure property of the basis functions ϕ_k can be employed so that

$$\begin{aligned} \sum_{k=J+1}^{\infty} \exp[\mathrm{i}\varepsilon_k(t_1 - t)]h_{ik}(t)h_{kj}(t_1) \approx \exp\{\mathrm{i}\bar{\varepsilon}(t_1 - t)\}[\langle i|V_{\mathrm{Ae}}(t)V_{\mathrm{Ae}}(t_1)|j\rangle \\ - \sum_{k=1}^{J} \langle i|V_{\mathrm{Ae}}(t)|k\rangle\langle k|V_{\mathrm{Ae}}(t_1)|j\rangle]. \end{aligned} \tag{8.40}$$

This closure approximation has been employed to discuss excitation by proton (and electron) impact of atomic hydrogen and helium (Bransden *et al.* 1972; Bransden and Issa 1975; Berrington *et al.* 1973; Sullivan *et al.* 1972) using a small set of equations with $J = 2$ or $J = 4$. The results for energies between 15 and 500 keV show a distinct improvement over the coupled channel equations without the optical potential, but large-scale applications have not been carried out. By using the approximate optical potential wave function in an integral expression for the charge exchange probability amplitude, cross sections for rearrangement can also be calculated (Begum *et al.* 1973).

An alternative method of evaluating the infinite sum in (8.39) is to employ a discrete pseudostate basis and explicitly perform the sum over k, until convergence is reached. The resulting amplitudes are then approximations to the solutions of the full set of coupled channel equations obtained with the same basis. This technique has been used successfully for electron scattering (Bransden *et al.* 1981), but has not yet been applied to the heavy particle case.

If a single-centre basis is used, either in conjunction with the closure approximation or in conjunction with pseudostates, the method will suffer the usual limitation that it is only applicable at energies for which the probability of charge exchange is small.

The local approximation

In the second-order potential approximation the coupling between different states belonging to the Q-space is completely ignored. This coupling can be included, to some extent, in a third-order approximation. This has been done successfully for electron scattering (Vučić *et al.* 1987) but has not been investigated for heavy particles. An alternative procedure which allows for coupling amongst the states spanning the Q-subspace has been introduced by Lüdde *et al.* (1987, 1988). It is based on a high-velocity approximation in which it is supposed that the successive interactions within Q-space take place instantaneously, so that in its simplest form the approximation is made that

$$\mathbf{V}(t)Q\mathbf{V}(t') \approx \mathbf{V}(t)Q\mathbf{V}(t)\delta(t-t'), \tag{8.41}$$

which neglects time ordering completely. Making this approximation on the right-hand side of (8.35), it is found that

$$\mathbf{W}P\mathbf{A} \approx P\mathbf{V}(t)\sum_{n=1}^{\infty}(-\mathrm{i})^n\int_{-\infty}^{t}[Q\mathbf{V}(t_1)]^n P\mathbf{A}(t_1)\,\mathrm{d}t_1. \tag{8.42}$$

Inserting (8.42) into (8.32), we obtain

$$\mathrm{i}P\dot{\mathbf{A}}(t) = P\mathbf{V}(t)\mathbf{G}_Q(t)P\mathbf{A}(t), \tag{8.43}$$

where $\tilde{\mathbf{G}}_Q$ is a Green's operator defined as

$$\tilde{\mathbf{G}}_Q = [1+\mathrm{i}Q\mathbf{V}(t)]^{-1} = \sum_{n=0}^{\infty}(-\mathrm{i})^n[Q\mathbf{V}(t)]^n. \tag{8.44}$$

Using the identity (G.23) given in Appendix G, $\tilde{\mathbf{G}}_Q$ can be put in the form

$$\tilde{\mathbf{G}}_Q = \mathbf{G} + \mathrm{i}\mathbf{G}P\mathbf{V}\tilde{\mathbf{G}}_Q, \tag{8.45}$$

where $\mathbf{G}$ is defined as the Green's operator

$$\mathbf{G} = [1+\mathrm{i}\mathbf{V}]^{-1}. \tag{8.46}$$

Using (8.45) and (8.46), the coupled equations for the P-space amplitudes can be expressed after a little manipulation as

$$\mathrm{i}P\dot{\mathbf{A}}(t) = [P\mathbf{U}P]P\mathbf{A}(t), \tag{8.47}$$

where

$$P\mathbf{U}P = P\mathrm{i}\{1-[P\mathbf{G}P]^{-1}\}. \tag{8.48}$$

The calculation of the P-subspace component of $\mathbf{G}$, $P\mathbf{G}P$, is not a simple matter because $\mathbf{G}$ contains the matrix $\mathbf{V}$, which is defined in the complete $(P+Q)$-space. The matrix elements of $\mathbf{V}$ are defined in (8.37) and (8.38). By introducing an average energy for the Q-space basis functions and using closure, Lüdde *et al.* obtain the approximation that

$$(P\mathbf{U}P)_{ij} = \mathrm{i}\,\mathrm{e}^{\mathrm{i}(\varepsilon_i-\varepsilon_j)t}\{\delta_{ij} - (P\bar{\mathbf{G}}P^{-1})_{ij}\} \tag{8.49}$$

$$(P\bar{\mathbf{G}}P)_{lm} = \langle l|\frac{1}{1+\mathrm{i}V_{\mathrm{Ae}}(t)+\mathrm{i}\bar{\varepsilon}}|m\rangle. \tag{8.50}$$

It is interesting to note that as the P-subspace is increased to the whole of function space, then (with $\bar{\varepsilon} = 0$), $(P\mathbf{U}P)_{ij} \to (\mathbf{V})_{ij}$ and the complete set of coupled equations (8.24) is regained.

The basic local approximation in which retardation effects are neglected is essentially the same as the omission of the second- (and higher-) order terms in the Magnus expansion (7.152). The calculation of Baye and Heenan (1973) shown in Table 7.4 show that this approximation may not be very accurate, even at high energies. However, in the present context it should be emphasized that retardation is only neglected for transitions between the Q-space functions and is fully taken into account within P-space. It is, however, possible to improve on the local approximation by scaling the basic relationship (8.41). This problem has been discussed in general by Lüdde *et al.* (1988) and by Lüdde and Dreizler (1989). In the latter paper it is pointed out that the interaction matrix $\mathbf{V}$ depends directly on the internuclear separation, rather than t. It is therefore preferable to reformulate the problem using the Z-component of $\mathbf{R}$ as the independent variable rather than t, where, with a straight line trajectory, $Z = v_0 t$. The local approximation is obtained by employing $\delta(Z - Z')$ in (8.42) rather than $\delta(t - t')$. A simple scaling relation is obtained by replacing (8.41) by

$$\begin{aligned} \mathbf{V}(t)Q\mathbf{V}(t') &\approx \alpha\mathbf{V}(t)Q\mathbf{V}(t)\delta(Z - Z') \\ &= \frac{\alpha}{v_0}\mathbf{V}(t)Q\mathbf{V}(t)\delta(t - t'), \end{aligned} \tag{8.51}$$

where α is a constant. One method of choosing α is so that results obtained with the local potential agree with those obtained with the second-order potential at some high energy, where the latter should be accurate. Further methods for choosing α are discussed in the papers cited.

The net effect of employing the scaling represented by (8.51) is to substitute $P\mathbf{U}P$ in (8.47) by

$$P\mathbf{U}^{s}P = \mathrm{i}\left(\frac{v_0}{\alpha}\right)\{1 - [P\mathbf{G}^{s}P]^{-1}\}, \tag{8.52}$$

where the scaled propagator $\mathbf{G}^{s}$ is defined as

$$\mathbf{G}^{s} = \left[1 + \frac{\alpha}{v_0}\mathbf{V}\right]^{-1}. \tag{8.53}$$

In the limit that the P-space becomes complete, the full coupled equations (8.24) are regained once more.

Applications A number of applications of the single-centre local potential have been reported (Ast *et al.* 1988; Stary *et al.* 1990). Of particular interest was the calculation of the excitation of atomic hydrogen to the $n = 2$ and $n = 3$ levels by proton impact (Ast *et al.* 1988). The P-space basis functions were taken to be the hydrogenic functions for $n = 1$, 2, and 3 (10 states) and the Q-space average energy was set at the ionization threshold ($\bar{\varepsilon} = 0$). This choice was motivated by noting that ionization is dominated by the ejection of low-energy electrons. Calculations were performed either with no scaling, or with a simple scaling law. Comparison with experiment (see Fig. 8.1) shows that reasonable agreement is obtained even in the region in which charge exchange is important ($E < 60$ keV), provided the scaling approximation is employed. In contrast, the second-order potential method, which has been applied to $n = 2$ excitation, fails below 60 keV, although second-order potential and local potential methods agree well at the higher energies.

In the single-centre optical potential model the elastic scattering and excitation channels are explicitly represented in P-space, while the charge exchange and ionization channels are implicitly represented in Q-space. It follows that separate charge exchange and ionization cross sections cannot be computed. On the other hand, by calculating the total probability flux loss from P-space to Q-space, the electron loss cross section (the sum of the capture and ionization cross sections) can be estimated. For the homonuclear system $H^+ + H$, the electron loss cross section formed in this way is shown in Fig. 8.4 for both scaled and unscaled potentials. Since below about 60 keV the loss cross section is dominated by capture and above this energy, by ionization, it is possible to compare the computed cross section with the

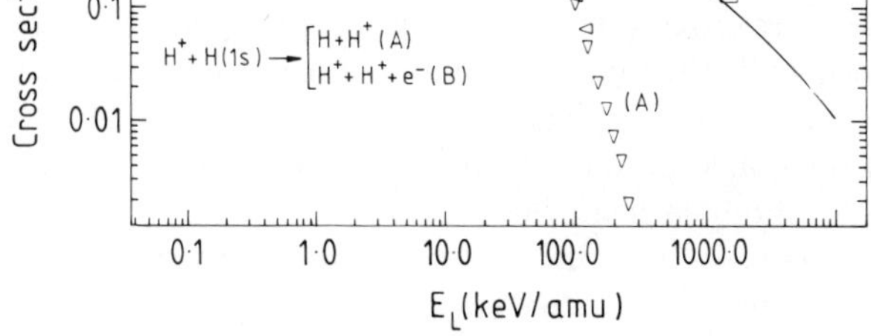

Fig. 8.4. Capture, ionization, and loss cross sections for H^+ impact on H(1s) (after Ast *et al.* (1988)). Experimental data: (a) capture ▽▽ Tawara *et al.* (1983); ▷▷ Gealy and van Zyl (1987); ◁◁ McClure (1966); (b) ionization □□ Shah and Gilbody (1981); △△ Shah *et al.* (1987). Theoretical cross-section for electron loss in one-centre optical potential model, Ast *et al.* (1988): ——— scaled potential; – – – unscaled potential.

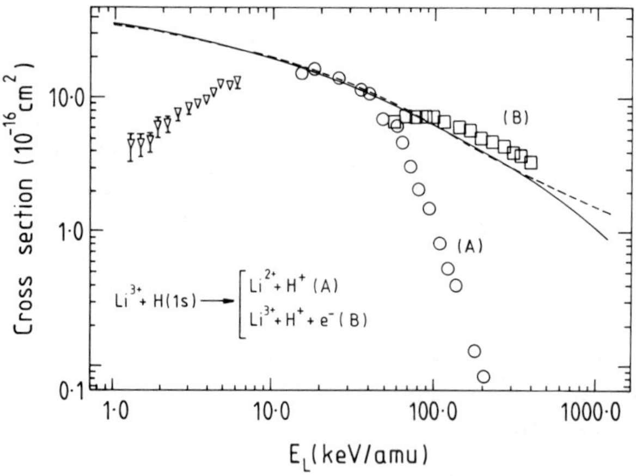

Fig. 8.5. Capture, ionization, and loss cross section for Li^{3+} impact on H(1s) (after Ast *et al.* (1988). Experimental data: (a) capture ▽ Seim *et al.* (1980); ○○ Shah *et al.* (1978); (b) ionization □□ Shah and Gilbody (1981). The theoretical data are as in Fig. 8.4.

experimental data for capture and for ionization. It is seen from Fig. 8.4 that the cross section obtained from the scaled potential is in good agreement with the experimental data, and is a considerable improvement over the cross section obtained from the unscaled potential. For heteronuclear systems the situation is not so favourable. For the system Li^{3+} + H(1s), it can be seen from Fig. 8.5 that below $E_L \approx 10$ keV/a.m.u. the computed loss cross section fails to represent the experimental data, increasing rather than decreasing with decreasing energy. At higher energies, the theoretical cross section is of the same order of magnitude as the measured ionization cross section (both for the scaled and unscaled potentials). For the He^{2+} + H(1s) system the calculated loss cross section is again in disagreement with the low-energy measurements, while the ionization data are better represented by the model. Both for Li^{3+} and He^{2+} projectiles the energy variation of the cross section provided by the scaled potentials appears to be in better agreement with the data than that obtained from the unscaled potential.

One of the reasons for the success of the local approximation in the H^+ + H(1s) system is that the continuum, which is part of Q-space, is comparatively unimportant. In contrast, it is known from coupled channel calculations that several positive energy pseudostates representing the continuum are required to give a reasonable description of systems such as $H^+ + He^+$ and $H^+ + Li^{2+}$, and in fact it is found that the unscaled local approximation is poor at all energies in these cases. The scaled approximation does provide cross sections which agree with the coupled channel pseudostate model at sufficiently high energies ($E > 100$ keV for $H^+ + He^+$ and $E > 500$ keV for the $H^+ + Li^+$ system), but also fails at lower energies.

Two-centre expansion and the local optical potential

In optical potentials formed from a single-centre basis, the Q-space should contain both the charge exchange and ionization channels. Since the major approximations are made within Q-space, one can infer that if charge exchange is important, the approximate excitation cross sections cannot be expected to be highly accurate. Furthermore, within such a model the charge exchange cross sections cannot be calculated directly. These defects can be remedied by representing both the important excitation and charge exchange channels explicitly in P-space. Some steps in this direction were taken by Potvliege *et al.* (1987) who investigated the elastic scattering of protons by atomic hydrogen at energies between 25 and 60 keV and of protons by helium at 100 keV. The second-order potential method employed did not completely avoid 'double-counting', that is overlap, between the charge exchange states explicitly represented in P-space and the Q-space wave functions, but nevertheless, very good agreement between the calculated and experimental differential cross sections for the hydrogen target was obtained. In the case of the helium target a large and unexplained discrepancy was found between theory and experiment, which is also found when comparing the results of the coupled channel calculations of Slim *et al.* (1991b) with the experimental data.

The work of Lüdde and Dreizler (1989) (based on the local approximation) avoids any overlap between P- and Q-space basis functions. The P-space is defined as the subspace spanned by a set of $N + M$ functions $y_i(\mathbf{s}, t)$ where

$$y_i(\mathbf{s}, t) = \begin{cases} \Phi_i(\mathbf{s}, t) & i = 1, 2, \ldots, N \\ \mathrm{X}_i(\mathbf{s}, t) & i = N+1, N+2, \ldots, N+M, \end{cases} \tag{8.54}$$

and Φ_i, χ_i are the travelling atomic orbitals (5.84) centred on B and A, respectively. A projection operator P on this space can be introduced so that if $f(\mathbf{s}, t)$ is any wave function

$$Pf(\mathbf{s}, t) \equiv \sum_{i=1}^{N+M} \sum_{j=1}^{N+M} y_i(\mathbf{s}, t)(\mathbf{N}^{-1})_{ij} \int y_j^*(\mathbf{s}', t) f(\mathbf{s}', t)\, d\mathbf{s}', \tag{8.55}$$

where N is the usual overlap matrix (see (3.207))

$$N_{ij}(t) = \int y_i^*(\mathbf{s}, t) y_j(\mathbf{s}, t)\, d\mathbf{s}. \tag{8.56}$$

Equation (8.55) can be written more compactly as

$$P|f\rangle = \sum_{i=1}^{N+M} \sum_{j=1}^{N+M} |i\rangle (\mathbf{N}^{-1})_{ij} \langle j|f\rangle. \tag{8.57}$$

It is easy to verify that

$$P^2 = P. \tag{8.58}$$

The set of target eigenfunctions is complete, so the functions Φ_i with $i > N$ can be employed to represent the Q-space provided they are first orthogonalized to the M projectile-centred functions X_m. This orthogonalization is achieved by defining a new infinite set of functions $Y_\alpha(\mathbf{s}, t)$ by

$$Y_\alpha(\mathbf{s}, t) = (1 - P)\Phi_{\alpha+N}(\mathbf{s}, t), \qquad \alpha = 1, 2, \ldots, \tag{8.59}$$

where P is defined by (8.57). It follows immediately from (8.59) and (8.57) that the Q-space basis functions Y_α and the P-space basis functions y_i are orthogonal. The functions Y_j are, however, not mutually orthogonal and the overlap matrix,

$$\begin{aligned} \mathcal{N}_{\alpha\beta}(t) &= \langle \alpha \mid \beta \rangle \\ &= \int Y_\alpha^*(\mathbf{s}, t) Y_\beta(\mathbf{s}, t)\, d\mathbf{s}, \end{aligned} \tag{8.60}$$

is not diagonal.

The projection operator on to Q-space,

$$Q = 1 - P, \tag{8.61}$$

can be expressed in a form similar to (8.57)

$$Q \mid f \rangle = \sum_\alpha \sum_\beta |\alpha\rangle (\mathcal{N}^{-1})_{\alpha\beta} \langle \beta \mid f \rangle, \tag{8.62}$$

and the relationships,

$$Q^2 = Q, \qquad QP = 0, \qquad \text{and} \qquad P + Q = 1 \tag{8.63}$$

follow from the definitions. It should be emphasized that the Q-space basis is not unique. For example the set of projectile-centred functions $X_n(\mathbf{s}, t)$ $(n > M)$ could be employed in place of the Φ_i $(i > N)$ functions, or some other choices could be envisaged.

Having defined the P- and Q-subspaces and the optical potential representing the effect of the Q-space in the P-space, coupled equations can be constructed along the lines discussed for the single-centre case. Again, if the local approximation is made, a simplified model can be obtained and numerical solutions found in the closure approximation. Applications to the scattering of protons and α-particles by atomic hydrogen have been reported by Lüdde and Dreizler (1989), and to the scattering of protons and antiprotons by lithium and sodium atoms by Stary *et al.* (1990). As an example, we shall take the capture and ionization reactions

$$He^{2+} + H(1s) \rightarrow He^{+} + H^{+} \qquad (8.64a)$$

$$He^{2+} + H(1s) \rightarrow He^{2+} + H^{+} + e^{-}. \qquad (8.64b)$$

For this system Lüdde and Dreizler represented the direct channels containing the $n = 1, 2$, and 3 levels of atomic hydrogen and the rearranged channels containing the $n = 1, 2$, and 3 levels of He^{+} in P-space. Since these are the most important excitation and capture channels, the Q-space is dominated by the ionization channels.

The total capture and ionization cross sections are shown in Fig. 8.6 compared with the experimental data. It is seen that the capture cross section lies significantly below the experimental data, except at the highest energy to which the data extend (100 keV). This lack of accuracy at low energies is seen even more clearly in the partial cross section for capture into the 2s level of He^{+}, shown in Fig. 8.7, where calculated cross section curves lie well below the experimental curve at energies from 15 keV–100 keV, by up to a factor of two. Turning to ionization (Fig. 8.6), it is seen that good agreement is obtained with the experimental data for $E_L \gtrsim 150$ keV/a.m.u., but below this energy agreement is lacking. In general, from this and other examples it can be concluded that capture and ionization cross sections can be determined with reasonable accuracy at energies above 100 keV/a.m.u., but at lower energies the accuracy is not high. This can be attributed to the

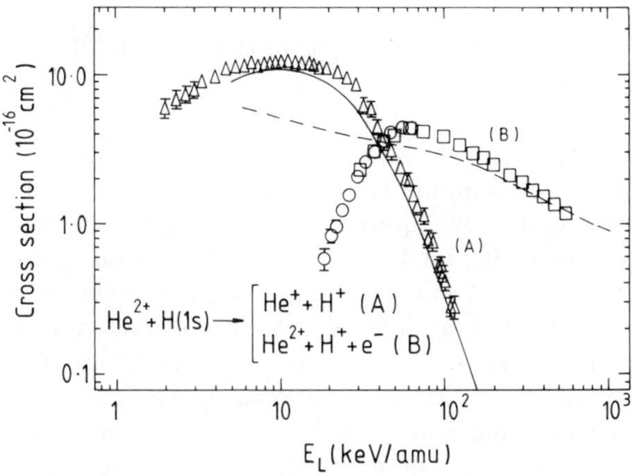

Fig. 8.6. Total capture and ionization cross sections for the scattering of He^{2+} by H(1s) (after Lüdde and Dreizler (1989)). Experimental data: (a) capture △ Shah and Gilbody (1978); (b) ionization □ Shah and Gilbody (1981), ○ Shah *et al.* (1988). Theoretical cross section in the two-centre optical potential model, Lüdde and Dreizler (1989): ——— capture; – – – ionization.

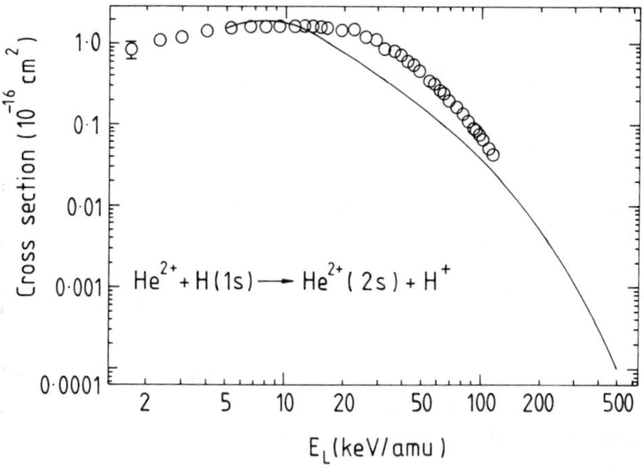

Fig. 8.7. Total cross section for capture into the 2s state of He^+ by He^{2+} impact on H(1s) (after Lüdde and Dreizler (1989)). Experimental data: ○○ Shah and Gilbody (1978). Theoretical cross section in the two-centre optical potential model: ——— Lüdde and Dreizler (1989).

inadequacy of the local approximations. It is clear that the optical potential approach is very promising and if the local approximation employed can be improved, it will provide a useful alternative to coupled channel calculations based on very large pseudostate sets.

8.4 THE CLASSICAL TRAJECTORY MONTE CARLO MODEL

Fully quantal and semi-classical calculations of cross sections require large-scale numerical computations which may not always be practicable because of the size of basis required. For this reason there has always been an interest in constructing models for ion–atom collisions based on classical mechanics, which can be evaluated comparatively rapidly and which can be expected to provide cross sections of at least moderate accuracy. The most successful of these models, the classical trajectory Monte Carlo (CTMC) method, was developed by Abrines and Percival (1966) and was first applied extensively by Olson and Salop (1977) to calculate ionization and capture cross sections for positive ions interacting with atomic hydrogen at intermediate energies.

In the approach of Abrines and Percival for scattering of a fully stripped ion of charge Z by atomic hydrogen, initially in the ground state, the target atom is described by a microcanonical ensemble of classical Kepler orbitals all corresponding to the ground-state energy ε_0. The classical equations of motion are then integrated numerically for a given set of initial conditions, obtained by statistically sampling the ensemble of target orbits. For a given incident velocity v_0, the calculations are repeated over a range of impact parameters $0 \leq b \leq b_m$, where b_m is chosen in such a way that for $b > b_m$ the contribution to the cross section is less than the statistical error of the calculations.

The integration of the classical equations of motion can be carried out either in the centre-of-mass system of coordinates or in the laboratory system. If the laboratory system is chosen, the position vectors of the incident ion A, the target proton B, and the electron e can be denoted by $\mathbf{r}_A$, $\mathbf{r}_B$, and $\mathbf{r}_e$, respectively, taking the origin of coordinates as the initial position of B. The equations of motion in Newton's form are

$$\begin{aligned} M_A \ddot{\mathbf{r}}_A &= Z_A(\mathbf{R}/R^3) + Z_A(\mathbf{x}/x^3) \\ M_B \ddot{\mathbf{r}}_B &= -Z_A(\mathbf{R}/R^3) + (\mathbf{r}/r^3) \\ \ddot{\mathbf{r}}_e &= -Z_A(\mathbf{x}/x^3) - (\mathbf{r}/r^3), \end{aligned} \tag{8.65}$$

where, in our usual notation, the relative position vectors $\mathbf{R}$, $\mathbf{r}$, and $\mathbf{x}$ are defined as

$$\mathbf{R} = \mathbf{r}_A - \mathbf{r}_B; \qquad \mathbf{r} = \mathbf{r}_e - \mathbf{r}_B; \qquad \mathbf{x} = \mathbf{r}_e - \mathbf{r}_A. \tag{8.66}$$

The simple form of equations (8.65) makes it possible to construct extremely fast and efficient computer programs to calculate the trajectories $\mathbf{r}_A(t)$, $\mathbf{r}_B(t)$, and $\mathbf{r}_e(t)$ starting from some specified initial set of positions and velocities, denoted by $\mathbf{r}_A^0$, $\mathbf{r}_B^0$, $\mathbf{r}_e^0$ and $\dot{\mathbf{r}}_A^0$, $\dot{\mathbf{r}}_B^0$, $\dot{\mathbf{r}}_e^0$.

If the Z-axis is the direction of incidence and if A is initially in the XZ-plane, the components of $\mathbf{r}_A^0$ are:

$$x_A^0 = b, \qquad y_A^0 = 0, \qquad z_A^0 = -(R_0^2 - b^2)^{\frac{1}{2}}, \tag{8.67}$$

where R_0 is the initial distance of A from B. In the work of Olson and Salop it was found sufficient to set $R_0 \simeq 10\, Z_A$. The corresponding initial velocity components are

$$\dot{x}_A^0 = \dot{y}_A^0 = 0, \qquad \dot{z}_A = v_0. \tag{8.68}$$

Since B is initially at the origin, $\mathbf{r}_B^0 = 0$, and to a good approximation (neglecting the mass of the electron), $\dot{\mathbf{r}}_B^0 = 0$.

When considering the initial position and velocity of the electron, it is convenient to introduce spherical polar coordinates (r_e, θ, ϕ) such that

$$x_e^0 = r_e^0 \sin\theta \cos\phi, \qquad y_e^0 = r_e^0 \sin\theta \sin\phi, \qquad z_e^0 = r_e^0 \cos\theta. \tag{8.69}$$

If v_e^0 is the magnitude of the initial velocity of the electron, and ξ is the angle between the plane of the electron Kepler orbit and the plane containing both the Z-axis and the major axis of the orbit, the components of $\dot{\mathbf{r}}_e^0$ are

$$\begin{aligned}\dot{x}_e^0 &= -v_e^0(\cos\theta\cos\phi\cos\xi - \sin\phi\sin\xi)\\ \dot{y}_e^0 &= -v_e^0(\cos\theta\cos\phi\cos\xi - \cos\phi\sin\xi)\\ \dot{z}_e^0 &= v_e^0\sin\theta\cos\xi.\end{aligned} \tag{8.70}$$

The eccentric angle u can be introduced by the substitution. See, for example, Landau and Lifshitz (1960))

$$r_e^0 = a(1 - \eta\cos u) \tag{8.71}$$

where a is the semi-major axis of the ellipse and η is the eccentricity. The velocity v_e^0 can be expressed in terms of η and u by

$$v_e^0 = a^{\frac{1}{2}}(1 - \eta^2\cos^2 u)^{\frac{1}{2}}(1 - \eta\cos u)^{-1}, \tag{8.72}$$

and a is determined by the energy ε_0 since

$$a = \frac{1}{2\varepsilon_0}. \tag{8.73}$$

The orbit itself is given by Kepler's equation

$$\theta_n = u - \eta\sin u \tag{8.74}$$

where θ_n is the angle swept by the radius vector as the electron traverses the orbit.

The five quantities θ, ϕ, ξ, η, and θ_n specify the initial position and velocity of the electron. It should be noted that, for given values of θ_n and η, the eccentric angle u must be calculated by an iterative solution of (8.74). The values of r_e^0 and v_e^0 then follow from (8.71) and (8.72), respectively.

Abrines and Percival showed that in the microcanonical ensemble of orbits with energy ε_0, the quantities $\cos\theta$, ϕ, ξ, η^2, and θ_n are distributed uniformly over the intervals $-1 \le \cos\theta \le 1$, $0 \le \phi \le 2\pi$, $0 \le \xi \le 2\pi$, $0 \le \eta^2 \le 1$, $0 \le \theta_n \le 2\pi$. Furthermore, the distribution in velocity of the electron given by this ensemble is the same as that predicted by the quantum theory for the hydrogenic ground state.

The calculations proceed by selecting sets of the six numbers $\cos\theta$, ϕ, ξ, η^2, θ_n, and b at random within the specified intervals. The equations of motion are then integrated for each such set, until the distance between A and B is sufficiently large (usually until $R > 10Z_A$). If after the collision the electron remains in orbit about B, then excitation or elastic scattering has occurred. Alternatively, if the electron is found in orbit about A the collision is recorded as charge transfer, or if the electron becomes free the collision is counted as one leading to ionization.

If N is the total number of trajectories and N_{CE} is the number leading to capture by A, the cross section for charge exchange is

$$\sigma_{\mathrm{CE}} = \frac{N_{\mathrm{CE}}}{N} \pi b_m^2 \left[1 \pm \left(\frac{N - N_{\mathrm{CE}}}{N N_{\mathrm{CE}}} \right)^{\frac{1}{2}} \right]. \tag{8.75}$$

The second term is the standard deviation of the estimated cross section. The ionization cross section, or that for elastic scattering and excitation, can be found in a similar way. In order to obtain good statistical accuracy several thousand trajectories must be computed for total cross sections, and several million for differential cross sections.

A further classification of the trajectories can be made to provide estimates of cross sections for reactions leading to particular final states, or for differential cross sections. For example, to find the partial cross section for capture in a particular final level of the atom $(\mathrm{A} + \mathrm{e}^-)$ with integer quantum numbers nl, one can proceed as follows. If the Kepler orbit of the electron about A, after the collision, is of energy E and angular momentum L then continuous numbers n_c, l_c can be defined such that

$$E = -\frac{Z_{\mathrm{A}}^2}{2n_c^2}, \qquad L = [l_c(l_c + 1)]^{\frac{1}{2}}.$$

If the numbers n_c, l_c lie in the intervals

$$n - \tfrac{1}{2} \le n_c \le n + \tfrac{1}{2}, \qquad l \le l_c \le l + 1$$

then the corresponding trajectory is counted as one leading to the level with quantum numbers nl. The partial cross section σ_{nl} can then be calculated in a similar way to the total cross section. Clearly this procedure can only be expected to be accurate if n and l are sufficiently large. Improved methods of binning to ensure a $(2l + 1)$ statistical distribution among the sublevels have been given by Olson (1981).

As an example of the computation of a differential cross section, we can take the double differential cross section for ionization (Schiwietz and Fritsch 1987). This can be computed from the expression

$$\frac{\mathrm{d}^2\sigma}{\mathrm{d}E\,\mathrm{d}\Omega} = 2\pi\lambda \frac{b_m}{\Delta E\,\Delta\Omega} \frac{1}{N} \sum_{i=1}^{N} b_i, \tag{8.76}$$

where N is the total number of trajectories and $b_i = b$ if the ith trajectory leads to the ejection of an electron within the energy interval ΔE and within the element of solid angle $\Delta\Omega$, or else $b_i = 0$. The factor λ is employed to correct for the finite size of ΔE, which must be large enough to ensure that a sufficient number of ejected electrons are found within this interval to obtain good statistical accuracy. The factor λ can be computed by assuming the shape of the energy distribution of the ejected electron, taken to be exponential in the work of Schiwietz and Fritsch.

Validity of the CTMC model

The accuracy to be expected from the CTMC model and its domain of validity is difficult to assess. At low impact velocities, charge exchange occurs largely by quantum tunnelling through the potential barrier encountered by the electron in transferring from the direct to the rearranged configuration, and in this region classical methods are inapplicable. At high velocities, the ionization cross section should be given correctly by the first-order Born approximation, which behaves like $[v_0^{-2} \log v_e]$, whereas all classical methods predict a v_0^{-2} variation of the cross section. For these reasons, and taking into account comparison of the model cross sections with experiment, the velocity range of validity of the CTMC model has been estimated as $1 \leq v_0/v_e \leq 4$. Within these limits, among the reasons for the success of a purely classical model are the following. First we have already shown, in deriving the semi-classical impact parameter model, that the relative motion of the heavy particles A and B can be described classically and that the interaction between A and B cannot contribute to total cross sections for inelastic processes. Further, in swift collisions the main role of the potential binding the electron to the target is to generate an initial electronic momentum (or velocity) distribution. The collision can then be considered as a two-body scattering between the first incident ion A and an electron with this initial velocity distribution, and for Coulomb interactions the two-body cross section is the same in classical and in quantum mechanics. This 'impulse parameter' picture requires modification in the case of charge exchange at high velocities because, as will be explained in Chapter 10, double scattering in which the electron interaction with the projectile A is followed by a subsequent scattering between the electron and B is important. Of course, since the classical three-body problem is solved exactly such processes are also represented in the CTMC model. The CTMC model can be viewed as an approximation to the exact quantum theory, in a similar way that the semi-classical impact parameter model is an approximation to the exact theory. This subject has been discussed by Reinhold and Falcon (1988*a*) who identify the wave functions associated with the CTMC model and other classical approximations, but the analysis has not, as yet, been carried to the point where the accuracy of the CTMC model in a given application can be determined.

Although the microcanonical distributions provide the same velocity distributions as in the quantum theory, the spatial distribution of the charge density is very poor, vanishing identically for $r > 2a_0$ rather than decreasing exponentially. Following earlier work by Eichenauer *et al.* (1981) and Hardie and Olson (1983) to correct this situation, Cohen (1985) has derived a new phase space distribution which exactly reproduces the spatial density of ground-state hydrogen. The momentum distribution is no longer exact, but remains within a few per cent of the exact distribution except for large $v_e (v_e > 9)$, which is, in any case, an unimportant region for the scattering problem. Cross sections obtained with the microcanonical distribution (CTMC-p) can be connected by an integral transform to those obtained by the new distribution (CTMC-r). It is found that ionization cross sections are markedly improved, while the charge exchange cross sections show little change.

One-electron systems

To illustrate the results of the CTMC model for the one-electron system, total cross sections for capture and ionization in collisions between fully stripped ions of charge Z and H(1s) computed by Reinhold and Falcon (1988*b*) are compared with experimental data in Figs 8.8 and 8.9 using both

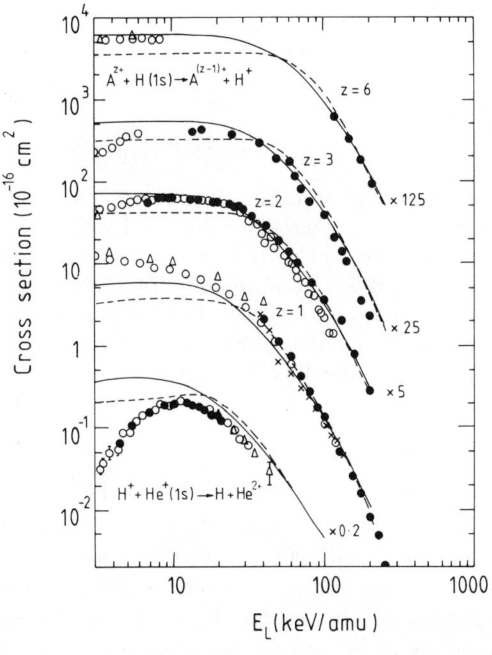

Fig. 8.8. Total cross section for electron capture by fully stripped ions of charge Z from H(1s) and by protons from He^+(1s) (after Reinhold and Falcon (1988b)). Experimental data: $Z = 6$: ● Goffe *et al.* (1979); ○ Meyer *et al.* (1985); △ Dijkkamp *et al.* (1985*a*, *b*). $Z = 3$: ● Shah *et al.* (1978); ○ Seim *et al.* (1980). $Z = 2$: △ Nutt *et al.* (1978); ● Olson *et al.* (1977); ○ Shah and Gilbody (1978). $Z = 1$: ● Wittkower *et al.* (1966); ○ McClure (1966); × Gilbody and Ryding (1966); △ Fite *et al.* (1960, 1962). For $H^+ + He^+$: ● Peart *et al.* (1983); ○ Rinn *et al.* (1985); △ Watts *et al.* (1986). Classical trajectory Monte Carlo cross-sections from Reinhold and Falcon (1988*b*): CTMC-r ———; CTMC-p — — —.

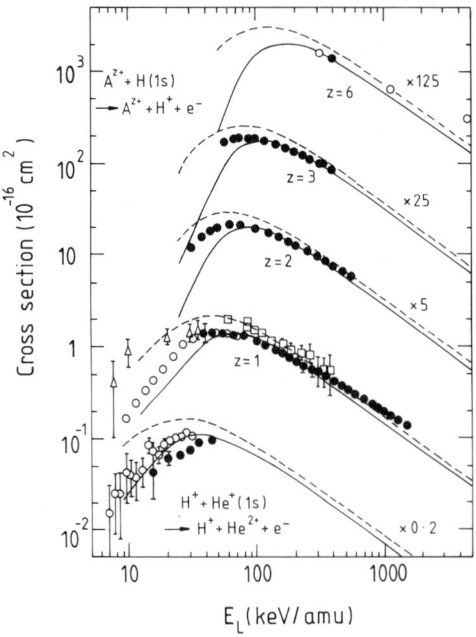

Fig. 8.9. Total cross section for ionizaton of H(1s) by fully stripped ions of charge Z and of He^+(1s) by protons (after Reinhold and Falcon (1988b)). Experimental data: $Z = 6$: ● Shah and Gilbody (1983); ○ Schlachter *et al.* (1981). $Z = 3$: ● Shan and Gilbody (1983). $Z = 2$: ● Shah and Gilbody (1981). $Z = 1$: ○ Shah *et al.* (1987); ● Shah and Gilbody (1981); △ Fite *et al.* (1960); □ Gilbody and Ireland (1963). For $H^+ + He^+$: ○ Rinn *et al.* (1986); ● Watts *et al.* (1986). Classical trajectory Monte Carlo cross sections from Reinnold and Falcon (1988*b*): CTMC-r ———; CTMC-p – – –.

the microcanonical (CTMC-p) and the Cohen (CTMC-r) distributions. The results show the general superiority of the CTMC-r model and show that the computed total cross sections for both processes compare well with the data provided the incident energy is above ~25 keV/a.m.u. ($v_0 = 1$). Also included in the figures are results for the system $H^+ + He^+$(1s). It is also clear that the accuracy of the model depends on the ratio Z_A/Z_B as well as the velocity. For example, for ionization the computed cross section for $H^+ + He^+$(1s) is in better agreement with experiment for small velocities than for systems with $Z_A/Z_B > 1$. For capture, on the other hand, the CTMC method fails to reproduce the data for the $H^+ + He^+$(1s) system, and the position is similar for $H^+ + Li^{2+}$(1s) (Ermolaev and McDowell 1987). This result is in harmony with the impulse approximation picture, which would be expected to be more accurate with increasing values of the ratio Z_A/Z_B.

Reinhold and Falcon (1988*a*) show that for a single-electron system the

CTMC cross sections satisfy the scaling law

$$\sigma(v_0, \lambda Z_A, \lambda Z_B) = \lambda^{-2}\sigma(v_0/\lambda, Z_A, Z_B)$$

where Z_A and Z_B are the charges in the two nuclei. It is easy to verify that the same law applies to cross sections calculated in the semi-classical impact parameter formalism. The scaling law allows us to relate scattering by hydrogenic ions with $Z_B > 1$ to the scattering of a particle of charge Z_A/Z_B by neutral atomic hydrogen. For example, the system $C^{+6} + He^{+}(1s)$ ($Z_B = 6$, $Z_A = 2$) can be related to the scattering of Li^{3+} by H(1s).

The impact parameter CTMC model

It is well established that the relative motion of the heavy particles A and B can be described by a predetermined classical trajectory. It follows that as in the semi-classical impact parameter formalism, the nuclear and electronic motion can be separated in any classical model (Toshima *et al.* 1989). In the impact parameter model the classical equations of motion are determined by the time-dependent Hamiltonian

$$H(t) = \tfrac{1}{2}\mathbf{P}^2 - \frac{Z_A}{|\mathbf{s} + p\mathbf{R}(t)|} - \frac{Z_B}{|\mathbf{s} - q\mathbf{R}(t)|} \tag{8.77}$$

where **s** is the electronic position vector with respect to the centre of mass of A and B, while

$$p = \frac{M_B}{M_A + M_B}, \quad q = \frac{M_A}{M_A + M_B}, \tag{8.78}$$

and **P** is the momentum of the electron. For the purposes of calculating total cross sections the trajectory $\mathbf{R} = \mathbf{R}(t)$ can be taken to be a straight line provided $Z_A Z_B$ is small. In this simplified scheme the CTMC total cross sections agree with those found from the full three-body treatment to within 0.01 per cent. For differential cross sections Toshima *et al.* calculated the trajectory $\mathbf{R}(t)$ from the static potential:

$$V_{AB}(R) = \frac{Z_A(Z_B - 1)}{R} + \left(Z_A Z_B + \frac{Z_A}{R}\right)e^{-2Z_B R}. \tag{8.79}$$

Differential cross sections can then be calculated from the expression

$$\frac{d\sigma}{d\Omega} = \left.\frac{d\sigma}{d\Omega}\right|_{cl} P(b), \tag{8.80}$$

where $d\sigma/d\Omega|_{cl}$ is the differential cross section for scattering by V_{AB} at angle θ, and $P(b)$ is the transition probability calculated from the CTMC model at the impact parameter corresponding to θ.

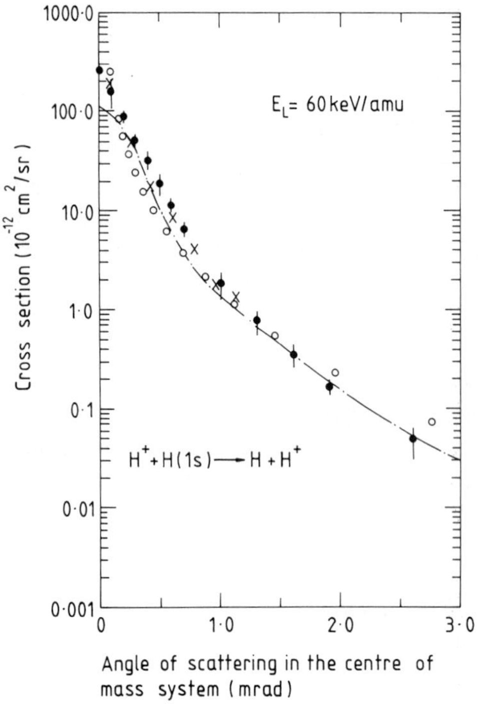

Fig. 8.10. The differential cross section for electron capture by H^+ ions from atomic hydrogen in the ground state at $E_L = 60$ keV/a.m.u. (after Toshima *et al.* (1989)). Experimental cross section: ●● Martin *et al.* (1981). Theoretical cross sections: ○○ impact parameter CTMC; × × CTMC model, Olson (1983); —·—·— eikonal model (see Chapter 10), Toshima *et al.* (1989).

As an example, the differential cross section for charge exchange is shown in Fig. 8.10 for the reaction $H^+ + H(1s) \rightarrow H + H^+$ at $E_L = 60$ keV/a.m.u. calculated both in the impact parameter CTMC model and in the usual CTMC model. The results are in good agreement with each other and also with the measured cross section, except near the forward direction. It seems safe to conclude that the use of a straight line trajectory to describe the relative nuclear motion is equally well justified for classical as for quantal models.

Extensions of the CTMC model

By using effective potentials between the active electron and the ionic cores, the CTMC method can be extended to many-electron systems (McDowell and Janev 1984; Peach *et al.* 1985; Reinhold and Falcon 1986). The main uncertainty lies in the preparation of the ensemble representing the target atom, since it is only for a pure Coulomb interaction that the microcanonical ensemble produces the exact velocity distribution. Detailed comparisons between experiment, quantal calculations, and the results of the CTMC, model† show that the CTMC cross sections for charge exchange can be expected to be within ± 30 per cent of the correct values. For ionization the CTMC cross sections are generally in better agreement with experiment.

There is no difficulty in integrating the classical equations of motion for a system containing more than one electron together with two heavy particles, although the time increases rapidly with increasing number of particles, so the extension of the CTMC model to systems of two or three electrons depends on defining classical models for two- or three-electron atoms. Although earlier attempts to construct a classical Bohr model of He were unsatisfactory, because the classical two-electron system is unstable against autoionization, by introducing suitable constraints suggested by the Heisenberg uncertainty principle, Kirschbaum and Wilets (1980) have succeeded in constructing a suitable classical model of the helium atom. This model has been employed by McKenzie and Olson (1987) to calculate cross sections for the interaction of highly charged ions with He, including specifically two-electron reactions such as simultaneous capture and ionization.

† As an example, see the detailed comparisons made by Ermolaev *et al.* (1987*a*, *b*) for the He^{2+}–Li systems.

9

MANY-ELECTRON SYSTEMS

9.1 INTRODUCTION

As we have already noted in Sections 4.7 and 5.9, the coupled channel methods developed in Chapter 3 for quasi-one-electron systems can be generalized, in principle, for systems with any number of electrons. The basis functions can be adiabatic molecular wave functions which diagonalize the electronic Hamiltonian, or atomic functions (and pseudostates) which diagonalize the unperturbed atomic Hamiltonians in the different arrangements. In the former case all the interactions including the electron–electron interaction are taken into account and in the latter all the internal interactions in the separate atoms are allowed for in the basis. Calculating such basis functions for a many-electron system with sufficient accuracy is already a demanding task; nevertheless there have been a growing number of investigations of two-electron or quasi-two-electron systems, some of which have been illustrated in Chapter 6. To some extent, the difficulties of constructing many-particle molecular wave functions can be overcome by using a basis in which electron–electron correlation is absent. Correlation can then be taken into account when forming the coupled equations. This approach, which is due to Harel and Salin, will be discussed below. Apart from the calculation of basis functions, the computational effort required to treat two-electron systems can be very great, compared with one-electron systems, for several reasons. For example, the elements of the interaction matrix become more difficult to compute. In particular, with an atomic basis composed of travelling Slater orbitals the computation of the matrix elements of the electron–electron interaction are particularly time consuming. If specifically two-electron processes are important, for example double capture, double excitation or ionization, simultaneous capture and excitation, or simultaneous capture and ionization, the size of the basis required to represent the two-electron wave function is much greater than that required for one-electron systems.

As the number of electrons increases, a detailed treatment is not generally feasible and simplifications are required. In many cases, if the principal reaction channels involve the capture or ionization of a single electron, the one-active-electron model described in Chapter 3 may be sufficiently accurate, as we have already seen in Chapters 4 and 6. In discussing the one-active-electron model, it has been assumed that the effective potentials V_{Ae}, V_{Be}

between the ion cores and the active electron are fixed and are determined semi-empirically to provide the correct energy level structure of the molecule $A + B + e^-$, or of the atoms $(A + e^-)$ and $(B + e^-)$. A different approach, which will be considered later in this chapter, is the calculation of an average potential by the time-dependent Hartree–Fock (TDHF) approximation. We shall also discuss briefly how the one-electron amplitudes can be combined to deduce approximate transition rates for double capture and other two-electron transitions, using the *independent particle model.*

9.2 THE HAREL–SALIN OEDM MODEL

To avoid the difficulty of constructing molecular basis functions for many-particle systems, a basis composed from the products of one-electron diatomic molecular orbitals (OEDM functions) can be employed (Harel and Salin 1980). As an example let us take the case of a fully stripped ion A, of charge Z_A, interacting with a helium atom with a nucleus B of charge $Z_B = 2$. The position vectors of the two electrons will be denoted by $\mathbf{s}_1, \mathbf{s}_2$ taking the mid-point of the internuclear line as the origin of coordinates. In the impact parameter formalism an approximate solution of the time-dependent Schrödinger equation,

$$\mathrm{i}\frac{\partial\psi(\mathbf{s}_1, \mathbf{s}_2, t)}{\partial t} = H_e\psi(\mathbf{s}_1, \mathbf{s}_2, t), \tag{9.1}$$

is sought, where the electronic Hamiltonian H_e is

$$H_e = -\tfrac{1}{2}(\nabla_1^2 + \nabla_2^2) - \frac{Z_A}{x_1} - \frac{Z_A}{x_2} - \frac{Z_B}{r_1} - \frac{Z_B}{r_2} + \frac{1}{|\mathbf{s}_1 - \mathbf{s}_2|} + \frac{Z_A Z_B}{R(t)}, \tag{9.2}$$

with

$$\mathbf{x}_i = \mathbf{s}_i - \tfrac{1}{2}\mathbf{R}(t), \qquad \mathbf{r}_i = \mathbf{s}_i + \tfrac{1}{2}\mathbf{R}(t). \tag{9.3}$$

As usual $\mathbf{R} = \mathbf{R}(t)$ is a classical trajectory which can be taken to be rectilinear except at low energies ($E_L \leq 1$ keV/a.m.u.).

A set of orthonormal one-electron orbitals $\phi_j(\mathbf{s}_i, \mathbf{R})$ is defined which is composed of eigenfunctions of the one-electron Hamiltonians

$$h_0(i) = -\tfrac{1}{2}\nabla_i^2 - \frac{Z_A}{x_i} - \frac{Z_B}{r_i}, \tag{9.4}$$

so that

$$h_0(i)\phi_j(\mathbf{s}_i, \mathbf{R}) = \varepsilon_j(R)\phi_j(\mathbf{s}_i, \mathbf{R}). \tag{9.5}$$

A basis for the expansion of the total wave function can be obtained by forming suitably symmetrized products of the one-electron functions $\phi_j(\mathbf{s}_i, \mathbf{R})$.

In the present example the helium is in the singlet ($S = 0$) state, so that the spatial wave function is symmetrical in the coordinates of the two electrons, and the basis function $\chi_{ij}(\mathbf{s}_1, \mathbf{s}_2, t)$ are†

$$\chi_{ij}(\mathbf{s}_1, \mathbf{s}_2, t) = N_{ij}[\phi_i(\mathbf{s}_1, \mathbf{R})\phi_j(\mathbf{s}_2, \mathbf{R}) + \phi_i(\mathbf{s}_2, \mathbf{R})\phi_j(\mathbf{s}_1, \mathbf{R})], \tag{9.6}$$

where

$$N_{ij} = \begin{cases} \dfrac{1}{\sqrt{2}} & i \neq j \\ \dfrac{1}{2} & i = j. \end{cases}$$

It is interesting to note that by forming independent particle product wave functions from *diabatic* orbitals and obtaining the corresponding correlation diagrams, a nearly complete qualitative understanding of many-particle systems can be obtained. The correlation rules have been obtained by Barat and Lichten (1972) and by Eichler *et al.* (1976). This topic will not be discussed here but reference may be made to the reviews of Barat (1983) and Briggs (1983).

To first order in the electron–electron interaction, the energy levels of the system $E_{ij}^{(1)}(R)$ are

$$\begin{aligned} E_{ij}^{(1)}(R) &= \langle \chi_{ij} | H_e | \chi_{ij} \rangle \\ &= \langle \chi_{ij} | \frac{1}{\mathbf{s}_1 - \mathbf{s}_2} | \chi_{ij} \rangle + \frac{2Z_A}{R} + \varepsilon_i(R) + \varepsilon_j(R). \end{aligned} \tag{9.7}$$

The first-order energies for the case $Z_A = 8$, the O^{8+} + He collision, are shown in Fig. 9.1 for the configurations listed in Table 9.1 (Bliman *et al.* 1983).

The solution $\psi(\mathbf{s}_1, \mathbf{s}_2, t)$ of the impact parameter equation (9.1) can be expanded in terms of the basis function χ_{ij} as

$$\psi(\mathbf{s}_1, \mathbf{s}_2, t) = \sum_J A_J(\mathbf{b}, t)\chi_J(\mathbf{s}_1, \mathbf{s}_2, t) \exp[i\gamma_J(\mathbf{s}_1, \mathbf{s}_2, t)] \tag{9.8}$$

where the index J denotes the combination (ij). The phase γ_J can be written as

$$\gamma_J = -[\varepsilon_i(\infty) + \varepsilon_j(\infty)]t + \pi_J(\mathbf{s}_1, \mathbf{s}_2, t) \tag{9.9}$$

where $\exp(i\pi_J)$ is a translation factor. If a common translation factor is

† If the incident ion A is an α-particle the expression (9.6) for the basis functions χ_{ij} can be modified to allow for the extra gerade–ungerade symmetry. Using the correlation diagram for homonuclear one-electron systems, the appropriate combinations leading to the separated atom wave functions can be obtained (Harel and Salin 1980).

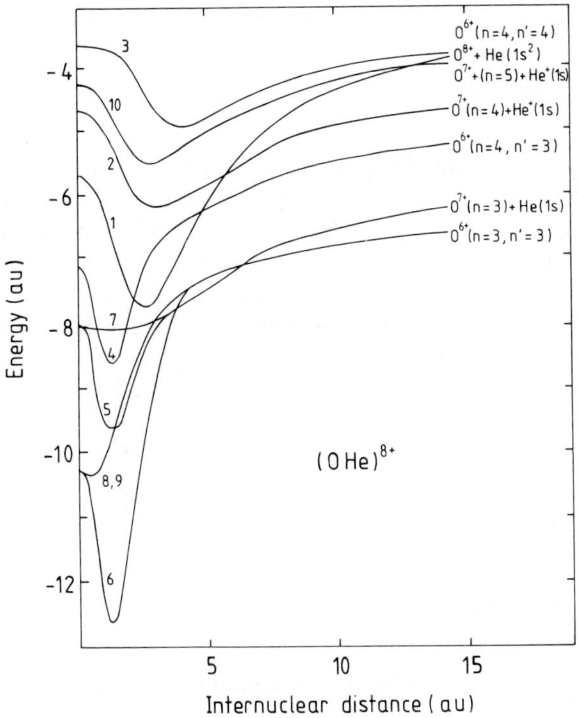

Fig. 9.1. The energies of the $(OHe)^{8+}$ system treating the electron–electron interaction to first order. The numbers attached to the curves correspond to the levels listed in Table 9.1.

Table 9.1. Configurations in the O^{8+} + He system.

	United atom designation	Separated atom limit
1	$4f\sigma^2$	$O^{8+} + He(1s^2)$
2	$4f\sigma\ 5g\sigma$	$O^{7+}(n = 4) + He^+(1s)$
3	$5g\sigma^2$	$O^{6+}(n = 4)^2 + He^{2+}$
4	$5g\sigma\ 3d\sigma$	$O^{6+}(n = 4,\ n' = 3) + He^{2+}$
5	$4f\sigma\ 3d\sigma$	$O^{7+}(n = 3) + He^+(1s)$
6	$3d\sigma^2$	$O^{6+}(n = 3)^2 + He^{2+}$
7	$4f\sigma\ 3d\pi$	$O^{7+}(n = 3) + He^+(1s)$
8	$3d\pi^2\ \Sigma$	$O^{6+}(n = 3)^2 + He^{2+}$
9	$3d\pi^2\ \Delta$	$O^{6+}(n = 3)^2 + He^{2+}$
10	$4f\sigma\ 6h\sigma$	$O^{7+}(n = 3) + He^+(1s)$

employed π_J can be expressed in terms of a switching function $f(\mathbf{s}_i, \mathbf{R})$ by

$$\pi_J(\mathbf{s}_1, \mathbf{s}_2, t) = \tfrac{1}{2} \sum_{i=1}^{2} f(\mathbf{s}_i, \mathbf{R})\mathbf{v}_0 \cdot \mathbf{s}_i - \tfrac{1}{4}v_0^2 t \tag{9.10}$$

where $f(\mathbf{s}_i, \mathbf{R})$ can be one of the forms discussed in Chapter 5.

The amplitudes $A_J(\mathbf{b}, t)$ satisfy coupled equations of the standard form (3.205). The initial boundary conditions to be satisfied, in the limit $t \to -\infty$, require some discussion because none of the basis functions coincide exactly with the ground-state wave function of helium in the separated atomic limit. This difficulty can be overcome by diagonalizing the Hamiltonian for the helium atom with respect to the basis functions χ_J in the separated atom limit, thus providing an approximate energy E_{He} and an approximate wave function ϕ_{He} where

$$\phi_{\mathrm{He}}(\mathbf{s}_1, \mathbf{s}_2) = \sum_J \alpha_J \chi_J(\mathbf{s}_1, \mathbf{s}_2, t \to -\infty). \tag{9.11}$$

The requirement

$$\lim_{t \to -\infty} \psi(\mathbf{s}_1, \mathbf{s}_2, t) = \phi_{\mathrm{He}}(\mathbf{s}_1, \mathbf{s}_2) \exp(-\mathrm{i}E_{\mathrm{He}}t) \exp(\mathrm{i}\pi_J) \tag{9.12}$$

is satisfied by imposing the boundary conditions

$$\lim_{t \to -\infty} A_J(\mathbf{b}, t) = \alpha_J \exp[-\mathrm{i}E_{\mathrm{He}}t + \mathrm{i}(\varepsilon_i(\infty) + \varepsilon_j(\infty))]. \tag{9.13}$$

In the limit $t \to \infty$, the amplitudes for the various possible reactions are obtained by projecting the wave function by the approximate atomic wave functions, again formed from the finite basis of functions χ_J.

The Harel–Salin model has been applied to a number of cases including He^{2+} + He (Harel and Salin 1980), and O^{8+} + He (Bliman *et al.* 1983). In both of these examples, no translational factors were employed; nevertheless the results were very encouraging. To take the case of the He^{2+} + He system, good agreement was obtained with both experiment and with other theoretical calculations over the energy range $2 < E_L < 50$ keV/a.m.u. In particular, as shown in Figs 9.2 and 9.3, the cross sections for both single and double capture agree closely with those of Kimura (1988) who employed an adiabatic configuration interaction molecular wave function basis, including translation factors, and with those of Gramlich *et al.* (1989) who employed a two-electron pseudostate basis formed from Gaussian orbitals with plane-wave translation factors.

In the O^{8+} + He system the principal source of charge exchange is a transition near 5.6 a.u. between the $4\mathrm{f}\sigma^2$ incident state (see Fig. 9.1) and the $4\mathrm{f}\sigma\, 5\mathrm{g}\sigma$ state which is asymptotic to $O^{7+}(n = 4) + He^{+}(1s)$. The cross section for double capture, leading to doubly excited states of O^{6+}, is

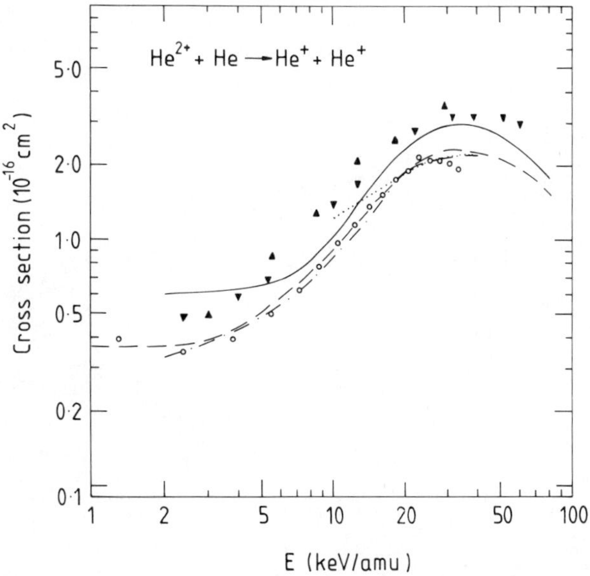

Fig. 9.2. Total cross sections for single charge exchange in collisions of He^{2+} with He. Experimental data: ▲ Bayfield and Khayrallah (1975a, b); ○ Afrosimov *et al.* (1975); ▼ Berkner *et al.* (1968). Theoretical cross sections: —·—·— OEDM basis, Harel and Salin (1980); ----- configuration interaction MO basis, Kimura (1988); ——— Gaussian pseudostate basis, Gramlich *et al.* (1989); ······ time-dependent Hartree–Fock cross section, Sandhya Devi and Garcia (1983).

comparable in magnitude to that for single capture, the final levels of importance being $O^{6+}(n = 3, n = 4)$ and $O^{6+}(n = 3, n' = 3)$. In the experiments of Bliman *et al.* the production of O^{7+} ions is measured, and these may arise both from single electron capture and from double capture followed by the autoionization of the O^{6+} doubly excited states. With this in mind it can be seen from Fig. 9.4 that the agreement with the theoretical model is reasonable.

9.3 THE TIME-DEPENDENT HARTREE–FOCK MODEL

All converged expansion methods in which the total wave function for a many-electron system is expanded in a formally complete set allow fully for electron–electron correlation. Nevertheless correlation effects are not always important in which case the electron–electron interaction can often be taken into account in an average way. When transitions involving a single electron are the only ones of importance, it is possible to use the one-active-

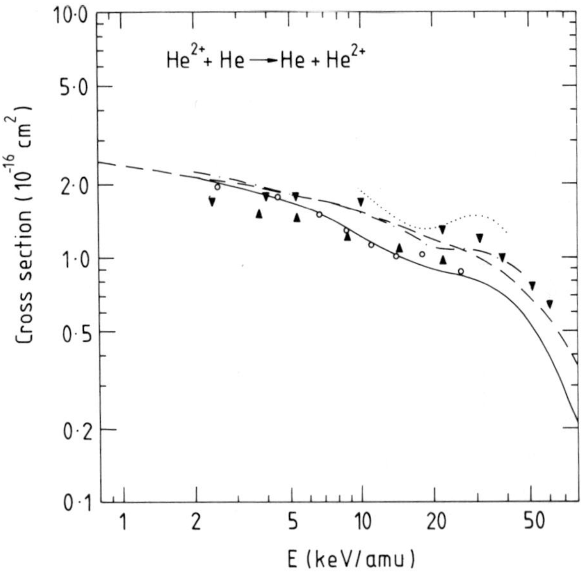

Fig. 9.3. Total cross sections for double charge exchange in collisions of He^{2+} with He. The symbols are as in Fig. 9.2.

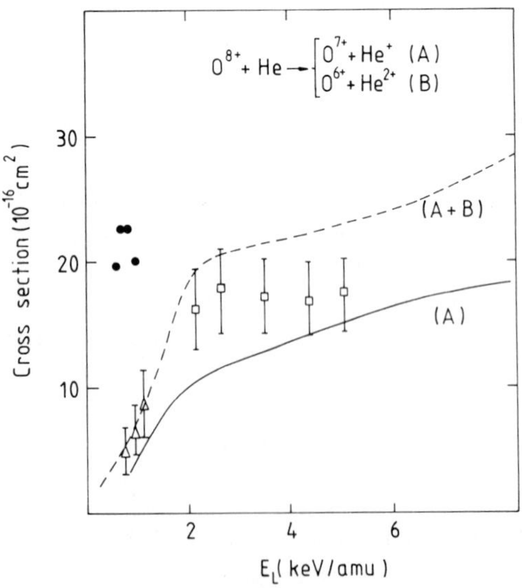

Fig. 9.4. Total cross sections for charge exchange in collisions of O^{8+} with He (after Bliman *et al.* (1983)). Experimental data for the production of O^{7+} ions: ●●● Kaneko *et al.* (1981); △△△ Iwai *et al.* (1982), as corrected by Bliman *et al.* (1983); □□ Bliman *et al.* (1983). Theoretical cross sections from Bliman *et al.* (1983): (a) ——— for single charge exchange; (a + b) – – – – – for the sum of single and double charge exchange.

electron model in which the electron moves in predetermined effective potentials V_{Ae}, V_{Be}. In such models, many-electron effects are ignored completely. Without making such an extreme approximation it is possible to calculate a potential for each electron in many-particle systems through the Hartree–Fock method, which retains some of the many-body aspects of the collision. The potential experienced by each electron depends on the charge density due to all the other electrons, but the electron–electron interaction is represented only on the average.

Let us consider the N-electron generalization of the two-electron system represented by the time-dependent Schrödinger equation (9.1). It is convenient to remove the internuclear potential (see (3.164)), and to work with the electronic Hamiltonian $\tilde{H}_e$ where $H = \tilde{H}_e + Z_A Z_B/R$. The Hamiltonian $\tilde{H}_e$ can be expressed as:

$$\tilde{H}_e = \sum_{i=1}^{N} h_0(i) + \sum_{i>j}^{N} \frac{1}{|\mathbf{s}_i - \mathbf{s}_j|} \tag{9.14}$$

where $h_0(i)$ is defined as (see (9.4))

$$h_0(i) = -\tfrac{1}{2}\nabla_i^2 - \frac{Z_A}{x_i} - \frac{Z_B}{r_i}, \tag{9.15}$$

and $\sum_{i>j}^{N}$ denotes a sum over both i and j subject to the restriction $i > j$.

The many-electron wave function $\psi(q_1, q_2, \ldots, q_N, t)$ is a function of both position vectors $\mathbf{s}_i$ and the spin coordinates σ_i of each electron, collectively denoted by q_i. In the time-dependent Hartree–Fock method the wave function is approximated by products of one-electron spin orbitals $U_\lambda(q_i, t)$ where λ is an index specifying the one-electron state. Each spin orbital is of the form

$$U_\lambda(q_i, t) = \phi_\lambda(\mathbf{s}_i, t)\chi_{\frac{1}{2}m_\lambda}(i) \tag{9.16}$$

where $\phi_\lambda(\mathbf{s}_i, t)$ is a spatial orbital and $\chi_{\frac{1}{2}m_\lambda}(i)$ is a spin wave function with $m_\lambda = \pm\frac{1}{2}$. To allow for the identity of the electrons the wave function $\psi(q_1, \ldots, q_N, t)$ must be antisymmetrized under an interchange of any pair of particles. This can be achieved by writing the wave function as a Slater determinant of N spin orbitals with $\lambda = \alpha, \beta, \ldots, \nu$, so that

$$\psi(q_1, q_2, \ldots, q_N, t) = \frac{1}{\sqrt{N!}} \begin{vmatrix} U_\alpha(q_1, t) & U_\beta(q_1, t) & \cdots & U_\nu(q_1, t) \\ U_\alpha(q_2, t) & U_\beta(q_2, t) & \cdots & U_\nu(q_2, t) \\ \vdots & & & \vdots \\ U_\alpha(q_N, t) & U_\beta(q_N, t) & \cdots & U_\nu(q_N, t) \end{vmatrix} \tag{9.17}$$

If the spin orbitals are taken to be orthonormal

$$\langle U_\alpha(t)|U_\beta(t)\rangle = \delta_{\alpha\beta}, \tag{9.18}$$

the total wave function (9.17) is normalized to unity. In the time-dependent Hartree–Fock method the wave function ψ is approximated by only one Slater determinant as in (9.17). This is the principal approximation, and it is clear that to represent the wave function exactly requires the linear superposition of an infinite number of Slater determinants each differing in the spin orbitals occupied by the N electrons.

To obtain the best wave function of the approximate form (9.17), the variational principle (3.181) is employed by forming the functional I where

$$I = \int_{-\infty}^{\infty} \mathrm{d}t \, \langle \psi(q_1, q_2, \ldots, q_N, t) | \tilde{H}_e - \mathrm{i}\frac{\partial}{\partial t} | \psi(q_1, q_2, \ldots, q_N, t) \rangle, \quad (9.19)$$

and requiring $\delta I = 0$ under independent first-order variations of each spin orbital:

$$U_\lambda(q_i, t) \to U_\lambda(q_i, t) + \delta U_\lambda(q_i, t). \quad (9.20)$$

It is convenient to write the approximate wave function (9.17) in the equivalent form (Bransden and Joachain 1983)

$$\psi(q_1, q_2, \ldots, q_N, t) = (N!)^{-\frac{1}{2}} \sum_P (-1)^P P U_\alpha(q_1, t) U_\beta(q_2, t) \cdots U_\nu(q_N, t), \quad (9.21a)$$

where the sum is over the $N!$ permutations of the electronic coordinates $q_1, q_2, \ldots, q_N$ and where $(-1)^P$ is equal to $+1$ when P is an even and to -1 when P is an odd permutation.

The wave function ψ can also be written in the form

$$\psi(q_1, q_2, \ldots, q_N, t) = (N!)^{\frac{1}{2}} A \psi_H, \quad (9.21b)$$

where ψ_H is a single product of spin orbitals

$$\psi_H(q_1, q_2, \ldots, q_N, t) = U_\alpha(q_1, t) U_\beta(q_2, t) \cdots U_\nu(q_N, t), \quad (9.22)$$

and the operator A is the antisymmetrization operator:

$$A = (N!)^{-1} \sum_P (-1)^P P. \quad (9.23)$$

Using the fact that $\tilde{H}_e$ is completely symmetrical in the space–spin coordinates $q_1, \ldots, q_N$, it is easy to see that

$$I = N! \int_{-\infty}^{\infty} \mathrm{d}t \, \langle \psi_H(q_1, q_2, \ldots, q_N, t) | \tilde{H}_e - \mathrm{i}\frac{\partial}{\partial t} | A \psi_H(q_1, q_2, \ldots, q_N, t) \rangle. \quad (9.24)$$

Let us write

$$I = I_1 + I_2 \quad (9.25)$$

where I_1 contains all the one-electron operators in $\tilde{H}_e$ and I_2 contains the interaction between the electrons. We then have:

$$I_1 = N! \int_{-\infty}^{\infty} \mathrm{d}t \, \langle \psi_H(q_1, q_2, \ldots, q_N, t) | \sum_{i=1}^{N} h_0(i) - \mathrm{i} \frac{\partial}{\partial t} | A\psi_H(q_1, q_2, \ldots, q_N, t) \rangle \tag{9.26}$$

and

$$I_2 = N! \int_{-\infty}^{\infty} \mathrm{d}t \, \langle \psi_H(q_1, q_2, \ldots, q_N, t) | \sum_{i>j} \frac{1}{|\mathbf{s}_i - \mathbf{s}_j|} | A\psi_H(q_1, q_2, \ldots, q_N, t) \rangle . \tag{9.27}$$

By using the orthonormality conditions (9.18), I_1 can be reduced to the form

$$I_1 = \int_{-\infty}^{\infty} \mathrm{d}t \sum_{\lambda} \langle U_\lambda(q_1, t) | h_0(i) - \mathrm{i} \frac{\partial}{\partial t} | U_\lambda(q_1, t) \rangle \tag{9.28}$$

where the sum over λ runs over the N different spin orbitals $U_\alpha, U_\beta, \ldots, U_\nu$. The matrix element I_2 can be expressed, using (9.23), as

$$\begin{aligned} I_2 &= \sum_{i>j} \sum_{P} (-1)^P \int_{-\infty}^{\infty} \mathrm{d}t \, \langle \psi_H | \frac{1}{|\mathbf{s}_i - \mathbf{s}_j|} P | \psi_H \rangle \\ &= \sum_{i>j} \int_{-\infty}^{\infty} \langle \psi_H | \frac{1}{|\mathbf{s}_i - \mathbf{s}_j|} (1 - P_{ij}) | \psi_H \rangle \, \mathrm{d}t \end{aligned} \tag{9.29}$$

where P_{ij} is the operator that interchanges the space and spin coordinates of electrons i and j. Taking into account the orthonormality conditions (9.18) and the definition of ψ_H, (9.22), we find that

$$\begin{aligned} I_2 = \frac{1}{2} \sum_{\lambda} \sum_{\mu} \int_{-\infty}^{\infty} \mathrm{d}t \Big[&\langle U_\lambda(q_1, t) U_\mu(q_2, t) | \frac{1}{|\mathbf{s}_1 - \mathbf{s}_2|} | U_\lambda(q_1, t) U_\mu(q_2, t) \rangle \\ &- \langle U_\lambda(q_1, t) U_\mu(q_2, t) | \frac{1}{|\mathbf{s}_1 - \mathbf{s}_2|} | U_\mu(q_1, t) U_\lambda(q_2, t) \rangle \Big] \end{aligned} \tag{9.30}$$

where the sums run over the N spin orbitals $\lambda, \mu = \alpha, \beta, \ldots, \nu$.

Under the independent arbitrary variations (9.20) we have, from (9.25), (9.28), and (9.30):

$$\begin{aligned} \delta I = \sum_{\lambda} \int_{-\infty}^{\infty} \mathrm{d}t \Big[&\langle \delta U_\lambda(q_1, t) | h_0^{(1)} - \mathrm{i} \frac{\partial}{\partial t} | U_\lambda(q_1, t) \rangle \\ &+ \sum_{\mu} \langle \delta U_\lambda(q_1, t) U_\mu(q_2, t) | \frac{1}{|\mathbf{s}_1 - \mathbf{s}_2|} (1 - P_{12}) | U_\lambda(q_1, t) U_\mu(q_2, t) \rangle \Big]. \end{aligned} \tag{9.31}$$

The condition $\delta I = 0$ requires that the coefficient of each of the functions δU_λ should vanish, leading to the system of equations

$$\left[-\tfrac{1}{2}\nabla_1^2 - \frac{Z_\mathrm{A}}{x_1} - \frac{Z_\mathrm{B}}{r_1} - \mathrm{i}\frac{\partial}{\partial t}\right] U_\lambda(q_1, t)$$

$$+ \sum_\mu \left[\langle U_\mu(q_2, t)| \frac{1}{|\mathbf{s}_1 - \mathbf{s}_2|} |U_\mu(q_2, t)\rangle U_\lambda(q_1, t) \right.$$

$$\left. - \langle U_\mu(q_2, t)| \frac{1}{|\mathbf{s}_1 - \mathbf{s}_2|} |U_\lambda(q_2, t)\rangle U_\mu(q_1, t) \right] = 0, \qquad \lambda + \alpha, \beta, \ldots, \nu. \tag{9.32}$$

These are the time-dependent Hartree–Fock equations. Expressing the spin orbitals in the form (9.16) and using the orthonormal property of the spin functions $\chi_{\frac{1}{2}m_\lambda}$

$$\langle \chi_{\frac{1}{2}m_\lambda} | \chi_{\frac{1}{2}m_\mu} \rangle = \delta_{m_\lambda m_\mu}, \tag{9.33}$$

the Hartree–Fock equations can be written in terms of the spatial orbitals $\phi_\lambda(\mathbf{s}_i, t)$. We have

$$\left[-\tfrac{1}{2}\nabla_1^2 - \frac{Z_\mathrm{A}}{x_1} - \frac{Z_\mathrm{B}}{r_1} - \mathrm{i}\frac{\partial}{\partial t}\right] \phi_\lambda(\mathbf{s}_1, t)$$

$$+ \sum_\mu \int \mathrm{d}\mathbf{s}_2\, \phi_\mu^*(\mathbf{s}_2, t) \frac{1}{|\mathbf{s}_1 - \mathbf{s}_2|} \{\phi_\mu(\mathbf{s}_2, t)\phi_\lambda(\mathbf{s}_1, t) - \delta_{m_\lambda m_\mu} \phi_\lambda(\mathbf{s}_2, t)\phi_\mu(\mathbf{s}_1, t)\} = 0,$$

$$\lambda = \alpha, \beta, \ldots, \nu. \tag{9.34}$$

We notice that there is no contribution to the sum over μ from the term $\mu = \lambda$. Taking this fact into account, it is useful to define a direct potential $V_\lambda(\mathbf{s}_1, t)$ together with an exchange kernel $K_\lambda(\mathbf{s}_1, \mathbf{s}_2, t)$ such that

$$V_\lambda(\mathbf{s}_1, t) = \sum_{\mu \neq \lambda} \int \frac{|\phi_\mu(\mathbf{s}_2, t)|^2}{|\mathbf{s}_1 - \mathbf{s}_2|} \mathrm{d}\mathbf{s}_2 \tag{9.35}$$

and

$$K_\lambda(\mathbf{s}_1, \mathbf{s}_2, t) = -\sum_{\mu \neq \lambda} \frac{\phi_\mu^*(\mathbf{s}_2, t)\phi_\mu(\mathbf{s}_1, t)}{|\mathbf{s}_1 - \mathbf{s}_2|} \delta_{m_\lambda m_\mu}. \tag{9.36}$$

In terms of V_λ and K_λ, the Hartree–Fock equations become

$$\mathrm{i}\frac{\partial}{\partial t}\phi_\lambda(\mathbf{s}_1, t) = \left[-\tfrac{1}{2}\nabla_1^2 - \frac{Z_\mathrm{A}}{x_1} - \frac{Z_\mathrm{B}}{r_1} + V_\lambda(\mathbf{s}_1, t)\right] \phi_\lambda(\mathbf{s}_1, t)$$

$$+ \int K_\lambda(\mathbf{s}_1, \mathbf{s}_2, t)\phi_\lambda(\mathbf{s}_2, t)\, \partial\mathbf{s}_2, \qquad \lambda = \alpha, \beta, \ldots, \nu. \tag{9.37}$$

The equations have the form of a one-electron two-centre system modified by the extra potentials V_λ and K_λ. The direct potential V_λ represents the average electrostatic potential of one electron due to the remaining $N - 1$ electrons and the non-local exchange potential takes into account exchange effects between one electron and the other $N - 1$ electrons.

Solving the time-dependent Hartree–Fock equations

The set of equations (9.37) is rather complicated since there are N coupled equations containing the non-linear terms V_λ and K_λ. However, the general methods we have discussed for the one-electron problem can still be employed. These can be divided into two main groups: methods in which the unknown functions $\phi_\lambda(\mathbf{s}, t)$ are expanded in a basis set, the expansion coefficients then satisfy coupled first-order *non-linear* equations; and methods in which a direct numerical solution, based on finite-difference or finite-element techniques, is attempted.

Examples of the expansion technique have been given for the two-electron system by Stich *et al.* (1983, 1985) and by Henne *et al.* (1986). These authors worked in the body-fixed frame and they employed an expansion by Hylleraas functions of the form described in Section 5.8, and given by equation (5.152). An alternative expansion in terms of Gaussian orbitals, with plane-wave translation functions, has been investigated by Gramlich *et al.* (1986), who also worked in the body-fixed coordinate frame. A convenient computer program based on a predictor–corrector method published by Shampine and Gordon (1975) was used by both sets of authors to solve the coupled equations for the amplitudes (expansion coefficients) $A_i(\mathbf{b}, t)$ which are of the non-linear form

$$\mathbf{A}(\mathbf{b}, t) = N^{-1}\{\mathbf{V} + \mathbf{E}(\mathbf{A})\}\mathbf{A}(\mathbf{b}, t), \tag{9.38}$$

where $\mathbf{N}$ and $\mathbf{V}$ are overlap and interaction matrices for the one-electron problem and the additional term $\mathbf{E}(\mathbf{A})$ represents the electron–electron interaction and depends on the unknown amplitudes $A_i(\mathbf{b}, t)$.

Finite-difference methods to solve the time-dependent Hartree–Fock equations for the two-electron system have been discussed by Kulander *et al.* (1982) and Sandhya Devi and Garcia (1983, 1984), and finite-element methods have been developed by Bottcher and his co-workers (Bottcher and Stayer 1987; Bottcher *et al.* 1988; Bottcher 1985).

To obtain transition probabilities for individual processes, such as elastic scattering, capture, and excitation, the Slater determinant approximating the wave function must be projected by the wave functions for each final state in the limit of large t. This gives rise to some difficulty because the probabilities $P(b, t)$ do not converge in the limit $t \to \infty$, but, as a result of the non-linear character of the equations, oscillate. The amplitude of these oscillations is rather small for total capture probabilities, but can be considerable for the probabilities for forming particular final states. It is likely that these oscillations become less important as the basis is increased towards completeness, or correspondingly if the accuracy of the finite-difference methods is increased. To allow for the oscillations, time averaged probabilities must be calculated, a computationally intensive procedure.

Applications

The time-dependent Hartree–Fock method offers an approach to systems with a large number of electrons for which the interaction of one electron with the remainder should be well represented by an average. Although this is well understood, most of the applications have been to systems containing only two electrons. In this case the smaller computational effort required makes it feasible to develop and test different approaches to the solution of the equations, for example the convergence of different kinds of basis set can be evaluated.

For the case of He^{2+} incident on $He(1s^2)$, results have been reported by Sandhya Devi and Garcia (1983, 1984), by Stich *et al.* (1985), and more recently by Schaudt *et al.* (1991). Since the system is in the singlet state the wave function must be of the form

$$\psi(q_1, q_2, t) = \phi(\mathbf{s}_1, t)\phi(\mathbf{s}_2, t)\frac{1}{\sqrt{2}}[\chi_{\frac{1}{2}\frac{1}{2}}(1)\chi_{\frac{1}{2}-\frac{1}{2}}(2) - \chi_{\frac{1}{2}\frac{1}{2}}(2)\chi_{\frac{1}{2}-\frac{1}{2}}(1)],$$

containing a single orbital $\phi(\mathbf{s}_1, t)$. There is no exchange term and, from (9.37), $\phi(\mathbf{s}_1)$ satisfies

$$\mathrm{i}\frac{\partial}{\partial t}\phi(\mathbf{s}_1, t) = \left[-\tfrac{1}{2}\nabla^2 - \frac{Z_A}{x_1} - \frac{Z_B}{r_1} + V(\mathbf{s}_1, t)\right]\phi(\mathbf{s}_1, t) \tag{9.40}$$

where

$$V(\mathbf{s}_1, t) = \int \frac{\rho(\mathbf{s}_2, t)}{|\mathbf{s}_2 - \mathbf{s}_1|}\,\mathrm{d}\mathbf{s}_2 \tag{9.41}$$

with

$$\rho(\mathbf{s}_2, t) = |\phi(\mathbf{s}_2, t)|^2. \tag{9.42}$$

It should be noted that it is usually more efficient to calculate $V(\mathbf{s}, t)$ from Poisson's equation

$$\frac{1}{4\pi}\nabla^2 V(\mathbf{s}, t) = -\rho(\mathbf{s}, t) \tag{9.43}$$

than to evaluate the integrals (9.41).

The computed cross sections for the capture of one or two electrons by He^{2+} from He are shown in Figs 9.2 and 9.3. The results are in rough agreement with those computed by Harel and Salin (1980) using the OEDM expansion, and also with experiment. A further example, this time of a

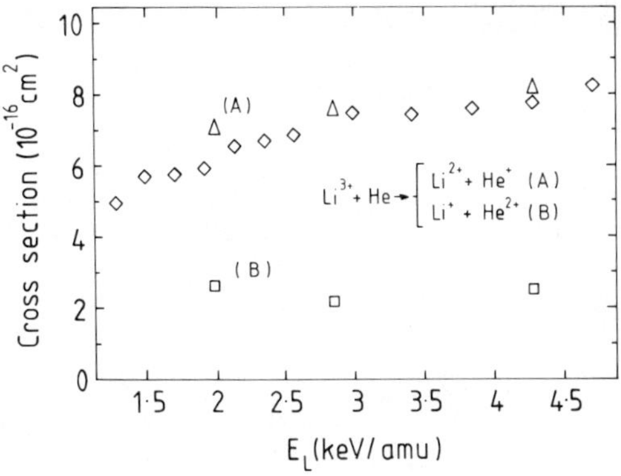

Fig. 9.5. Total cross sections for charge exchange in collisions between Li^{3+} ions and helium. Experimental data for one electron capture: ◇ Wirkner-Bott *et al.* (1981). Theoretical cross sections employing the time-dependent Hartree–Fock approximation from Stich *et al.* (1985): △, △ one electron capture; □, □ two electron capture.

heteronuclear collision, is single and double electron capture by Li^{3+}ions from helium. The calculation of Stich *et al.* (1985) at the three energies $E_L = 2$, 2.86, and 4.29 keV/a.m.u. predict single electron capture cross sections which agree well with the measured cross section as shown in Fig. 9.5. Although these results appear reasonably encouraging, the limitations of the TDHF method are exposed when one consider the incident channel $He^+(1s) + He^+(1s)$. In this case, we either have (a) $\chi_{\frac{1}{2}m}(1) = \pm\frac{1}{2}$ and $\chi_{\frac{1}{2}m}(2) = \pm\frac{1}{2}$, or (b) $\chi_{\frac{1}{2}m}(1) = \pm\frac{1}{2}$ and $\chi_{\frac{1}{2}m}(2) = \mp\frac{1}{2}$. The first set (a) gives rise to eigenfunctions of spin with $S = 1$, $M_s = \pm 1$, and

$$\psi^1(\mathbf{s}_1, \mathbf{s}_2, t) = \frac{1}{\sqrt{2}}[\phi_1^1(\mathbf{s}_1, t)\phi_2^1(\mathbf{s}_2, t) - \phi_1^1(\mathbf{s}_2, t)\phi_2^1(\mathbf{s}_1, t)]\chi_{\frac{1}{2}\pm\frac{1}{2}}(1)\chi_{\frac{1}{2}\pm\frac{1}{2}}(2). \tag{9.44}$$

The second set of spin functions (b) belongs to $M_s = 0$ and we have

$$\psi^0(\mathbf{s}_1, \mathbf{s}_2, t) = \frac{1}{\sqrt{2}}[\phi_1^0(\mathbf{s}_1, t)\phi_2^0(\mathbf{s}_2, t)\chi_{\frac{1}{2}\pm\frac{1}{2}}(1)\chi_{\frac{1}{2}\mp\frac{1}{2}}(2) - \phi_1^0(\mathbf{s}_2, t)\phi_2^0(\mathbf{s}_1, t)\chi_{\frac{1}{2}\pm\frac{1}{2}}(2)\chi_{\frac{1}{2}\mp\frac{1}{2}}(1)]. \tag{9.45}$$

The spin functions represent equal mixtures of the two states $S = 0$, $M_S = 0$ and $S = 1$, $M_S = 0$. The transition to a particular final state can be computed by assuming the four spin states are equally populated, so that

$$\sigma = \tfrac{1}{2}\sigma(M_S = 0) + \tfrac{1}{2}\sigma(M_S = 1). \tag{9.46}$$

This procedure is actually incorrect for this system, since the two states with

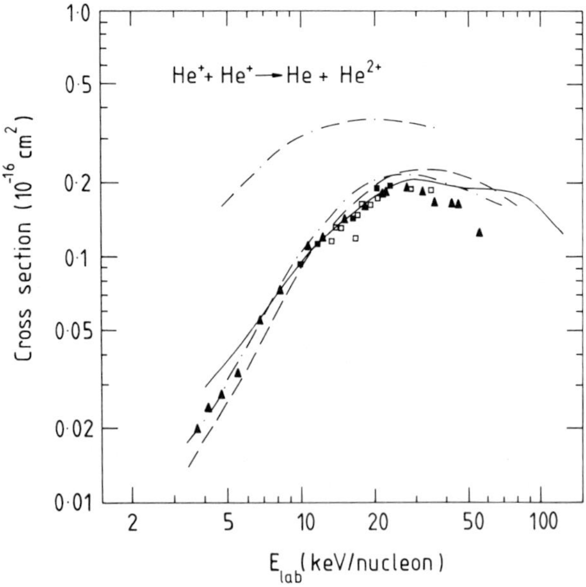

Fig. 9.6. Total cross sections for charge exchange between He^+ ions. Experimental data: ▲, □ Melchert *et al.* (1987); Peart *et al.* (1983). Theoretical cross sections: ——·—— time-dependent Hartree–Fock model, Henne *et al.* (1986); ----- MO basis, Kimura (1988); —·—·— AO^+ basis, Fritsch and Lin (1987); ——— Gaussian pseudostate basis, Gramlich *et al.* (1989).

$S = 0$ and $S = 1$ should propagate independently and we should have $\sigma = \frac{1}{4}\sigma(S = 0) + \frac{3}{4}\sigma(S = 1)$ rather than (9.46). However, the use of the wave function (9.45) mixes these two spin states, and this difficulty is inherent in the approximation in which the wave function is represented by a single determinant. Some computed cross sections for this system (Henne *et al.* 1986) are shown in Fig. 9.6. Not surprisingly the TDHF results are very poor compared with those of the more realistic MO or AO^+ (two-electron) expansions, in which singlet and triplet wave functions of the correct symmetry are employed.

As an example of the application of the TDHF method to a many-electron system, it is interesting to cite the calculation of Bottcher (1985) on the ten-electron system

$$Fe^{9+} + Ne \rightarrow Fe^{(9-m)+} + Ne^{n+} + (n - m)e^- \qquad (9.47)$$

at energies of 25 and 100 keV/a.m.u. In this work, the direct potential V_λ was replaced by V^D, an average over all the states λ, and a simple local approximation was employed for the exchange potential, namely $V^E = -CV^D$ where C is a constant. With these simplifications it proved possible to calculate total probabilities for capture and ionization as a function of impact parameter.

9.4 CROSS SECTIONS WITHIN THE INDEPENDENT PARTICLE MODEL

In an independent particle model the wave function is expressible as an antisymmetrical product of single-particle spin orbitals, as in the expressions (9.17) and (9.21). Various approximations can be employed to calculate the individual spin orbitals, for instance the time-dependent Hartree–Fock approximation, or by solving the time-dependent Schrödinger equation for a single electron moving in effective predetermined potentials V_{Ae}, V_{Be}, as described in Chapters 3 and 5.

To obtain transition probabilities for excitation or capture, the approximate wave function of the form (9.21) must be projected in turn by each possible final state of the system in the limit $t \to \infty$. Within the independent particle model, the final states of the system can also be represented by Slater determinants composed of atomic orbitals centred on either the target nucleus B or the projectile nucleus A. The spin orbitals centred on A and B will be denoted by $\omega_\lambda^{\mathrm{A}}(q_j, t)$ and $\omega_\lambda^{\mathrm{B}}(q_i, t)$, respectively, where

$$\begin{aligned}\omega_\lambda^{\mathrm{A}}(q_j, t) &= \mathrm{X}_{nlm}^{\mathrm{A}}(\mathbf{s}_j, t)\chi_{\frac{1}{2}m_s}(j)\\ \omega_\lambda^{\mathrm{B}}(q_j, t) &= \Phi_{nlm}^{B}(\mathbf{s}_j, t)\chi_{\frac{1}{2}m_s}(j).\end{aligned} \tag{9.48}$$

The functions $\mathrm{X}_{nlm}^{\mathrm{A}}$, Φ_{nlm}^{B} are atomic orbitals modified by plane-wave translational factors, as defined in (3.175) and (3.170); however, the contracted notation of Chapter 3 has been expanded to display the three quantum numbers *nlm* explicitly. The index λ denotes the set of quantum numbers *nlm* and m_s and so defines each atomic spin orbital completely. The spin orbitals are taken to be orthonormal

$$\langle \omega_\lambda^{\mathrm{A}} | \omega_\mu^{\mathrm{A}} \rangle = \delta_{\lambda\mu}; \qquad \langle \omega_\lambda^{\mathrm{B}} | \omega_\mu^{\mathrm{B}} \rangle = \delta_{\lambda\mu} \tag{9.49}$$

and the overlap integral between atomic orbitals centred on A and B vanishes in the limit $|t| \to \infty$

$$\lim_{|t| \to \infty} \langle \omega_\lambda^{\mathrm{A}} | \omega_\mu^{\mathrm{B}} \rangle = 0. \tag{9.50}$$

For simplicity let us consider a two-particle system. At large separations, $R \to \infty$, $|t| \to \infty$, there are three types of asymptotic state: those in which both electrons are attached to A, those in which both electrons are attached to B, and those in which one electron is attached to A and one to B. The three arrangements are described by the antisymmetric wave functions

$$\begin{aligned}\Gamma^{AA}_{\lambda\mu}(q_1, q_2, t) &= 2^{-\frac{1}{2}}[\omega^{A}_{\lambda}(q_1, t)\omega^{A}_{\mu}(q_2, t) - \omega^{A}_{\lambda}(q_2, t)\omega^{A}_{\mu}(q_2, t)]\\ \Gamma^{BB}_{\lambda\mu}(q_1, q_2, t) &= 2^{-\frac{1}{2}}[\omega^{B}_{\lambda}(q_1, t)\omega^{B}_{\mu}(q_2, t) - \omega^{B}_{\lambda}(q_2, t)\omega^{B}_{\mu}(q_2, t)] \qquad (9.51)\\ \Gamma^{AB}_{\lambda\mu}(q_1, q_2, t) &= 2^{-\frac{1}{2}}[\omega^{A}_{\lambda}(q_1, t)\omega^{B}_{\mu}(q_2, t) - \omega^{A}_{\lambda}(q_2, t)\omega^{B}_{\mu}(q_2, t)].\end{aligned}$$

Since the overlap matrix (9.50) vanishes as $|t| \to \infty$, the three sets of functions $\Gamma^{AA}_{\lambda\mu}$, $\Gamma^{BB}_{\lambda\mu}$, and $\Gamma^{AB}_{\lambda\mu}$ are mutually orthogonal in the asymptotic region.

The probability amplitude $A^{AA}_{\lambda\mu}(b)$ for finding both electrons attached to A in the final state in the levels λ and μ is found by projecting the approximate wave function ψ given in the form (9.17) or (9.21) by $\Gamma^{AA}_{\lambda\mu}$, in the limit, $t \to \infty$

$$A^{AA}_{\lambda\mu}(\mathbf{b}) = \lim_{t\to\infty} \langle \Gamma^{AA}_{\lambda\mu} | \psi \rangle. \qquad (9.52)$$

The probability amplitudes $A^{BB}_{\lambda\mu}(\mathbf{b})$ and $A^{AB}_{\lambda\mu}(\mathbf{b})$ for the other arrangements are found similarly.

Frequently the number of final states of importance for a many-electron system is very large, so that to obtain total cross sections for capture may be laborious. To avoid this problem, direct methods to obtain the *global probabilities* for capture or for excitation have been developed (Bottcher 1982; Reading and Ford 1980; Lüdde and Dreizler 1983).

Global probabilities

In the approach of Lüdde and Dreizler use is made of the fact that wave functions for the jth particle centred about A in the asymptotic region are effectively zero outside the half-space defined by the inequality $x_j < r_j$, where, as usual, $\mathbf{x}_j$ is the position vector of the electron with respect to A and $\mathbf{r}_j$ is the corresponding position vector with respect to B. Integrals over this half-space are denoted by

$$\langle f | g \rangle_A = \int_A f^*(\mathbf{s}^j) g(\mathbf{s}_j)\, d\mathbf{s}_j. \qquad (9.53)$$

Similarly wave functions centred about B in the asymptotic region are confined to the half-space $r_j < x_j$, and integrals over this half-space are denoted by

$$\langle f | g \rangle_B = \int_B f^*(\mathbf{s}_j) g(\mathbf{s}_j)\, d\mathbf{s}_j. \qquad (9.54)$$

In the same way multiple integrals over the half-spaces A and B can be defined so that, for example,

$$\langle f|g\rangle_{\mathrm{ABA}\ldots} = \int_{\mathrm{A}} \mathrm{d}\mathbf{s}_1 \int_{\mathrm{B}} \mathrm{d}\mathbf{s}_2 \int_{\mathrm{A}} \mathrm{d}\mathbf{s}_3 \ldots f^*(\mathbf{s}_1, \mathbf{s}_2, \mathbf{s}_3 \ldots) g(\mathbf{s}_1, \mathbf{s}_2, \mathbf{s}_3 \ldots). \quad (9.55)$$

Now consider a two-particle system described by the approximate wave function $\psi(q_1, q_2, t)$. Let us assume that ionization is unimportant so that asymptotically the electrons are either bound to A or to B. In that case, in the asymptotic region ψ can be represented as a linear combination of the wave functions $\Gamma^{\mathrm{AA}}_{\lambda\mu}$, $\Gamma^{\mathrm{BB}}_{\lambda\mu}$, and $\Gamma^{\mathrm{AB}}_{\lambda\mu}$ defined by (9.51)

$$\psi(q_1, q_2, t) \underset{t\to\infty}{\sim} \sum_{\lambda,\mu} \{A^{\mathrm{AA}}_{\lambda\mu}(\mathbf{b})\Gamma^{\mathrm{AA}}_{\lambda\mu}(q_1, q_2, t) + A^{\mathrm{BB}}_{\lambda\mu}(\mathbf{b})\Gamma^{\mathrm{BB}}_{\lambda\mu}(q_1, q_2, t)$$

$$+ A^{\mathrm{AB}}_{\lambda\mu}(\mathbf{b})\Gamma^{\mathrm{AB}}_{\lambda\mu}(q_1, q_2, t)\}. \quad (9.56)$$

Using the expression (9.56), we see that

$$\lim_{t\to\infty} \langle\psi|\psi\rangle_{\mathrm{AA}} = \sum_{\lambda,\mu} |A^{\mathrm{AA}}_{\lambda\mu}(\mathbf{b})|^2, \quad (9.57)$$

since both $\Gamma^{\mathrm{BB}}_{\lambda\mu}$ and $\Gamma^{\mathrm{AB}}_{\lambda\mu}$ vanish when both $\mathbf{s}_1$ and $\mathbf{s}_2$ lie within the half-space A. The total probability, $P_{20}(\mathbf{b})$, for finding both electrons centred on A after the collision is thus

$$P_{20}(\mathbf{b}) = \sum_{\lambda\mu} |A^{\mathrm{AA}}_{\lambda\mu}(\mathbf{b})|^2 = \lim_{t\to\infty} \langle\psi|\psi\rangle_{\mathrm{AA}}. \quad (9.58)$$

Similarly the probability for finding both electrons centred about B is

$$P_{02}(\mathbf{b}) = \lim_{t\to\infty} \langle\psi|\psi\rangle_{\mathrm{BB}}. \quad (9.59)$$

The calculation of $P_{11}(\mathbf{b})$, the probability of finding one electron bound to A and one to B, is slightly different. We have

$$\begin{aligned}
\lim_{t\to\infty} \langle\psi|\psi\rangle_{\mathrm{AB}} &= \sum_{\mathrm{spin}} \int_{\mathrm{A}} \mathrm{d}\mathbf{s}_1 \int_{\mathrm{B}} \mathrm{d}\mathbf{s}_2 \sum_{\lambda\mu}\sum_{\lambda'\mu'} [A^{\mathrm{AB}}_{\lambda\mu}(\mathbf{b})]^* A^{\mathrm{AB}}_{\lambda'\mu'}(\mathbf{b}) \\
&\quad \times \tfrac{1}{2}\{\omega^{\mathrm{A}}_{\lambda}(q_1, t)\omega^{\mathrm{B}}_{\mu}(q_2, t) - \omega^{\mathrm{A}}_{\lambda}(q_2, t)\omega^{\mathrm{B}}_{\mu}(q_1, t)\}^* \\
&\quad \times \{\omega^{\mathrm{A}}_{\lambda'}(q_1, t)\omega^{\mathrm{B}}_{\mu'}(q_2, t) - \omega^{\mathrm{A}}_{\lambda'}(q_2, t)\omega^{\mathrm{B}}_{\mu'}(q_1, t)\} \\
&= \tfrac{1}{2} \sum_{\mathrm{spin}} \int_{\mathrm{A}} \mathrm{d}\mathbf{s}_1 \int_{\mathrm{B}} \mathrm{d}\mathbf{s}_2\, [A^{\mathrm{AB}}_{\lambda\mu}(\mathbf{b})]^* A^{\mathrm{AB}}_{\lambda'\mu'}(\mathbf{b}) \{\omega^{\mathrm{A}}_{\lambda}(q_1, t)\omega^{\mathrm{B}}_{\mu}(q_2, t)\}^* \\
&\quad \times \omega^{\mathrm{A}}_{\lambda'}(q_1, t)\omega^{\mathrm{B}}_{\mu'}(q_2, t) \\
&= \tfrac{1}{2} \sum_{\lambda\mu} |A^{\mathrm{AB}}_{\lambda\mu}(\mathbf{b})|.
\end{aligned} \quad (9.60)$$

It follows that

$$P_{11}(\mathbf{b}) = \sum_{\lambda\mu} |A^{AB}_{\lambda\mu}(\mathbf{b})|^2 = \lim_{t\to\infty} 2\langle\psi|\psi\rangle_{AB}, \tag{9.61}$$

and it is also easy to see that

$$\langle\psi|\psi\rangle_{AB} = \langle\psi|\psi\rangle_{BA}. \tag{9.62}$$

As an example (Lüdde and Dreizler 1983), we may take the case of scattering of a fully stripped ion by ground-state helium. In the time-dependent Hartree–Fock approximation, the wave function $\psi(q_1, q_2, t)$ is given by (9.39) in terms of a single spatial orbital $\phi(\mathbf{s}_j, t)$. The probability for elastic scattering or excitation $P_{02}(\mathbf{b})$ is given by

$$P_{02}(\mathbf{b}) = \lim_{t\to\infty} \langle\psi|\psi\rangle_{BB} = \lim_{t\to\infty} \left[\int_B \mathrm{d}\mathbf{s}\, |\phi(\mathbf{s}, t)|^2\right]^2, \tag{9.63}$$

while for single and double electron capture we have

$$P_{11}(\mathbf{b}) = \lim_{t\to\infty} 2\langle\psi|\psi\rangle_{AB} = \lim_{t\to\infty} 2\left[\int_A \mathrm{d}\mathbf{s}\, |\phi(\mathbf{s}, t)|^2 \int_B \mathrm{d}\mathbf{s}\, |\phi(\mathbf{s}, t)|^2\right], \tag{9.64}$$

and

$$P_{20}(\mathbf{b}) = \lim_{t\to\infty} \langle\psi|\psi\rangle_{AB} = \lim_{t\to\infty} \left[\int_A \mathrm{d}\mathbf{s}\, |\phi(\mathbf{s}, t)|^2\right]^2. \tag{9.65}$$

We note the interesting relationships

$$P_{11} = 2[P_{20}P_{02}]^{\frac{1}{2}} \tag{9.66}$$

and

$$P_{02} = (1 - P_{20}^{\frac{1}{2}})^2. \tag{9.67}$$

The generalization of the formulae (9.63), (9.64), and (9.65) to an N-electron system has been given by Lüdde and Dreizler. They show that the probability for finding n electrons bound to the target B and $N - n$ electrons bound to the projectile A after the collision is given by

$$P_{N-n,n}(\mathbf{b}) = {}_N C_n \langle\psi|\psi\rangle_{A^{N-n}B^n} \tag{9.68}$$

where ${}_N C_n$ is the binomial coefficient.

Inclusive reactions

Frequently experiments are concerned with the cross section for a reaction (excitation, capture, or ionization) in which one or more electrons are transferred to specific final states, while transitions undergone by the remaining electrons are not observed. The probabilities for such *inclusive reactions* can be obtained by projecting the approximate wave function on to all asymptotic states of importance and summing over all the unobserved transitions. As with the calculation of global probabilities, more direct procedures are possible and these have been considered by a number of authors (McGuire and Weaver 1977; Reading 1973; Reading and Ford 1980). We shall follow the approach of McGuire and Weaver and consider the capture by a fully stripped ion of a single electron from an atomic shell containing N electrons. Let $P(\mathbf{b})$ be the probability of capture calculated in the single active electron model. It follows that $\{1 - P(\mathbf{b})\}$ is the probability that an electron is not captured during a collision. If one electron is captured and $N - 1$ electrons are not captured, from a shell of N equivalent electrons, the probability for single electron capture, without observation of the transitions undergone by the remaining $N - 1$ electrons is

$$P(\mathbf{b}) = NP(\mathbf{b})\{1 - P(\mathbf{b})\}^{N-1}, \tag{9.69}$$

where the factor N arises because there are N ways of choosing the captured electron.

Arguing along similar lines, the creation of q vacancies in the N-electron shell by excitation, capture, or ionization can be expressed as

$$P_{\mathrm{v}}(\mathbf{b}) = {}_{N}C_{\mathrm{q}}[P_{\mathrm{v}}(\mathbf{b})]^{q}\{1 - P_{\mathrm{v}}(\mathbf{b})\}^{N-q}, \tag{9.70}$$

where $P_{\mathrm{v}}(\mathbf{b})$ is the probability for the creation of a vacancy in the single electron model and ${}_{N}C_{q}$ represents the number of ways in which the q electrons can be chosen from the N in the shell.

This simple model, although often employed, is not quite sufficient for considering the multiple capture of electrons, or for capture by a partially ionized system, because it takes no acount of the restrictions imposed by the Pauli principle. The extension of the model to take these restrictions into account has been discussed by Becker *et al.* (1984) and by Lüdde and Dreizler (1985).

Limitations of the independent particle model

Studies of multiple capture and ionization in ion–atom collisions have shown that, with some exceptions, the independent particle model can provide a useful guide to cross sections. (See, for example, Gayet (1989).) The most favourable conditions under which the independent particle model can be expected to be accurate is when the electron–electron interaction is small compared with the interaction between the electrons and the nucleus, which occurs, for instance, in capture or ionization from the inner shells of atoms or ions which have a large nuclear charge. It follows that the severest test of the model is presented by systems in which helium is the target atom. We have already seen in Chapter 6 that for *single* electron excitation, capture, and ionization the one-active-electron model (which is, of course, an independent particle model) usually provides cross sections of reasonable accuracy. In particular, studies of the $H^+ + He$ system in this approximation (Jain *et al.* 1987*a, b*, 1988; Shingal and Lin 1991*b*) predict cross sections which are close, but not identical, to those obtained from a specifically two-electron formulation allowing fully for correlation (Slim *et al.* 1990, 1991*a, b*). In the standard one-electron model the active electron moves in an effective potential chosen so that the binding energy of the ground state and first few excited states are reproduced as accurately as possible. For a helium target such a potential is (Bransden *et al.* 1984, see also Shingal 1988*a, b*)

$$V_E(r) = -\frac{1.0}{r} - (1.0 + 0.4143r)\exp(-2.499r), \tag{9.71}$$

which provides an ionization potential of 0.9 a.u., compared with the experimental value of 0.904 a.u. Shingal (1990*a, b*) and Shingal and Lin (1991*a*) have employed the potential V_E to calculate probabilities in the one-active-electron approximation employing the standard coupled channel model with a two-centre basis of states and pseudostates and plane-wave translation factors. Within this model probability is conserved so that

$$P_{EL}(b) + P_E(b) + P_C(b) + P_I(b) = 1, \tag{9.72}$$

where P_{EL}, P_E, P_C, and P_I are the calculated probabilities for elastic scattering, excitation, capture, and ionization, respectively. To obtain the probabilities within the independent particle model for a helium target containing two equivalent electrons we note that

$$[P_{EL}(b) + P_E(b) + P_C(b) + P_I(b)]^2 = 1 \tag{9.73}$$

and identify the probabilities for single electron capture, ionization, or excitation as $\mathscr{P}_C(b) = 2P_{EL}(b)P_C(b)$, $\mathscr{P}_I(b) = 2P_{EL}(b)P_I(b)$, and $\mathscr{P}_E(b) = 2P_{EL}(b)P_E(b)$, respectively. Correspondingly the probabilities for double capture, ionization, or excitation are $\mathscr{P}_{2C}(b) = P_C^2(b)$, $\mathscr{P}_{2I}(b) = P_I^2(b)$, $\mathscr{P}_{2E}(b) = P_E^2(b)$, while $\mathscr{P}_{CI}(b) = 2P_C(b)P_I(b)$, $\mathscr{P}_{IE}(b) = 2P_I(b)P_E(b)$, and $\mathscr{P}_{CE}(b) = 2P_C(b)P_E(b)$ represent the probabilities for simultaneous capture and ionization (transfer ionization), ionization and excitation, and capture and excitation (transfer excitation), respectively. Within this model Shingal (1990*a, b*) and Shingal and Lin (1991*a*) have calculated cross sections for both single and double electron transitions with H^+, He^{2+}, Li^{3+}, C^{6+}, and F^{9+} projectiles incident on helium in the ground state, in various energy intervals ranging from $E_L = 20$–100 keV/a.m.u. for H^+ projectiles to $E_L = 250$–1500 keV/a.m.u. for F^{9+} projectiles. The overall agreement with the experimental data for the single electron processes was good, but in contrast, the cross sections for the double electron transitions agreed poorly with the data, and it is clear that the independent particle model in the form outlined above is defective.

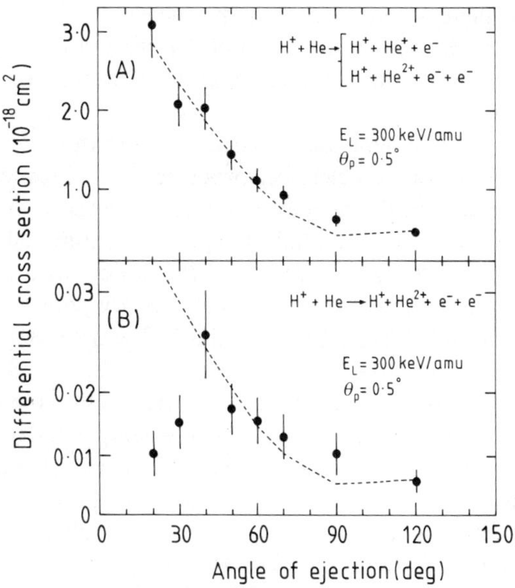

Fig. 9.7. Differential cross sections for the ejection of electrons in proton collisions with helium at $E_L = 300$ keV/a.m.u. at a proton scattering angle of $\theta_p = 0.5°$ (after Skogvall and Schiwietz (1990)): (a) cross sections for the sum of single and double ionization; (b) cross sections for double ionisation. Experimental data: ⧫⧫ Skogvall and Schiwietz (1990). Theoretical cross sections: independent particle + coupled channel model, Schiwietz (1990).

The failure of the independent particle model has also been noted by Skogvall and Schiwietz (1990) who measured absolute differential cross sections for single and double ionization of helium by proton impact at 300 keV. The data for single ionization was well represented by coupled channel calculations using a large single-centred basis (Schiwietz 1990), but the cross sections for double ionization predicted by the independent particle model did not agree with the measured data. Irrespective of which method is used to calculate the probability $P_I(b)$, the single and double ionization probabilities are given in the independent particle model by $\mathscr{P}_I(b) = 2P_{EL}(b)$, $P_I(b) \approx 2P_I(b)$ and $\mathscr{P}_{2I}(b) = P_I^2(b)$ and this implies that the angular distributions of the ejected electrons are identical, apart from a constant factor, for both single and double ionization. The angular distributions for the sum of single and double ionization and of double ionization are illustrated in Fig. 9.7, where it is seen that the two measured distributions behave quite differently for ejection angles of less than 40°. Also included in the figure are the results obtained from the independent particle model in which the single electron ionization probability $P_I(\mathbf{b})$ has been calculated from a coupled channel calculation using a large pseudostate basis. It is seen that good agreement is obtained with the measured total ionization cross section and also with the measured double ionization cross section at large angles of ejection, but the model fails to describe double ionization at small angles of ejection.

The failure of the independent particle approximation to provide a complete description of two-electron processes in helium can, in part, be ascribed to the change in the field experienced by the second electron on the ionization of the first; the potential changes from some effective potential such as V_E given by (9.71) to a pure Coulomb potential $V_C = -2/r$. One consequence of this is that if the two electrons are considered to move in the field V_E, the double ionization potential is $2 \times 0.9 = 1.8$ a.u., which can be contrasted with the true value of $0.9 + 2 = 2.9$ a.u. Since, to obtain reasonable cross sections, the energies of the levels concerned must be represented accurately, the inaccuracy of the simple independent particle model for double ionization or capture is not particularly surprising.

While staying within the general framework of the independent particle model, it is possible to find a better description of double electron processes. One way of doing this is to continue to use a model in which both electrons move in the same potential, but to adjust the potential so that the correct double ionization potential is obtained. This is the course followed by Shingal and Lin (1991*a*), who have introduced the effective Coulomb potential $V_m(r) = -Z_m/r$, with $Z_m = 1.703$ so that the ionization potential per electron is 1.45.

Of course, if this potential is employed to calculate cross sections for single electron transitions erroneous results will be obtained, but when cross sections for double electron transitions are calculated reasonable agreement with the experimental measurements is found. This is illustrated in Fig. 9.8 for the case of the Li^{3+} + He system.

The use of different effective potentials to describe single and double electron transitions is, perhaps, a little unsatisfactory. A different model which also produces cross sections for double transitions of reasonable accuracy has been discussed by Shingal (1990*a*, *b*) and, in a different form, by Deb and Crothers (1990). In this model the two-electron processes are viewed as occurring through successive independent events. For example in double capture the first electron is captured with a probability $P_C(b)$, which in the model of Shingal is calculated from the one-active-electron model with the potential V_E. Subsequently the ion formed in this way $(A + e^-)$ interacts with the residual $He^+(1s)$ ion and the second electron is captured. The second electron interacts with the helium nucleus with the Coulomb interaction $V_c = -2/r$ and with the screened nucleus A with an effective potential. If the probability for capture in the second event is $\bar{P}_C(b)$ the double capture probability is $\mathscr{P}_{2C} = P_C(b)\bar{P}_C(b)$. In the model of Deb and Crothers, which was applied to double ionization of helium by He^{2+}, the probability of the first event was calculated from the continuum distorted wave (CDW) model,

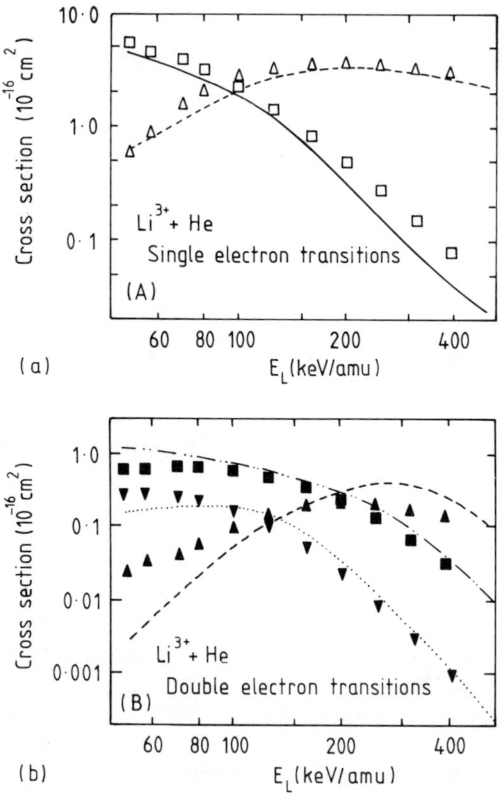

Fig. 9.8. Cross sections for Li^{3+} + He interactions (after Shingal and Lin (1991*a*)): (a) single electron transitions. (1) Ionization: ----- theoretical cross section (Shingal and Lin 1991*a*) (potential 9.71) △△△ experimental data (Shah and Gilbody 1985). (2) Capture: ——— theoretical cross section (Shingal and Lin 1991*a*) (potential 9.71) □□□ experimental data (Shah and Gilbody 1985). (b) Double electron transitions. (1) Double ionization: — — — theoretical cross section (Shingal and Lin 1991*a*) (modified potential) ▲▲▲ experimental data (Shah and Gilbody 1985). (2) Double capture: ······ theoretical cross section (Shingal and Lin 1991*a*) (modified potential) ▼▼▼ experimental data. (3) Transfer ionization: —··—·· theoretical cross section (Shingal and Lin 1991*a*) (modified potential) ■■■ experimental data (Shah and Gilbody 1985).

to be discussed in the next chapter, employing a correlated two-electron description of helium. The second event is viewed as ionization of $He^{+}(1s)$ by He^{2+} and again was calculated in a version of the CDW approximation. Very good agreement with the measurements was obtained at high energies ($E_L > 200$ keV/a.m.u.), for which the CDW approximation is expected to be accurate.

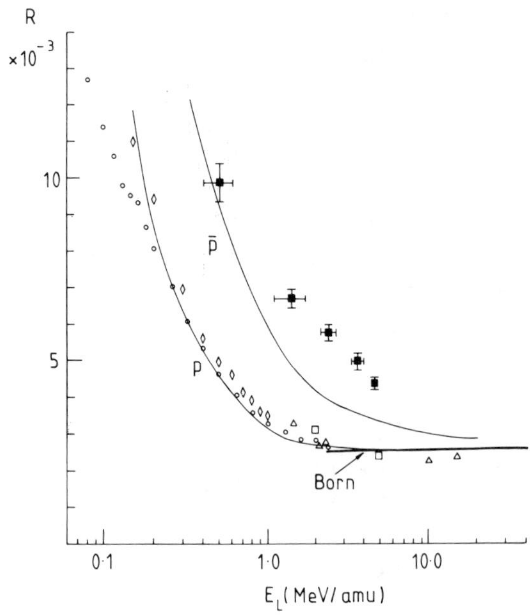

Fig. 9.9. The ratio R between double and single ionization of He by p and p̄ impact. Experimental data: p̄: ■ Andersen *et al.* (1986); p: □ Andersen *et al.* (1986); △ Knudsen *et al.* (1984); ○ Shah and Gilbody (1985); ◇ Puckett and Martin (1970). Theoretical data: ——— from the forced impulse approximation model scaled by a factor of 1.35 (Reading and Ford 1987*a*, *b*); ——— first-order Born approximation (Ford and Reading 1988).

The failure of the independent particle approximation for double ionization of helium by fully stripped ions at high energies has been recognized for some time. At the higher energies, it might be supposed that the first-order Born approximation would be accurate for which the probability of single ionization is proportional to Z_A^2, where Z_A is the charge on the incident ion. The independent particle approximation then predicts that the probability for double ionization should be proportional to Z_A^4, whereas in fact the dependence of the cross section at high energies on Z_A is Z_A^2. Another sign of the breakdown of the independent particle approximation is that while the single ionization cross section of He by proton impact is not very different from that due to antiproton impact at energies above ~100 keV/a.m.u., the corresponding double ionization cross sections differ widely, even at E_L = 1 MeV/a.m.u. This is illustrated in Fig. 9.9 where the ratio of double to single ionization of helium by proton and antiproton impact is plotted as a function of energy. It is difficult to see how these results can be reconciled with a modified independent particle model, and models allowing more fully for correlation would seem to be indicated.

9.5 TWO-ELECTRON TRANSITIONS

In Chapters 4, 5, and 6 we saw that the coupled channel description of many-electron systems can be extended to systems of two (or more) electrons, provided the expansion of the wave function is in terms of two-electron basis sets which diagonalize the molecular, or two-centre atomic, Hamiltonians. Such basis sets allow fully for correlation, and enable the calculation of both single and double electron transitions. In this section some studies of double electron transitions are described which either allow a simplification of the coupled channel method or which give some insight into the specific mechanisms leading to two-electron transitions.

Double excitation and double ionization at high energies

At energies which are sufficiently high for charge exchange to be unimportant the excitation and ionization of a two-electron ion can be described by a coupled channel model obtained from an expansion in terms of a two-electron basis centred on the target nucleus. An interesting study of the mechanisms for excitation of doubly excited states of helium by fully stripped ions using a single-centred basis of this kind has been carried out by Fritsch and Lin (1990). By diagonalizing the helium Hamiltonian on a basis of products of single electron hydrogenic functions with an effective charge, a set of correlated helium wave functions was obtained. These functions were then employed to form a six-state basis for the dynamical calculation, which included the ground state $1s^2\,{}^1S$, the two singly excited states $1s\,2s\,{}^1S^e$, $1s\,2p\,{}^1P^0$, and the four doubly excited states $2s^2\,{}^1S^e$, $2p^2\,{}^1D^e$, $2s\,2p\,{}^1P^0$, and $2p^2\,{}^1S^e$. From the coupled equations, formed by using this basis, cross sections for excitation by fully stripped ions of charge Z_A can be computed. It should be noted that since the doubly excited states decay by autoionization, comparison with experiment is obtained indirectly by calculating the spectrum of ejected electrons which contains a contribution both from direct ionization and from autoionization. An interesting question which can be answered without appeal to experiment is whether direct excitation of the doubly excited states is more or less important than a two-step mechanism in which the singly excited levels form intermediate states. To answer this question Fritsch and Lin compared the cross sections calculated from the full model with those obtained by omitting the singly excited $1s\,2s\,{}^1S^e$ and $1s\,2p\,{}^1P^0$ states from the basis. A selection of their results for projectiles of both positive and negative charge is shown in Table 9.2. For small values, $Z_A \leq 2$, of the charge on the projectile the cross sections obtained by omitting the singly excited states from the basis are close to those obtained for the full basis, the largest deviation being for the ${}^1D^e$ level. In contrast, as Z_A increases the two sets of cross sections become very different. It can be concluded that for small Z_A the single step process is of primary importance, but as Z_A increases

Table 9.2. Cross sections for the excitation of doubly excited states of helium at $E_L = 1.5$ MeV/a.m.u. (in units of 10^{-20} cm^2). A is the coupled channel calculation of Fritsch and Lin (1990) using a basis of two singly excited and four doubly excited states together with the ground state of helium; B is as in A, but with the omission of the singly excited states from the basis.

	$2s^2\ ^1S^e$		$2p^2\ ^1D^e$		$2s\,2p\ ^1P^0$		$2p^2\ ^1S^e$	
Z_A	A	B	A	B	A	B	A	B
1	0.74	0.71	0.48	0.34	3.0	3.0	0.45	0.39
2	3.2	2.7	3.6	1.7	12.6	11.4	22.0	15.0
6	51.6	20.6	156.0	35.0	162.0	68.6	48.6	10.8
−1	0.73	0.74	0.27	0.22	3.3	3.2	0.41	0.40

the two-step mechanism becomes more and more significant. Further studies show that for a given Z_A the two-step mechanism is of greater importance at low energies and decreases in importance as the energy increases. This analysis demonstrates conclusively the importance of correlation, since with a basis of uncorrelated wave functions, a single step mechanism is not possible.

By extending the basis to include both singly and doubly excited pseudostates, cross sections for the ionization of one or of two electrons can, in principle, be calculated. A major difficulty is that because the pseudostates are not exact eigenstates of the helium Hamiltonian, the amplitude for excitation of a given two-electron pseudostate contains contributions to both single and double ionization. Because the probability of double ionization compared with single ionization is small (see Fig. 9.9) the projection of the double ionization contribution from the amplitudes for pseudostate excitation is not straightforward. These difficulties have been carefully analysed by Reading and Ford (1987*a*, *b*) who have developed a satisfactory, but rather complicated, method to overcome them. A second practical difficulty is that to obtain converged results the basis set must be rather large, with the result that the calculations become too time consuming. To overcome this problem Reading *et al.* (1984) have developed the forced impulse approximation. This is based on the idea that in fast collisions the interchange of energy between electrons, that is correlation, is slow compared with the response of a single electron to the collision with the projectile. It is therefore possible to divide the time during which the projectile interacts with the target into a number of intervals within which correlation can be neglected. The calculation then proceeds as follows. At some time $t = -t_0$, the correlated two-particle wave function for the target is formed as a linear combination of products of single-particle wave functions. Within the first time interval $-|t_0| < t < t_1$, each single-particle wave function is propagated independently, using the standard coupled channel model. An antisymmetrical two-body wave function is formed from the products of the one-particle functions. At the end of the interval this basis of uncorrelated functions is used to diagonalize the target Hamiltonian, obtaining a fresh set of correlated wave functions. These again define the coefficients of linear combinations of one-electron functions each of which is propagated in the second interval independently. The procedure is repeated for each time interval in turn. In this way, Reading and Ford (1987*a*, *b*) have successfully explained the differing ratios of double to single ionization for p and $\bar{\text{p}}$ interactions with helium (see Fig. 9.9). In this work the single-particle basis consisted of 36 s and p states, from which 423 two-particle states were formed. The ratio of the cross section obtained has the correct energy variation but is some 35 per cent too small. At very high energies, both the p and $\bar{\text{p}}$ cross sections should become the same and equal to that given by the first-order Born approximations. Ford and Reading (1988) have shown that to obtain a converged Born approximation (as shown in Fig. 9.9), d states are required in the basis, and it seems possible that if the d states are included in the basis of the forced impulse approximation calculations the correct magnitude of the cross section will be obtained. In later work (Ford and Reading 1990*a*, *b*; Reading *et al.* 1989), the impact parameter dependence and angular dependence of the double ionization cross section has been studied and also the double ionization of two-electron atoms of varying nuclear charge Z_B.

Simultaneous excitation and capture (transfer excitation)

A two-electron transition of particular interest in an ion–atom collision is the capture of an electron from the target and the simultaneous excitation of an electron initially bound to the projectile, to produce a doubly excited projectile state

$$A^{Z+} + B \rightarrow (A^{(Z-1)+})^{**} + B^{+}. \tag{9.74}$$

The doubly excited projectile state will subsequently decay by the emission of photons, or by autoionization

$$\begin{aligned}(A^{(Z-1)+})^{**} &\rightarrow (A^{(Z-1)+})^{*} + h\nu \\ &\rightarrow A^{Z+} + e^{-}.\end{aligned} \tag{9.75}$$

Experimental studies have been made of a number of such processes including the interaction of S^{13+} with a helium target (Tanis *et al.* 1985). The doubly excited state of S^{12+} formed contains a K-shell vacancy and subsequently decays by the emission of X-rays or Auger electrons. The cross section for the transition can be measured by detecting the charge state of the projectile in coincidence with the emitted X-rays. For the S^{13+} + He

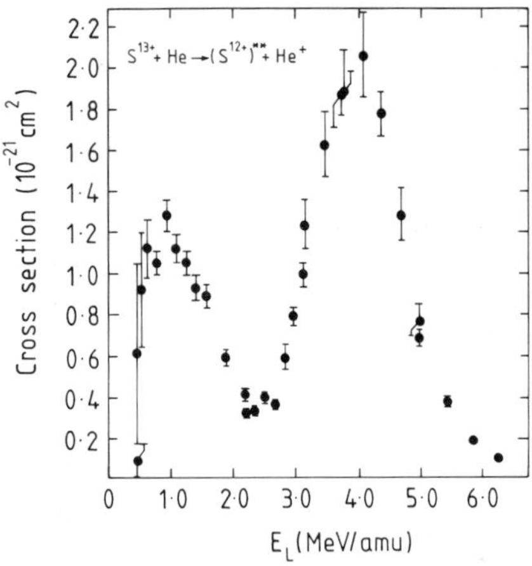

Fig. 9.10. Cross section for production of a K-shell vacancy together with single electron capture in collisions of S^{13+} with He. Experimental data from Tanis *et al.* (1985).

collision the experimental data are shown in Fig. 9.10. The cross section as a function of energy displays two well-defined peaks which have been interpreted in terms of two different mechanisms. The peak centred at the lower energy is considered to be due to capture of a target electron by the interaction with the Coulomb potential of the projectile nucleus A, with the excitation of a projectile electron by the Coulomb potential of the target nucleus B. This is termed non-resonant transfer excitation (NTE) and is basically a two-step process, with each process occuring through uncorrelated independent events. The peak at the higher energy is considered to be due to a single step process through the interaction of a target electron with a projectile electron. In the rest frame of the projectile the target atom B can be viewed as supplying a beam of quasi-free electrons, and the processes termed resonant transfer excitation (RTE) can be compared with dielectronic recombination in electron–atom collisions which proceeds through the formation of an autoionizing state

$$
\begin{aligned}
e^- + A^{Z+} &\to (A^{(Z-1)+})^{**} \\
&\to (A^{(Z-1)+})^{*} + h\nu .
\end{aligned}
\tag{9.76}
$$

The RTE reaction is resonant in the sense that if the captured electron can be treated as being free with kinetic energy E, energy conservation requires that the reaction should only proceed when E is equal to the difference in energy of the doubly excited ion $(A^{(Z-1)+})^{**}$ and the projectile ion A^{Z-1}. The division of the amplitude for transfer excitation into NTE and RTE contributions is approximate, but it is possible to attempt to distinguish between the two mechanisms experimentally by studying the dependence of the cross section on Z_A and Z_B. The structure of the transfer excitation amplitude has been studied by high-energy methods based on the impulse and related approximations (see particularly, Brandt (1983), Feagin *et al.* (1984), and Hahn (1989)), but rather than pursue these ideas, we shall discuss briefly an application of the standard coupled channel model to the problem.

Fritsch and Lin (1988) have studied the system $He^+ + H(1s)$ using the standard coupled channel model with a two-centre basis and plane-wave translation factors. The scattering equations uncouple into two sets, one for the singlet ($S = 0$) states and one for the triplet states ($S = 1$). For each spin symmetry a different basis set can be employed. For singlet systems Fritsch and Lin took atomic wave functions representing the incident channel $He^+(1s) \,.\, H(1s)$ together with the six excitation channels $He^+(1s) \,.\, H(nlm)$ and $He^+(nlm) \,.\, H(1s)$ with $n = 2, l = 0, m = 0$ and $l = 1, m = 0, \pm 1$. Pseudo-states representing the same configurations, but with the addition of united atom orbitals with $Z = 3$, were included to improve the representation of the wave function at small internuclear separations. In addition, the capture channels were taken into account by adding to the basis $1s^2\ {}^1S$, $1s\,2s\ {}^1S$, and $1s\,2p_m\ {}^1P$ terms on the helium centre as well as the doubly excited states $2s^2\ {}^1S$, $2p^2\ {}^1D$, $2s\,2p\ {}^1P$, and $2p^2\ {}^1S$. For the triplet system the terms representing the entrance and excitation channels in the basis were similar, but capture channels were represented in the basis set by $1s\,2s\ {}^3S$, $1s\,2p\ {}^3P$, $2s\,2p\ {}^3P$, and $2p^2\ {}^3P$ terms on the helium centre. The basis was constructed to diagonalize the helium Hamiltonian and so allowed fully for correlation.

In Fig. 9.11, the cross sections calculated by Fritsch and Lin for the

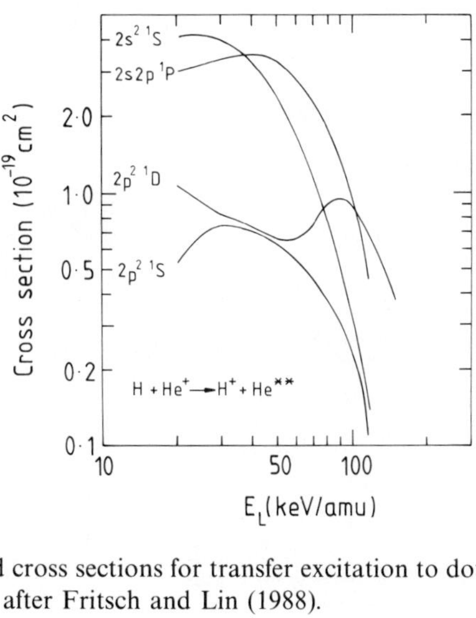

Fig. 9.11. Calculated cross sections for transfer excitation to doubly excited states in $He^+ + H$ collisions, after Fritsch and Lin (1988).

formation of doubly excited states by transfer excitation are shown as a function of energy. From a knowledge of the energies of the doubly excited states it can be estimated that, if the RTE mechanism were to dominate transfer excitation amplitudes, the cross sections would be expected to peak in the region of $E_L = 86$–96 keV/a.m.u. As can be seen only the $2p^2\ {}^1D$ cross section exhibits such a peak and for the other transitions the RTE mechanism is not in evidence. By removing all the single excitation and capture terms from the basis, calculations can be carried out, as in the corresponding investigations of double excitation (see above), in which the two-step NTE mechanism is suppressed. Using this technique a detailed analysis has been carried out by Fritsch and Lin in which the relative contributions of NTE and RTE can be assessed as a function of impact parameter. The general conclusion is that the simple predictions of high-energy models are not valid, at least for this particular system, and that the more detailed coupled channel model is required for an adequate description of transfer excitation.

10

NON-RELATIVISTIC MODELS AT HIGH ENERGIES

10.1 INTRODUCTION

In this chapter, we shall be concerned with charge exchange at high impact velocities for which methods based on perturbation or distorted wave series are appropriate, but for which a non-relativistic formulation remains accurate. This high-velocity region cannot be precisely defined, but as a guide we can take $v_0/v_e \geq 3$–4 where v_0 is the incident velocity and v_e the Bohr velocity of the active electron in the target. Relativistic corrections for collisions between light ions become significant when $v_0 \gtrsim 60$ a.u. and we shall consider the modifications required to the theory to take account of these in the next chapter.

It is interesting to consider the mechanisms which allow the capture of an electron from a target atom $(B + e^-)$ by a very fast ion A. (A very full analysis has been given by Shakeshaft and Spruch (1979)). In a reference frame centred on the ion A, the initial electronic energy (kinetic) of $v_0^2/2$ is large, while the final electronic energy, after capture, of the atom $(A + e^-)$ is small. The different mechanisms for capture correspond to different means of effecting this change of energy. The first possibility is radiative capture, $A + (B + e^-) \rightarrow (A + e^-) + B + h\nu$. At sufficiently high energies, the ion B acts as a spectator and the cross section is effectively that for radiative capture of a free electron by the moving ion A. This process will be discussed in Chapter 11. All that needs to be noted here is that although the cross section for radiative capture is small at low impact velocities it decreases less rapidly (like E^{-1} or v_0^{-2}) than non-radiative capture cross sections and is the dominant process at high energies. This occurs at impact energies in excess of ~ 10 MeV/a.m.u. for collisions between light ions.

Turning to non-radiative mechanisms, we saw in Chapter 7 that at high energies the cross section for capture provided by a coupled channel model becomes equal to that calculated from first-order perturbation theory. The corresponding amplitude describes a single scattering mechanism which depends on the momentum distribution of the bound electron in the target containing components which 'match' the incident velocity v_0. Since the atomic momentum distribution decreases rapidly with increasing momentum, the capture cross section also decreases rapidly and explicit calculation shows that the first-order cross section decreases like v_0^{-12} for large v_0.

A second non-radiative mechanism, not completely described by the coupled channel model with a finite basis, was first described by Thomas (1927), who put forward a purely classical model of a double scattering process. Thomas considered the reaction $H^+ + H \rightarrow H + H^+$, so that the heavy particles A and B are protons and the interactions are Coulombic. In the first collision the incident proton A strikes the bound electron (assumed stationary) which recoils at an angle close to 60° towards the target proton B with a speed equal to v_0. In the second collision, between the electron and B, the electron is scattered, without change of speed, again through 60° so that it emerges in a direction parallel to the particle A, when capture is assumed to take place (as shown in Fig. 10.1). At each collision, the classical Rutherford differential cross section is employed (which coincides with the quantum cross section for a strictly two-body interaction). It is assumed that the electron is initially to be found on the surface of a sphere of a given radius, a, which is treated as a parameter of the model. After the collision, the target nucleus B remains essentially stationary, while the projectile A continues with constant speed v_0, but is deflected through a small angle $\Theta_c = \sqrt{3}\,(m_e/M_p)/2$, called the Thomas angle, where m_e and M_p are the electron and proton masses. Classically a double scattering mechanism is the simplest that allows for the simultaneous conservation of momentum and energy in the reaction. Much later Drisko (1955) showed that the corresponding double scattering was contained in the second-order terms of the quantum perturbation series. Both the classical and quantum double scattering mechanisms provide cross sections which decrease like v_0^{-11} and so are ultimately more important than the single scattering mechanism. The direct observation in the 1s–1s capture reaction of the v_0^{-11} variation of the cross section is not possible, because this term only dominates at energies for which relativistic corrections become important ($E \gtrsim 100$ MeV/a.m.u.). It is possible that for capture into states of non-zero angular momentum this energy variation may be observable at lower energies. The double scattering process also gives rise to a peak in the differential cross section

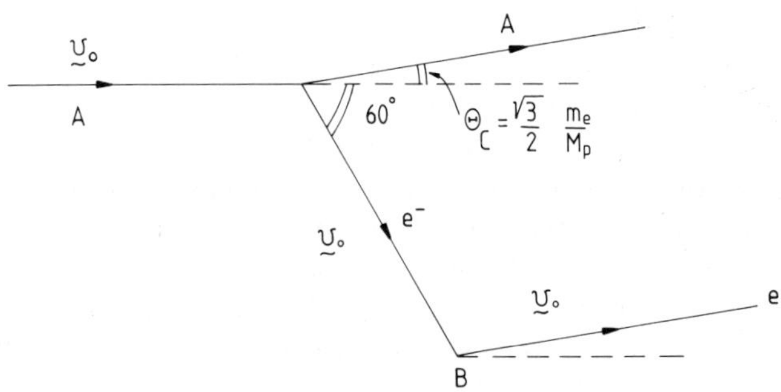

Fig. 10.1. Thomas double scattering.

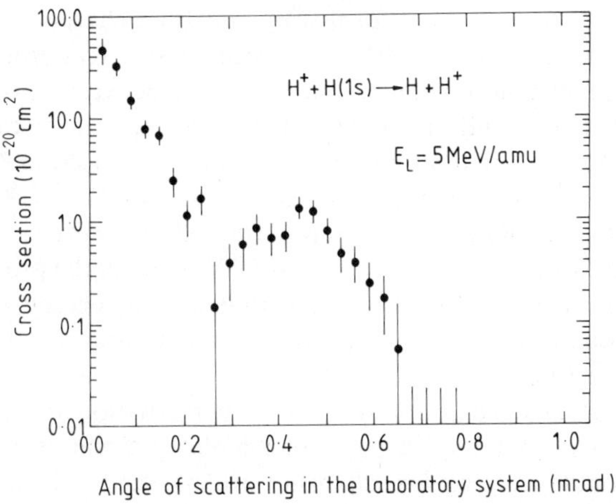

Fig. 10.2. The differential cross section in the laboratory system for electron capture from H(1s) by H^+ impact at $E_L = 5$ MeV/a.m.u. (after Vogt *et al.* (1986)). Experimental data from Vogt *et al.* (1986).

at high energies at the Thomas angle and indeed such a peak has been observed in H^+ + He and H^+ + H collisions (Horsdal-Pedersen *et al.* 1983*a*; McGuire *et al.* 1984; Vogt *et al.* 1986). Figure 10.2 shows an example of the peak for H^+ + H at an incident energy of 5.0 MeV. In addition to the peak in the differential cross section near the forward direction due to the Thomas mechanism, further peaks may arise at very large angles. These are again due to a double scattering mechanism in which the first collision is between the projectile A and the electron, as before, and in the second collision the projectile is scattered through 60° by the interionic interaction $V_{AB}(R)$ (Shakeshaft and Spruch 1979; Briggs 1986).

Since both the single and double scattering mechanisms described above are important close to the forward direction, the theory can be accurately formulated in terms of the impact parameter formalism employing a straight line trajectory. There is a non-relativistic mechanism which, in contrast, is important only in the backward direction. This is the 'knock-on' mechanism (Mapleton 1964) in which the incident nucleus A makes a head-on collision with the target nucleus B, which is initially at rest in the laboratory frame. The nucleus B is ejected with velocity v_0, and nucleus A takes its place. In this case, the electron acts as spectator, but the net effect is the transfer of the electron from an orbit about B to one about A. Kinematically the interchange of A and B in this way can only take place if A and B are of equal mass. If A and B are distinguishable a capture cross section can be defined, but if A and B are identical, as in the reaction $H^+ + H(1s) \rightarrow H(1s) + H^+$, the atoms emerging in the backward direction in the centre-of-mass system (as shown in Fig. 1.2), can either be undeflected atoms belonging to the incident beam, or those formed by knock-on charge exchange. In this case no distinction can be made between elastic scattering and charge exchange. Similarly no distinction can be made between knock-on capture into an excited state and direct excitation. The large-angle scattering is due to the internuclear potential and, as we shall see in Section 10.7, the Born approximation shows that the cross section decreases as v_0^{-6} at asymptotically high velocities. Knock-on is unimportant at low energies, since the cross section is proportional to $1/M^2$ where M is the mass of the proton. However, because of its relatively slow decrease with increasing energy, it becomes the largest non-radiative charge exchange cross section in the asymptotic high-energy region.

In this chapter, we shall first discuss the perturbation and distorted wave approximations to the impact parameter model, keeping in mind that both single and double scattering mechanisms are expected to be important. Our attention will then be turned to an account of the continuum distorted wave and symmetrical eikonal models, both of which aim to include a large part of the interaction in the basis states, and finally the wave formulation of perturbation theory for charge exchange will be considered briefly, which will allow a treatment of the large angle 'knock-on' process.

10.2 PERTURBATION THEORY AND THE IMPACT PARAMETER MODEL

Although perturbation theory can be developed equally well for the impact parameter model and for the full Schrödinger equation, it is somewhat simpler to work in terms of the impact parameter model, which is, as we have seen earlier, highly accurate for fast collisions. By doing so we exclude from the discussion the 'knock-on' process, which will be considered separately in Section 10.7. In Chapter 7, the perturbation expansion of the solution of the coupled channel equations was obtained, and it might be thought this would provide a sufficient basis for calculation at the higher energies. In fact, the high-energy limit of a finite set of coupled channel equations is just the perturbation solution of the two-state approximation in which the initial and final states of interest are retained. The total cross sections for elastic scattering and for excitation calculated in this way are indeed accurate at high energies, but the double scattering mechanism, which becomes increasingly important in charge exchange at high energies, is not included in such a first-order theory; a second-order theory is required with intermediate states lying in the continuum. As we have seen, at low energies the effect of the continuum can be represented by the introduction of pseudostates, but the high-energy limit of the cross section cannot be obtained from a *finite* set of pseudostates of the usual kind. It will therefore be necessary to develop the required perturbation theory using complete, rather than truncated, basis sets.

Let us consider a one-electron system for which the impact parameter equation is, in the notation of Chapter 3, (see 3.163a),

$$\mathrm{i}\frac{\partial\psi(\mathbf{s}, t)}{\partial t} = H_\mathrm{e}\psi(\mathbf{s}, t) \tag{10.1}$$

where

$$H_\mathrm{e} = -\tfrac{1}{2}\nabla_s^2 + V_{\mathrm{Ae}}(x) + V_{\mathrm{Be}}(r) + V_{\mathrm{AB}}(R). \tag{10.2}$$

At high energies, it is accurate to employ the straight line, constant velocity, trajectory (with $W(R) = 0$):

$$\mathbf{R}(t) = \mathbf{b} + \mathbf{v}_0 t; \qquad \mathbf{b}\cdot\mathbf{v}_0 = 0. \tag{10.3}$$

If the charges on the ionic cores A and B are Z_A and Z_B, respectively, the effective interactions V_{Ae}, V_{Be}, and V_{AB} are asymptotic to the Coulomb potentials

$$V_{\mathrm{Ae}}(x) \sim -\frac{Z_\mathrm{A}}{x}; \qquad V_{\mathrm{Be}}(r) \sim -\frac{Z_\mathrm{B}}{r}; \qquad V_{\mathrm{AB}}(R) \sim \frac{Z_\mathrm{A}Z_\mathrm{B}}{R}. \tag{10.4}$$

As shown in Chapter 5, provided the appropriate translation factors are included in the wave functions, any point on the line AB can be used as the origin of the coordinate system. Here it is convenient to take the origin to be on the target nucleus B, so that the independent variables are $\mathbf{r}$ and t and the operation $\partial/\partial t$ is to be taken with $\mathbf{r}$ held fixed.

The phase transformation (3.164) can be used to remove the interaction $V_{\mathrm{AB}}(R)$ from the Schrödinger equation (10.1). Setting

$$\psi(\mathbf{r}, t) = \tilde{\psi}(\mathbf{r}, t)\exp\left[-\mathrm{i}\int^t \mathrm{d}t'\, V_{\mathrm{AB}}(R(t'))\right] \tag{10.5}$$

the transformed equation becomes

$$\mathrm{i}\frac{\partial\tilde{\psi}(\mathbf{r}, t)}{\partial t} = \tilde{H}_\mathrm{e}\tilde{\psi}(\mathbf{r}, t) \tag{10.6}$$

with

$$\tilde{H}_\mathrm{e} = -\tfrac{1}{2}\nabla_r^2 + V_{\mathrm{Ae}}(x) + V_{\mathrm{Be}}(r). \tag{10.7}$$

The electronic Hamiltonian depends on t through x, since $\mathbf{x}(t) = \mathbf{r} - \mathbf{R}(t)$.

Asymptotic solutions

In Chapter 7, it was shown that divergences could occur in the higher-order terms of the perturbation series unless the long-range Coulomb interactions acting in each arrangement were allowed for properly. Having removed the internuclear potential, we see that (10.7) describes a system in which, in the direct channels, A + (B + e^-), there is an overall Coulomb interaction in the asymptotic region for $x \gg r$, $|t| \to \infty$, of $-Z_A/x$ and this is asymptotically equal to $-Z_A/R$. Likewise in the rearranged channels (A + e^-) + B, there is, for $r \gg x$, $|t| \to \infty$, an overall Coulomb interaction $-Z_B/r$, asymptotically equal to $-Z_B/R$. Allowing for these overall interactions, unperturbed wave functions $\Phi_j^B(\mathbf{r}, t)$ can be defined for the direct channels satisfying the equation

$$\left(\mathrm{i}\frac{\partial}{\partial t} - H_B\right)\Phi_j^B(\mathbf{r}, t) = 0 \tag{10.8}$$

with

$$H_B = -\tfrac{1}{2}\nabla_r^2 + V_{Be}(r) - \frac{Z_A}{R}. \tag{10.9}$$

Similarly the unperturbed wave functions $X_n^A(\mathbf{r}, t)$ in the rearranged channels can be defined which satisfy

$$\left(\mathrm{i}\frac{\partial}{\partial t} - H_A\right)X_n^A(\mathbf{r}, t) = 0 \tag{10.10}$$

with

$$H_A = -\tfrac{1}{2}\nabla_r^2 + V_{Be}(x) - \frac{Z_B}{R}. \tag{10.11}$$

Solutions of (10.8) differ slightly from the corresponding functions given in Chapter 3, (3.170), because of the different choice of the origin of coordinates, now taken to be at B, and also because of the extra term $-Z_A/R$ representing the overall Coulomb interaction. We have

$$\Phi_j^B(\mathbf{r}, t) = \phi_j(\mathbf{r}) \exp[\mathrm{i}\alpha_j(t)] \tag{10.12}$$

where

$$\alpha_j(t) = -\varepsilon_j t - \sigma(t). \tag{10.13}$$

The Coulomb phase $\sigma(t)$ is (cf. (3.178)),

$$\sigma(t) = \frac{Z_A}{v_0}\log[R(t) - v_0 t]. \tag{10.14}$$

The functions Φ_j^B are orthonormal and, provided the continuum states of the atom (B + e^-) are included, form a complete set for each value of t.

The time-reversed functions $\Phi_j^{BT}(\mathbf{r}, t)$, where

$$\Phi_j^{BT}(\mathbf{r}, t) = \Phi_j^{B*}(\mathbf{r}, -t) = \phi_j(\mathbf{r}) \exp[\mathrm{i}\alpha_j^T(t)] \tag{10.15}$$

and

$$\alpha_j^T(t) = -\varepsilon_j t + \sigma(-t) \tag{10.16}$$

are also solutions of (10.8), and again form a complete orthonormal set. In obtaining (10.15), we have taken the internal wave functions ϕ_j and χ_n of the atoms $(B + e^-)$ and $(A + e^-)$ to be real, which is always possible. The corresponding solutions of (10.10) are (cf. 3.175)

$$X_n^A(\mathbf{r}, t) = \chi_n(\mathbf{x}) \exp[i\beta_n(t)] \tag{10.17}$$

with

$$\beta_n(t) = -\eta_n t - \delta(t) + \mathbf{v}_0 \cdot \mathbf{r} - \tfrac{1}{2}v_0^2 t. \tag{10.18}$$

The plane-wave translation factor $\exp[i\mathbf{v}_0 \cdot \mathbf{r} - \frac{1}{2}iv_0^2 t]$ allows for the fact that the atom $(A + e^-)$ travels with a velocity $\mathbf{v}_0$ with respect to the origin at B. The Coulomb phase $\delta(t)$ is given by

$$\delta(t) = \frac{Z_B}{v_0} \log[R(t) - v_0 t]. \tag{10.19}$$

Time-reversed solutions of (10.10) are defined as

$$X_n^{AT}(\mathbf{v}_0; \mathbf{r}, t) = X_n^{A*}(-\mathbf{v}_0; \mathbf{r}, -t) \tag{10.20}$$

where the dependence on $\mathbf{v}_0$ has been displayed explicitly.

Probability amplitudes and cross sections

If, in the initial state of the system, A is incident on the ion $(B + e^-)$, in the ith level the corresponding solution $\tilde{\psi}_i(\mathbf{r}, t)$ of (10.6) must satisfy the boundary condition

$$\tilde{\psi}_i(\mathbf{r}, t) \xrightarrow[t \to -\infty]{} \Phi_i^B(\mathbf{r}, t). \tag{10.21}$$

Another solution of (10.6), $\tilde{\psi}_j^T(\mathbf{r}, t)$ can be found for which, in the limit $t \to \infty$, the system evolves to the jth level of the ion $(B + e^-)$

$$\tilde{\psi}_j^T(\mathbf{r}, t) \xrightarrow[t \to \infty]{} \Phi_j^{BT}(\mathbf{r}, t). \tag{10.22}$$

As shown in Chapter 3, the probability amplitude $\mathscr{A}_{ji}(\mathbf{b})$ for excitation (or ionization) of the atom $(B + e^-)$ can then be expressed as

$$\begin{aligned} \mathscr{A}_{ji}(\mathbf{b}) &= \langle \tilde{\psi}_j^T(t) \,|\, \tilde{\psi}_i(t) \rangle \\ &= \int [\tilde{\psi}_j^T(\mathbf{r}, t)]^* \tilde{\psi}_i(\mathbf{r}, t)\, d\mathbf{r}. \end{aligned} \tag{10.23}$$

Since the matrix element $\langle \tilde{\psi}_j^T(t) \,|\, \tilde{\psi}_i(t) \rangle$ is independent of the time t, taking the limits $t \to \infty$ or $t \to -\infty$ provides alternative expressions for $\mathscr{A}_{ji}(\mathbf{b})$:

$$\mathscr{A}_{ji}(\mathbf{b}) = \lim_{t \to -\infty} \langle \tilde{\psi}_j^{\mathrm{T}}(t) \,|\, \Phi_i^{\mathrm{B}}(t) \rangle \tag{10.24a}$$

$$= \lim_{t \to \infty} \langle \Phi_j^{\mathrm{BT}}(t) \,|\, \tilde{\psi}_i(t) \rangle. \tag{10.24b}$$

The cross section for excitation (or ionization) is obtained by integrating over impact parameter space:

$$\sigma_{ji}(v_0) = \int |\mathscr{A}_{ji}(\mathbf{b})|^2 \, \mathrm{d}^2\mathbf{b}. \tag{10.25}$$

If j represents a continuum state of $(\mathrm{B} + \mathrm{e}^-)$, the total ionization cross section is

$$\sigma_i^{\pm}(v_0) = \sum_j{}' \, \sigma_{ji}(v_0) \tag{10.26}$$

where the prime denotes a sum over continuum states. If these states are characterized by $\mathbf{q}$ the momentum of the electron, a suitable normalization of the continuum wave function ϕ_j is

$$\langle \phi_j \,|\, \phi_{j'} \rangle \equiv \langle \phi_{\mathbf{q}} \,|\, \phi_{\mathbf{q}'} \rangle = \delta(\mathbf{q} - \mathbf{q}') \tag{10.27}$$

in which case $\sum_j' \to \int \mathrm{d}\mathbf{q}$.

To calculate the cross section for charge exchange, a solution of (10.6) can be found, $\tilde{\psi}_n^{\mathrm{T}}(\mathbf{r}, t)$ for which the system evolves to the nth level of the ion $(\mathrm{A} + \mathrm{e}^-)$. The probability amplitude for charge exchange to this level from the ith level of the atom $(\mathrm{B} + \mathrm{e}^-)$ is then

$$\mathscr{A}_{ni}(\mathbf{b}) = \langle \tilde{\psi}_n^{\mathrm{T}}(t) \,|\, \tilde{\psi}_i(t) \rangle \tag{10.28a}$$

$$= \lim_{t \to -\infty} \langle \tilde{\psi}_n^{\mathrm{T}}(t) \,|\, \Phi_i^{\mathrm{B}}(t) \rangle \tag{10.28b}$$

$$= \lim_{t \to \infty} \langle \mathrm{X}_n^{\mathrm{AT}}(t) \,|\, \tilde{\psi}_i(t) \rangle. \tag{10.28c}$$

Total cross sections for charge exchange are obtained in the usual way by integrating $|\mathscr{A}_{ni}(\mathbf{b})|^2$ over impact parameter space. Since the level n can be in the continuum of the atom $(\mathrm{A} + \mathrm{e}^-)$, ionization cross sections can, in principle, be calculated from $|\mathscr{A}_{ni}(\mathbf{b})|^2$ as well as from $|\mathscr{A}_{ji}(\mathbf{b})|^2$. With exact wave functions, and recollecting that the wave functions ϕ_j and χ_n both form complete sets, it is clear that the total ionization cross sections calculated from either expression should be identical. Experimentally it is observed that a distinction can be made between ejected electrons correlated with the motion of B and those correlated with the motion of A. In the former case we speak of 'direct' ionization and in the latter 'charge transfer to the continuum', and it is possible to devise approximations in which each contribution to the ionization cross sections can be calculated separately. We return to this topic in Section 10.8.

Differential cross sections for excitation or charge exchange can be calculated as described in Chapter 3, provided the phase generated by the internuclear potential $V_{AB}(R)$ is restored. For example, when A and B are bare nuclei and $V_{AB} = Z_A Z_B/R$, the phase-corrected amplitude $\mathscr{A}^c_{ji}(\mathbf{b})$ is

$$\mathscr{A}^c_{ji}(\mathbf{b}) = b^{2i\nu} A_{ji}(\mathbf{b}) \qquad (10.29)$$

with $\nu = Z_A Z_B/v_0$, and where the result

$$\int^t \frac{1}{R(t')}\,\mathrm{d}t' = -\frac{1}{v_0}\log(R - v_0 t] \qquad (10.30)$$

has been employed. The differential cross section, for excitation or charge exchange, is, from (3.231) and (3.232),

$$\frac{\mathrm{d}\sigma}{\mathrm{d}\Omega} = \left|\mu v_0 \int_0^\infty b^{2i\nu+1} J_{\Delta m}(\mu v_0 b\Theta)\mathscr{A}_{ji}(\mathbf{b})\,\mathrm{d}b\right|^2, \qquad j \neq i. \qquad (10.31)$$

Integral equations and perturbation series

The solutions of the Schrödinger equation (10.6) satisfy integral equations which automatically ensure that the specified boundary conditions are satisfied. For example the solution $\tilde{\psi}_i(\mathbf{r}, t)$ satisfying the boundary condition (10.21) can be written in terms of a Green's function $g_B^+(\mathbf{r}, t; \mathbf{r}', t')$ as

$$\tilde{\psi}_i(\mathbf{r}, t) = \Phi_i^B(\mathbf{r}, t) + \int_{-\infty}^{\infty} \mathrm{d}t' \int \mathrm{d}\mathbf{r}' g_B^+(\mathbf{r}, t; \mathbf{r}', t') U_A(\mathbf{r}', t')\tilde{\psi}_i(\mathbf{r}', t') \qquad (10.32)$$

where

$$U_A(\mathbf{r}', t') = V_{Ae}(x'(t)) + \frac{Z_A}{R(t)}; \qquad \mathbf{x}'(t) = \mathbf{r}' - \mathbf{R}(t'). \qquad (10.33)$$

The explicit form of g_B^+, which is obtained in Appendix G, is

$$g_B^+(\mathbf{r}, t; \mathbf{r}', t') = -i\sum_k \Phi_k^{BT}(\mathbf{r}, t)[\Phi_k^{BT}(\mathbf{r}', t')]^*\theta(t - t') \qquad (10.34)$$

where the sum over k runs over all states of the atom $(B + e^-)$, both discrete and continuous. The step function $\theta(t - t')$ in (10.34) ensures that the second term on the right-hand side of (10.32) vanishes in the limit $t \to -\infty$, so that the boundary condition (10.21) is satisfied. Using the completeness of the set of functions Φ_j^{BT} it is easy to verify by substitution that (10.32) is a solution of (10.6).

From (10.14) it can be seen that the sum of the phases $\sigma(t)$ and $\sigma(-t)$ is independent of time:

$$\sigma(t) + \sigma(-t) = \frac{Z_A}{v_0}\log[(R(t) - v_0 t)(R(t) + v_0 t)]$$

$$= \frac{2Z_A}{v_0}\log(b). \tag{10.35}$$

A similar relation holds for the phases $\delta(t)$ and $\delta(-t)$. Using the definitions of Φ_j^{BT} and Φ_j^{B} given by (10.11) and (10.12), respectively, it is then easy to see that

$$\Phi_k^{\mathrm{BT}}(\mathbf{r}, t)[\Phi_k^{\mathrm{BT}}(\mathbf{r}', t')]^* = \Phi_k^{\mathrm{B}}(\mathbf{r}, t)[\Phi_k^{\mathrm{B}}(\mathbf{r}', t')]^* \tag{10.36a}$$

so that g_{B}^{+} can equally well be expressed as

$$g_{\mathrm{B}}^{+}(\mathbf{r}, t; \mathbf{r}', t') = -\mathrm{i}\sum_k \Phi_k^{\mathrm{B}}(\mathbf{r}, t)[\Phi_k^{\mathrm{B}}(\mathbf{r}', t')]^*\theta(t - t'). \tag{10.36b}$$

Whether (10.34) or (10.37) is employed is a matter of convenience.

For elastic scattering, direct excitation, or ionization, an equation for the probability amplitude $\mathscr{A}_{ji}(\mathbf{b})$ can be obtained from (10.24b), (10.32), and (10.34):

$$\mathscr{A}_{ji}(\mathbf{b}) = \lim_{t\to\infty} \langle \Phi_j^{\mathrm{BT}}(t) \,|\, \tilde{\psi}_i(t)\rangle$$

$$= \lim_{t\to\infty} \langle \Phi_j^{\mathrm{BT}}(t) \,|\, \Phi_i^{\mathrm{B}}(t)\rangle - \mathrm{i}\int_{-\infty}^{\infty} \langle \Phi_j^{\mathrm{BT}}(t) | U_{\mathrm{A}} | \tilde{\psi}_i(t)\rangle \,\mathrm{d}t. \tag{10.37}$$

Using the relationship (10.35), we have

$$\lim_{t\to\infty} \langle \Phi_j^{\mathrm{BT}}(t) \,|\, \Phi_i^{\mathrm{B}}(t)\rangle = \delta_{ij}\exp\left[-\mathrm{i}\frac{2Z_A}{v_0}\log(b)\right]. \tag{10.38}$$

This term contributes to the probability of elastic scattering and represents scattering by the overall Coulomb potential. For inelastic processes, (10.37) reduces to

$$\mathscr{A}_{ji}(\mathbf{b}) = -\mathrm{i}\int_{-\infty}^{\infty} \langle \Phi_j^{\mathrm{BT}}(t) | U_{\mathrm{A}} | \tilde{\psi}_i(t)\rangle \,\mathrm{d}t, \qquad i \neq j. \tag{10.39}$$

The perturbation series for the exact wave function $\tilde{\psi}_i$ can be obtained from (10.32) by iteration. Using an operator notation, we obtain

$$\tilde{\psi}_i = (I + g_{\mathrm{B}}^{+} U_{\mathrm{A}} + g_{\mathrm{B}}^{+} U_{\mathrm{A}} g_{\mathrm{B}}^{+} + \cdots)\Phi_i^{\mathrm{B}}, \tag{10.40}$$

and the corresponding series expansion of $\mathscr{A}_{ji}$ is obtained by inserting (10.40) into (10.39):

$$\mathscr{A}_{ji}(\mathbf{b}) = -\mathrm{i}\int_{-\infty}^{\infty} \langle \Phi_j^{\mathrm{BT}} | U_{\mathrm{A}} + U_{\mathrm{A}} g_{\mathrm{B}}^{+} U_{\mathrm{A}} + U_{\mathrm{A}} g_{\mathrm{B}}^{+} U_{\mathrm{A}} g_{\mathrm{B}}^{+} U_{\mathrm{A}} + \cdots | \Phi_i^{\mathrm{B}} \rangle \, \mathrm{d}t. \tag{10.41}$$

Charge exchange To find a corresponding series for the probability amplitude for charge exchange, it is convenient to start from (10.28c). This can be written as

$$\mathscr{A}_{ni}(\mathbf{b}) = \int_{-\infty}^{\infty} \frac{\mathrm{d}}{\mathrm{d}t} \langle \mathrm{X}_n^{\mathrm{AT}}(t) | \tilde{\psi}_i(t) \rangle \, \mathrm{d}t + \lim_{t \to -\infty} \langle \mathrm{X}_n^{\mathrm{AT}}(t) | \tilde{\psi}_i(t) \rangle. \tag{10.42}$$

In view of the boundary condition (10.21)

$$\lim_{t \to -\infty} \langle \mathrm{X}_n^{\mathrm{AT}}(t) | \tilde{\psi}_i(t) \rangle = \lim_{t \to -\infty} \langle \mathrm{X}_n^{\mathrm{AT}}(t) | \Phi_i^{\mathrm{B}}(t) \rangle \tag{10.43}$$

and this is zero because the overlap between $\phi_i(\mathbf{r})$ and $\chi_n(\mathbf{x})$ vanishes as $R \to \infty$. The remaining term in (10.42) can be transformed using equations (10.6) and (10.10), satisfied by $\tilde{\psi}_i$ and $\mathrm{X}_n^{\mathrm{AT}}$, respectively, and recollecting that $H_{\mathrm{A}}^{\dagger} = H_{\mathrm{A}}$:

$$\begin{aligned} \mathscr{A}_{ni}(\mathbf{b}) &= -\mathrm{i}\int_{-\infty}^{\infty} \langle \mathrm{X}_n^{\mathrm{AT}}(t) | H_{\mathrm{e}} - H_{\mathrm{A}} | \tilde{\psi}_i(t) \rangle \, \mathrm{d}t \\ &= -\mathrm{i}\int_{-\infty}^{\infty} \langle \mathrm{X}_n^{\mathrm{AT}}(t) | U_{\mathrm{B}} | \tilde{\psi}_i(t) \rangle \, \mathrm{d}t \end{aligned} \tag{10.44}$$

where

$$U_{\mathrm{B}}(\mathbf{r}, t) = V_{\mathrm{Be}}(r) + \frac{Z_{\mathrm{B}}}{R(t)}. \tag{10.45}$$

Inserting (10.40) into (10.44), one form of perturbation series for $\mathscr{A}_{ni}(\mathbf{b})$ is obtained:

$$\mathscr{A}_{ni}(\mathbf{b}) = -\mathrm{i}\int_{-\infty}^{\infty} \mathrm{d}t \langle \mathrm{X}_n^{\mathrm{AT}} | U_{\mathrm{B}} + U_{\mathrm{B}} g_{\mathrm{B}}^{+} U_{\mathrm{A}} + U_{\mathrm{B}} g_{\mathrm{B}}^{+} U_{\mathrm{A}} g_{\mathrm{B}}^{+} U_{\mathrm{A}} + \cdots | \Phi_i^{\mathrm{B}} \rangle. \tag{10.46}$$

This series expansion of $\mathscr{A}_{ni}(\mathbf{b})$ in terms of powers of U_{A} is evidently not unique, since a corresponding series containing successive interactions by the potential U_{B} must exist. To find such a series, we can start from a solution $\tilde{\psi}_n^{\mathrm{T}}$ of (10.6) which satisfies the boundary condition

$$\tilde{\psi}_n^{\mathrm{T}}(\mathbf{r}, t) \xrightarrow[t \to +\infty]{} \mathrm{X}_n^{\mathrm{AT}}(\mathbf{r}, t). \tag{10.47}$$

This solution satisfies the integral equation (in operator form)

$$\tilde{\psi}_n^{\rm T} = {\rm X}_n^{\rm AT} + g_{\rm A}^- U_{\rm B}\tilde{\psi}_n^{\rm T} \qquad (10.48)$$

where the explicit expression for $g_{\rm A}^-$ is

$$g_{\rm A}^-(\mathbf{r}, t; \mathbf{r}', t') = {\rm i}\theta(t'-t)\sum_m {\rm X}_m^{\rm AT}(\mathbf{r}, t)[{\rm X}_m^{\rm AT}(\mathbf{r}', t')]^*, \qquad (10.49)$$

and the sum over m includes an integration over the continuum states of $({\rm A} + {\rm e}^-)$. It is again straightforward to verify that (10.48) satisfies (10.6) with the boundary condition (10.47), by using the fact that the functions ${\rm X}_m^{\rm AT}$ form a complete orthonormal set. From (10.48) the series expansion of $\tilde{\psi}_n^{\rm T}$ in terms of $U_{\rm B}$ is obtained immediately:

$$\tilde{\psi}_n^{\rm T} = [I + g_{\rm A}^- U_{\rm B} + g_{\rm A}^- U_{\rm B} g_{\rm A}^- U_{\rm B} + \cdots]{\rm X}_n^{\rm AT}. \qquad (10.50)$$

To find the corresponding series for $\mathscr{A}_{ni}(\mathbf{b})$, the probability amplitude, an integral expression analogous to (10.44) but in terms of $\tilde{\psi}_n^{\rm T}$ is required. This may be found by writing (10.28b) as

$$\begin{aligned}\mathscr{A}_{ni}(\mathbf{b}) &= \lim_{t\to-\infty} \langle \tilde{\psi}_n^{\rm T}(t) | \Phi_i^{\rm B}(t)\rangle \\ &= -\int_{-\infty}^{\infty} \frac{\rm d}{{\rm d}t}\langle \tilde{\psi}_n^{\rm T}(t) | \Phi_i^{\rm B}(t)\rangle \,{\rm d}t + \lim_{t\to\infty} \langle \tilde{\psi}_n^{\rm T}(t) | \Phi_i^{\rm B}(t)\rangle. \end{aligned} \qquad (10.51)$$

The last term vanishes since $\psi_n^{\rm T} \to {\rm X}_n^{\rm AT}$ as $t \to \infty$, and using equations (10.6) and (10.8) satisfied by $\tilde{\psi}_n^{\rm T}$ and $\Phi_i^{\rm B}$, $\mathscr{A}_{ni}(\mathbf{b})$ can be put in the form:

$$\mathscr{A}_{ni}(\mathbf{b}) = -{\rm i}\int_{-\infty}^{\infty} \langle \tilde{\psi}_n^{\rm T} | U_{\rm A} | \Phi_i^{\rm B}\rangle \,{\rm d}t. \qquad (10.52)$$

Finally by substituting $\tilde{\psi}_n^{\rm T}$ given by (10.50) into (10.52) and using the relationship $g_{\rm A}^{-\dagger} = g_{\rm A}^+$, the desired expansion for $\mathscr{A}_{ni}(\mathbf{b})$ is:

$$\mathscr{A}_{ni}(\mathbf{b}) = -{\rm i}\int_{-\infty}^{\infty} \langle {\rm X}_n^{\rm AT} | U_{\rm A} + U_{\rm B} g_{\rm A}^+ U_{\rm A} + U_{\rm B} g_{\rm A}^+ U_{\rm B} g_{\rm A}^+ U_{\rm A} + \cdots | \Phi_i^{\rm B}\rangle \,{\rm d}t \qquad (10.53)$$

The series (10.41), (10.46), and (10.53) are examples of what are often called Coulomb Born series, in which the unperturbed wave functions are modified by Coulomb interactions. Since in these particular series the Coulomb phases are introduced in order to satisfy the correct boundary conditions,† the

† Other forms of Coulomb Born series have been proposed; see, for example, the work of Mukherjee *et al.* (1979) and Mandal *et al.* (1981). The relationships between different versions of the Coulomb Born approximations have been analysed by Datta *et al.* (1990).

terminology of Dewangan and Eichler (1985) will be followed and the series will be called *boundary corrected Born* series. If the series for the charge exchange amplitude are truncated at the Nth term, two different approximations are obtained; ${}^{\mathrm{I}}\mathscr{A}_{ni}^{(N)}(\mathbf{b})$ based on the expansion (10.46) in terms of powers of U_{A}, and ${}^{\mathrm{II}}\mathscr{A}_{ni}^{(N)}(\mathbf{b})$ based on the expansion (10.53) in terms of powers of U_{B}:

$$ {}^{\mathrm{I}}\mathscr{A}_{ni}^{(N)}(\mathbf{b}) = -\mathrm{i}\int_{-\infty}^{\infty} \langle \mathrm{X}_n^{\mathrm{AT}}| U_{\mathrm{B}} \sum_{k=1}^{N} [g_{\mathrm{B}}^{+} U_{\mathrm{A}}]^{k-1} |\Phi_i^{\mathrm{B}}\rangle \, \mathrm{d}t \tag{10.54} $$

$$ {}^{\mathrm{II}}\mathscr{A}_{ni}^{(N)}(\mathbf{b}) = -\mathrm{i}\int_{-\infty}^{\infty} \langle \mathrm{X}_n^{\mathrm{AT}}| \sum_{k=1}^{N} [U_{\mathrm{B}} g_{\mathrm{A}}^{+}]^{k-1} U_{\mathrm{A}} |\Phi_i^{\mathrm{B}}\rangle \, \mathrm{d}t. \tag{10.55} $$

These approximations to the amplitude will be called Nth-order boundary corrected Born approximations and are denoted by the symbol BNB with $N = 1, 2, \ldots$

Calculations beyond the first-order term of the sequences ${}^{\mathrm{I}}\mathscr{A}_{ni}^{(N)}(\mathbf{b})$ and ${}^{\mathrm{II}}\mathscr{A}_{ni}^{(N)}(\mathbf{b})$ are extremely laborious because of the infinite sums over the intermediate states contained in the Green's functions $g_{\mathrm{A,B}}^{+}$. Some simplification results by expanding g_{B}^{+} in a series of powers of V_{Be} (or by expanding g_{A}^{+} in powers of V_{Ae}) as in equation (G.28) of Appendix G. The second-order approximation called B2B0 by Decker and Eichler (1989*b*) is then

$$ \mathscr{A}_{ni}^{(2)}(\mathbf{b}) \simeq -\mathrm{i}\int_{-\infty}^{\infty} \langle \mathrm{X}_n^{\mathrm{AT}}| U_{\mathrm{B}} + U_{\mathrm{B}} g_0^{+} U_{\mathrm{A}} |\Phi_i^{\mathrm{B}}\rangle \, \mathrm{d}t. \tag{10.56} $$

Post–prior discrepancies

Provided the internal wave functions χ_n and ϕ_i are exact, it is straightforward to show that the two first-order amplitudes ${}^{\mathrm{I}}\mathscr{A}_{ni}^{(1)}(\mathbf{b})$ and ${}^{\mathrm{II}}\mathscr{A}_{ni}^{(1)}(\mathbf{b})$ are identical, so there is no 'post–prior discrepancy'. This is not the case for higher-order terms in the boundary corrected Born series, and in general ${}^{\mathrm{I}}\mathscr{A}_{ni}^{(N)}(\mathbf{b}) \neq {}^{\mathrm{II}}\mathscr{A}_{ni}^{(N)}(\mathbf{b})$. In the B2B0 approximation the second-order terms obtained by expanding g_{B}^{+} are identical with those obtained by expanding g_{A}^{+}, so there is no post–prior discrepancy in this case.

Since, in general, the sequences of approximations ${}^{\mathrm{I}}\mathscr{A}_{ni}^{(N)}(\mathbf{b})$ and ${}^{\mathrm{II}}\mathscr{A}_{ni}^{(N)}(\mathbf{b})$ are not the same, the question arises as to which should be preferred for a particular system. Clearly for a nearly symmetrical collision, in the sense that U_{B} and U_{A} are potentials of comparable magnitude (an example is the symmetrical reaction $\mathrm{H}^{+} + \mathrm{H}(1s) \to \mathrm{H}(1s) + \mathrm{H}^{+}$), both sequences converge equally well, or equally badly. On the other hand, as pointed out by Briggs (1977) and by Macek and Shakeshaft (1980), in a highly asymmetrical situation for which $Z_{\mathrm{B}} \ll Z_{\mathrm{A}}$, expansions in terms of the weaker perturbation U_{B} might be expected to converge more rapidly, so that the sequence ${}^{\mathrm{II}}\mathscr{A}_{ni}^{(N)}(\mathbf{b})$ would be preferred. If it is decided to retain all first-order terms and neglect all higher-order terms in the weaker potential U_B, the resulting (second-order) approximation is given by ${}^{\text{II}}\mathscr{A}_{ni}^{(2)}(\mathbf{b})$. This approximation to the amplitude has been termed the *strong potential Born approximation.*

Convergence

The perturbations U_A and U_B both vanish at large R faster than $1/R$, and this is sufficient for each term in the series (10.46) or (10.53) to exist. This property is due to the proper treatment of the Coulomb boundary conditions as emphasized by Cheshire (1964) and Belkić *et al.* (1979), and we saw in Chapter 7 that if the Coulomb boundary conditions are ignored, second- and higher-order terms in the perturbation series diverge† (Mapleton 1972, Dewangan and Eichler 1985). Even if each term in the perturbation series is well defined, the series may not be convergent, and, indeed, a proof of convergence has not been given; however some important steps have been taken. It has been shown (Dettmann and Liebfried 1969) for a wide class of potentials, including the Coulomb potential, that as $v_0 \to \infty$ the dominant term in the Born series for the ground-state capture amplitude is the first-order term, when $M_A = M_B$ (knock-on collisions), and the second-order term when $M_A \neq M_B$ (Thomas scattering). This work was carried out within the full three-body quantum mechanical formulation. A further advance by Shakeshaft and Spruch (1973) was to show that if the exact amplitude $\mathscr{A}_{ni}(\mathbf{b})$, for ground-state capture, is written as

$$\mathscr{A}_{ni}(\mathbf{b}) = \mathscr{A}_{ni}^{(N)}(\mathbf{b}) + R^{(N)} \tag{10.57}$$

where $\mathscr{A}_{ni}^{(N)}(\mathbf{b})$ is the Nth-order perturbation approximation, the remainder $R^{(N)}$ is bounded ($N > 2$). This result was obtained within the impact parameter formulation with a straight line trajectory and for a class of short-range potentials, not including the Coulomb potential. These two results show that for the short-range potentials considered by Shakeshaft and Spruch, the scattering amplitude for rearrangement in the high energy limit is given by the second-order term in the Born series, whether or not the series converges. It is greatly believed that this result is also correct for Coulomb potentials, and it appears likely, but not certain, that the sequences (10.54) and (10.55) do converge at sufficiently high energies.

A historical digression

The history of the development of perturbation methods for charge exchange is interesting and instructive. Oppenheimer (1928), using the wave formulation, and Brinkman and Kramers (1930), using the impact parameter method,

† It should be noted that it is possible to employ a basis of plane-wave states (that is to ignore the overall Coulomb interactions $-Z_A/R$ or $-Z_B/R$) provided the divergences are removed using a systematic renormalization procedure. This has been achieved by Macek (1988), who discusses the problem in detail.

obtained equivalent first-order expressions for the charge exchange amplitude. It was realized that the potential between the heavy particles $V_{AB}(R)$ could not contribute to the probability for charge exchange in the impact parameter model, and could only contribute small terms of the order $(1/\mu)$ in a full three-body quantum mechanical treatment. Accordingly it was supposed that the perturbing potential could either be taken to be V_{Ae} (prior form) or V_{Be} (post form). It was then tacitly assumed that as charge exchange takes place only at small distances at which the atomic wave functions centred on A and B have an appreciable overlap, the Coulomb boundary conditions consistent with the Schrödinger equation (10.6) were unimportant. It followed that the first-order perturbation approximation to the charge exchange amplitude, now usually known as the OBK approximation, was (in the prior form)

$$^{\mathrm{OBK}}\mathscr{A}^{(1)}_{ni}(\mathbf{b}) = -\mathrm{i}\int_{-\infty}^{\infty} \langle \tilde{\mathrm{X}}^{\mathrm{A}}_{n}(t)|V_{\mathrm{Ae}}(x)|\tilde{\Phi}^{\mathrm{B}}_{i}(t)\rangle\,\mathrm{d}t \qquad (10.58)$$

where $\tilde{\mathrm{X}}^{\mathrm{A}}_{n}$ and $\tilde{\Phi}^{\mathrm{B}}_{i}$ differ from $\mathrm{X}^{\mathrm{A}}_{n}$ and Φ^{B}_{i} in the omission of the Coulomb phases $\delta(t)$ and $\sigma(t)$:

$$\tilde{\Phi}^{\mathrm{B}}_{i}(\mathbf{r}, t) = \phi_{j}(\mathbf{r})\exp[-\mathrm{i}\varepsilon_{j}t] \qquad (10.59)$$

$$\tilde{\mathrm{X}}^{\mathrm{A}}_{n}(\mathbf{r}, t) = \chi_{n}(\mathbf{x})\exp[-\mathrm{i}\eta_{j}t + \mathrm{i}\mathbf{v}_{0}\cdot\mathbf{r} - \tfrac{1}{2}\mathrm{i}v_{0}^{2}t]. \qquad (10.60)$$

When, some twenty-five years later, experimental measurements were made, it became apparent that the OBK approximation was very poor; for instance the calculated cross section for the reaction $H^{+} + H(1s) \to H + H^{+}$ exceeded the data by a factor of five or more at energies about 50 keV. Faced with this difficulty, Bates and Dalgarno (1952, 1953) and, independently Jackson and Schiff (1953), suggested that for the $H^{+} + H$ system, the internuclear potential should be retained in the first-order matrix element, on the grounds that the signs of $V_{Ae} = 1/x$ and $V_{AB} = 1/R$ are opposite and the errors in the contributions to the amplitude from each term might be expected to cancel. In fact, it turns out that what has become known as the JS (Jackson-Schiff) approximation using $(V_{Ae} + V_{AB})$ as the perturbations,

$$^{\mathrm{JS}}\mathscr{A}^{(1)}_{ni}(\mathbf{b}) = -\mathrm{i}\int_{-\infty}^{\infty} \langle \tilde{\mathrm{X}}^{\mathrm{A}}_{n}(t)|V_{\mathrm{AB}}(R) + V_{\mathrm{Ae}}(x)|\tilde{\Phi}^{\mathrm{B}}_{i}(t)\rangle\,\mathrm{d}t, \qquad (10.61)$$

provided cross sections that agree well with the data for $E \gtrsim 100$ keV for the reaction $H^{+} + H \to H + H^{+}$. This was also the case for the reaction $H^{+} + He \to H + He^{+}$ for which there is no overall Coulomb interaction in either the initial or final channels (Bransden *et al.* 1954). Unfortunately when applied to other systems the JS approximation failed to agree with experiment, so that neither the OBK or JS approximations could be considered satisfactory in first order. It was then assumed that the situation would be improved with the calculation of higher-order terms. Since the matrix element of V_{AB} in the JS approximation was large and as V_{AB} cannot contribute to the exact amplitude (in the impact parameter model) this contribution should be cancelled in higher order. However, explicit calculation of the second-order OBK perturbation series, using plane-wave states to construct the Green's function, produces cross sections which are even further from the data than the first-order OBK cross sections (Miraglia *et al.* 1981; Simony and McGuire 1981), and if the Green's functions $g_{\mathrm{B}}^{\pm}$ or $g_{\mathrm{A}}^{\pm}$ are employed the second-order matrix element is singular, as we have seen. These difficulties are eliminated when the boundary corrected Born series or other methods respecting the Coulomb boundary conditions are employed. It is now possible to understand why the JS cross section gave such good results for $H^+ + H$ (and for $H^+ + He$), but not for other systems. For the $H^+ + H$ system the assumed perturbations $U_{\mathrm{A}} = -1/x + 1/R$ for the boundary corrected and the JS approximations are identical. Furthermore, the two phase $\sigma(t)$ and $\delta(t)$ are equal, so that (see (10.35))

$$\begin{aligned}{}^{\mathrm{I}}\mathscr{A}_{ni}^{(1)}(\mathbf{b}) &= -\mathrm{i}\int_{-\infty}^{\infty}\langle \mathrm{X}_n^{\mathrm{AT}}(t)|-\frac{1}{x}+\frac{1}{R}|\Phi_i^{\mathrm{B}}(t)\rangle\,\mathrm{d}t\\ &= -\mathrm{i}\exp\left[-\frac{2\mathrm{i}}{v_0}\log(b)\right]\int_{-\infty}^{\infty}\langle \tilde{\mathrm{X}}_n^{\mathrm{A}}(t)|-\frac{1}{x}+\frac{1}{R}|\tilde{\Phi}_i^{\mathrm{B}}(t)\rangle\,\mathrm{d}t\\ &= \exp\left[-\frac{2\mathrm{i}}{v_0}\log(b)\right]{}^{\mathrm{JS}}\mathscr{A}_{ni}^{(1)}(\mathbf{b}).\end{aligned} \tag{10.62}$$

Since ${}^{\mathrm{I}}\mathscr{A}_{ni}^{(1)}(\mathbf{b})$ and ${}^{\mathrm{JS}}\mathscr{A}_{ni}^{(1)}(\mathbf{b})$ only differ by a phase the resulting total cross sections are identical. It is easy to establish that this relationship holds to all orders, so that for the particular case of $H^+ + H \rightarrow H + H^+$, the Nth-order approximations to the cross sections arising from the JS and boundary corrected methods are the same. In addition, if the differential cross section is computed the additional phase due to the internuclear potential has to be included in the boundary corrected amplitude as in (10.29), and when this is done the JS and boundary corrected amplitudes agree both in magnitude and phase, so that the differential cross sections are also the same in both approximations for this system.

High-energy behaviour

First-order cross sections It is useful to compare the high-energy behaviour of various approximations with that obtained from the first-order OBK approximation. Let us consider a true one-electron system in which the heavy particles A and B are bare nuclei of charges Z_{A} and Z_{B}, respectively. The OBK cross section has been evaluated in closed form by Omidvar (1967) for capture from an arbitrary initial state to an arbitrary final state with quantum numbers n_n, l_n, m_n. At asymptotic velocities the cross section varies like $v_0^{-12-2l_i-2l_n}$. The capture cross section, in this approximation, from the 1s ground state into the nth level is given at high velocities by the expression

$$^{\mathrm{OBK}}\sigma(1\mathrm{s} \to n) \underset{v_0 \to \infty}{\sim} \frac{2^{18}(Z_A Z_B)^5}{5n^3 v_0^{12}}. \tag{10.63}$$

The B1B cross section also varies like $v_0^{-12-2l_i-2l_n}$ at high velocities (Omidvar 1975; Karnakov 1982).

In the case of the symmetrical resonance reaction

$$H^+ + H(1s) \to H(1s) + H^+, \tag{10.64}$$

a closed form for the B1B cross section was obtained by Jackson and Schiff (1953) and this has been extended by Dewangan and Chakraborty (1989) (see also Dubé *et al.* 1990) to 1s → 1s capture but with arbitrary charges Z_A and Z_B. For reaction (10.64), the asymptotic cross section in the B1B (or JS) approximation is

$$^{\mathrm{B1B}}\sigma(1\mathrm{s} \to 1\mathrm{s}) \underset{v_0 \to \infty}{\sim} 0.661\ ^{\mathrm{OBK}}\sigma(1\mathrm{s} \to 1\mathrm{s}). \tag{10.65}$$

It can be seen from (10.63) that for capture from the ground state to the nth level, the OBK cross section is proportional to n^{-3} at high impact velocities. This '$1/n^3$' rule was originally suggested by Oppenheimer (1928), and has been shown to apply also to the B1B cross section for the $H^+ + H$ system (Karnakov 1982; Omidvar 1975; Roy *et al.* 1980).

Second-order cross sections For the symmetrical resonance reaction (10.64), Drisko (1955) employed the B2B0 approximation defined earlier to obtain the high velocity result

$$^{\mathrm{B2B0}}\sigma(1\mathrm{s} \to 1\mathrm{s}) \sim \left[0.294 + \frac{5\pi v_0}{2^{12}}\right] {}^{\mathrm{OBK}}\sigma^{(1)}_{(1\mathrm{s} \to 1\mathrm{s})}. \tag{10.66}$$

As expected from our previous discussion the high-energy behaviour is modified from v_0^{-12} to v_0^{-11}, and in addition the coefficient of the v_0^{-12} term is modified from that of the first-order B1B approximation. Drisko also showed that in the third-order approximation, B3B0, the coefficient of the v_0^{-12} term is again altered, but the v_0^{-11} term remains unchanged, (see also Shakeshaft 1978a),

$$^{\mathrm{B3B0}}\sigma(1\mathrm{s} \to 1\mathrm{s}) \sim \left[0.319 + \frac{5\pi v_0}{2^{12}}\right] {}^{\mathrm{OBK}}\sigma^{(1)}_{(1\mathrm{s} \to 1\mathrm{s})}. \tag{10.67}$$

The asymptotic form of the second-order OBK cross section can be obtained provided a free-particle Green's function is employed, since the use of either g_A^+ or g_B^+ leads to a divergent integral. For capture from the n_i, l_i, m_i state into the state n_n, l_n, m_n, the asymptotic cross section has been calculated by Shakeshaft (1974*a*) and shows that the second-order cross sections behave like v_0^{-11} at high energies, independently of the quantum numbers of the initial and final states. At high energies, the second-order, as well as the first-order approximations, all display the $1/n^3$ rule of Oppenheimer (Dubé and Briggs 1981; Shakeshaft 1974*a*; Karnakov 1982).

In the case of 1s → 1s capture direct observation of the asymptotic v_0^{-11} behaviour of the cross section is not possible, because the v_0^{-11} term only dominates at energies in excess of ~100 MeV, for which relativistic corrections become important. The first-order Born approximation contribution decreases with increasing velocity like $v_0^{-12-2l_i-2l_n}$, so that for transitions to states with non-zero angular momentum the rapid decrease of the first-order contribution offers a chance of observing the v_0^{-11} tail of the cross section at much lower energies. For example, Dewangan (1988) has suggested that this behaviour may be observable in the reaction $H^+ + H(1s) \rightarrow H(2p) + H^+$ at energies as low as 10 MeV. The double scattering which gives rise to the v_0^{-11} term in the cross section gives rise to a peak in the angular distribution, the Thomas peak, which, as we saw in Section 10.1, has been observed in the differential cross sections for the $H^+ + He$ and $H^+ + H$ systems.

The precise connection between the double scattering mechanism contained in second-order perturbation theory and the classical model of Thomas has been discussed by Shakeshaft (1974*b*) and by Shakeshaft and Spruch (1979), and it has been shown by Spruch (1978) that for capture involving atomic states with large values of n (Rydberg states), the classical and quantal cross sections become identical, as might be expected from the correspondence principle.

10.3 DISTORTED WAVE SERIES

Rather than base perturbation expansions on the functions Φ_j^B and X_n^A which satisfy (10.8) and (10.10), respectively, *distorted waves* can be introduced to represent some physical features of the system, for example, to allow for the effective interaction in each channel. There are clearly infinitely many ways in which to choose distorted waves, but for the moment we shall introduce functions $\mathscr{L}_j^B(\mathbf{r}, t)$ and $\mathscr{M}_n^A(\mathbf{r}, t)$ which satisfy the equations

$$\mathrm{i}\frac{\partial}{\partial t}\mathscr{L}_j^B(\mathbf{r}, t) = [H_B + W_A(R)]\mathscr{L}_j^B(\mathbf{r}, t), \tag{10.68}$$

and

$$\mathrm{i}\frac{\partial}{\partial t}\mathscr{M}_n^A(\mathbf{r}, t) = [H_A + W_B(R)]\mathscr{M}_n^A(\mathbf{r}, t). \tag{10.69}$$

In (10.68) and (10.69), W_A and W_B are distorting potentials in the direct and rearranged channels, respectively, which depend on the interionic separation R only and hence are functions of $\mathbf{b}$ and t. The condition will be imposed that W_A and W_B are of short range, and decrease as $|t| \to \infty$ faster than $1/R$. The two sets of functions $\mathscr{L}_j^B$ and $\mathscr{M}_n^A$ are each complete and have the explicit form

$$\mathscr{L}_j^B(\mathbf{r}, t) = \psi_j(\mathbf{r}) \exp[i\alpha_j(t) - i\omega_j(t)] \tag{10.70}$$

$$\mathscr{M}_n^A(\mathbf{r}, t) = \chi_n(\mathbf{r}) \exp[i\beta_n(t) - i\Omega_n(t)] \tag{10.71}$$

where α_j and β_n are defined by (10.13) and (10.18), respectively. The phases ω_j and Ω_n represent the distortion by the the short-range potentials W_A and W_B and have the form

$$\omega_j(t) = \int_{-\infty}^{t} W_A(R)\, dt', \tag{10.72}$$

$$\Omega_n(t) = \int_{-\infty}^{t} W_B(R)\, dt'. \tag{10.73}$$

The time-reversed functions $\mathscr{L}_j^{BT}$ and $\mathscr{M}_n^{AT}$, defined as

$$\mathscr{L}_j^{BT}(\mathbf{r}, t) = [\mathscr{L}_j^B(\mathbf{r}, t)]^*; \qquad \mathscr{M}_n^{AT}(\mathbf{r}, t) = [\mathscr{M}_n^A(\mathbf{r}, -t)]^*, \tag{10.74}$$

also satisfy equations (10.68) and (10.69), respectively. The analogous relation to (10.35) is that

$$\sigma(t) + \omega_j(t) + \sigma(-t) + \omega_j(-t) = \frac{2Z_A}{v_0} \log(b) + \int_{-\infty}^{\infty} W_A(R(\mathbf{b}, t'))\, dt' \tag{10.75}$$

and a similar relationship holds for the phases $\eta_i(t)$ and $\Omega_n(t)$. The right-hand side of (10.75) is independent of t and it follows that (cf. (10.36))

$$\mathscr{L}_j^{BT}(\mathbf{r}, t)[\mathscr{L}_j^{BT}(\mathbf{r}', t')]^* = \mathscr{L}_j^B(\mathbf{r}, t)[\mathscr{L}_j^B(\mathbf{r}', t')]^* \tag{10.76}$$

with a similar relation for the functions $\mathscr{M}^{AT}$ and $\mathscr{M}^A$.

Following the procedures of the previous section, it is straightforward to show that for excitation (or ionization) the probability amplitude $\mathscr{A}_{ji}(\mathbf{b})$ can be expressed as

$$\mathscr{A}_{ji}(\mathbf{b}) = -i \int_{-\infty}^{\infty} \langle \mathscr{L}_j^{BT} | U_A - W_A | \tilde{\psi}_i \rangle\, dt, \qquad j \neq i, \tag{10.77}$$

and for charge exchange

$$\mathscr{A}_{ni}(\mathbf{b}) = -\mathrm{i}\int_{-\infty}^{\infty} \langle \mathscr{M}_n^{\mathrm{AT}}|U_{\mathrm{B}} - W_{\mathrm{B}}|\tilde{\psi}_i\rangle\, \mathrm{d}t \tag{10.78}$$

$$= -\mathrm{i}\int_{-\infty}^{\infty} \langle \tilde{\psi}_n^{\mathrm{T}}|U_{\mathrm{A}} - W_{\mathrm{A}}|\mathscr{L}_i^{\mathrm{B}}\rangle\, \mathrm{d}t. \tag{10.79}$$

Integral equations for $\tilde{\psi}_i$ and $\tilde{\psi}_n^{\mathrm{T}}$ similar in structure to (10.32) and (10.48) can be found. In operator form these are:

$$\tilde{\psi}_i = \mathscr{L}_i^{\mathrm{B}} + \mathscr{G}_{\mathrm{B}}^{+}[U_{\mathrm{A}} - W_{\mathrm{A}}]\tilde{\psi}_i \tag{10.80}$$

$$\tilde{\psi}_n^{\mathrm{T}} = \mathscr{M}_n^{\mathrm{AT}} + \mathscr{G}_{\mathrm{A}}^{-}[U_{\mathrm{B}} - W_{\mathrm{B}}]\tilde{\psi}_n^{\mathrm{T}} \tag{10.81}$$

where

$$\mathscr{G}_{\mathrm{B}}^{\pm} = \lim_{\varepsilon\to 0^+} \frac{1}{\mathrm{i}\dfrac{\partial}{\partial t} \pm \mathrm{i}\varepsilon - H_{\mathrm{B}} - W_{\mathrm{A}}} \tag{10.82}$$

and

$$\mathscr{G}_{\mathrm{A}}^{\pm} = \lim_{\varepsilon\to 0^+} \frac{1}{\mathrm{i}\dfrac{\partial}{\partial t} \pm \mathrm{i}\varepsilon - H_{\mathrm{A}} - W_{\mathrm{B}}}. \tag{10.83}$$

Explicit forms of the distorted wave Green's functions are

$$\mathscr{G}_{\mathrm{B}}^{+}(\mathbf{r}, t; \mathbf{r}', t') = -\mathrm{i}\theta(t - t')\sum_j \mathscr{L}_j^{\mathrm{BT}}(\mathbf{r}, t)[\mathscr{L}_j^{\mathrm{BT}}(\mathbf{r}', t')]^* \tag{10.84}$$

$$\mathscr{G}_{\mathrm{A}}^{-}(\mathbf{r}, t; \mathbf{r}', t') = \mathrm{i}\theta(t' - t)\sum_n \mathscr{M}_n^{\mathrm{AT}}(\mathbf{r}, t)[\mathscr{M}_n^{\mathrm{AT}}(\mathbf{r}', t')]^*. \tag{10.85}$$

The sums in (10.84) and (10.85) include an integration over the continuum of the atoms $(\mathrm{B} + \mathrm{e}^-)$ or $(\mathrm{A} + \mathrm{e}^-)$, respectively. The relations $\mathscr{G}_{\mathrm{B}}^{-\dagger} = \mathscr{G}_{\mathrm{B}}^{+}$ and $G_{\mathrm{A}}^{-\dagger} = \mathscr{G}_{\mathrm{A}}^{+}$ should be noted.

Combining the iterative solutions of (10.80) or (10.81) with (10.78) or (10.79) the distorted wave Born (DWB) series for the charge exchange amplitude are obtained. The Nth-order amplitude for charge exchange can be written either in terms of $\mathscr{G}_{\mathrm{A}}^{+}$ or of $\mathscr{G}_{\mathrm{B}}^{+}$, as:

$${}^{\mathrm{I}}\mathscr{A}_{ni}^{(N)}(\mathbf{b}) = -\mathrm{i}\int_{-\infty}^{\infty} \langle \mathscr{M}_n^{\mathrm{AT}}|(U_{\mathrm{B}} - W_{\mathrm{B}})\sum_{k=1}^{N}[\mathscr{G}_{\mathrm{B}}^{+}(U_{\mathrm{A}} - W_{\mathrm{A}})]^{k-1}|\mathscr{L}_i^{\mathrm{B}}\rangle\, \mathrm{d}t, \tag{10.86}$$

$${}^{\mathrm{II}}\mathscr{A}_{ni}^{(N)}(\mathbf{b}) = -\mathrm{i}\int_{-\infty}^{\infty} \langle \mathscr{M}_n^{\mathrm{AT}}|\sum_{k=1}^{N}[(U_{\mathrm{B}} - W_{\mathrm{B}})\mathscr{G}_{\mathrm{A}}^{+}]^{k-1}(U_{\mathrm{A}} - W_{\mathrm{A}})|\mathscr{L}_i^{\mathrm{B}}\rangle\, \mathrm{d}t. \tag{10.87}$$

As for the boundary corrected perturbation series, it can be seen that provided the internal wave functions $\phi_i(\mathbf{r})$ and $\chi_n(\mathbf{x})$ are exact there is no first-order post–prior discrepancy, although in general ${}^{\mathrm{I}}\mathscr{A}_{ni}^{(N)} \neq {}^{\mathrm{II}}\mathscr{A}_{ni}^{(N)}$, for $N > 1$.

The potentials W_{A} and W_{B} can be chosen to represent as much as possible of the effective interaction between A and the atom $(\mathrm{B} + \mathrm{e}^-)$ in the initial state, or between B and $(\mathrm{A} + \mathrm{e}^-)$ in the final state. The simplest choice is to set $W_{\mathrm{A}} - Z_{\mathrm{A}}/R$ equal to the static interaction in the incident channel, and $W_{\mathrm{B}} - Z_{\mathrm{B}}/R$ to the static interaction in the final channel

$$W_{\mathrm{A}} - \frac{Z_{\mathrm{A}}}{R} = \langle \mathscr{L}_i^{\mathrm{B}} | V_{\mathrm{Ae}} | \mathscr{L}_i^{\mathrm{B}} \rangle \tag{10.88}$$

$$W_{\mathrm{B}} - \frac{Z_{\mathrm{B}}}{R} = \langle \mathscr{M}_n^{\mathrm{A}} | V_{\mathrm{Be}} | \mathscr{M}_n^{\mathrm{A}} \rangle . \tag{10.89}$$

The first-order DW approximation (DW1B) becomes

$${}^{\mathrm{II}}\mathscr{A}_{ni}^{(1)}(\mathbf{b}) = {}^{\mathrm{I}}\mathscr{A}_{ni}^{(1)}(\mathbf{b}) = -\mathrm{i} \int_{-\infty}^{\infty} \langle \mathscr{M}_n^{\mathrm{T}} | V_{\mathrm{Ae}} - \langle \mathscr{L}_i^{\mathrm{B}} | V_{\mathrm{Ae}} | \mathscr{L}_i^{\mathrm{B}} \rangle | \mathscr{L}_i^{\mathrm{B}} \rangle \, \mathrm{d}t, \tag{10.90}$$

which is closely related to the first-order approximation to the pair of coupled equations obtained in the two-state approximation, obtained by Bates (1958) and discussed in Chapter 7. Referring to equation (7.123) and (7.124), it is seen that in the approximation in which $(1 - |S_{\mathrm{if}}|^2)$ is replaced by unity in the denominators, the first-order amplitude $\mathscr{D}_{\mathrm{f}}(\mathbf{b})$ defined by (7.123) is equal to the first-order expression (10.90).

The first-order distorted wave cross section (DWB1) provided by (10.90) is asymptotic to the first-order OBK cross section and varies like v_0^{-12} for $1\mathrm{s} \to 1\mathrm{s}$ capture

$${}^{\mathrm{DWB1}}\sigma \underset{v_0 \to \infty}{\sim} {}^{\mathrm{OBK1}}\sigma . \tag{10.91}$$

As usual the second-order amplitude contains the double scattering mechanism which provides the dominant v_0^{-11} behaviour at large velocities.

10.4 APPLICATIONS OF THE BOUNDARY CORRECTED AND DISTORTED WAVE BORN APPROXIMATIONS

First-order approximations

Although as $v_0 \to \infty$, the second-order term in the boundary corrected or distorted wave perturbation series must become increasingly important, this by no means implies that the first-order approximations will not be useful at non-asymptotic velocities, and extensive calculations have shown that first order *total* cross sections are in remarkable agreement with experimental data over a considerable velocity range extending from $v_0 \approx v_e$ to $v_0 \approx 10v_e$. For the boundary corrected series, the first B1B calculations were for capture by H^+ from the K-shell of argon (1–10 MeV), by Dewangan and Eichler (1986); for capture by H^+, He^{2+}, and Li^{3+} from the K-shell of carbon (0.1–5 MeV/a.m.u.), by H^+ from He (0.1–1 MeV) and by H^+ from Ar by Belkić *et al.* (1986*a*). As an example, the cross section for electron capture from the K-shell of argon by proton impact is shown in Fig. 10.3. In these examples the target is complex, but of particular significance is a series of calculations for capture by fully stripped ions from the ground state of atomic hydrogen; a true one-electron system (Belkić *et al.* 1987*b*). Excellent agreement was obtained with the total cross section measurements (where these exist) over an energy range from ~100 keV/a.m.u. to ~4 MeV/a.m.u. for the incident ions H^+, He^{2+}, Li^{3+}, Be^{4+}, Be^{5+}, and C^{6+}. The results for H^+,

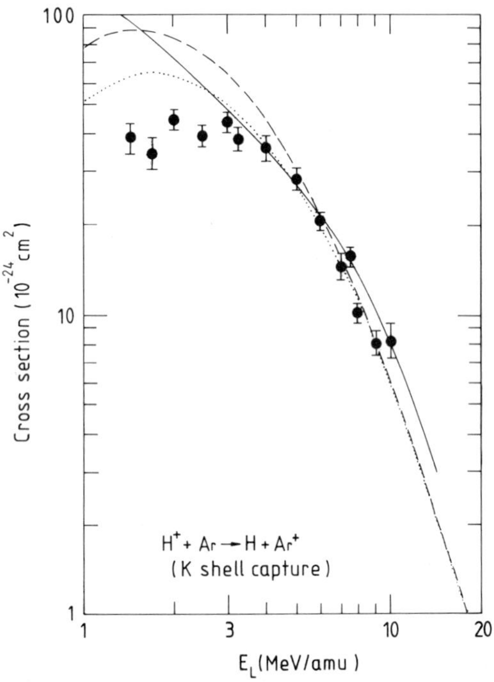

Fig. 10.3. Total cross section for electron capture by protons from the K-shell of argon (after Taulbjerg *et al.* (1990)). Experimental data: ● from Horsdal-Pedersen *et al.* (1983*b*). Theoretical cross sections: in the B1B approximation ———, from Dewangan and Eichler (1986); in the distorted wave strong potential Born approximation ······, from Taulbjerg *et al.* (1990); in the impulse approximation – – –, Alston (1983).

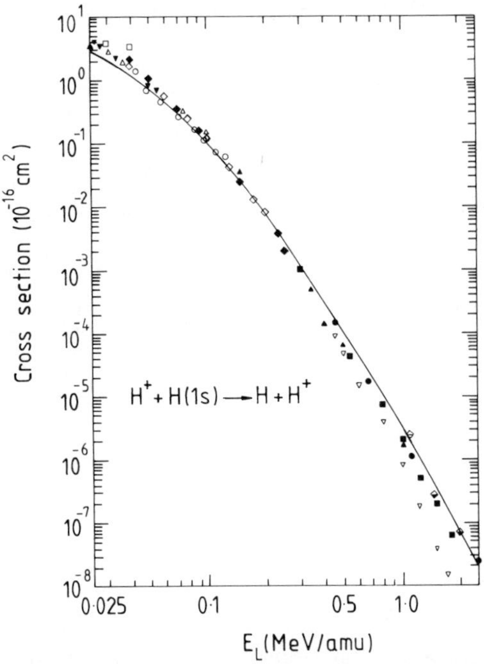

Fig. 10.4. Total cross section for electron capture from H(1s) by protons (after Belkić *et al.* (1987*b*). Experimental data: atomic hydrogen target: ▼ Bayfield (1969); □ Fite *et al.* (1960); ○ Gilbody and Ryding (1966); △ McClure (1966); ◆ Wittkower *et al.* (1966). Molecular hydrogen target: equivalent atomic hydrogen cross section (see Belkić and Gayet 1977): ▲ Barnett and Reynolds (1958); ⬙ Schryber (1967); ◇ Stier and Barnett (1956); ■ Toburen *et al.* (1968); ● Welsh *et al.* (1967); ▽ Williams (1967). Theoretical cross sections: ——— B1B and JS models, from Belkić *et al.* (1987*b*).

Li^{3+}, and C^{6+} are illustrated in Figs 10.4, 10.5, and 10.6, which also show the first-order JS cross section. This cross section is the same as that of the B1B model in the case of the H^+ projectile, but increasingly diverges from the B1B cross section, and the data, as the projectile charge is increased. Other interesting results have been obtained by Belkić *et al.* (1986*b*, 1987*a*) for capture from Li by He^{2+} (250 keV–2 MeV) using an independent particle model and by Deco *et al.* (1986*c*), who calculated the cross sections for capture by H^+ into H(2s) and H(2p) from H(1s) and He(1s^2) targets. In the latter case, the agreement with the experimental data is not as good as for the overall total cross sections, but the data do not extend to very high energies (<250 keV for 2s capture and <40 keV for 2p capture from H(1s); <200 keV for 2s capture and <150 keV for 2p capture from He(1s^2)).

A further, and more stringent, test of the first-order boundary corrected model is to compare the computed differential capture cross section with

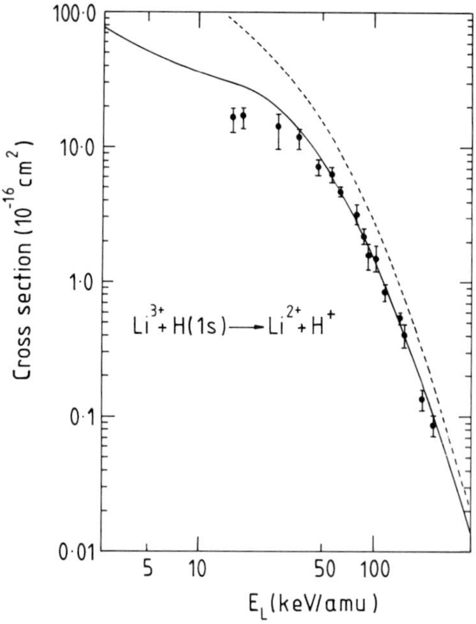

Fig. 10.5. Total cross section for electron capture from H(1s) by Li^{3+} ions (after Belkić *et al.* (1987*b*)). Experimental data: ● Shah *et al.* (1978). Theoretical cross sections from Belkić *et al.* (1987*b*): ——— B1B model; ----- JS model.

experiment. A clear-cut case is the symmetrical reaction $H^+ + H(1s) \rightarrow H(1s) + H^+$ (Belkić *et al.* 1986*b*; Toshima and Ishihara 1989; Decker and Eichler 1989*a, b*). The computed cross section for capture into the ground state shows a deep unphysical dip at an angle which varies with energy. This is shown clearly in Figs 10.7 and 10.8 where the results at 125 keV and 5 MeV increased by 20 per cent to allow for capture into excited states are compared with the experimental data. It is seen that while the data near the forward direction are well represented by the theoretical cross sections, in the vicinity of the dip the B1B cross sections fail completely. This is not, of course, too surprising since even for direct excitation higher-order terms in the Born series are known to be significant.

Certain sequences of higher-order terms, namely those representing elastic scattering in the initial and final channels, are summed in the first-order distorted wave approximation. The DWB1 approximation does not contain the double scattering Thomas mechanism and cannot improve on the B1B approximation in that respect but, nevertheless, it is of interest to examine how the total and differential cross sections predicted compare with experiment. As far as the total cross section is concerned the work of Toshima *et al.* (1987) on capture from $He(1s^2)$ by H^+ in the energy range 100 keV to

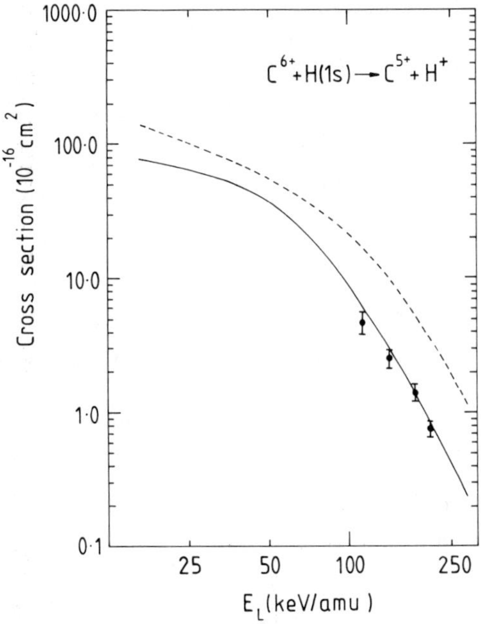

Fig. 10.6. Total cross sections for electron capture from H(1s) by C^{6+} ions (after Belkić *et al.* (1987*b*)). Experimental data: ● Goffe *et al.* (1979). Theoretical cross sections from Belkić *et al.* (1987*b*): ——— B1B model; ----- JS model.

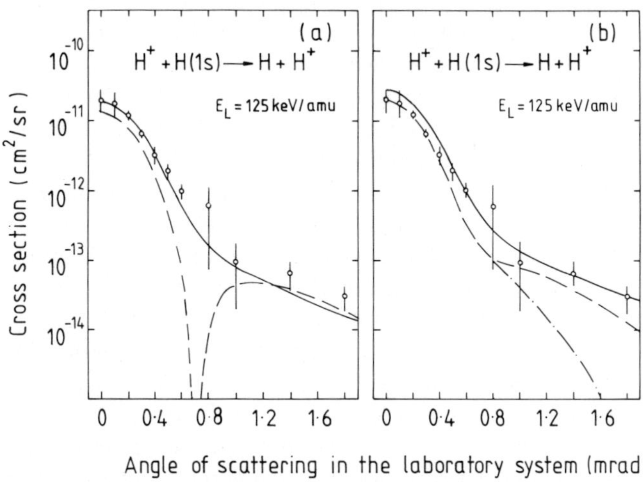

Fig. 10.7. Differential cross section for electron capture by protons from atomic hydrogen at $E_L = 125$ keV/a.m.u. Experimental data from Martin *et al.* (1981). Theoretical cross sections: (a) first order: ——— DWB1, Decker and Eichler (1989*b*); ----- B1B; (b) second order: ——— DWB2, Decker and Eichler (1989*b*); ----- B2B0, Belkić (1988); —·—·— B2B, Dewangan and Bransden (1988).

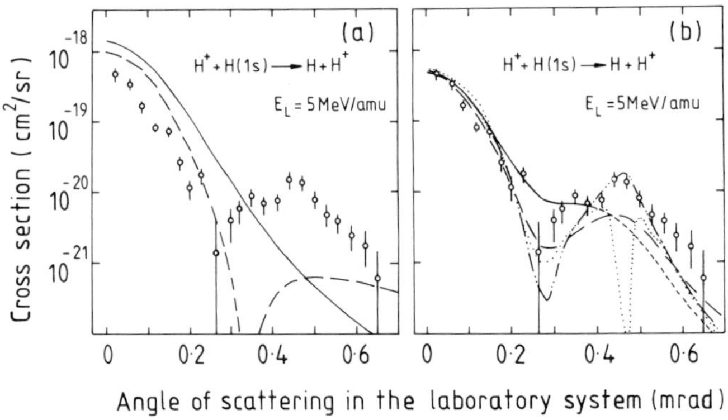

Fig. 10.8. Differential cross section for electron capture by protons from atomic hydrogen at $E_L = 5$ MeV/a.m.u. Experimental data from Vogt *et al.* (1986). Theoretical cross sections: (a) first-order—see Fig. 10.7(a); (b) second order: ——— DWB2, the prolongation of this curve for $\theta_L > 0.4$ mrad, denoted by -----, indicates that the numerical error on the calculated cross-section may exceed 20 per cent, Decker and Eichler (1989*b*); ——— B2B0, Belkić (1988); ····· CDW, Crothers and Dunseath (1990); —··—··— TCDW, Crothers and Dunseath (1990).

10 MeV shows that the DWB1 cross section is too large by a factor of two compared with the data, except at energies below ~150 keV. In contrast the B1B total cross section appears to be in good accord with the data. On the other hand, the general shape of the differential cross section for capture from H(1s) by H^+ is given rather well by the DWB1 approximation at 60 keV and 125 keV, but at 5 MeV the agreement is poor although the dip displayed by the B1B cross section is filled in (Toshima and Ishihara 1989; Decker and Eichler 1989*a*, *b*), as can be seen in Figs 10.7 and 10.8. Too much should not be read into the relative success of the B1B compared with the DWB1 approximation for the *total* charge exchange cross section. While empirically it seems safe to compute total cross sections using the B1B approximation, it is quite clear that second- and higher-order terms must be included to obtain a sound theoretical description of charge exchange. It is salutary to recall that (see Chapter 7) the first-order Born approximation for the total excitation cross section is often reasonably accurate under circumstances in which a number of higher-order terms are individually significant, but happen to cancel. In that case also the first-order Born approximation fails to provide an adequate description of the differential cross section.

Second-order approximations

The second term in the boundary corrected Born distorted wave series is very difficult to evaluate because of the infinite sum over intermediate states

Table 10.1. Total cross sections for the reaction $H^+ + H(1s) \rightarrow H(1s) + H^+$. The numbers in brackets are the powers of ten by which the entry has to be multiplied. The DW1B and DW2B0 cross sections are from Decker and Eichler (1989*a*, *b*); the B2B0 cross sections are from Kramer (1972), Decker and Eichler (1989*a*, *b*), and Belkić (1988); the B2B cross section is from Dewangan and Bransden (1988). The experimental data are from Stier and Barnett (1956) (at 125 keV and 2 MeV), Bayfield (1969) (at 500 keV), Barnett and Reynolds (1958) (at 1 MeV), and Schwab *et al.* (1987) (at 5 MeV). The experimental cross sections are for capture into all states of atomic hydrogen.

	Cross sections (10^{-16} cm^2)				
E_L (keV/a.m.u.)	B1B	B2B0(B2B)	DW1B	DW2B0	Experiment
125	3.1(−2)	4.0(−2)(4.0(−2))	4.8(−2)	7.4(−2)	4.2(−2)
500	8.8(−5)	6.9(−5)	–	–	6.6(−5)
1000	2.6(−6)	1.9(−6)	5.1(−6)	3.5(−6)	2.0(−6)
2000	5.9(−8)	4.0(−8)	–	–	7.2(−8)
5000	3.3(−10)	1.6(−10)	6.1(−10)	2.4(−10)	3.1(−10)

which must be carried out numerically. Considerable simplification results if the Green's functions $g^{\pm}_{A,B}$ or $G^{\pm}_{A,B}$ are replaced by the phase modified free-particle Green's function $g^{\pm}_{0}$ and the most extensive calculations have been made in the B2B0 and DW2B0 approximation in which this substitution has been made. In Table 10.1, total cross sections for the reaction $H^+ + H(1s) \rightarrow H(1s) + H^+$ are shown for both the first- and second-order boundary corrected Born and distorted wave approximations. At 125 keV, results are also available for a full B2B calculation in which all the important Coulomb intermediate states were summed and integrated over. Bound states with $n \leq 6$ and $l \leq 3$ made a significant contribution as did continuum states with $l \leq 3$. The integration over the continuum was effected numerically.

At the single energy for which comparison is possible (125 keV), the B2B and B2B0 total capture cross sections agree well, which supports the hope that the use of $g^{\pm}_{0}$ rather than $g^{\pm}_{A,B}$ is generally a good approximation, for symmetrical or near symmetrical systems. The change in the cross section in going from the first-order to second-order theories is small enough, except at the highest energy shown, to suggest that the third-order corrections are not substantial. Keeping in mind that the theoretical cross sections have to be increased to allow for capture into excited states (by $\sim$20 per cent if the n^{-3} rule holds), agreement with experiment is reasonable (Table 10.1 and

Fig. 10.4), with rather better agreement being shown for the B2B0 than the DW2B0 cross sections, except at 5 MeV.

The corresponding B2B0 and DW2B0 differential cross sections have been computed at a number of energies (B2B0: Belkić (1988) and Decker and Eichler (1989*a*); DW2B0: Decker and Eichler 1989*b*) and the B2B differential cross section has been given by Dewangan and Bransden (1988) at 125 keV. The computed cross sections at 125 keV and 5 MeV are shown together with the experimental data in Figs. 10.7 and 10.8. At 125 keV, both the B2B0 and DW2B0 cross sections agree with the data, on the other hand the B2B cross section, while in good agreement at small angles falls off too rapidly with increasing angle. This is probably due to numerical difficulties in the evaluation of the second-order term containing $g_{\mathrm{B}}^{\pm}$ at small impact parameters. At 5 MeV, the B2B0 cross section reproduces the Thomas peak rather well, but the DW2B0 cross section displays a shoulder rather than a peak near $\Theta_{\mathrm{L}} = 0.47$ mrad.

10.5 THE STRONG POTENTIAL BORN AND IMPULSE APPROXIMATIONS

It has already been remarked that if one of the post or prior perturbations is much stronger than the other, an expansion in powers of the weaker potential is appropriate. For example, if $U_{\mathrm{B}} \ll U_{\mathrm{A}}$, the boundary corrected second-order Born approximation (see (10.55))

$$^{\mathrm{II}}\mathscr{A}_{ni}^{(2)}(\mathbf{b}) = -\mathrm{i}\int_{-\infty}^{\infty} \langle \mathrm{X}_n^{\mathrm{AT}}|\{1 + U_{\mathrm{B}}g_{\mathrm{A}}^{+}\}U_{\mathrm{A}}|\Phi_i^{\mathrm{B}}\rangle\,\mathrm{d}t \qquad (10.92)$$

$$= -\mathrm{i}\int_{-\infty}^{\infty} \langle \mathrm{X}_n^{\mathrm{AT}}|U_{\mathrm{B}}\{1 + g_{\mathrm{A}}^{+}U_{\mathrm{A}}\}|\Phi_i^{\mathrm{B}}\rangle\,\mathrm{d}t \qquad (10.93)$$

contains all the first-order terms in the weaker potential U_{B}. In the second line the fact that there is no 'post–prior discrepancy' in the first-order term has been employed. Correspondingly, if a distorted wave formulation is employed and if $U_{\mathrm{B}} - W_{\mathrm{B}} \ll U_{\mathrm{A}} - W_{\mathrm{A}}$, the DW2B amplitude (10.87) contains all first-order terms in the weaker potential. This observation is the basis of the strong potential Born approximation (SPB), originally introduced by Macek and Taulbjerg (1981) and Macek and Alston (1982) and developed in a distorted wave formulation by Macek and Taulbjerg (1989*a*, *b*) and Taulbjerg *et al.* (1990). The evaluation of the B2B or DW2B approximations is extremely laborious because of the sum over intermediate states and, in particular, the integration over those states which lie in the continuum. For this reason, in the SPB approximation the matrix elements (10.54) or (10.55) are put in a form in which certain approximations can be made, leading to more tractable results.

Let us consider the particular case of a one-electron system in which A and B are bare nuclei of charges Z_A and Z_B, with $Z_B \ll Z_A$. The probability amplitude (10.93) can be written as

$$^{II}\mathscr{A}_{ni}^{(2)}(\mathbf{b}) = -\mathrm{i}\int_{-\infty}^{\infty} \langle X_n^{AT}|U_B|\Psi_{SPB}^{+}\rangle\, \mathrm{d}t \tag{10.94}$$

where the wave function Ψ_{SPB}^{+} is defined as

$$|\Psi_{SPB}^{+}(t)\rangle = \{1 + g_A^{+}U_A\}|\Phi_i^B\rangle. \tag{10.95}$$

Recollecting that $g_A^{+} = \lim_{\varepsilon\to 0^+}[\mathrm{i}\,\partial/\partial t + \mathrm{i}\varepsilon - H_A]^{-1}$, using the result (Dodd and Greider 1966) $\lim_{\varepsilon\to 0^+}[\mathrm{i}\varepsilon g_A^{+}] = 0$, and noting that equation (10.8) is satisfied by Φ_i^B, it follows that

$$|\Psi_{SPB}^{+}(t)\rangle = g_A^{+}\{H_B - H_A + U_A\}|\Phi_i^B\rangle \tag{10.96}$$

$$= g_A^{+}U_B|\Phi_i^B\rangle. \tag{10.97}$$

It is convenient to evaluate the amplitude (10.94) in a frame of reference with the projectile A as origin. In this frame $\mathbf{x}$ and t are the independent variables, and the initial and final Coulomb distorted wave functions are

$$\Phi_i^B(\mathbf{x}, t) = \phi_i(\mathbf{r})\exp[-\mathrm{i}\{\varepsilon_i t + \mathbf{v}_0\cdot\mathbf{x} + \tfrac{1}{2}v_0^2 t + \sigma(t)\} \tag{10.98}$$

and

$$X_n^A(\mathbf{x}, t) = \chi_n(\mathbf{x})\exp[-\mathrm{i}\eta_n t - \mathrm{i}\delta(t)] \tag{10.99}$$

where $\sigma(t)$ and $\delta(t)$ are defined by (10.14) and (10.19), respectively. Equation (10.97) can be written explicitly in the form

$$\Psi_{SPB}^{+}(\mathbf{x}, t) = \lim_{\varepsilon\to 0^+}\left[\mathrm{i}\frac{\partial}{\partial t} + \mathrm{i}\varepsilon - H_A\right]^{-1}\int \mathrm{d}\mathbf{x}'\int_{-\infty}^{\infty}\mathrm{d}t'$$
$$\times\, \delta(\mathbf{x} - x')\delta(t - t')U_B(\mathbf{x}', t')\Phi_i^B(\mathbf{x}', t'). \tag{10.100}$$

Using the closure property of plane waves† to write

$$\delta(\mathbf{x} - \mathbf{x}')\delta(t - t') = (2\pi)^{-4}\int \mathrm{d}\mathbf{q}\int_{-\infty}^{\infty}\exp[\mathrm{i}\mathbf{q}\cdot(\mathbf{x} - \mathbf{x}') - \mathrm{i}\omega(t - t')]\,\mathrm{d}\omega, \tag{10.101}$$

expression (10.100) can be written in the product form as

$$\Psi_{SPB}^{+}(\mathbf{x}, t) = \int_{-\infty}^{\infty}\mathrm{d}t'\int \mathrm{d}\mathbf{q}N(t, t')M_i(t'), \tag{10.102}$$

† The treatment adopted here should be contrasted with tht in Appendix G, (G.4) to (G.7), where an expansion in terms of the eigenfunctions of the atomic subsystem is employed.

where

$$N(t, t') = (2\pi)^{-4} \int_{-\infty}^{\infty} [\omega + i\varepsilon - H_A]^{-1} \exp[i\mathbf{q}\cdot\mathbf{x} - i\omega(t - t')]\, d\omega, \quad (10.103)$$

and

$$M_i(t') = \int e^{-i\mathbf{q}\cdot\mathbf{x}} U_B(\mathbf{x}', t')\Phi_i(\mathbf{x}', t')\, d\mathbf{x}'. \quad (10.104)$$

To evaluate $M_i(t')$, we note that $\mathbf{x}' = \mathbf{r} + \mathbf{R}(t')$ and hence obtain the result

$$M_i(t') = \exp\left[i\left\{\frac{\mathbf{q} + \mathbf{v}_0}{2}\right\}^2 t' - \varepsilon_i t' - \sigma(t') + i\mathbf{q}\cdot\mathbf{b}\right]$$
$$\times \int \exp[-i(\mathbf{q} + \mathbf{v}_0)\cdot\mathbf{r}]\left(-\frac{Z_B}{r} + \frac{Z_B}{R}\right)\phi_i(\mathbf{r})\, d\mathbf{r}. \quad (10.105)$$

This expression can be simplified (Macek and Taulbjerg 1989*a*) by dropping the weak distorting potential Z_B/R in the matrix element. This would seem not to lead to any divergent behaviour because in (10.97) we are dealing with the product of Green's function and a perturbation belonging to different arrangements. Using the wave equation (3.15) satisfied by $\phi_i(\mathbf{r})$, it follows immediately that

$$\int \exp[-i(\mathbf{q} + \mathbf{v}_0)\cdot\mathbf{r}]\left(-\frac{Z_B}{r}\right)\phi_i(\mathbf{r})\, d\mathbf{r} = [\varepsilon_i - \tfrac{1}{2}(\mathbf{q} + \mathbf{v}_0)^2](2\pi)^{\frac{3}{2}}\tilde{\phi}_i(\mathbf{q} + \mathbf{v}_0)$$
$$(10.106)$$

where $\tilde{\phi}_i$ is the Fourier transform of the wave function ϕ_i:

$$\tilde{\phi}_i(\mathbf{k}) = (2\pi)^{-\frac{3}{2}} \int \exp(-i\mathbf{k}\cdot\mathbf{r})\phi_i(\mathbf{r})\, d\mathbf{r}. \quad (10.107)$$

Finally,

$$M_i(t') = [\varepsilon_i - \tfrac{1}{2}(\mathbf{q} + \mathbf{v}_0)^2](2\pi)^{\frac{3}{2}}\tilde{\phi}_i(\mathbf{q} + \mathbf{v}_0)$$
$$\times \exp\left[i\left\{(\mathbf{q} + \mathbf{v}_0)^2 \frac{t'}{2} - \varepsilon_i t' - \sigma(t') + \mathbf{q}\cdot\mathbf{b}\right\}\right] \quad (10.108)$$

To evaluate $N(t, t')$ defined by (10.103), the weak distorting Coulomb potential Z_B/R can be dropped if we work to first order in Z_B/Z_A, so that H_A can be approximated by $H'_A = -\frac{1}{2}\nabla_x^2 - Z_A/x$. Using the result obtained by applying the identity (G.22) of Appendix G,

$$\frac{1}{\omega + i\varepsilon - H'_A} - \frac{1}{\omega + i\varepsilon - H_0} = \frac{1}{\omega + i\varepsilon - H'_A}(H_A - H_0)\frac{1}{\omega + i\varepsilon - H_0}$$
$$(10.109)$$

where $H_0 = -\frac{1}{2}\nabla_x^2$, we find that

$$[\omega + \mathrm{i}\varepsilon - H'_{\mathrm{A}}]^{-1}\,\mathrm{e}^{\mathrm{i}\mathbf{q}\cdot\mathbf{x}} = \left[\omega + \mathrm{i}\varepsilon - \frac{q^2}{2}\right]^{-1} \psi_{\mathbf{q},\omega}(\mathbf{x}) \tag{10.110}$$

with

$$\psi_{\mathbf{q},\omega}(\mathbf{x}) = \left\{1 + \lim_{\varepsilon\to 0^+} \frac{1}{\omega + \mathrm{i}\varepsilon - H'_{\mathrm{A}}}\left(-\frac{Z_{\mathrm{A}}}{x}\right)\right\}\mathrm{e}^{\mathrm{i}\mathbf{q}\cdot\mathbf{x}}. \tag{10.111}$$

The expression (10.111) for $\psi_{\mathbf{q},\omega}(\mathbf{x})$ can be taken to be the definition of the Coulomb wave function off the energy shell, that is with $\omega \neq q^2/2$. A peculiarity of the Coulomb interaction is that the off-shell function does not converge uniformly to the ordinary Coulomb function $\psi_{\mathbf{q}}^{\mathrm{c}}(\mathbf{x})$, where (McDowell and Coleman 1970; Bransden 1983)

$$\psi_{\mathbf{q}}^{\mathrm{c}}(\mathbf{x}) = \mathrm{e}^{-\nu_{\mathrm{A}}\pi/2}\Gamma(1 + i\nu_{\mathrm{A}})\,\mathrm{e}^{\mathrm{i}\mathbf{q}\cdot\mathbf{x}}\,{}_1F_1(-\mathrm{i}\nu_{\mathrm{A}}, 1, \mathrm{i}(qx - \mathbf{q}\cdot\mathbf{x})), \qquad \nu_{\mathrm{A}} = -Z_{\mathrm{A}}/q, \tag{10.112}$$

in the on-shell limit with $\omega = q^2/2$. Analysis by Mapleton (1961) (see also Pradhan (1957) and Macek and Alston (1982)) shows that close to the energy shell, the off-shell function $\psi_{\mathbf{q},\omega}(\mathbf{x})$ is of the form

$$\psi_{\mathbf{q},\omega}(\mathbf{x}) = S\left(\omega, \frac{q^2}{2}\right)\psi_{\mathbf{q}}^{\mathrm{c}}(\mathbf{x}) \tag{10.113}$$

where

$$S(\omega, \omega') = (4\omega)^{-\mathrm{i}\nu}(\omega' - \omega - \mathrm{i}\varepsilon)^{\mathrm{i}\nu}\Gamma(1 - \mathrm{i}\nu)\,\mathrm{e}^{-\pi\nu/2}, \qquad \nu = -Z_{\mathrm{A}}/q. \tag{10.114}$$

Combining (10.113) with (10.110) it is seen that the integral over ω in (10.110) is of the form

$$J(\tau) = \int_{-\infty}^{\infty} \frac{\exp[-\mathrm{i}\omega\tau]}{\omega + \mathrm{i}\varepsilon - q^2/2} S\left(\omega, \frac{q^2}{2}\right)\mathrm{d}\omega, \qquad \text{with } \tau = t - t'. \tag{10.115}$$

This integral can be evaluated by closing the path of integration by an infinite semicircle in the upper or lower half-planes according to whether $\tau < 0$ or $\tau > 0$, and it is found (Dettmann 1971) that

$$J(\tau) = -\mathrm{i}\theta(\tau)\tau^{-\mathrm{i}\nu}\exp\left[-\mathrm{i}q^2\frac{\tau}{2}\right]. \tag{10.116}$$

Using this result in (10.103) together with the expression (10.108) for M_i, the wave function $\Psi^+_{\text{SPB}}(\mathbf{x}, t)$ becomes in this near-shell approximation

$$\Psi^+_{\text{SPB}}(\mathbf{x}, t) = (2\pi)^{\frac{5}{2}} \int d\mathbf{q}\, \psi^c_{\mathbf{q}}(\mathbf{x}) \exp\left[i\left\{ \frac{(\mathbf{q} + \mathbf{v}_0)^2}{2} - \varepsilon_i \right\} t + i\mathbf{q}\cdot\mathbf{b} \right]$$

$$\times\, \tilde{\phi}_i(\mathbf{q} + \mathbf{v}_0)\gamma\left(\frac{(\mathbf{q} + \mathbf{v}_0)^2}{2} \right) \tag{10.117}$$

where

$$\gamma(\Omega) = -i \int_{-\infty}^{\infty} \theta(\tau)(\varepsilon_i - \Omega)\tau^{-iv} \exp[-i\{\Omega - \varepsilon_i\}\tau - i\sigma(t')]\, dt'. \tag{10.118}$$

The wave function Ψ^+_{SPB} expressed in (10.117) is the near-shell form of the strong potential Born approximation, which can be used in (10.94) to provide an approximate amplitude for charge exchange. The numerical evaluation of (10.94) requires some further approximations (see Macek and Alston 1982), in particular to the factor $\gamma(\Omega)$ (Macek and Taulbjerg 1989*a*, Taulbjerg *et al.* 1990); nevertheless the aim of providing an approximation to the second-order boundary corrected Born or of the second-order distorted wave Born amplitudes which retains the essential features of these amplitudes, but which is more tractable, has been achieved for asymmetrical collisions ($Z_B \ll Z_A$).

The impulse approximation

It has been argued (Macek and Taulbjerg 1989*b*) that as the region of closest approach is the most important in the evaluation of the probability (10.94), $\gamma(\Omega)$ can be approximated by setting $t = 0$. If in addition the small b limit is taken then

$$\sigma(t') = Z_A \int^{t'} \frac{1}{R(t'')}\, dt'' \approx \frac{Z_A}{v_0} \int^{t'} \frac{1}{|t''|}\, dt'' = \frac{Z_A}{v_0} \log(t'). \tag{10.119}$$

With these approximations the factor τ^{-iv} cancels with $\exp[-i\sigma(t')]$ and $\gamma(\Omega) = 1$. In this case, the wave function Ψ^+_{SPB} reduces to the *impulse approximation* wave function in the form due to Jakubassa-Amundsen and Amundsen (1980)

$$\Psi_{\text{IMP}}(\mathbf{x}, t) = \int d\mathbf{q}\psi^c_{\mathbf{q}}(\mathbf{x})\, e^{i\mathbf{q}\cdot\mathbf{b}} \tilde{\phi}_i(\mathbf{q} + \mathbf{v}_0) \exp\left[i\left\{ \frac{(\mathbf{q} + \mathbf{v}_0)^2}{2} \right\} t - \varepsilon_i t \right]. \tag{10.120}$$

The impulse approximation, introduced for short-range potentials by Chew (1950) and Chew and Wick (1952) represents the collision as a two-body collision between the incident nucleus A and the target electron, which has a momentum given by the momentum distribution of the target atom. In previous derivations (see for example Bransden and Cheshire (1963), Cheshire (1963), Coleman and McDowell (1965), and Briggs (1977)) a renormalization procedure was employed which has the effect of setting the off-shell factor $S(\omega, \omega')$ in (10.113) equal to unity. Okubo and Feldman (1960) have shown that this procedure was necessary in order to obtain a convergent (on-shell) Coulomb scattering amplitude from an off-shell T-matrix constructed from the wave function $\psi_{\mathbf{q},\omega}(\mathbf{x})$, and it was argued that this renormalization was necessary in the present context to obtain the correct limit in the case of vanishing binding of the electron to the target. The derivation of Macek and Taulbjerg is more satisfactory and clearly identifies the crucial role of the distortion of the wave function by the overall Coulomb potential.

In Fig. 10.3, calculated cross sections for the capture of K-shell electron from argon by proton impact are shown for the distorted wave strong potential Born approximation and for the impulse approximation ($\gamma = 1$). The target K-shell wave function is determined by taking the interaction V_{Be} to be a Coulomb potential with an effective charge Z_{B}. It is seen that in the high-velocity region, $v_0/v_{\mathrm{e}} > 1$, corresponding to $E > 8$ MeV, both cross sections agree well with the total cross section data and are very close to the total cross section obtained from the B1B approximation. At lower energies neither approximation is satisfactory as would be expected.

10.6 THE CONTINUUM DISTORTED WAVE AND EIKONAL MODELS

If the potential V_{Ae}, V_{Be} acting between the active electron and the ionic cores A and B are unmodified Coulomb potentials the *continuum distorted wave* model (CDW) introduced by Cheshire (1964) can be employed, in which the effect of the potentials is entirely included in the unperturbed basis, leaving part of the kinetic energy as the corresponding perturbation. Following Crothers (1982) it is convenient to work in terms of variables $\mathbf{r}$, $\mathbf{x}$, and t, where differentiations with respect to t are taken with both $\mathbf{r}$ and $\mathbf{x}$ constant, rather than the variables $\mathbf{r}$, t that we have been using in this chapter.

The electronic Schrödinger equation (10.6) can then be expressed as (Crothers 1982)

$$\left[\tilde{H}_{\mathrm{e}} - \mathrm{i}\frac{\mathrm{d}}{\mathrm{d}t}\bigg|_{\mathbf{r}}\right]\tilde{\psi}(\mathbf{r}, t)$$

$$= \left[-\tfrac{1}{2}\nabla_r^2 - \tfrac{1}{2}\nabla_x^2 - \boldsymbol{\nabla}_r\cdot\boldsymbol{\nabla}_x - \frac{Z_{\mathrm{A}}}{x} - \frac{Z_{\mathrm{B}}}{r} + \mathrm{i}\mathbf{v}_0\cdot\boldsymbol{\nabla}_x - \frac{\partial}{\partial t}\right]\tilde{\psi}(\mathbf{r}, \mathbf{x}; t) = 0 \tag{10.121}$$

where the wave function $\tilde{\psi}$ is expressed explicitly as a function of $\mathbf{r}$, $\mathbf{x}$, and t. With our choice of origin of coordinates on B, the position vector $\mathbf{x}$ is, of course, a function of time $\mathbf{x}(t) = \mathbf{r} - \mathbf{R}(t)$, and in obtaining (10.121) we have used the result

$$\left.\frac{\mathrm{d}}{\mathrm{d}t}\right|_{\mathbf{r}} \tilde{\psi}(\mathbf{r}, t) = \left[\dot{\mathbf{x}}\cdot\boldsymbol{\nabla}_x + \frac{\partial}{\partial t}\right]\tilde{\psi}(\mathbf{r}, \mathbf{x}; t), \tag{10.122}$$

where, on the right-hand side, the derivative $\partial/\partial t$ is taken with both $\mathbf{r}$ and $\mathbf{x}$ held constant.

If the target electron were free, its wave function in the field of the projectile A would satisfy the Schrödinger equation

$$\left[-\tfrac{1}{2}\nabla^2 - \frac{Z_{\mathrm{A}}}{x(t)} - \mathrm{i}\left.\frac{\mathrm{d}}{\mathrm{d}t}\right|_{\mathbf{r}}\right]\mathscr{F}_{\mathrm{A}}(\mathbf{x}) = 0. \tag{10.123}$$

Since $\mathscr{F}_{\mathrm{A}}(\mathbf{x})$ is a function of $\mathbf{x}$ alone, $\mathrm{d}/\mathrm{d}t|_{\mathbf{r}} = -\mathbf{v}_0\cdot\boldsymbol{\nabla}_x$, and the solution of (10.123) is

$$\mathscr{F}^{+}_{\mathrm{A}}(\mathbf{x}) = N_{\mathrm{A}}\, {}_1F_1[\mathrm{i}\nu_{\mathrm{A}}; 1; \mathrm{i}v_0 x + \mathrm{i}\mathbf{v}_0\cdot\mathbf{x}], \qquad \nu_{\mathrm{A}} = Z_{\mathrm{A}}/v_0. \tag{10.124}$$

That this is the case is easily seen if it is recognized that with the normalization

$$N_{\mathrm{A}} = \Gamma(1 - \mathrm{i}\nu_{\mathrm{A}})\exp(\pi\nu_{\mathrm{A}}/2), \tag{10.125}$$

$\psi^{\mathrm{c}}_{\mathbf{v}_0}(\mathbf{x}) = \exp[\mathrm{i}\mathbf{v}_0\cdot\mathbf{x}]\mathscr{F}^{+}_{\mathrm{A}}(\mathbf{x})$ is an ordinary on-shell Coulomb wave function satisfying the Schrödinger equation

$$\left[-\tfrac{1}{2}\nabla^2 - \frac{Z_{\mathrm{A}}}{x} - \tfrac{1}{2}v_0^2\right]\psi^{\mathrm{c}}_{\mathbf{v}_0}(\mathbf{x}) = 0. \tag{10.126}$$

The asymptotic form of $\mathscr{F}^{+}_{\mathrm{A}}$ for $|t| \to -\infty$, with the normalization (10.125), is

$$\mathscr{F}^{+}_{\mathrm{A}}(\mathbf{x}) \xrightarrow[t\to-\infty]{} \exp[-\mathrm{i}\nu_{\mathrm{A}}\log(v_0 R - v_0^2 t)] = \exp[-\mathrm{i}\sigma(t) - \mathrm{i}\xi_{\mathrm{A}}] \tag{10.127}$$

where ξ_{A} is the constant phase $\xi_{\mathrm{A}} = \nu_{\mathrm{A}}\log v_0$, and where $\sigma(t)$ is defined by (10.14). It follows that the distorted wave function

$$L_i(\mathbf{r}; \mathbf{x}; t) = \phi_i(\mathbf{r})\exp[-\mathrm{i}\varepsilon_i t + \mathrm{i}\xi_{\mathrm{A}}]\mathscr{F}^{+}_{\mathrm{A}}(\mathbf{x}) \tag{10.128}$$

coincides with the unperturbed functions Φ^{B}_i in the limit $t \to -\infty$ and contains the effect of the distorting potential $-Z_{\mathrm{A}}/x$, not just at large separations, but at all distances. The function L_i is not, of course, an exact solution of the Schrödinger equation (10.121); indeed we have

$$\left(\tilde{H}_{\mathrm{e}} - \mathrm{i}\left.\frac{\mathrm{d}}{\mathrm{d}t}\right|_{\mathbf{r}}\right)L_i(\mathbf{r}, \mathbf{x}, t) = -\exp[-\mathrm{i}\varepsilon_i t + \mathrm{i}\xi_{\mathrm{A}}][\boldsymbol{\nabla}_x\mathscr{F}^{+}_{\mathrm{A}}(\mathbf{x})\cdot\boldsymbol{\nabla}_r\phi_i(\mathbf{r})]. \tag{10.129}$$

If an operator P is introduced so that

$$PL_i \equiv \exp[-\mathrm{i}\varepsilon_i t + \mathrm{i}\xi_\mathrm{A}][\mathbf{V}_x \mathscr{F}_\mathrm{A}^+(\mathbf{x}) \cdot \mathbf{V}_r \phi_i(\mathbf{r})], \tag{10.130}$$

we can, following the methods of Section 10.1, show that the probability for charge exchange can be expressed as

$$\begin{aligned} \mathscr{A}_{ni}(\mathbf{b}) &= \lim_{t \to -\infty} \langle \tilde{\psi}_n^\mathrm{T} | L_i \rangle \\ &= -\mathrm{i} \int_{-\infty}^{\infty} \langle \tilde{\psi}_n^\mathrm{T} | P | L_i \rangle \, \mathrm{d}t. \end{aligned} \tag{10.131}$$

Similarly a second set of distorted wave functions can be defined as

$$M_n(\mathbf{r}, \mathbf{x}, t) = \chi_n(\mathbf{x}) \exp\left[-\mathrm{i}\eta_n t + \mathrm{i}\xi_\mathrm{B} + \mathrm{i}\mathbf{v}_0 \cdot \mathbf{r} - \frac{\mathrm{i}v_0^2 t}{2}\right] \mathscr{F}_\mathrm{B}^+(\mathbf{r}) \tag{10.132}$$

where

$$\mathscr{F}_\mathrm{B}^+(\mathbf{r}) = N_\mathrm{B}\, {}_1F_1[\mathrm{i}\nu_\mathrm{B}; 1, \mathrm{i}v_0 r + \mathrm{i}\mathbf{v}_0 \cdot \mathbf{r}], \qquad \nu_\mathrm{B} = Z_\mathrm{B}/v_0 \tag{10.133}$$

and

$$N_\mathrm{B} = \Gamma(1 - \mathrm{i}\nu_\mathrm{B}) \exp\left(\frac{\pi \nu_\mathrm{B}}{2}\right). \tag{10.134}$$

If the constant phase ξ_B is set equal to $[\nu_\mathrm{B} \log v_0]$, the distorted wave functions M_n coincide with the functions $\mathrm{X}^\mathrm{A}(\mathbf{r}, t)$ in the limit $t \to -\infty$. It follows that the charge exchange transition probability can also be written in the form

$$\mathscr{A}_{ni}(\mathbf{b}) = -\mathrm{i} \int_{-\infty}^{\infty} \langle M_n^\mathrm{T} | P | \tilde{\psi}_i \rangle \, \mathrm{d}t \tag{10.135}$$

where $M_n^\mathrm{T}(\mathbf{v}_0, t) = M_n^*(-\mathbf{v}_0, -t)$ and the action of P is defined as

$$PM_n \equiv \exp\left[-\mathrm{i}\eta_n t + \mathrm{i}\xi_\mathrm{B} + \tfrac{1}{2}\mathrm{i}\mathbf{v}_0 \cdot \mathbf{r} - \frac{\mathrm{i}v_0^2 t}{2}\right][\mathbf{V}_x \chi_n(\mathbf{x}) \cdot \mathbf{V}_r \mathscr{F}_\mathrm{B}(\mathbf{r})]. \tag{10.136}$$

It is clear from (10.131) and (10.135) that the constant phases ξ_A, ξ_B have no effect on cross sections and they will be omitted from now on.

The first-order CDW approximation

The original work of Cheshire was based on a first-order approximation to (10.131) or (10.135). Setting $\tilde{\psi}_i \approx L_i$ in (10.135), or $\tilde{\psi}_n \approx M_n$ in (10.131), the CDWI approximation is obtained

$$\mathscr{A}_{ni}^\mathrm{I}(\mathbf{b}) = -\mathrm{i} \int_{-\infty}^{\infty} \langle M_n^\mathrm{T} | P | L_i \rangle \, \mathrm{d}t. \tag{10.137}$$

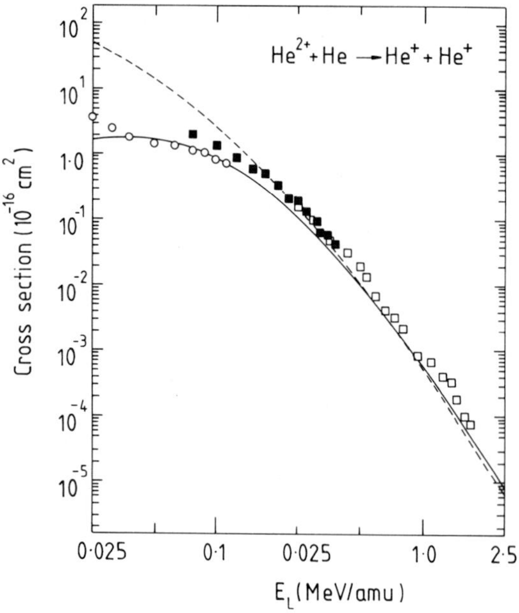

Fig. 10.9. Total cross section for the capture of a single electron from He by He^{2+} ions (after Deco *et al.* (1984)). Experimental data: ○ Allison (1958); ■ Pivovar *et al.* (1962); □ Hvelplund *et al.* (1976). Theoretical cross section in the CDW approximation: – – – from Belkić and Gayet (1977); ——— Symmetrical eikonal (SE) approximation, Deco *et al.* (1984).

This approximation has been shown to be rather successful in reproducing the measured total charge exchange cross sections at high energies where, in the case of a composite target, the interaction between the active electron and the core is taken to be a Coulomb potential $-Z_B/r$ where Z_B is an effective charge chosen to obtain the correct binding energy. As an example the capture cross section by α-particles from He is shown in Fig. 10.9. It is seen that agreement between theory and experiment is obtained for $E \gtrsim 250$ keV/a.m.u. As a second example, the total capture cross section calculated in the CDW approximation for Li^{3+} ions interacting with atomic hydrogen in the ground state has been included in Fig. 8.3. It is seen that the experimental data are well represented for $E_L \gtrsim 100$ keV/a.m.u. In contrast, the cross sections predicted from the two-centre expansion model with no pseudostates, and from the OHC expansion method, are in poor agreement with the high-energy data. As pointed out in Chapter 8, this must be because these particular coupled channel calculations do not contain terms representing the Li^{2+} continuum, which is represented in the CDW wave function.

Detailed comparisons with experiment for a number of systems have been given by Belkić *et al.* (1979) and by Rivarola (1985), and satisfactory agreement is obtained with *total* cross section data for a wide range of values of Z_A and Z_B at high energies. From the symmetrical nature of the approximation there is no post–prior discrepancy and the method is particularly attractive for nearly symmetrical systems for which Z_A and Z_B are comparable. In the case of asymmetrical systems with $Z_A \ll Z_B$, a simplification is possible in which the initial distorted wave function L_i is replaced by an undistorted atomic orbital $\phi_i(\mathbf{r}) \exp(-i\varepsilon_i t)$. This approximation was originally introduced by Cheshire (1964), and termed the continuum intermediate state (CIS) approximation by Belkić (1977). It is equivalent to the target continuum distorted wave approximation (TCDW) discussed by Crothers and McCann (1984) (see also Crothers and Dunseath (1987, 1990), Dunseath *et al.* (1988), Deb (1988), Deb and Drothers (1989) which has been shown to be remarkably accurate, both for total and differential cross sections.

In Fig. 10.8, a comparison is made between the calculated CDW and TCDW differential cross sections for electron capture by H^+ from H(1s) at 5 MeV/a.m.u., and the experimental data together with the B2B0 and DWB2 cross section. The CDW cross section exhibits an unphysical dip near $\Theta_L = 0.5$ mrad which is not present in any of the other models. For this case $Z_A = Z_B$, so that it is rather surprising that the TCDW approximation predicts a cross section which appears to be in best overall agreement with the measurements, particularly in the region of the Thomas peak.

Second-order approximations

At high velocities the CDWI cross section for capture by H^+ from H(1s) into s states is asymptotically the same as that given by the second-order boundary corrected amplitude. However, this does not extend to capture into states with non-zero angular momentum. This implies that while the CDWI approximation contains contributions from each term in the perturbation series, it does not fully contain all second-order effects. This is reflected in the differential cross section which exhibits an unphysical dip at the Thomas angle. However, in the small-angle region which makes the most important contribution to total cross sections at high, but not asymptotically high, energies the corrections to the CDWI approximation are small. The CDW cross section can be avoided by going to a second-order approximation (Crothers and McCann 1984; Crothers 1985). The second-order amplitude is of the familiar form

$$\mathscr{A}^{\text{II}}_{ni}(\mathbf{b}) = \mathscr{A}^{\text{I}}_{ni}(\mathbf{b}) + (\text{i})^2 \int_{-\infty}^{\infty} \text{d}t \int_{-\infty}^{\infty} \text{d}t' \langle M^{\text{T}}_n | P \dagger g^+_x P | L_i \rangle. \qquad (10.138)$$

The Green's function g^+_x describing propagation in intermediate states can be chosen in different ways, as in the case of the distorted wave series we have discussed earlier. For example, a free-particle Green's function g_0 can be employed, or alternatively one based on either the target Hamiltonian containing the interaction $-Z_B/r$, or the final state Hamiltonian containing $-Z_A/x$. Crothers (1985, 1989) has shown that these second-order approximations contain the Thomas double scattering mechanism completely. In the case of 1s-1s capture, the full CDWII cross section behaves asymptotically like

$$\sigma^{\text{CDWII}} = \left[0.319 + \frac{5\pi v_0}{2^{11}(Z_A + Z_B)}\right]\sigma^{\text{OBK}}, \tag{10.139}$$

where σ^{OBK} is the first-order OBK cross section behaving like v_0^{-12}. This result agrees completely with the asymptotic *third-order* OBK cross section (Shakeshaft 1978*a*) based on free-particle propagation in intermediate states.

The second-order terms of the TCDW model have been investigated by Crothers and McCann (1984) and Crothers and Dunseath (1987, 1990). The position is complicated. If the second-order term is based on the free-particle Green's function, the correction to the first-order cross section is large near the Thomas angle. However, if a modified target Coulomb Green's function is employed the second-order terms are small. The modification of the Coulomb Green's function referred to is to remove the divergence that would otherwise occur because the initial state wave function employed does not satisfy the appropriate long-range boundary conditions. Despite the success of the TCDW model in reproducing the experimental data, when Z_A and Z_B are of comparable magnitude, more symmetrical approximations such as the CDW2 would appear to be theoretically more satisfactory.

Coupled channel CDW models

In common with most perturbation methods the CDW approximation fails at low energies. This is in part due to the fact that the functions M_n and L_i are not normalized to unity at finite values of t, and also to the fact that the approximation does not ensure the automatic conservation of probability and for some ranges of b the calculated probabilities can exceed unity at low energies (Crothers 1982). Both difficulties can be solved by using the functions L_i and M_n as basis functions in a coupled channel model. Crothers and McCann (1985) have discussed a two-channel system, and applied it to a number of cases. The wave function is represented as

$$\tilde{\psi}_i(\mathbf{r}, t) = A_1(\mathbf{b}, t)L_i(\mathbf{r}, \mathbf{x}, t) + A_2(\mathbf{b}, t)M_n^{\text{T}}(\mathbf{r}, \mathbf{x}, t). \tag{10.140}$$

Coupled equations of the standard form (3.185) or (5.92) are obtained, where the overlap and potential coupling matrices are defined by (3.184) and (3.186) with the distorted wave functions L_i and M_n^{T} replacing the free wave functions Φ_i and χ_n, respectively. The model is unitary within the two-dimensional basis, and should provide accurate cross sections at intermediate and high energies. At low energies a realistic approximation certainly requires a larger basis set, because the asymptotic channels representing capture into different states, together with excitation and ionization, will all be strongly coupled. As with the usual two-state coupled channel equations, the Bates distorted wave approximation can be written down (Crothers 1982), retaining the essential improvements over the simple CDWI approximation at the higher energies.

The symmetrical eikonal and eikonal approximation

The CDWII, and in many circumstances the CDWI or TCDWI approximations, provide an accurate description of high-energy charge exchange. A simplification of the CDW approach, which retains the essential features, is to employ the eikonal approximation to the Coulomb distorted wave functions. The basis functions then are of the form of (10.124) and (10.133), with $\mathscr{F}_{\mathrm{A}}^{+}$ and $\mathscr{F}_{\mathrm{B}}^{+}$ replaced by $\mathscr{F}_{\mathrm{AE}}^{+}$, $\mathscr{F}_{\mathrm{BE}}^{+}$ where

$$\mathscr{F}_{\mathrm{AE}}^{+} = \exp[-\mathrm{i}\nu_{\mathrm{A}} \log(v_0 x + \mathbf{v}_0 \cdot \mathbf{x})], \tag{10.141}$$

$$\mathscr{F}_{\mathrm{BE}}^{+} = \exp[-\mathrm{i}\nu_{\mathrm{B}} \log(v_0 r + \mathbf{v}_0 \cdot \mathbf{r})]. \tag{10.142}$$

Using these expressions the first-order symmetrical eikonal approximation is given by equation (10.137), where the perturbating operator P is now (cf. (10.129), (10.130))

$$PL_i = \exp[-\mathrm{i}\varepsilon_i t]\{\phi_i(\mathbf{r})\tfrac{1}{2}\nabla_x^2 \mathscr{F}_{\mathrm{AE}}^{+}(\mathbf{x}) + \nabla_r \phi_i(\mathbf{r}) \cdot \nabla_x \mathscr{F}_{\mathrm{AE}}^{+}(\mathbf{x})\}. \tag{10.143}$$

Both the first- and second-order symmetrical eikonal amplitudes have been developed and applications made to excitation and ionization as well as to charge exchange (Maidagen and Rivarola 1984; Deco and Rivarola 1985; Deco *et al.* 1984, 1986*a*, *b*; Fainstein and Rivarola 1987; Rivarola *et al.* 1985).

Naturally, unsymmetrical versions of the eikonal approximation for charge exchange can be employed in which Coulomb distortion is allowed for in either the initial or final channels, but not in both. This is achieved by setting one of the functions $\mathscr{F}_{\mathrm{AE}}^{+}$ or $\mathscr{F}_{\mathrm{BE}}^{+}$ equal to unity, with the other being given by (10.141) or (10.142). The unsymmetrical form of the eikonal approximation for charge exchange was introduced by Dewangan (1975, 1977) and developed by, among others, Chan and Eichler (1979) and Eichler (1981).

At high energies the symmetrical eikonal (SE) total cross section for charge exchange approaches the CDW cross section. This is because the SE cross section is accurate at small angles. At large angles the SE and CDW differential cross sections disagree and the SE model does not contain the Thomas mechanism. This is illustrated for the $\mathrm{He}^{2+} + \mathrm{He}$ system in Fig. 10.9, where it can be seen that the two cross sections coincide for $E_{\mathrm{L}} \gtrsim 500$ keV/a.m.u. Despite the fact that the CDW wave functions account for the Coulomb interactions in a much more detailed way, it can be seen that at lower energies the SE cross section continues to represent the experimental data rather well, while the CDW cross section is too large. The reason for the failure of the CDW approximation at low energies is due, in part, as discussed above, to the lack of normalization of the approximate wave functions at all values of the time t. The symmetrical eikonal wave functions remain normalized to unity for all t because the distorted waves $\mathscr{F}^{+}_{\mathrm{AE}}$ and $\mathscr{F}^{+}_{\mathrm{BE}}$ are purely phase factors, and this accounts for the relative success of this approximation at intermediate energies and for large charges Z_A, Z_{B}.

10.7 QUANTUM PERTURBATION EXPANSION FOR CHARGE EXCHANGE AND KNOCK-ON COLLISIONS

The various distorted wave and perturbation models discussed in the earlier sections of this chapter can be obtained from a complete quantum mechanical formulation; however, the impact parameter method adopted is accurate for forward scattering, which dominates both excitation and charge exchange, apart from knock-on collisions, in which the incident heavy particle exchanges with that of the target, leaving the active electron as a 'spectator'. The quantum perturbation theory of rearrangement collisions has been discussed in some detail by McDowell and Coleman (1970) and by Bransden (1965, 1969, 1983) and only a brief account will be given here.

To avoid complication of notation let us consider systems in which there is no overall Coulomb interaction in either the direct or rearranged channels, such as in collisions between protons and a neutral atom. The generalization to an overall Coulomb interaction requires the introduction of Coulomb rather than plane waves in the asymptotic wave functions along the lines indicated in Chapter 3, equations (3.32), (3.33), and (3.36).

The full Hamiltonian for a one-active-electron system is

$$H = T + V_{\mathrm{Ae}}(x) + V_{\mathrm{Be}}(r) + V_{\mathrm{AB}}(R), \tag{10.144}$$

where T is the kinetic energy operator expressed in one of the sets of Jacobi coordinates introduced in Section 3.1. The complete wave function Ψ_i^+, satisfying the boundary conditions discussed in Section 3.1, satisfies the Schrödinger equation

$$(H - E)\Psi_i^+ = 0. \tag{10.145}$$

To find an integral equation for the scattering amplitude g_{ni}, defined by (3.29a), Ψ_i^+ can be expanded in the complete set of eigenfunctions χ_m for the atom $(A + e^-)$

$$\Psi_i^+(\mathbf{x}, \boldsymbol{\rho}) = \sum_m \chi_m(\mathbf{x}) S_m(\boldsymbol{\rho}), \tag{10.146}$$

where the sum over m includes an integration over the continuum states. The functions $S_m(\boldsymbol{\rho})$ are asymptotically equal to the functions $\mathscr{G}_m(\boldsymbol{\rho})$ defined by (3.26), and hence have the asymptotic form (3.29a)

$$S_m(\boldsymbol{\rho}) \sim \rho^{-1} g_{mi}\, e^{i\kappa_m \rho}. \tag{10.147}$$

Projecting the Schrödinger equation with the function $\chi_n(\mathbf{x})$ and using the wave equation (3.24), we obtain

$$(\nabla_\rho^2 + \kappa_n^2) S_n(\boldsymbol{\rho}) = 2\mu \int \mathrm{d}\mathbf{x}\, \chi_n^*(\mathbf{x})[V_{\mathrm{Be}}(r) + V_{\mathrm{AB}}(R)]\Psi_i^+(\mathbf{x}, \boldsymbol{\rho}). \tag{10.148}$$

A solution with outgoing waves corresponding to the asymptotic form (3.29a) is

$$S_n(\boldsymbol{\rho}) = \int \mathrm{d}\mathbf{x} \int \mathrm{d}\boldsymbol{\rho}\, g_0(\kappa_n; \boldsymbol{\rho}, \boldsymbol{\rho}')\chi_n^*(\mathbf{x})[V_{\mathrm{Be}}(r) + V_{\mathrm{AB}}(R)]\Psi_i(\mathbf{x}, \boldsymbol{\rho}) \tag{10.149}$$

where g_0 is the free-particle Green's function (Bransden 1983)

$$g_0(\kappa_n; \boldsymbol{\rho}, \boldsymbol{\rho}') = -\left(\frac{\mu}{2\pi}\right)\frac{\exp(i\kappa_n|\boldsymbol{\rho} - \boldsymbol{\rho}'|)}{|\boldsymbol{\rho} - \boldsymbol{\rho}'|}. \tag{10.150}$$

For large ρ, $(\rho \gg \rho')$

$$g_0(\kappa_n; \boldsymbol{\rho}, \boldsymbol{\rho}') \approx -\frac{\mu}{2\pi}\rho^{-1}\exp[i(\kappa_n\rho - \boldsymbol{\kappa}_n \cdot \boldsymbol{\rho}')], \tag{10.151}$$

where the direction of $\boldsymbol{\kappa}_n$ is taken to be that of the vector $\boldsymbol{\rho}$. Combining (10.151), (10.149) with (10.147), we obtain

$$g_{ni}(\bar{\theta}, \bar{\phi}) = -\frac{\mu}{2\pi}\int \mathrm{d}\mathbf{x} \int \mathrm{d}\boldsymbol{\rho}\, \exp(-i\boldsymbol{\kappa}_n \cdot \boldsymbol{\rho})\chi_n^*(\mathbf{x})[V_{\mathrm{Be}}(r) + V_{\mathrm{AB}}(R)]\Psi_i^+(\mathbf{x}, \boldsymbol{\rho}). \tag{10.152}$$

This can be written as

$$g_{ni} = -\frac{\mu}{2\pi} T_{ni}, \tag{10.153}$$

where the transition matrix T_{ni} is defined as

$$T_{ni} = \langle X_n | V_{\mathrm{Be}} + V_{\mathrm{AB}} | \Psi_i^+ \rangle, \tag{10.154}$$

and X_n is the unperturbed plane-wave final state wave function

$$X_n(\mathbf{x}, \boldsymbol{\rho}) = \exp(i\boldsymbol{\kappa}_n \cdot \boldsymbol{\rho})\chi_n(\mathbf{x}), \tag{10.155}$$

which is a solution of (3.23).

Plane-wave unperturbed functions for the initial configuration A + (B + e$^-$) satisfy (3.14) and are of the form

$$\Phi_j(\mathbf{r}, \boldsymbol{\sigma}) = \exp(i\mathbf{k}_j \cdot \boldsymbol{\sigma})\phi_j(\mathbf{r}). \tag{10.156}$$

If A is incident on the state of (B + e$^-$) labelled i, then $\Psi_{\text{out}} = \Psi_i^+ - \Phi_i$ contains only outgoing waves in all channels. The function Ψ_{out} satisfies the equation

$$(H - E)\Psi_{\text{out}} = -(V_{\text{Ae}} + V_{\text{AB}})\Phi_i. \tag{10.157}$$

It follows that

$$\Psi_{\text{out}} = G(E)(V_{\text{Ae}} + V_{\text{AB}})\Phi_i \tag{10.158}$$

and

$$\Psi_i^+ = \Phi_i + G(E)(V_{\text{Ae}} + V_{\text{AB}})\Phi_i \tag{10.159}$$

where $G(\eta)$ is the Green's operator

$$G(\eta) = \frac{1}{\eta - H}. \tag{10.160}$$

This is an analytic function of η except on the energy spectrum of H; however the singularity when $\eta = E$ can be avoided by defining

$$G^+(E) = \lim_{\varepsilon \to 0^+} \frac{1}{E + i\varepsilon - H}. \tag{10.161}$$

The prescription $E \to E + i\varepsilon$ ensures that G^+ contains only outgoing waves (see Bransden 1983), the proof being similar to that given for the impact parameter Green's functions in Appendix G.

Combining (10.161) and (10.156), an integral equation for T_{ni} is obtained:

$$T_{ni} = \langle X_n | (V_{\text{Be}} + V_{\text{AB}})\{1 + G^+(E)(V_{\text{Ae}} + V_{\text{AB}})\} | \Phi_i \rangle. \tag{10.162}$$

The Green's function G^+ can be expanded by using the identity (G.22) in a variety of forms. In particular we have

$$G^+ = G_0^+\{1 + (V_{\text{Ae}} + V_{\text{Be}} + V_{\text{AB}})G^+\} \tag{10.163}$$

$$= G_{\text{A}}^+\{1 + (V_{\text{Be}} + V_{\text{AB}})G^+\} \tag{10.164}$$

$$= G_{\text{B}}^+\{1 + (V_{\text{Ae}} + V_{\text{AB}})G^+\} \tag{10.165}$$

where G_0^+, G_A^+, and G_B^+ are defined as

$$G_0^+(E) = \lim_{\varepsilon \to 0^+} (E + i\varepsilon - T)^{-1} \tag{10.166}$$

$$G_A^+(E) = \lim_{\varepsilon \to 0^+} (E + i\varepsilon - T - V_{Ae})^{-1} \tag{10.167}$$

$$G_B^+(E) = \lim_{\varepsilon \to 0^+} (E + i\varepsilon - T - V_{Be})^{-1}. \tag{10.168}$$

By iterating (10.164) or (10.165) and using (10.162), the Born series can be obtained which are the wave equivalents (for the case in which there is no overall Coulomb interaction) to the post and prior forms of the impact parameter boundary corrected Born series (10.54) or (10.55), respectively, while by iterating (10.163) the BNB0 approximations are obtained. Distorted wave approximations, with general distorting potentials in the initial and final arrangements, can be obtained by proceeding along very similar lines to those employed in the impact parameter approximation.

No proof of the convergence or lack of convergence of the Born series for T_{ni} has been given and there are some well-known difficulties which must be considered before such a proof can be constructed. To explain these, we will for the moment assume that V_{Ae}, V_{Be}, and V_{AB} are potentials which vanish faster than the Coulomb potential at large distances. Consider the expansion of G^+ given by iterating (10.163). We find

$$G^+(E) = G_0^+(E) + G_0^+(E)(V_{Ae} + V_{Be} + V_{AB})G_0^+(E) + G_0^+(E)(V_{Ae} + V_{Be} + V_{AB})G_0^+(E)(V_{Ae} + V_{Be} + V_{AB})G_0^+(E) + \cdots. \tag{10.169}$$

Among the various sequences of terms in (10.169) are those which contain one potential only, for example

$$G_0^+(E) + G_0^+(E)V_{Ae}G_0^+(E) + G_0^+(E)V_{Ae}G_0^+(E)V_{Ae}G_0^+(E) + \cdots. \tag{10.170}$$

By introducing a complete set of plane-wave states

$$\phi_{\mathbf{p},\mathbf{q}}(\mathbf{x}, \boldsymbol{\rho}) = (2\pi)^{-3} \exp[i(\mathbf{p}\cdot\mathbf{x} + \mathbf{q}\cdot\boldsymbol{\rho})] \tag{10.171}$$

the transition matrix (10.162) can be written in the form

$$T_{ni} = \langle X_n | V_{Be} + V_{AB} | \Phi_i \rangle + \int d\mathbf{p} \int d\mathbf{q} \int d\mathbf{p}' \int d\mathbf{q}' \, \langle X_n | V_{Be} + V_{AB} | \phi_{\mathbf{p},\mathbf{q}} \rangle \times \langle \phi_{\mathbf{p},\mathbf{q}} | G^+(E) | \phi_{\mathbf{p}',\mathbf{q}'} \rangle \langle \phi_{\mathbf{p}',\mathbf{q}'} | V_{Ae} + V_{AB} | \Phi_i \rangle. \tag{10.172}$$

The factor $\langle\phi_{\mathbf{p},\mathbf{q}}|G^+(E)|_{\mathbf{p}',\mathbf{q}'}\rangle \equiv \langle\mathbf{p},\mathbf{q}|G^+(E)|\mathbf{p}',\mathbf{q}'\rangle$ contains the sequence of terms (10.170). Since $G_0^+(E)$ is diagonal in the momentum $\mathbf{q}$, the contribution to the matrix element of $G^+(E)$ from the subseries (10.170) reduces to

$$\delta(\mathbf{q}-\mathbf{q}')\langle p|g_0(E-E_q)+g_0(E-E_q)V_{\mathrm{Ae}}g_0(E-E_q)+\cdots|p'\rangle, \qquad (10.173)$$

where $g_0(z)$ is a two-body free-particle Green's function

$$g_0(z)=\left(z-\frac{p^2}{2}\right)^{-1}, \qquad (10.174)$$

and where

$$E_q=\frac{q^2}{2\mu_{\mathrm{B}}}. \qquad (10.175)$$

Formally the series (10.173) can be summed to give

$$\delta(\mathbf{q}-\mathbf{q}')\langle\mathbf{p}|g_{\mathrm{A}}(E-E_q)|\mathbf{p}'\rangle, \qquad (10.176)$$

where $g_{\mathrm{A}}(z)$ is a two-body Green's function for scattering by the potential V_{Ae} alone

$$g_{\mathrm{A}}(z)=\left(z-\frac{p^2}{2}-V_{\mathrm{Ae}}\right)^{-1}. \qquad (10.177)$$

Since the potential V_{Ae} supports bound states the series expansion of g_{A} is known to converge only for $|z|>|\varepsilon_{\mathrm{A}}|$ where $|\varepsilon^{\mathrm{A}}|$ is the energy of the most strongly bound state. It follows (Aaron *et al.* 1961) that the expansion (10.170) is divergent for some part of the range of $\mathbf{q}$, and that, in general, the expansion of $G^+(E)$ itself must diverge for some ranges of $\mathbf{q}$. The divergence of the series expansion of the Green's function does not imply the divergence of the corresponding expansion of T_{ni} and indeed a one-dimensional counterexample has been given by Dettmann and Liebfried (1966).

The term (10.176) is said to be a disconnected part of the kernel $G_0(V_{\mathrm{Ae}}+V_{\mathrm{Be}}+V_{\mathrm{AB}})$ of the integral equation (10.163). Physically it represents scattering between the particles A and e while particle B propagates freely. Because disconnected terms contain a delta function in one of the momenta, the trace of the kernel $G_0(V_{\mathrm{Ae}}+V_{\mathrm{Be}}+V_{\mathrm{AB}})$ is not finite, so that a Fredholm solution of the integral equation does not exist. These difficulties apply equally to the corresponding series for excitation or ionization and are characteristic of the three-body problem.

While the presence of disconnected terms does not necessarily make the Born series for the transition amplitude diverge, methods in which the disconnected terms are removed are preferable. It should also be recollected that low-order Born approximations can be obtained from variational principles and may be useful approximations even if the Born series is not formally convergent. Several methods of removing disconnected terms have been developed. The most famous is that due to Faddeev (1960) in which the integral equations are recast in terms of the two-body transition matrices which replace the potentials. (Elementary accounts of the Faddeev equations are given by McDowell and Coleman (1970), Bransden (1983), and by Joachain (1975).) This approach has been attempted infrequently in atomic physics because of difficulties in handling the singularities in the Coulomb two-body transition matrices, but for charge exchange some progress has been reported by Alston (1989*a*, *b*).

An alternative approach due to Dodd and Greider (1966) is to develop a particular class of distorted wave series in which disconnected terms do not occur. It has been shown by Gayet (1972) that the CDWI model is the first term of such a series. Further work by Burgdörfer and Taulbjerg (1986) has shown that the wave version of the second-order distorted wave approximation, considered earlier, is equivalent to a second-order approximation to the Faddeev equations, and has shown how sequences of terms in the Born series can be re-summed to produce new series with connected kernels.

The Born approximtion and knock-on collisions

In order to explain how the knock-on contribution to charge exchange arises, the first Born approximation will be discussed for the case $Z_A = Z_B = 1$. There are no overall Coulomb interactions in either the direct, $A + (B + e^-)$, or exchange, $(A + e^-) + B$, channels and accordingly the first-order Born approximation (which coincides with the first-order boundary corrected approximation up to a phase factor) for the transition matrix T_{ni} is given by the first term in (10.162),

$$T_{ni} \approx \langle X_n | V_{Be} + V_{AB} | \Phi_i \rangle$$
$$= \int d\mathbf{x} \int d\boldsymbol{\rho}\, e^{-i\boldsymbol{\kappa}_n \cdot \boldsymbol{\rho}} \chi_n(\mathbf{x}) \left(-\frac{1}{r} + \frac{1}{R} \right) e^{i\mathbf{k}_i \cdot \boldsymbol{\sigma}} \phi_i(\mathbf{r}). \qquad (10.178)$$

Energy conservation requires that

$$\frac{k_i^2}{2\mu_A} - \frac{\kappa_n^2}{2\mu_B} = \eta_n - \varepsilon_i \qquad (10.179)$$

where μ_A and μ_B are defined by (3.4) and (3.6), respectively. From (3.49) and (3.50), $\boldsymbol{\sigma}$ and $\boldsymbol{\rho}$ can be expressed in terms of $\mathbf{x}$ and $\mathbf{r}$:

$$\boldsymbol{\sigma} = b\mathbf{r} - \mathbf{x}, \qquad \boldsymbol{\rho} = \mathbf{r} - a\mathbf{x} \qquad (10.180)$$

where

$$b = \frac{M_B}{M_B + 1} \quad \text{and} \quad a = \frac{M_A}{M_A + 1}. \qquad (10.181)$$

Hence changing the variables of integration to $\mathbf{r}$ and $\mathbf{x}$

$$T_{ni} = \int \mathrm{d}\mathbf{r} \int \mathrm{d}\mathbf{x}\, \mathrm{e}^{\mathrm{i}\boldsymbol{\alpha}\cdot\mathbf{r}} \chi_n(\mathbf{x}) \left(-\frac{1}{r} + \frac{1}{R} \right) \mathrm{e}^{\mathrm{i}\boldsymbol{\beta}\cdot\mathbf{x}} \phi_i(\mathbf{r})$$

$$\equiv T_{ni}^{\mathrm{OBK}} + T_{ni}^{\mathrm{R}}, \tag{10.182}$$

where

$$\boldsymbol{\alpha} = -\boldsymbol{\kappa}_n + b\mathbf{k}_i, \qquad \boldsymbol{\beta} = -\mathbf{k}_i + a\boldsymbol{\kappa}_n. \tag{10.183}$$

Let us consider the two terms in (10.182) separately. The first is the OBK amplitude containing the potential $-1/r$

$$T_{ni}^{\mathrm{OBK}} = -G_n(\boldsymbol{\beta}) g_i(\boldsymbol{\alpha}), \tag{10.184}$$

where

$$G_n(\boldsymbol{\beta}) = \int \mathrm{e}^{\mathrm{i}\boldsymbol{\beta}\cdot\mathbf{x}} \chi_n(\mathbf{x})\, \mathrm{d}\mathbf{x}, \tag{10.185}$$

and

$$g_i(\boldsymbol{\alpha}) = \int \mathrm{e}^{\mathrm{i}\boldsymbol{\alpha}\cdot\mathbf{r}} \frac{\phi_i(\mathbf{r})}{r}\, \mathrm{d}\mathbf{r}. \tag{10.186}$$

Since the hydrogenic wave function $\phi_i(\mathbf{r})$ satisfies the equation

$$\left(-\tfrac{1}{2}\nabla^2 - \frac{1}{r} - \varepsilon_i \right) \phi_i(\mathbf{r}) = 0, \tag{10.187}$$

we have that

$$g_i(\boldsymbol{\alpha}) = \int \mathrm{e}^{\mathrm{i}\boldsymbol{\alpha}\cdot\mathbf{r}} (-\tfrac{1}{2}\nabla^2 - \varepsilon_i) \phi_i(\mathbf{r})\, \mathrm{d}\mathbf{r}$$

$$= (-\tfrac{1}{2}\alpha^2 - \varepsilon_i) G_i(\boldsymbol{\alpha}), \tag{10.188}$$

and consequently

$$T_{ni}^{\mathrm{OBK}} = -(-\tfrac{1}{2}\alpha^2 - \varepsilon_i) G_i(\boldsymbol{\alpha}) G_n(\boldsymbol{\beta}). \tag{10.189}$$

For capture from the ground state to the ground state of atomic hydrogen, $\mathrm{H}^+ + \mathrm{H}(1s) \to \mathrm{H}(1s) + \mathrm{H}^+$,

$$G_{1s}(\boldsymbol{\alpha}) = \frac{8\pi^{\frac{1}{2}}}{\alpha^2 + 1^2}. \tag{10.190}$$

In this case $\eta_n = \varepsilon_i$ and $M_{\mathrm{A}} = M_{\mathrm{B}} = M$, where M is the mass of the proton, so that $a = b$ and $\beta^2 = \alpha^2$. It follows that

$$T_{1s,1s}^{\mathrm{OBK}} = -\frac{32\pi}{(\alpha^2 + 1)^3}. \tag{10.191}$$

The relationship between α^2 and the angle of scattering is

$$\alpha^2 \approx \frac{v_0^2}{4}\left(1 + 4M^2 \sin^2\left(\frac{\theta}{2}\right)\right). \tag{10.192}$$

It is seen that, except for very low velocities, $T^{\mathrm{OBK}}_{1s,1s}$ is only significant in the forward direction, being confined to angles of order $1/M$, as expected.

Turning to the second term in (10.182) containing the internuclear potential, the Fourier transform,

$$\frac{1}{R} = \frac{1}{2\pi^2}\int k^{-2} \exp(-i\mathbf{R}\cdot\mathbf{k})\, d\mathbf{k}, \tag{10.193}$$

can be used with $\mathbf{R} = \mathbf{x} - \mathbf{r}$ to obtain the result

$$T^{\mathrm{R}}_{ni} = \frac{1}{2\pi^2}\int k^{-2} G_n(\boldsymbol{\beta} - \mathbf{k}) G_i(\boldsymbol{\alpha} + \mathbf{k})\, d\mathbf{k}. \tag{10.194}$$

Thus for 1s to 1s capture, using (10.190), we have

$$T^{\mathrm{R}}_{ni} = \frac{32}{\pi}\int d\mathbf{k}\, k^{-2}\{|\boldsymbol{\beta} - \mathbf{k}|^2 + 1\}^{-2}\{|\boldsymbol{\alpha} + \mathbf{k}|^2 + 1\}^{-2}. \tag{10.195}$$

The integral in (10.195) can be evaluated analytically by a technique introduced by Jackson and Schiff (1953). The details have been given in the text of McDowell and Coleman (1970) and will not be repeated here. The result is

$$T^{\mathrm{R}}_{ni} = 32\pi\left[\frac{1}{(1+\beta^2)^3}\left\{\frac{1}{\omega} + \frac{\omega - 4}{2\omega^{\frac{3}{2}}}\tan^{-1}\left(\frac{\omega^{\frac{1}{2}}}{2}\right)\right\} + \frac{1}{(1+\beta^2)^2}\left\{\frac{2}{\omega^{\frac{3}{2}}}\tan^{-1}\left(\frac{\omega^{\frac{1}{2}}}{2}\right) - \frac{4}{\omega(4+\omega)}\right\} + \frac{1}{(1+\beta^2)}\frac{1}{(4+\omega)^2}\right], \tag{10.196}$$

where

$$\omega = (\boldsymbol{\alpha} + \boldsymbol{\beta})^2 \tag{10.197}$$

and for exact resonance with $M_{\mathrm{A}} = M_{\mathrm{B}}$

$$\omega = (1 - a)^2(\mathbf{k}_i + \boldsymbol{\kappa}_n)^2. \tag{10.198}$$

The major contribution to the cross section, with the exception to be noted, comes from the forward direction where β is smallest, $\beta_{\mathrm{min}} = v_0/2$ at $\theta = 0$ and $\beta_{\mathrm{max}} = Mv_0$ at $\theta = \pi$. Near the forward direction $\omega = v_0^2$, if terms of the order $1/M$ are neglected. When combined with T^{OBK}_{ni}, and the integration over angles is carried out, the cross section becomes, in the high-velocity limit, equal to 0.661 σ^{OBK}, as we have already noted, and

$$\sigma^{\mathrm{JS}} = 0.661\left(\frac{2^{18}}{5}\right)v_0^{-12}. \tag{10.199}$$

Near the backward direction, $\omega \to 0$, and in this case the last term in (10.196) provides the major contribution to the cross section in the high-velocity limit, becoming equal to σ^{K} where

$$\sigma^{\mathrm{K}} \sim \frac{16}{3M^2}v_0^{-6}. \tag{10.200}$$

This is the knock-on cross section described in Section 10.1. Because σ^{K} is proportional to M^{-2}, it is negligible compared with the contribution from forward scattering at energies below $\sim$40 MeV, but dominates the non-relativistic cross section at extremely high energies. Since relativistic cross sections become important above a few tens of MeV, this result is of theoretical interest. If the masses of A and B are not identical, for example if the incident particle is a deuteron or a triton, the condition $\boldsymbol{\alpha} + \boldsymbol{\beta} = 0$ at $\theta = \pi$ is not satisfied, and the knock-on process cannot take place.

10.8 IONIZATION AND CHARGE EXCHANGE INTO THE CONTINUUM

In Chapter 8 it was seen that at sufficiently high energies the total ionization cross section can be calculated by the first-order Born approximation, and at lower energies by the coupled channel method, with pseudostates, or by the classical trajectory Monte Carlo method. A different approach to ionization is the application of the CDW model (Belkić 1978, 1980). In a simplified version due to Crothers and McCann (1983) and Fainstein *et al.* (1988), the initial state is described by an eikonal distorted wave function and the final state by a CDW function (the CDW–EIS model). Applications include interesting studies of the difference between the single ionization cross section of helium by antiprotons and by protons (Fainstein *et al.* 1989).

The single differential cross section for the single ionization of helium by proton and antiproton impact is shown in Fig. 10.10 for impact energies of 200 keV/a.m.u. and 500 keV/a.m.u., calculated in the CDW–EIS model and (for 200 keV/a.m.u.) in the CTMC model. The experimental data for proton impact is well represented by the CDW–EIS model at both energies; however, the CTMC model is less successful, failing to reproduce the data at 200 keV/a.m.u. for large angles of ejection. In principle, the CTMC model should be particularly suitable to study three-body effects in ionization and despite this apparent failure at the impact energy of 200 keV/a.m.u., Reinhold

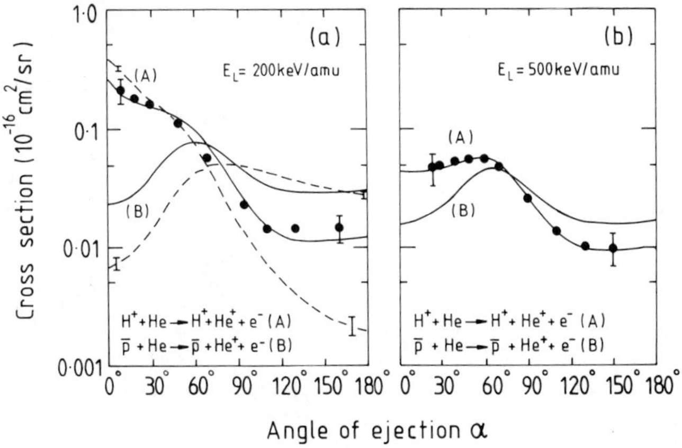

Fig. 10.10. The single differential cross section $d\sigma/d\Omega$ for the ejection of electrons at an angle α in the ionization of helium by proton and antiproton impact at (a) 200 keV/a.m.u. and (b) 500 keV/a.m.u. (after Fainstein *et al.* (1989)). Experimental data: ● Rudd *et al.* (1976). Theoretical cross sections: CDW–EIS model ———, Fainstein *et al.* (1989); CTMC model -----, Olson and Gay (1988).

and Olson (1989*a*, *b*) (see also Reinhold and Schulz 1989) have shown that the magnitude and shape of cross sections calculated from this model are in general agreement with the experimental data for the ionization of helium by the impact of H^+, He^+, and He^{2+} ions at $E_L = 50$ and 100 keV/a.m.u. It should however be pointed out that the measurements differ considerably among themselves and do not define, at present, the differential cross sections very closely. Figure 10.11 shows the experimental and calculated double differential cross sections for a number of angles of ejection, as a function of the energy of the ejected electron, in the ionization of helium by proton impact at 50 and 100 keV/a.m.u.

When the charges on the incident and target nuclei are comparable in size, the bulk of the ionized electrons are ejected with small velocities, as predicted by the Born approximation. However, there is a significant enhancement of the cross section above the predictions of the Born approximation for ejected electrons that have a velocity v_E close to that of the incident ion v_0, and in this region the cross section is cusp-like, behaving approximately like $|\mathbf{v}_0 - \mathbf{v}_E|^{-1}$ (Crooks and Rudd 1970; Harrison and Lucas 1970). Since the incident ion is not appreciably deflected during the collision, the electrons concerned are those that emerge close to the forward direction. This effect is clearly seen in the experiments of Manson *et al.* (1975) on the ionization of helium by protons of energies between 800 keV and 5 MeV, and in Fig. 10.12 the ratio of the experimental cross section to the Born cross section for electrons having the same speed as the incident proton is shown for a number of angles of ejection as a function of incident energy.

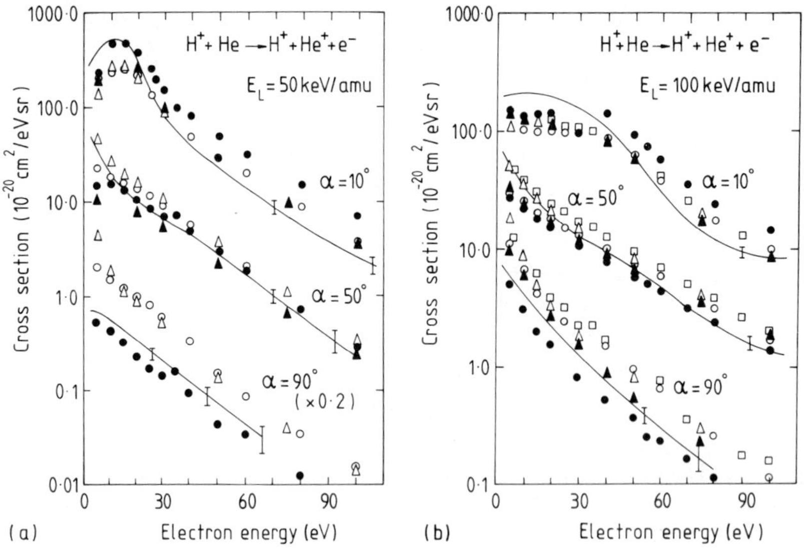

Fig. 10.11. The double differential cross section $d\sigma/d\Omega\, d\varepsilon$ for ejection of energy ε at an angle α in the ionization of helium by proton impact at (a) 50 keV/a.m.u. and (b) 100 keV/a.m.u. Experimental cross sections: ● Gibson and Reid (1986); ▲ Bernardi *et al.*, see Reinhold and Olson (1989*a*, *b*); △ Rudd and Madison (1976); ○ Rudd and Jurgensen (1963); □ Rudd *et al.* (1976). Theoretical cross sections: CTMC model, Reinhold and Olson (1989*a*, *b*).

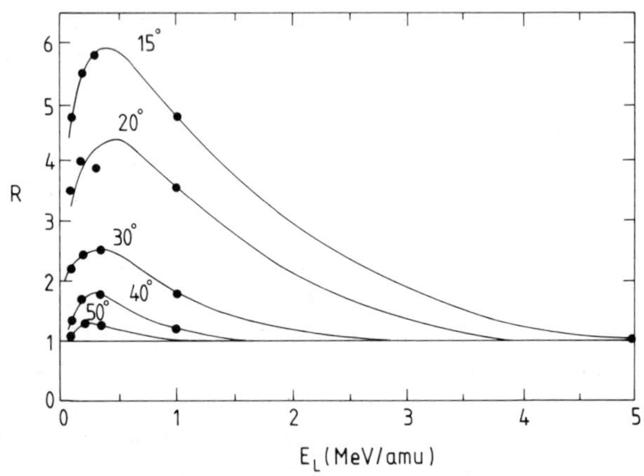

Fig. 10.12. The ratio R of the experimental cross section to the Born approximation cross section for the ionization of helium by proton impact, for the case in which the ejected electrons have the same speed as the incident proton.

One way of looking at this enhancement is as an electron capture process in which the electron is captured into the continuum rather than a discrete state of the incident ion. This explanation was put forward by Macek (1970) (see also Rudd and Macek (1972)) who coined the term charge exchange into the continuum. The fact that the discrepancies in the forward direction could be attributed to a final state interaction between the ejected electrons and the incident ion was first suggested by Oldham (1967). Following Rudd and Macek the following argument based upon analytic continuation across the ionization threshold of the ion $(\mathrm{A} + \mathrm{e}^-)$ can be made. Cross sections for electron capture to a Rydberg level with quantum numbers nlm follow the Oppenheimer n^{-3} rule:

$$\sigma_{nl} = \sigma_l \frac{1}{n^3} \tag{10.201}$$

where n is large. The energy of the ion $(\mathrm{A} + \mathrm{e}^-)$ in a frame with origin on A is

$$\eta_n = -\frac{1}{2n^2} \tag{10.202}$$

so that (10.201) can be expressed as

$$\sigma_{nl} = \sigma_l \left(\frac{\mathrm{d}\eta_n}{\mathrm{d}n}\right). \tag{10.203}$$

If σ_{nl} is now continued across the threshold at $\eta_n = 0$, the differential cross section for capture in continuum states with energies between η and $\eta + \mathrm{d}\eta$, where η is now positive, can be written as

$$\mathrm{d}\sigma = \bar{\sigma} \frac{1}{4\pi} \mathrm{d}\Omega\, \mathrm{d}\eta \tag{10.204}$$

where the angular quantum numbers l, m have been replaced by the element of solid angle $(4\pi)^{-1}\, \mathrm{d}\Omega$. Both $\mathrm{d}\Omega$ and $\mathrm{d}\eta$ are defined in the frame of reference with origin at A. Transforming to the laboratory frame with origin on B, the ionization cross section becomes

$$\mathrm{d}\sigma_\mathrm{L} = \frac{1}{4\pi} \frac{v_\mathrm{E}}{|\mathbf{v}_\mathrm{E} - \mathbf{v}_0|} \bar{\sigma}\, \mathrm{d}\eta_\mathrm{L}\, \mathrm{d}\Omega_\mathrm{L}. \tag{10.205}$$

Both Salin (1969) and Macek (1970) were able to obtain the forward enhancement by introducing a final state interaction into the Born approximation. A similar enhancement was found by Band (1974) who employed the Bates distorted wave approximation and also by Dettmann *et al.* (1974) who used a treatment based on the second-order Born approximation. These models have not been completely successful in reproducing the data quantitatively.

Clearly, in a proper theoretical treatment one should not attempt to describe the ionization cross section as a Born background with a charge exchange process superimposed; rather, a good model should contain both direct ionization and capture into the continuum automatically, as for example do coupled channel calculations with pseudostates on each heavy particle centre. The CDW generalization of Belkić (1978) is just such a model and above 50 keV satisfactory agreement is obtained with the experimental data of Park *et al.* (1977) for the ionization of atomic hydrogen by proton impact, provided the distortion of the wave function in both the initial and final channels is taken into account. More recently Crothers and McCann (1987) have developed a CDW model and have obtained good agreement with the ionization cross sections for the ionization of helium by fully stripped ions of carbon, helium, and hydrogen, but agreement with differential cross section data near the cusp is less stisfactory. In this work the CDW wave function for the initial state is of the usual form (10.128). In the final ionized state it is necessary to take into account the Coulomb interaction between the electron and both ions, and the wave function is written as

$$M(\mathbf{r}, t) = (2\pi)^{\frac{3}{2}}\omega_{\mathrm{A}}^{-}(\mathbf{v}_{\mathrm{E}}, \mathbf{x})\omega_{\mathrm{B}}^{-}(\mathbf{v}_{0}, \mathbf{r}) \exp[-\mathrm{i}\mathbf{v}_{0}\cdot\mathbf{r} - \eta t] \tag{10.206}$$

where

$$\omega_{\alpha}^{-}(\mathbf{v}, \mathbf{r}) = \mathrm{e}^{-\pi\nu_{\alpha}/2}\Gamma(1 + \mathrm{i}\nu_{\alpha})\,{}_{1}F_{1}(-\mathrm{i}\nu_{\mathrm{A}}, 1, -\mathrm{i}\mathbf{v}\cdot\mathbf{r} - \mathrm{i}\mathbf{v}\cdot\mathbf{r})$$

and

$$\nu_{\alpha} = \frac{Z_{\alpha}}{v}. \tag{10.207}$$

It is interesting to note that this wave function is closely related to that suggested for ionization by electron impact by Crothers (1982) and applied by Brauner *et al.* (1989) following the approach of Pluvinage (1980). The principle difference is that in the heavy particle case there is no contribution (to order $1/M$) from the interionic potential. For electron scattering a third Coulomb factor representing the interaction between the incident electron and the target nucleus B has to be introduced into the wave function.

11

CHARGE EXCHANGE AT RELATIVISTIC ENERGIES

In ion–atom collision theory relativistic effects enter from two directions. If the nuclear charges Z_A or Z_B are sufficiently large, the mean velocity of an electron in an atomic bound state ($\bar{v}_e = Z/n$) can become comparable with the velocity of light c. Accordingly relativistic corrections, which are negligible for small Z, $\alpha Z \lesssim 0.1$, where α is the fine structure constant, become significant for large Z and the atomic or molecular orbitals employed must be based on solutions of the single- or two-centre Dirac equation, or approximations to these. A different source of relativistic effects arises when the collision velocity is large compared with c, the velocity of light. Relativistic kinematics must then be employed, and, in addition, even for small values of Z_A and Z_B, relativistic atomic orbitals must be used since the momentum distribution of the bound electron, on which high-energy charge exchange depends explicitly, is very different in the relativistic and non-relativistic cases. For electron capture relativistic corrections to the Oppenheimer–Brinkman–Kramers perturbation approximation were first investigated by Mittleman (1964) for ground-state capture of electrons from atomic hydrogen

$$H^+ + H(1s) \rightarrow H(1s) + H^+. \tag{11.1}$$

He showed that at $E_L = 10$ MeV/a.m.u.,† the relativistic corrections were small ($\lesssim 3$ per cent), but became increasingly significant at higher energies, ultimately changing the energy variation of the OBK cross section from E_L^{-6} to E_L^{-1}. Subsequently Humphries and Moiseiwitsch (1984) (see also Moiseiwitsch (1988, 1989)) have shown that the corresponding second-order approximation also decreases like E_L^{-1}, in contrast to the $E_L^{-5.5}$ variation predicted by the non-relativistic theory. The development of the relativistic theory of charge exchange, particularly for heavy ions, has been stimulated by an experimental programme at Lawrence Berkeley Laboratory where the Bevalac accelerator (a combination of a heavy-ion linear accelerator and a synchrotron) can produce heavy-ion beams of energies up to 1 GeV/a.m.u. In this chapter, we shall discuss how the principal methods employed to calculate capture cross sections at non-relativisitc velocities can be generalized

† As in earlier chapters, E_L is the kinetic energy of the incident particle A in the laboratory system divided by the rest mass of A (see Appendix A).

to the relativistic regime, starting with the coupled channel approach and continuing with perturbative methods.

At very high energies non-radiative capture competes with radiative electron capture and also with the capture of an electron formed by electron–positron pair creation in the field of colliding heavy ions (Becker 1987; Rivarola 1989). It was noted in Section 4.8 that radiative electron capture can, under certain circumstances, be of importance at very low energies; however from energies of a few keV to a few MeV the radiative reaction does not contribute significantly to the total capture cross section. This situation changes as the impact energy is increased beyond a few MeV/a.m.u. For capture involving light ions, the non-radiative capture cross section continues to decrease very rapidly, $\sim E_L^{-6}$ or $E_L^{-5.5}$, not reaching the asymptotic decrease of E_L^{-1} predicted by the relativistic theory until energies of the order of 10^3 MeV/a.m.u. are reached. For this reason, the radiative capture cross section, which decreases at first like $E_L^{-5/2}$ and ultimately like E_L^{-1}, becomes relatively more important as the energy increases, and for light ions radiative capture is the dominant mechanism for $E_L \gtrsim 9$ MeV/a.m.u. (Briggs and Dettmann (1974)).

For heavy ions the situation is somewhat different and the radiative and non-radiative capture mechanisms can be of comparable importance. This is illustrated in Fig. 11.1, where some results of Becker (1987) are shown for K-shell capture by S^{16+} ions of 15 GeV/a.m.u. from targets with differing values of nuclear charge Z_B. It is seen that the non-radiative mechanism is unimportant for $Z_B < 100$, but for $Z_B > 100$ both the radiative and non-radiative mechanisms contribute to the capture reaction. Also shown in Fig. 11.1 are calculated cross sections for electron capture following pair creation, which can be seen to provide a larger cross section than non-radiative capture for $Z_B \gtrsim 85$ at this energy. At even higher energies this process is expected to be the dominant capture mechanism. The radiative capture mechanism will be discussed briefly in Section 11.4, and capture following pair creation in Section 11.5.

It will be recollected that at high but non-relativistic velocities, elastic scattering and ionization are the most important scattering mechanisms. As the energy is further increased, excitation becomes relatively less important and in ultra-high-energy heavy-ion collisions ionization is the most important inelastic reaction. At high energies the ionization cross section can be calculated in the first-order Born approximation if the perturbing charge is small. When the incident charge is large the distortion of the target wave function has to be taken into account and a distorted wave model employed. Both the Born approximation and distortion approximations can be modified to take into account relativistic effects (Anholt and Gould 1986; Moiseiwitsch 1980, 1985). To illustrate the relative importance of ionization and capture at high energies some cross sections are shown in Fig. 11.2(a) for the creation of vacancies in the K-shell of a target atom of nuclear charge

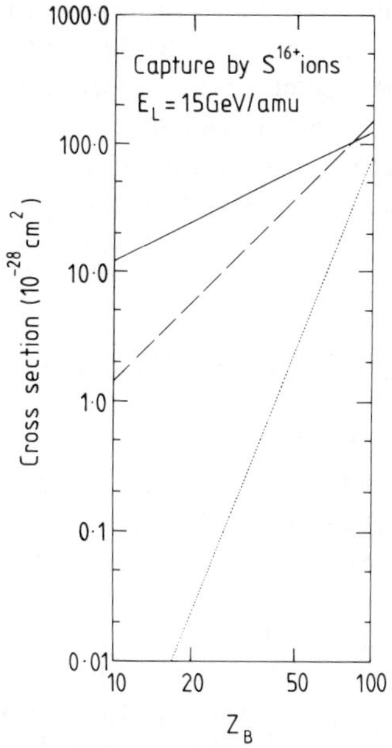

Fig. 11.1. Theoretical cross sections for electron capture by S^{16+} ions from targets with nuclear charge Z_B at 15 GeV/a.m.u. (after Becker (1987)): ——— radiative electron capture; ······ non-radiative electron capture; — — — capture following pair formation.

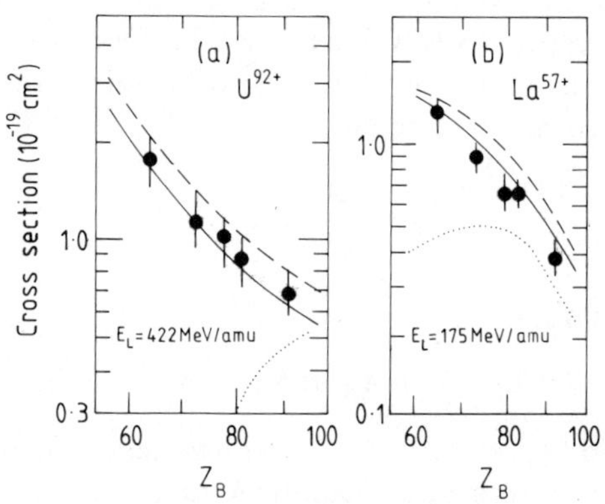

Fig. 11.2. Cross sections for the production of K-shell vacancies in targets of nuclear charge Z_B by the impact of (a) fully stripped uranium ions of energy 422 MeV/a.m.u., and (b) fully stripped Lanthanum ions of energy 175 meV/a.m.u. (after Deco *et al.* 1990): — — — total (ionization + capture) cross section in the relativistic Born approximation (Anholt *et al.* 1984); ——— total (ionization + capture) cross section in a relativistic distorted wave approximation (Deco *et al.* 1990); ······ capture cross section in a relativistic distorted wave approximation (Deco and Rivarola 1987*b*); ⧫⧫ Experimental data (Anholt *et al.* 1984).

Z_B under the impact of fully stripped uranium atoms at 422 MeV/a.m.u. Similar results are shown in Fig. 11.2(b) for fully stripped lanthanum projectiles of energy 175 MeV/a.m.u. The theoretical cross sections shown are obtained from a relativistic model which employs a CDW final state and an eikonal initial state distorted wave function. It is seen that ionization dominates for small Z_B, but capture becomes of considerable importance when $Z_B \gtrsim 85$.

11.1 THE RELATIVISTIC IMPACT PARAMETER MODEL

A relativistic generalization of the impact parameter model introduced in Section 3.7 can be developed straightforwardly for a one-electron system. Let us consider a fully stripped ion of charge Z_A incident on a one-electron ion† with a nucleus of charge Z_B. As we have seen, at high energies the relative motion of A and B can be described by a classical straight line trajectory traversed at constant velocity. Any point on the line AB can be taken to be the origin of an inertial frame, but it is most convenient to take a frame of reference with an origin at the position of the nucleus B, the rest frame of B. In this frame the position of A is described by the trajectory

$$\mathbf{R}(t) = \mathbf{b} + \mathbf{v}_0 t, \qquad \mathbf{b}\cdot\mathbf{v}_0 = 0, \tag{11.2}$$

where $\mathbf{b}$ is the impact parameter vector taken parallel to the X-axis and $\mathbf{v}_0$ is the incident velocity which is in the Z-direction. The position vector of the electron with respect to B will, as usual, be denoted by $\mathbf{r}$.

The electronic wave function $\Psi(\mathbf{r}, t)$ is a four-dimensional spinor which satisfies the two-centre Dirac equation:

$$\left(H_D - i\frac{\partial}{\partial t}\right)\Psi(\mathbf{r}, t) = 0, \tag{11.3}$$

where (in atomic units)

$$H_D = \boldsymbol{\alpha}\cdot(-ic\nabla + \mathbf{A}_A + \mathbf{A}_B) - \phi_A - \phi_B + \beta c^2, \tag{11.4}$$

where $(\mathbf{A}_A, \phi_A)$ and $(\mathbf{A}_B, \phi_B)$ are the electromagnetic potentials acting on the electron due to the charges Z_A and Z_B, respectively. The (4×4)-dimensional Dirac matrices $\boldsymbol{\alpha}$ and β which appear in (11.4) have the standard representation‡ in terms of the two-dimensional Pauli matrices $\boldsymbol{\sigma}$ and the

† Many-electron systems can usually be treated in the one-active-electron approximation. If the active electron belongs to the K-shell (the most important case), it is usual to allow for screening by the second electron by introducing the effective charge $Z^B_{\text{eff}} = Z^B - 0.3$, and to employ the measured K-shell binding energies.

‡ Useful accounts of the theory of the Dirac equation can be found in the standard texts by Sakurai (1967), Merzbacher (1970), and Greiner (1990).

(2×2)-dimensional unit matrix I

$$\boldsymbol{\alpha} = \begin{pmatrix} 0 & \boldsymbol{\sigma} \\ \boldsymbol{\sigma} & 0 \end{pmatrix}, \qquad \beta = \begin{pmatrix} I & 0 \\ 0 & -I \end{pmatrix}. \tag{11.5}$$

Both $\boldsymbol{\alpha}$ and β are Hermitian

$$\boldsymbol{\alpha}^\dagger = \boldsymbol{\alpha}, \qquad \beta^\dagger = \beta, \tag{11.6}$$

and satisfy the conditions

$$\alpha_x^2 = \alpha_y^2 = \alpha_z^2 = \beta^2 = 1, \tag{11.7a}$$

$$[\alpha_x, \alpha_y]_+ = [\alpha_y, \alpha_z]_+ = [\alpha_z, \alpha_x]_+ = 0, \tag{11.7b}$$

$$[\alpha_x, \beta]_+ = [\alpha_y, \beta]_+ = [\alpha_z, \beta]_+ = 0, \tag{11.7c}$$

where $[A, B]_+$ denotes the anticommutator

$$[A, B]_+ = AB + BA. \tag{11.8}$$

The interaction between the charges Z_A and Z_B has been omitted from the Dirac Hamiltonian (11.4) since, as in the non-relativistic theory, it can be removed by a phase transformation.

In the rest frame of B, the interaction between the electron and the nucleus B is a Coulomb potential, so that

$$\mathbf{A}_B = 0 \qquad \text{and} \qquad \phi_B = \frac{Z_B}{r}. \tag{11.9}$$

The moving charge Z_A gives rise to both magnetic and electric fields at the electron, which are most easily determined as follows. Let us introduce a frame of reference with origin at A, the rest frame of A, and let the position of the electron in this frame be $\mathbf{x}'$. The time as measured in this frame will be denoted by t'. The variables $(\mathbf{r}, t)$ and $(\mathbf{x}', t')$ are connected by the inhomogeneous Lorentz transformation:

$$\mathbf{x}' = \mathbf{r} - \mathbf{b} + (\gamma - 1)(\mathbf{r}\cdot\hat{\mathbf{v}}_0)\hat{\mathbf{v}}_0 - \gamma\mathbf{v}_0 t \tag{11.10a}$$

$$t' = \gamma\left[t - \frac{\mathbf{r}\cdot\mathbf{v}_0}{c^2}\right] \tag{11.10b}$$

where $\gamma = (1 - v_0^2/c^2)^{-\frac{1}{2}}$ and $\hat{\mathbf{v}}_0$ is the unit vector in the direction of $\mathbf{v}_0$. The corresponding inverse transformation is

$$\mathbf{r} = \mathbf{x}' + \mathbf{b} + (\gamma - 1)(\mathbf{x}'\cdot\hat{\mathbf{v}}_0)\hat{\mathbf{v}}_0 + \gamma\mathbf{v}_0 t' \tag{11.11a}$$

$$t = \gamma\left[t' + \frac{\mathbf{x}'\cdot\mathbf{v}_0}{c^2}\right]. \tag{11.11b}$$

In the rest frame of A, the potential between the nucleus A and the electron is electrostatic so that

$$\mathbf{A}'_{\mathrm{A}} = 0, \qquad \phi'_{\mathrm{A}} = \frac{Z_{\mathrm{A}}}{x'}. \tag{11.12}$$

Recollecting that $\mathbf{A}$ and ϕ form a four-vector which transforms under a homogeneous Lorentz transformation in the same way as $\mathbf{r}$ and ct, we can transform the potentials $(\mathbf{A}'_{\mathrm{A}}, \phi'_{\mathrm{A}})$ defined in the rest frame of A to corresponding potentials $(\mathbf{A}_{\mathrm{A}}, \phi_{\mathrm{A}})$ in the rest frame of B. We have

$$\mathbf{A}_{\mathrm{A}} = \mathbf{A}'_{\mathrm{A}} + (\gamma - 1)(\mathbf{A}'_{\mathrm{A}} \cdot \hat{\mathbf{v}}_0)\hat{\mathbf{v}}_0 + \frac{\gamma \mathbf{v}_0 \phi'_{\mathrm{A}}}{c}$$

$$\phi_{\mathrm{A}} = \gamma\left[\phi'_{\mathrm{A}} + \frac{\mathbf{A}'_{\mathrm{A}} \cdot \mathbf{v}_0}{c}\right] \tag{11.13}$$

and from (11.12), since $\mathbf{A}'_{\mathrm{A}} = 0$

$$\mathbf{A}_{\mathrm{A}} = \frac{\gamma \mathbf{v}_0 \phi'_{\mathrm{A}}(\mathbf{x}')}{c}$$

$$\phi_{\mathrm{A}} = \gamma \phi'_{\mathrm{A}}(\mathbf{x}') \tag{11.14}$$

where $\mathbf{x}'$ is to be expressed in terms of $\mathbf{r}$ and t by using (11.11).

Defining the operator L so that

$$L\phi'_{\mathrm{A}}(\mathbf{x}') = \phi_{\mathrm{A}} - \boldsymbol{\alpha} \cdot \mathbf{A}_{\mathrm{A}} \tag{11.15}$$

we see that

$$L = \gamma\left(1 - \frac{\mathbf{v}_0 \cdot \boldsymbol{\alpha}}{c}\right), \tag{11.16}$$

and that the Dirac equation (11.3) can be expressed in the frame of B as (from (11.4), (11.12), and (11.15))

$$\left[-\mathrm{i}c\boldsymbol{\alpha} \cdot \nabla - L\left(\frac{Z_{\mathrm{A}}}{x'}\right) - \frac{Z_{\mathrm{B}}}{r} + \beta c^2 - \mathrm{i}\frac{\partial}{\partial t}\right]\Psi(\mathbf{r}, t) = 0. \tag{11.17}$$

We shall now discuss the asymptotic solutions of this equation in the limit $|t| \to \infty$.

Asymptotic wave functions

The set of unperturbed energy eigenfunctions $\phi_j(\mathbf{r})$ for the target atom $(\mathrm{B} + \mathrm{e}^-)$ satisfy the Dirac equation

$$\left(-\mathrm{i}c\boldsymbol{\alpha} \cdot \nabla - \frac{Z_{\mathrm{B}}}{r} + \beta c^2 - E_j^{\mathrm{B}}\right)\phi_j(\mathbf{r}) = 0. \tag{11.18}$$

The functions ϕ_j are four-dimensional spinor functions and form an orthonormal set

$$\langle\phi_i(\mathbf{r})\,|\,\phi_j(\mathbf{r})\rangle = \int \mathrm{d}\mathbf{r}\phi_i^\dagger(\mathbf{r})\phi_j(\mathbf{r}) = \delta_{ij}. \tag{11.19}$$

A similar set of spinor functions $\chi'_n(\mathbf{x}')$ can be defined in the rest frame of A which describe the atom (A + e^-) and which satisfy

$$\left(-\mathrm{i}c\boldsymbol{\alpha}\cdot\boldsymbol{\nabla}' - \frac{Z_\mathrm{B}}{x'} + \beta c^2 - E_n^\mathrm{A}\right)\chi'_n(\mathbf{x}') = 0 \tag{11.20}$$

with

$$\langle\chi'_n(\mathbf{x}')\,|\,\chi'_m(\mathbf{x}')\rangle = \delta_{nm}. \tag{11.21}$$

The spinor functions $\chi'_m(\mathbf{x}')$ can be transformed to the rest frame of B by an operator S

$$\chi_n(\mathbf{r}, t) = S\chi'_n(\mathbf{x}') \tag{11.22}$$

where (Sakurai 1967; Greiner 1990)

$$S = \left(\frac{\gamma+1}{2}\right)^{\frac{1}{2}} + \left(\frac{\gamma-1}{2}\right)^{\frac{1}{2}}\boldsymbol{\alpha}\cdot\hat{\mathbf{v}}_0. \tag{11.23}$$

The inverse transformation is given by S^{-1} where

$$S^{-1} = \left(\frac{\gamma+1}{2}\right)^{\frac{1}{2}} - \left(\frac{\gamma-1}{2}\right)^{\frac{1}{2}}\boldsymbol{\alpha}\cdot\hat{\mathbf{v}}_0. \tag{11.24}$$

The relation

$$L = (S^{-1})^2 \tag{11.25}$$

should be noted, which follows directly from (11.16) and (11.23). In fact the requirement that the two-centre Dirac equation should take the same form in either the rest frame of A or of B, leads directly to the relation

$$SLS = 1 \tag{11.26}$$

and using (11.16), the expression (11.24) for S^{-1} can be deduced (Moiseiwitsch 1985).

The asymptotic solutions of equation (11.17) in the limit $|t| \to \infty$ can now be found in terms of the functions $\phi_j(\mathbf{r})$ or $\chi'_n(\mathbf{x}')$. Let us first consider terms asymptotic to the initial arrangement A + (B + e^-). In the limit $|t| \to \infty$, $x' \to R'$, where R' is the distance AB as seen from the rest frame of A, so that the asymptotic solutions $\Phi_j^\mathrm{B}(\mathbf{r}, t)$ of (11.17) satisfy the equation

$$\left[-\mathrm{i}c\boldsymbol{\alpha}\cdot\boldsymbol{\nabla} - L\left(\frac{Z_\mathrm{A}}{R'}\right) - \frac{Z_\mathrm{B}}{r} + \beta c^2 - \mathrm{i}\frac{\partial}{\partial t}\right]\Phi_j^\mathrm{B}(\mathbf{r}, t) = 0. \tag{11.27}$$

In the rest frame of B, we have, from (11.2) that

$$R(t) = (b^2 + v_0^2 t^2)^{\frac{1}{2}} \tag{11.28a}$$

while in the rest frame of A

$$R'(t') = (b^2 + v_0^2 t'^2)^{\frac{1}{2}}. \tag{11.28b}$$

Recollecting that Z_A/R' is the asymptotic form of the electronic interaction Z_A/x', it is seen that t' is the time at which the electron is at the point $\mathbf{x}'$ so that in terms of $\mathbf{r}$ and t, t' is given by (11.10b).

To determine the form of the function $\Phi_j^B(\mathbf{r}, t)$, we first transform (11.27) to the rest frame of A, giving

$$\left[-ic\boldsymbol{\alpha}\cdot\boldsymbol{\nabla}' - \frac{Z_A}{R'} - L^{-1}\frac{Z_B}{r} + \beta c^2 - i\frac{\partial}{\partial t'}\right]\Phi_j'^B(\mathbf{x}', t') = 0 \tag{11.29}$$

where

$$S\Phi_j'^B(\mathbf{x}, t') = \Phi_j^B(\mathbf{r}, t). \tag{11.30}$$

In this frame the interaction Z_A/R' is a pure Coulomb potential which can be removed by the phase transformation

$$\Phi_j'^B(\mathbf{x}', t') = \exp[-i\nu_A \log\{R'(t') - v_0 t'\}]\tilde{\Phi}_j(\mathbf{x}', t') \tag{11.31}$$

where $\tilde{\Phi}_j(\mathbf{x}', t')$ satisfies the equation

$$\left[-ic\boldsymbol{\alpha}\cdot\boldsymbol{\nabla}' - L^{-1}\frac{Z_B}{r} + \beta^2 c^2 - i\frac{\partial}{\partial t'}\right]\tilde{\Phi}_j(\mathbf{x}', t') = 0. \tag{11.32}$$

Equation (11.32) is the Dirac equation for the atom $(B + e^-)$ *written in the rest frame of A*, so that $\tilde{\Phi}_j(\mathbf{x}', t')$ must be of the form

$$\tilde{\Phi}_j(\mathbf{x}', t') = S^{-1}[\phi_j(\mathbf{r}) \exp(-iE_j^B t)]. \tag{11.33}$$

Combining (11.30), (11.31), and (11.33), the asymptotic wave function Φ_j^B can be written as

$$\Phi_j^B(\mathbf{r}, t) = \exp[-i\nu_A \log\{R'(t') - v_0 t'\}]\phi_j(\mathbf{r}) \exp(-iE_j^B t). \tag{11.34}$$

From (11.10b) it is seen that the phase factor describing the Coulomb distortion depends both on the time t and on the spatial coordinate $\mathbf{r}$, and this dependence is necessary for the interaction Z_A/R' to be eliminated completely. Nevertheless in the asymptotic region of the arrangement $A + (B + e^-)$, with $x' \approx R'$ and $r \ll x$, the quantity $(\mathbf{r}\cdot\mathbf{v}_0)/c^2$ is very small compared with $|t|$ and in the calculation of matrix elements the approximations can be made that

$$t' \approx \gamma t \tag{11.35a}$$

$$R'(t') \approx (b^2 + \gamma^2 v_0^2 t^2)^{\frac{1}{2}}. \tag{11.35b}$$

The asymptotic solutions of (11.17) in the arrangement $(\mathrm{A} + \mathrm{e}^-) + \mathrm{B}$ can be found in a similar way. In this arrangement $r \to R$ as $|t| \to \infty$ and $x \ll r$. The corresponding asymptotic functions $\mathrm{X}_n^{\mathrm{A}}(\mathbf{r}, t)$ written in the rest frame of B satisfy:

$$\left[-\mathrm{i}c\boldsymbol{\alpha}\cdot\boldsymbol{\nabla} - L\left(\frac{Z_{\mathrm{A}}}{x'}\right) - \frac{Z_{\mathrm{B}}}{R} + \beta c^2 - \mathrm{i}\frac{\partial}{\partial t}\right]\mathrm{X}_n^{\mathrm{A}}(\mathbf{r}, t) = 0. \tag{11.36}$$

The Coulomb potential Z_{B}/R can be removed by setting

$$\mathrm{X}_n^{\mathrm{A}}(\mathbf{r}, t) = \bar{\mathrm{X}}_n(\mathbf{r}, t)\exp[-\mathrm{i}\nu_{\mathrm{B}}(R - v_0 t)] \tag{11.37}$$

where $\nu_{\mathrm{B}} = Z_{\mathrm{B}}/v_0$. Inserting (11.37) into (11.36), $\bar{\mathrm{X}}_n$ is found to satisfy

$$\left[-\mathrm{i}c\boldsymbol{\alpha}\cdot\boldsymbol{\nabla} - L\left(\frac{Z_{\mathrm{A}}}{x'}\right) + \beta c^2 - \mathrm{i}\frac{\partial}{\partial t}\right]\bar{\mathrm{X}}_n(\mathbf{r}, t) = 0. \tag{11.38}$$

Transformed to the rest frame of A, this equation reads

$$\left[-\mathrm{i}c\boldsymbol{\alpha}\cdot\boldsymbol{\nabla}' - \frac{Z_{\mathrm{A}}}{x'} + \beta c^2 - \mathrm{i}\frac{\partial}{\partial t'}\right]\bar{\mathrm{X}}_n'(x', t') = 0 \tag{11.39}$$

with the solution

$$\bar{\mathrm{X}}_n'(\mathbf{x}', t') = \chi_n'(\mathbf{x}')\exp(-\mathrm{i}E_n^{\mathrm{A}}t'). \tag{11.40}$$

Since, from (11.22), $\bar{\mathrm{X}}_n(\mathbf{r}, t) = S\bar{\mathrm{X}}_n'(\mathbf{x}', t)$, we have from (11.37) that

$$\mathrm{X}_n^{\mathrm{A}}(\mathbf{r}, t) = \exp[-\mathrm{i}\nu_{\mathrm{B}}\log(R - v_0 t)]S[\chi_n'(\mathbf{x}')\exp(-\mathrm{i}E_n^{\mathrm{A}}t')]. \tag{11.41}$$

Time-reversed functions Φ_j^{BT} and $\mathrm{X}_n^{\mathrm{AT}}$ can be obtained from Φ_j^{B} and $\mathrm{X}_n^{\mathrm{A}}$ (see Chapter 10) by the relations

$$\Phi_j^{\mathrm{BT}}(\mathbf{r}, t) = \Phi_j^{\mathrm{B}*}(\mathbf{r}, -t), \qquad \mathrm{X}_n^{\mathrm{AT}}(\mathbf{r}, t) = X_n^{\mathrm{A}*}(\mathbf{r}, t). \tag{11.42}$$

Probability amplitudes and cross sections

The determination of probability amplitudes and cross sections for inelastic processes from solutions of the two-centre Dirac equation (11.17) proceeds along the lines already discussed in Chapters 3 and 10. If the initial state of the system is one in which the ion A is incident on the ith state of the atom $\mathrm{B} + \mathrm{e}^-$, a solution of (11.17) is sought satisfying the boundary condition

$$\Psi_i(\mathbf{r}, t) \underset{t \to -\infty}{\sim} \Phi_i^{\mathrm{B}}(\mathbf{r}, t). \tag{11.43}$$

Then the probability amplitude for excitation to the jth level of the atom $\mathrm{B} + \mathrm{e}^-$ is given by (see (10.39) and (10.33))

$$\mathscr{A}_{ji}(\mathbf{b}) = -\mathrm{i}\int_{-\infty}^{\infty} \langle\Phi_j^{\mathrm{BT}}(\mathbf{r}, t)|U_{\mathrm{A}}^{\mathrm{R}}|\Psi_i(\mathbf{r}, t)\rangle\, \mathrm{d}t, \qquad i \neq j \tag{11.44}$$

where

$$U_{\mathrm{A}}^{\mathrm{R}}(\mathbf{r}, t) = L\left(-\frac{Z_{\mathrm{A}}}{x'} + \frac{Z_{\mathrm{A}}}{R'}\right). \tag{11.45}$$

The probability amplitude for ionization also takes the form (11.44), where $\phi_j(\mathbf{r})$ is taken to be a Dirac–Coulomb continuum function. Likewise, the probability amplitude for electron capture to the nth level of the atom $(\mathrm{A} + \mathrm{e}^-)$ is given by (see (10.44) and (10.45))

$$\mathscr{A}_{ni}(\mathbf{b}) = -\mathrm{i}\int_{-\infty}^{\infty} \langle \mathrm{X}_n^{\mathrm{AT}}(\mathbf{r}, t)|U_{\mathrm{B}}^{\mathrm{R}}|\Psi_i(\mathbf{r}, t)\rangle \, \mathrm{d}t \tag{11.46}$$

with

$$U_{\mathrm{B}}^{\mathrm{R}} = \left(-\frac{Z_{\mathrm{B}}}{r} + \frac{Z_{\mathrm{B}}}{R}\right). \tag{11.47}$$

In equations (11.44) and (11.46) all the functions Ψ_i, Φ_j^{B}, and $\mathrm{X}_n^{\mathrm{A}}$ are understood to be expressed in the rest frame of target atom B. The inelastic total cross sections are found by integrating over the impact parameter in the usual way:

$$\sigma_{ni} = \int \mathrm{d}^2\mathbf{b}|\mathscr{A}_{ni}(\mathbf{b})|^2. \tag{11.48}$$

To calculate differential cross sections the phase generated by the omitted internuclear potential $Z_{\mathrm{A}}Z_{\mathrm{B}}/R'$ must be restored. This can be achieved by including the additional interaction in equations (11.17), (11.29) and (11.36). By following the procedure outlined in Chapter 3 (see particularly (3.239)–(3.242)) the phase-corrected amplitude $\mathscr{A}_{ni}^{\mathrm{c}}(\mathbf{b})$ is found by multiplying the amplitude $\mathscr{A}_{ni}(\mathbf{b})$, calculated from (11.44) or (11.46) by the phase factor $b^{2\mathrm{i}\nu'}$, where $\nu' = Z_{\mathrm{A}}Z_{\mathrm{B}}/(\gamma v_0)$

$$\mathscr{A}_{ni}^{\mathrm{c}}(\mathbf{b}) = b^{2\mathrm{i}\nu'}\mathscr{A}_{ni}(\mathbf{b}). \tag{11.49}$$

The differential cross section in the rest frame of the target ion B is

$$\frac{\mathrm{d}\sigma}{\mathrm{d}\Omega} = |f_{ni}(\Theta, \Phi)|^2 \tag{11.50}$$

where the scattering amplitude is given by an expression similar to (3.226)

$$f_{ni}(\Theta, \Phi) = -\frac{1}{2\pi}\frac{M_{\mathrm{A}}\gamma}{v_0}\int \mathrm{d}^2\mathbf{b}\,\exp(\mathrm{i}\mathbf{K}\cdot\mathbf{b})\mathscr{A}_{ji}^{\mathrm{c}}(\mathbf{b}), \qquad j \neq i. \tag{11.51}$$

The magnitude of the momentum transfer $\mathbf{K}$ is (cf. (3.223))

$$K \approx K_{\mathrm{A}}\Theta \tag{11.52}$$

where $\mathbf{K}_A$ is the momentum of the incident ion in the rest frame of B:

$$\mathbf{K}_A = \gamma M_A \mathbf{v}_0. \tag{11.53}$$

11.2 COUPLED CHANNEL MODELS

Approximate solutions of the two-centred Dirac equation (11.17) can be obtained by expanding the wave function $\Psi(\mathbf{r}, t)$ in an appropriate basis. If $v_0/v_e < 1$, where v_0 is the incident velocity and v_e characterizes the electronic velocity in the bound states, an appropriate basis can be formed from a set of molecular orbitals determined by solving the two-centre Dirac equation at fixed values of the internuclear separation. For example, if Z_A (or Z_B) > 20 and $v_0 \lesssim 20$ this approach will be useful, since conditions are nearly adiabatic. In contrast, as v_0 approaches the velocity of light, conditions are far from adiabatic even for large values of Z_A and Z_B so that an expansion based on the Dirac atomic wave functions ϕ_j or χ_n would be more appropriate.

Molecular orbital expansions and superheavy molecules

In the collisions of heavy ions under conditions in which nuclei approach to within distances comparable to the nuclear radius, the united atom will be formed for a brief period ($\sim 10^{-17}$ s) with nuclear charge $Z_A + Z_B$. Such collisions have been studied intensively both experimentally and theoretically, because values of the combined charge $(Z_A + Z_B)$ are attained which are far beyond the limits imposed by the periodic table. Typical systems which have been studied are Pb + Pb, $(Z_A + Z_B = 164)$ and U + U, $(Z_B + Z_A = 184)$. In the intense field generated during the collision, pair creation takes place and its study may form the basis of tests of quantum electrodynamics. In this connection, the collision energies of interest are those in which in a head-on collision $(b = 0)$ the nuclei are just brought into contact. At these energies $E_L \sim 6$ MeV/a.m.u., $v_0/c \simeq 0.05$ and the relative motion of the ions can be treated non-relativistically. Under these circumstances the time-dependent Dirac equation (11.17) can be approximated by setting $L = I$ and $x' = x$ so that Ψ satisfies the equation (for a one-electron system)

$$\left[-ic\boldsymbol{\alpha}\cdot\boldsymbol{\nabla} - \frac{Z_A}{x} - \frac{Z_B}{r} + \beta c^2 - i\frac{\partial}{\partial t}\right]\Psi(\mathbf{r}, t) = 0 \tag{11.54}$$

where

$$\mathbf{x} = \mathbf{r} - \mathbf{R}(t).$$

The relativistic molecular orbitals $\chi_j(\mathbf{r}, R)$ are solutions of the eigenvalue equation

$$\left[-\mathrm{i}c\boldsymbol{\alpha}\cdot\boldsymbol{\nabla}-\frac{Z_{\mathrm{A}}}{x}-\frac{Z_{\mathrm{B}}}{r}+\beta c^2-E_j(R)\right]\chi(\mathbf{r},R)=0 \tag{11.55}$$

for fixed values of R. A solution of this equation has been given by Müller and Greiner (1973, 1976) in terms of spheroidal coordinates in the body-fixed frame (see also Fricke *et al.* (1975)) and this method is discussed in the text of Greiner (1990).

Provided the relative motion can be treated non-relativistically the equations for the probability amplitudes $A_j(\mathbf{b}, t)$ can be derived by the methods discussed in Chapter 5. If translational factors are neglected these equations can be put into the simple form

$$\dot{A}_j(\mathbf{b},t)=\sum_k \langle\chi_j|\frac{\partial}{\partial t}|\chi_k\rangle \exp\left[-\mathrm{i}\int^t (E_j(t)-E_k(t))\,\mathrm{d}t\right]A_k(\mathbf{b},t). \tag{11.56}$$

The operator $\partial/\partial t$ in the space-fixed frame can be expressed, as shown in Section 5.2, in terms of radial and angular couplings

$$\frac{\partial}{\partial t}=\dot{R}(t)\frac{\partial}{\partial R}-\mathrm{i}\Theta(t)L_y. \tag{11.57}$$

An account of the application of this simple molecular orbital model to the interpretation of the ionization and pair production experiments is outside the scope of this book, but details may be found in the article of Müller (1989).

Two-centre atomic basis sets

We now turn to the case of fast collisions for which the fully relativistic equation (11.17) must be employed. The perturbation methods described later in the chapter preserve the relativistic invariance of the theory, but are not in general gauge invariant because of the lack of orthogonality of the basis sets in the direct and rearranged channels. This difficulty is eliminated in a coupled channel approach in which the equations satisfied by the probability amplitudes $A_j(\mathbf{b}, t)$ are obtained from a variational principle. In the method developed by Toshima and Eichler (1988*a*, *b*) the trial wave function is expressed as

$$\Psi(\mathbf{r},t)=\sum_{j=1}^{N}A_j(\mathbf{b},t)\phi_j(\mathbf{r})\exp[\mathrm{i}\alpha_j(t)]+\sum_{j=N+1}^{N+m}A_n(\mathbf{b},t)S\chi_n(\mathbf{x}')\exp[\mathrm{i}\beta_n(t')] \tag{11.58}$$

where the phases α_j and β_n are chosen to be

$$\alpha_j(t) = -E_j^{\mathrm{B}}t \tag{11.59}$$

$$\beta_n(t') = -E_n^{\mathrm{A}}t' = -E_n^{\mathrm{A}}\gamma\left[t - \frac{\mathbf{r}\cdot\mathbf{v}_0}{c^2}\right]. \tag{11.60}$$

As in the non-relativistic case (see the remarks below (3.179) and (3.180)) the Coulomb phase factors which appear in the asymptotic functions Φ_j^{B} and χ_n^{A} have been absorbed into the amplitudes A_j and A_n.

In the non-relativistic limit

$$\begin{aligned} E_j^{\mathrm{B}} &\to mc^2 + \varepsilon_j \\ E_n^{\mathrm{A}} &\to mc^2 + \eta_n \\ \gamma &\to \left(1 + \frac{v - v_0}{2c^2}\right) \end{aligned} \tag{11.61}$$

where ε_j and η_n are the non-relativistic binding energies of the atoms $(\mathrm{B} + \mathrm{e}^-)$ and $(\mathrm{A} + \mathrm{e}^-)$, respectively, and m $(m = 1)$ is the mass of the electron. It follows that in this limit

$$\exp \mathrm{i}\alpha_j(t) \to \exp(-imc^2t)\exp(-\mathrm{i}\varepsilon_j t) \tag{11.62a}$$

$$\exp \mathrm{i}\beta_n(t) \to \exp(-imc^2t)\exp(-\mathrm{i}\eta_n t)\exp[-\tfrac{1}{2}\mathrm{i}v_0^2 t + \mathrm{i}\mathbf{v}_0\cdot\mathbf{r}]. \tag{11.62b}$$

The factor $\exp(-imc^2t)$ is an overall phase and can be removed, and the last factor in (11.62b) will be recognized as the proper translational factor representing the motion of the electron when attached to A, relative to the chosen origin B.

As in the non-relativistic case, the probability amplitudes satisfy coupled equations of the form (3.185), which can be written together as the matrix equations (3.205). These equations can be solved numerically with the boundary conditions (3.187*a*, *b*) to find the probability of excitation or charge transfer and the corresponding cross sections given by (3.188) and (3.156).

By using the equations satisfied by the Dirac spinor wave functions for the atoms $(\mathrm{B} + \mathrm{e}^-)$ and $(\mathrm{A} + \mathrm{e}^-)$ given by (11.18) and (11.20), respectively, explicit expressions can be obtained for the submatrices $\mathbf{S}$, $\mathbf{H}$, $\bar{\mathbf{H}}$, $\mathbf{K}$, and $\bar{\mathbf{K}}$ appearing in (3.186). It is found that:

$$S_{jn}(t) = \mathrm{e}^{\mathrm{i}(E_j^{\mathrm{B}} - \gamma E_n^{\mathrm{A}})t}\int \phi_j^{\dagger}(\mathbf{r})\,\mathrm{e}^{\mathrm{i}\gamma E_n^{\mathrm{A}}(\mathbf{r}\cdot\mathbf{v}_0)/c^2} S\chi_n(\mathbf{x}')\,\mathrm{d}\mathbf{r} \tag{11.63a}$$

$$\bar{S}_{nk}(t) = \mathrm{e}^{\mathrm{i}(\gamma E_n^{\mathrm{A}} - E_k^{\mathrm{B}})t}\int \chi_n^{\dagger}(\mathbf{x}')S\phi_k(\mathbf{r})\,\mathrm{e}^{-\mathrm{i}\gamma E_n^{\mathrm{A}}(\mathbf{r}\cdot\mathbf{v}_0)/c^2}\,\mathrm{d}\mathbf{r} \tag{11.63b}$$

$$H_{jk}(t) = \int \phi_j^\dagger(\mathbf{r})\, \mathrm{e}^{-\mathrm{i}E_j^{\mathrm{B}}t}\left(H_{\mathrm{D}} - \mathrm{i}\frac{\partial}{\partial t}\right)\phi_k(\mathbf{r})\, \mathrm{e}^{-\mathrm{i}E_k^{\mathrm{B}}t}\, \mathrm{d}\mathbf{r}$$

$$= -\mathrm{e}^{\mathrm{i}(E_j^{\mathrm{B}} - E_k^{\mathrm{B}})t} \int \phi_j^\dagger(\mathbf{r}) L\left(\frac{Z_{\mathrm{A}}}{x'}\right)\phi_k(\mathbf{r})\, \mathrm{d}\mathbf{r} \tag{11.63c}$$

$$\bar{H}_{nm}(t) = \int (S\chi_n(\mathbf{x}'))^\dagger\, \mathrm{e}^{+\mathrm{i}E_n^{\mathrm{A}}t'}\left(H_{\mathrm{D}} - \mathrm{i}\frac{\partial}{\partial t}\right)S\chi_m(\mathbf{x}')\, \mathrm{e}^{-\mathrm{i}E_m^{\mathrm{A}}t'}\, \mathrm{d}\mathbf{r}$$

$$= -\mathrm{e}^{\mathrm{i}\gamma(E_n^{\mathrm{A}} - E_m^{\mathrm{A}})t} \int \chi_n^\dagger(\mathbf{x}') L^{-1}\left(\frac{Z_{\mathrm{B}}}{r}\right)\chi_m(\mathbf{x}')\, \mathrm{e}^{\mathrm{i}\gamma(E_n^{\mathrm{A}} - E_m^{\mathrm{A}})(\mathbf{r}\cdot\mathbf{v}_0)/c^2}\, \mathrm{d}\mathbf{r} \tag{11.63d}$$

$$K_{jm}(t) = \int \phi_j^\dagger(\mathbf{r})\, \mathrm{e}^{\mathrm{i}E_j^{\mathrm{B}}t}\left(H_{\mathrm{D}} - \mathrm{i}\frac{\partial}{\partial t}\right)S\chi_m(\mathbf{x}')\, \mathrm{e}^{-\mathrm{i}E_m^{\mathrm{A}}t'}\, \mathrm{d}\mathbf{r}$$

$$= -\mathrm{e}^{\mathrm{i}(E_j^{\mathrm{B}} - \gamma E_m^{\mathrm{A}})t} \int \phi_j^\dagger(\mathbf{r}) S\left(\frac{Z_{\mathrm{B}}}{r}\right)\chi_m(\mathbf{x}')\, \mathrm{e}^{-\mathrm{i}\gamma E_m^{\mathrm{A}}(\mathbf{r}\cdot\mathbf{v}_0)/c^2}\, \mathrm{d}\mathbf{r}. \tag{11.63e}$$

$$\bar{K}_{nk}(t) = \int (S\chi_n(\mathbf{x}'))^\dagger\, \mathrm{e}^{\mathrm{i}E_n^{\mathrm{A}}t'}\left(H_{\mathrm{D}} - \mathrm{i}\frac{\partial}{\partial t}\right)\phi_j(\mathbf{r})\, \mathrm{e}^{-\mathrm{i}E_j^{\mathrm{B}}t}\, \mathrm{d}\mathbf{r}$$

$$= -\mathrm{e}^{\mathrm{i}(\gamma E_n^{\mathrm{A}} - E_j^{\mathrm{B}})t} \int \chi_n{}^\dagger(\mathbf{x}') S^{-1}\left(\frac{Z_{\mathrm{A}}}{x'}\right)\phi_j(\mathbf{r})\, \mathrm{e}^{\mathrm{i}\gamma E_n^{\mathrm{A}}(\mathbf{r}\cdot\mathbf{v}_0)/c^2}\, \mathrm{d}\mathbf{r} \tag{11.63f}$$

To reduce the matrix elements (11.63) to a form suitable for numerical computation the sums over the spinor indices must be carried out. The explicit form of the Dirac wave functions $\phi_i(\mathbf{r})$ (the functions $\chi_n(\mathbf{x}')$ obviously have the same form) in the standard representation is (Merzbacher 1970; Greiner 1990)

$$\phi_i(\mathbf{r}) = \begin{pmatrix} f_i(r) & \mathcal{Y}_{Jl}^m \\ -\mathrm{i}g_i(r) & \mathcal{Y}_{Jl'}^m \end{pmatrix}, \tag{11.64}$$

where $l = J \mp \frac{1}{2}$ and $l' = J \pm \frac{1}{2}$. The functions $\mathcal{Y}_{Jl}^m$ are the two-dimensional spinors

$$\mathcal{Y}_{Jl}^m = \sum_{m_l m_s} C_{l\frac{1}{2}}(JM, m_l m_s) Y_{l,m_l}(\hat{\mathbf{r}})\chi_{\frac{1}{2}m_s}, \tag{11.65}$$

where $C_{ll'}(JM, mm')$ is a Clebsch–Gordan coefficient.

The hydrogenic wave functions $\phi_i(\mathbf{r})$ are eigenfunctions of $\mathbf{J}^2$, J_Z, and the parity operator and can be labelled by the quantum numbers J, l, and M as well as a principal quantum number n ($i \equiv Jlmn$). We shall use the designation $nl_J(M)$ when referring to a given state. Since the radial functions $f_i(r)$, $g_i(r)$ are combinations of Slater functions, when the spinor operations have been carried out, the matrix elements are combinations of integrals of the same general type as in the non-relativistic case and the same analytical and numerical methods can be employed for their evaluation.

In the coupled channel model represented by equations (3.186) and (11.63) there are certain difficulties. The first (Toshima and Eichler 1989) is that although the exact theory is completely independent of which inertial reference frame is employed, this is not the case when the total wave function Ψ is represented by a truncated expansion. However, in practice this may not be an important objection to the method if the basis employed is large enough. A more serious difficulty is that if the basis is confined to functions ϕ_j, χ_n which are exact bound state solutions of Dirac equations, the ionization channels are omitted. In the non-relativistic coupled channel model this deficiency is corrected by the addition of pseudostates to the basis. This can also be done in the relativistic theory but this in turn gives rise to further problems which have been discussed by Toshima and Eichler (1989). For instance, the diagonalization of the target or projectile Hamiltonians on some discrete basis to produce pseudostates will, in general, introduce negative energy solutions of the Dirac equation with $E < -2mc^2$, since the Dirac energy spectrum is not bounded from below. This difficulty can be overcome in practice by imposing additional conditions, for example by ensuring that the ratio of the large and small components of the Dirac spinor is fixed to some reasonable value typical of states with $E > 0$. A second difficulty is that the exact wave functions $\chi_n(\mathbf{x}')$ are orthonormal in the rest frame of A, and the transformed functions $S\chi_n(\mathbf{x}')$ are orthonormal in the rest frame of B. This is not the case with pseudostate functions. Toshima and Eichler discuss various ways of minimizing this problem, for example by employing pseudostates only in the frame in which the calculation is carried out, that is, in the formulation given above, only those centred on B.

Applications

Toshima and Eichler (1988*a*, *b*) have made a thorough investigation of the single-electron system $U^{92+} + U^{91+}$ at an energy of 500 MeV/a.m.u., obtaining excitation and capture cross sections for both 20- and 36-state basis sets of bound state wave functions. In their later work (Toshima and Eichler 1989) this work was extended by the inclusion of pseudostates, thus allowing for an estimate of the ionization cross section in addition to excitation and capture. In Table 11.1, the calculated capture cross sections for the 36-state basis, and with the same basis augmented by the addition of 16 pseudostates, are compared with the results of the relativistic first-order boundary corrected approximation (R1B). The 36 bound state basis consisted of the 18 states $1s_{\frac{1}{2}}(\pm\frac{1}{2})$, $2s_{\frac{1}{2}}(\pm\frac{1}{2})$, $2p_{\frac{1}{2}}(\pm\frac{1}{2})$, $2p_{\frac{3}{2}}(\pm\frac{1}{2})$, $2p_{\frac{3}{2}}(\pm\frac{3}{2})$, $3s_{\frac{1}{2}}(\pm\frac{1}{2})$, $3p_{\frac{1}{2}}(\pm\frac{1}{2})$, $3p_{\frac{3}{2}}(\pm\frac{1}{2})$, $3p_{\frac{3}{2}}(\pm\frac{3}{2})$, on each of the nuclei A and B. The 16 pseudostate basis is distributed with eight states on each nucleus, designated $1\bar{s}_{\frac{1}{2}}(\pm\frac{1}{2})$, $2\bar{s}_{\frac{1}{2}}(\pm\frac{1}{2})$, $2\bar{p}_{\frac{1}{2}}(\pm\frac{1}{2})$, $2\bar{\bar{p}}_{\frac{1}{2}}(\pm\frac{1}{2})$; the corresponding energies after diagonalizing the target Hamiltonian lie in the continuum.

The results obtained from the earlier 20-state calculation agree to within

Table 11.1. Electron capture by U^{92+} from $U^{91+}(1s_{\frac{1}{2}})$ at 500 MeV/a.m.u. CCA are coupled channel calculations with a 36 bound state basis (Toshima and Eichler 1988*b*); CCB are coupled channel calculations with a 36 bound state + 16 pseudostate basis (Toshima and Eichler 1989); R1B is a relativistic first-order boundary corrected Born approximation (Eichler 1987, 1988).

Final level	Cross section (10^{-21} cm^2)		
	CCA	CCB	R1B
$1s_{\frac{1}{2}}$	4.12	4.47	3.39
$2s_{\frac{1}{2}}$	1.03	1.78	0.51
$2p_{\frac{1}{2}}$	0.391	0.535	0.595
$2p_{\frac{3}{2}}$	0.231	0.352	0.274
$3s_{\frac{1}{2}}$	0.333	0.632	0.151
$3p_{\frac{1}{2}}$	0.099	0.366	0.175
$3p_{\frac{3}{2}}$	0.071	0.086	0.091
Total	6.31	8.22	5.19

10 per cent with those of the 36-state calculation; furthermore the addition of pseudostates again only alters the capture and excitation by the order of 10 per cent. This suggests that, while perhaps not fully converged, the coupled channel model with basis sets of reasonable size can produce meaningful cross sections. The theoretical cross sections receive support from an experiment on the product of vacancies in the K-shell during U + U collisions at 422 MeV/a.m.u. At 500 MeV the cross section for vacancy production can be estimated by adding the excitation, capture, and ionization cross sections. The result is ~ 3–3.8×10^{-20} cm^2 which compares favourably with the measured value per K-electron of 3–3.5×10^{-20} cm^2 (Anholt *et al.* 1984).

In addition to these extensive calculations on the 'test-bed' system of $U^{91+} + U^{92+}$, Toshima and Eichler have discussed electron capture in collisions of fully stripped xenon (Xe^{54+}) with Ag and Au targets, which can be taken to be representative of medium heavy and heavy target atoms, respectively. Since the capture cross section receives contributions from all levels in the target atom, the one-electron coupled channel model is modified by performing a series of calculations for the capture from each one-electron target state in turn. The target wave functions are hydrogenic, using a different effective charge for each target subshell. As excitation into occupied

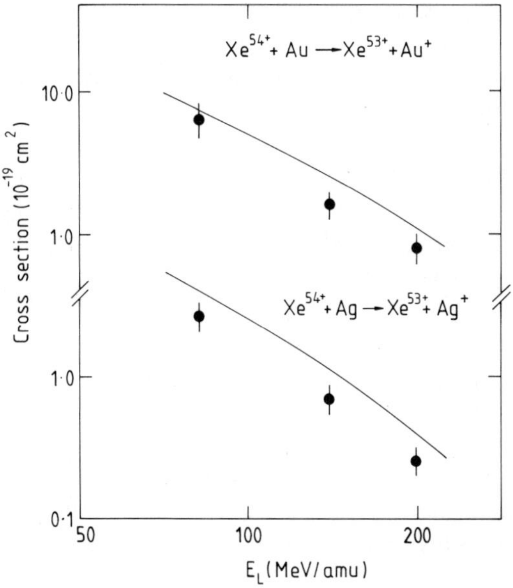

Fig. 11.3. The total cross section for electron capture by Xe^{54+} from neutral Au and Ag atoms (after Toshima and Eichler 1988*b*). Experimental data points from Meyerhof *et al.* (1985), Meyerhof (1988); ——— relativistic coupled channel calculations (Toshima and Eichler 1988*b*).

levels cannot take place, the basis for each calculation consists of a single target state representing the particular electron to be captured and a set of 28 projectile data containing all the *K*-, *L*-, and M-shell wave functions. The calculated total capture cross section shown in Fig. 11.3 for energies between 80 and 200 MeV/a.m.u. lies rather above the experimental data, but is of the correct order of magnitude.

It is probably safe to conclude that for a one-electron system results of comparatively high accuracy can be obtained from the relativistic coupled channel model. For targets containing many electrons cross sections of useful accuracy can be obtained, but there are limitations imposed by the restrictions to an effective charge description of the target electrons and absence of exchange and correlation effects.

11.3 RELATIVISTIC PERTURBATION AND DISTORTED WAVE APPROXIMATIONS

In Chapter 10 we saw that for light ions (Z_A, $Z_B \lesssim 20$) perturbation and distorted wave models can be employed to describe fast but non-relativistic ion–atom collisions. At energies of 100 MeV and above relativistic corrections need to be taken into account (see Table 11.2). Using the basic formulae outlined in Section 11.1, nearly all the methods discussed in Chapter 10 can be generalized for use at relativistic energies.

Excitation and ionization

For excitation, or ionization, the first-order perturbation approximation for the probability amplitude $\mathscr{A}^{\mathrm{I}}_{ji}(\mathbf{b})$ is obtained from (11.44) by replacing $\Psi_i(t)$ by the unperturbed wave function describing the initial state $\Phi_i^{\mathrm{B}}(\mathbf{r}, t)$. We have

$$\mathscr{A}^{\mathrm{I}}_{ji}(\mathbf{b}) = -\mathrm{i}\int_{-\infty}^{\infty} \langle \Phi_j^{\mathrm{BT}}(\mathbf{r}, t)|U_{\mathrm{A}}^{\mathrm{R}}|\Phi_i^{\mathrm{B}}(\mathbf{r}, t)\rangle \,\mathrm{d}t. \tag{11.66}$$

The unperturbed wave functions (11.34) allow for distortion by the overall Coulomb potential, but for excitation and ionization the phase factors for the incident and final states cancel and using (11.45) we obtain

$$\begin{aligned}\mathscr{A}^{\mathrm{I}}_{ji}(\mathbf{b}) &= -\mathrm{i}Z_{\mathrm{A}}\int_{-\infty}^{\infty} \exp[\mathrm{i}(E_j^{\mathrm{B}} - E_i^{\mathrm{B}})t]\langle\phi_j(\mathbf{r})|L\left(-\frac{1}{x'} + \frac{1}{R'}\right)|\phi_i(\mathbf{r})\rangle \,\mathrm{d}t \\ &= -\mathrm{i}Z_{\mathrm{A}}\gamma\int_{-\infty}^{\infty} \mathrm{d}t \exp[\mathrm{i}(E_j^{\mathrm{B}} - E_i^{\mathrm{B}})t]\int \phi_j(\mathbf{r})(1 - \mathbf{v}_0\cdot\boldsymbol{\alpha}/c) \\ &\qquad\times\left(-\frac{1}{x'} + \frac{1}{R'}\right)\phi_i(\mathbf{r})\,\mathrm{d}\mathbf{r}.\end{aligned} \tag{11.67}$$

Expressing R' in terms of the target rest frame quantities $\mathbf{r}$ and t we have

$$\begin{aligned}R'^2(t') &= b^2 + v_0^2 t'^2 \\ &= b^2 + \gamma^2 v_0^2\left(t - \frac{\mathbf{v}_0\cdot\mathbf{r}}{c^2}\right)^2,\end{aligned} \tag{11.68}$$

where in the second line (11.10b) has been employed. Since $R'(t')$ depends on the spatial coordinate $\mathbf{r}$, the matrix element $\langle\phi_j(\mathbf{r})|L(1/R')|\phi_i(\mathbf{r})\rangle$ does not vanish, despite the fact that ϕ_j and ϕ_i are orthogonal ($i \neq j$). However, by transforming into the rest frame of the projectile, Toshima and Eichler (1991) have shown that this term vanishes when the integration over t is carried out, so that

$$\mathscr{A}^{\mathrm{I}}_{ji}(\mathbf{b}) = Z_{\mathrm{A}}\gamma\int_{-\infty}^{\infty} \exp[\mathrm{i}(E_j^{\mathrm{B}} - E_i^{\mathrm{B}})t]\langle\phi_j(\mathbf{r})|L\frac{1}{x'}|\phi_i(\mathbf{r})\rangle \,\mathrm{d}t \tag{11.69}$$

where $\mathbf{x}'$ is given in terms of $\mathbf{r}$ and t and t by (11.10a). Details of the evaluation of the expression (11.69), particularly in the dipole (Bethe) approximation, can be conveniently consulted in a recent article by Eichler (1990).

Earlier studies of excitation and ionization employed the plane-wave Born approximation (see, for example, Anholt and Gould (1986)), but this has been shown to be entirely equivalent to the impact parameter formulation presented here (Amundsen and Aashamer 1981). A number of applications of the theory have been made, both in the first-order approximation represented by (11.69) and in corresponding distorted wave approximations in which eikonal or CDW wave functions are employed (Deco and Grün 1989*a*). The general nature of the calculated cross sections is illustrated by the particular examples shown in Fig. 11.2. At present the experimental data available in the high-energy region with $\gamma > 2$ is very limited. The calculated K-shell ionization cross sections are in reasonable agreement with the measurements where these exist, but the data is not sufficiently precise to allow meaningful distinctions to be made between the various theoretical calculations, which do not differ greatly among themselves.

Charge exchange amplitudes

Following the initial work of Mittleman (1964), the first-order OBK cross section was studied by Shakeshaft (1979) and by Moiseiwitsch and Stockman (1980) and subsequently the relativistic version of the second-order OBK approximation was obtained by Humphries and Moiseiwitsch (1984) and developed by Moiseiwitsch (1988, 1989). As discussed in Chapters 7 and 10 methods not allowing for the Coulomb distortion of the wave function in the initial and final channels cannot be considered to be satisfactory. The simplest method which does allow for this distortion is the boundary corrected Born approximation and the relativistic generalization has been given by Eichler (1987, 1988). Following work on a relativistic unsymmetrized eikonal approximation (Eichler 1985), a symmetrical form of the eikonal approximation has been derived by Moiseiwitsch (1986, 1987) and independently by Deco and Rivarola (1987*a*), while relativistic generalizations of the continuum distorted wave method have been developed by McCann (1985) and by Deco and Rivarola (1986, 1987*b*).

The relativistic generalization of the first-order boundary corrected method(R1B) can be obtained in the post form by making the approximation $\Psi_i \approx \Phi_i^{\mathrm{B}}(t)$ in (11.46) obtaining

$$\mathscr{A}_{ni}^{+}(\mathbf{b}) \approx -\mathrm{i} \int_{-\infty}^{\infty} \langle \mathrm{X}_n^{\mathrm{AT}}(\mathbf{r}, t)| U_{\mathrm{B}}^{\mathrm{R}} |\Phi_i^{\mathrm{B}}(\mathbf{r}, t)\rangle \, \mathrm{d}t \tag{11.70}$$

where $U_{\mathrm{B}}^{\mathrm{R}}$ is defined by (11.47). It is straightforward to obtain the prior form of the amplitude

$$\bar{\mathscr{A}}_{ni}(\mathbf{b}) \approx -\mathrm{i}\int_{-\infty}^{\infty} \langle \mathrm{X}_n^{\mathrm{AT}}(\mathbf{r}, t)|U_{\mathrm{A}}^{\mathrm{R}}|\Phi_i^{\mathrm{B}}(\mathbf{r}, t)\rangle\, \mathrm{d}t, \tag{11.71}$$

where $U_{\mathrm{A}}^{\mathrm{R}}$ is defined by (11.45), and to show that numerically both these amplitudes are equal: there is no post–prior discrepancy.

The amplitude for capture in the relativistic symmetrical eikonal approximation in the post form can be written as

$$A_{ni}^{+}(\mathbf{b}) \approx -\mathrm{i}\int_{-\infty}^{\infty} \langle \tilde{\mathrm{X}}_n^{\mathrm{AT}}(\mathbf{r}, t)|\, W_{\mathrm{B}}^{\mathrm{R}}|\tilde{\Phi}_i^{\mathrm{B}}(\mathbf{r}, t)\rangle\, \mathrm{d}t \tag{11.72}$$

$$\tilde{\Phi}_i^{\mathrm{B}}(\mathbf{r}, t) = \exp\left[-\mathrm{i}\frac{Z_{\mathrm{A}}}{v_0}\log(v_0 x' + \mathbf{v}_0\cdot\mathbf{x}') - \mathrm{i}E_j^{\mathrm{B}}t\right]\phi_j(\mathbf{r}) \tag{11.73a}$$

$$\tilde{\mathrm{X}}_n^{\mathrm{A}}(\mathbf{r}, t) = \exp\left[-\mathrm{i}\frac{Z_{\mathrm{B}}}{v_0}\log(v_0 r + \mathbf{v}_0\cdot r) - \mathrm{i}E_n^{\mathrm{A}}t\right]S\chi_n(\mathbf{x}'). \tag{11.73b}$$

The perturbation $W_{\mathrm{B}}^{\mathrm{R}}$ corresponding to this choice of distorted waves is

$$W_{\mathrm{B}}^{\mathrm{R}} = \left[1 - \frac{c}{v_0}\frac{\boldsymbol{\alpha}\cdot(\hat{\mathbf{r}} + \hat{\mathbf{v}}_0)}{1 + \hat{\mathbf{v}}_0\cdot\hat{\mathbf{r}}}\right]\frac{Z_{\mathrm{B}}}{r}. \tag{11.74}$$

The corresponding prior form of the amplitude is

$$\bar{\mathscr{A}}_{ni}(\mathbf{b}) = -\mathrm{i}\int_{-\infty}^{\infty} \langle \tilde{\mathrm{X}}_n^{\mathrm{AT}}(\mathbf{r}, t)|\, W_{\mathrm{A}}^{\mathrm{R}}|\tilde{\Phi}_i^{\mathrm{B}}(\mathbf{r}, t)\rangle\, \mathrm{d}t \tag{11.75}$$

with $\tilde{\chi}_n^{\mathrm{A}}$ and $\tilde{\Phi}_i^{\mathrm{B}}$ given by (11.73a, b) and with

$$W_{\mathrm{A}}^{\mathrm{R}} = -S^{-1}\left[1 - \frac{c}{v_0}\frac{\boldsymbol{\alpha}\cdot(\hat{\mathbf{x}}' + \hat{\mathbf{v}}_0)}{1 + \hat{\mathbf{v}}_0\cdot\hat{\mathbf{x}}'}\right]\frac{Z_{\mathrm{A}}}{x'}S^{-1}. \tag{11.76}$$

In the non-symmetrical eikonal approximation, if $Z_{\mathrm{B}} > Z_{\mathrm{A}}$ it is appropriate to allow for distortion in the final state, and in this case we have that

$$\bar{\mathscr{A}}_{ni}(\mathbf{b}) = -\mathrm{i}\int_{-\infty}^{\infty} \langle \tilde{\mathrm{X}}_n^{\mathrm{A}}(\mathbf{r}, t)|L\left(\frac{Z_{\mathrm{A}}}{x'}\right)|\phi_i(\mathbf{r})\rangle \exp(-\mathrm{i}E_i^{\mathrm{B}}t)\, \mathrm{d}t, \tag{11.77}$$

where $\tilde{\mathrm{X}}_n$ is given by (11.73b) and L by (11.16).

Both and R1B and the unsymmetrical eikonal amplitudes go over to the corresponding non-relativistic amplitudes in the limit $v_0/c \ll 1$. In contrast it has been pointed out by Toshima and Eichler (1990) that this is not the case with the symmetrical eikonal approximation in the form given by (11.72) or (11.75) and that in this limit the amplitudes (11.72) and (11.75) contain spin-flip contributions of comparable magnitude to the non-spin-flip contribution. Numerical studies show that for $H^+ + H(1s)$ collisions at 100 keV the symmetrical eikonal approximation predicts the ratio of the spin-flip to non-spin-flip contributions to be 0.12, in contrast to the ratio of the order 10^{-9} given by the R1B and unsymmetrical eikonal approximations. This unphysical feature of the symmetrical eikonal approximation appears to apply to the relativistic continuum distorted wave approximation when distortion is allowed for in both the initial and final channels. In contrast the one-sided CDW approximation (TCDW) is not open to this criticism. In both the CDW and symmetrical eikonal methods the effective perturbation contains contributions from the kinetic energy operator and it is possible that this is the source of the difficulty, which at the time of writing has not been resolved.

Applications

To illustrate the magnitude of relativistic effects in collisions between light ions, in Table 11.2 both the non-relativistic B1B and relativisitc R1B cross sections are shown for the symmetrical resonance reaction $H^+ + H(1s) \rightarrow H(1s) + H^+$. Also included are cross sections computed from the unsymmetric relativistic eikonal approximation. It is seen that the R1B cross

Table 11.2. Cross sections for the symmetrical resonance reaction $H^+ + H(1s) \rightarrow H(1s) + H^+$ at high energies. The number in brackets gives the power of ten multiplying the preceding number. A is the non-relativistic B1B approximation (Dewangan and Eichler 1986); B is the relativistic R1B approximation (Eichler 1988); C is the relativistic eikonal approximation (Anholt and Eichler 1985).

	Cross sections (10^{-28} cm^2)		
E_L (MeV/a.m.u.)	A	B	C
10	5.77	5.75	4.27
50	4.29 (−4)	4.60 (−4)	3.26 (−4)
100	6.92 (−6)	8.39 (−6)	5.61 (−6)
500	4.67 (−10)	9.22 (−10)	6.12 (−10)
1000	7.35 (−12)	–	1.65 (−11)

sections exceed the eikonal cross sections by about 40–50 per cent at the highest energies considered. As we have explained earlier, for systems with small Z, the non-radiative capture cross section is very small, far smaller than the corresponding radiative capture cross section. In the other extreme of high Z, the radiative and non-radiative capture cross sections are comparable. In the large Z case it is interesting to compare the R1B cross sections with those obtained by the coupled channel method for $U^{92+} + U^{91+}(1s_{\frac{1}{2}})$ collisions at 500 MeV/a.m.u. The results from Table 11.1 show that the total cross section provided by the first-order boundary corrected approximation compare well with those of the coupled channel approach. As at non-relativistic energies, this agreement would not be expected to extend to differential cross sections which certainly require a second-order approach to be successful.

In certain cases a comparison can be made with the results of experiment, and in Fig. 11.4 we show the theoretical cross sections for capture by fully

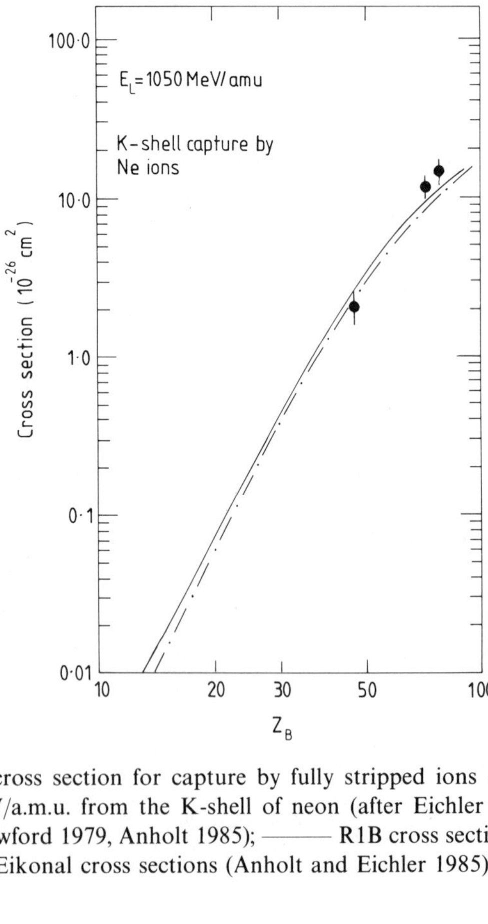

Fig. 11.4. Total cross section for capture by fully stripped ions of charge Z_A and energy 1050 MeV/a.m.u. from the K-shell of neon (after Eichler 1990): ⧫⧫ experimental data (Crawford 1979, Anholt 1985); ——— R1B cross sections (Eichler 1987, 1988); —·—·— Eikonal cross sections (Anholt and Eichler 1985).

stripped neon ions from the K-shell of ions with nuclear charge Z_B, compared with the measurements of Crawford (1979) and Anholt (1985) at an energy of $E_L = 1050$ MeV/a.m.u. The system is treated by the one-electron model, with the effective potential binding the target electron being of Coulomb form with an effective charge to take account of shielding. The computed cross sections are multiplied by two to compare with experiment since there are two K-electrons in the target. The contribution from L-shell electrons in the target is very small and can be ignored. It is seen that the agreement with experiment for both the R1B and eikonal cross sections is satisfactory.

11.4 RADIATIVE CHARGE TRANSFER

At non-relativistic energies, the cross section for radiative charge transfer is given by the expressions given in Chapter 4, (4.101) and (4.104), in terms of the matrix element of the dipole operator with respect to wave functions Ψ_i^+ and Ψ_f^-. These wave functions are the full solutions of the Schrödinger equation with boundary conditions appropriate to incident channels i and f, respectively. Above the very low-energy region where a full partial-wave treatment is required Ψ_i^+ and Ψ_f^- can be determined within a semi-classical impact parameter approximation, employing coupled channel methods at intermediate energies or a distorted wave type of approximation at high energies. For example, the approximation in which CDW wave functions are employed has been developed by Gonzáles and Miraglia (1984). It has been argued that at very high energies the initial binding of the electron to be captured can be ignored, in which case the cross section is just that for capture of a free electron by the incident ion. By detailed balancing this cross section differs only by a kinematical factor from that for photoionization, and in the relativistic region

$$\sigma_c = \frac{\gamma - 1}{\gamma + 1} \sigma_I \tag{11.78}$$

where σ_I is the photoionization cross section and $\gamma = (1 - v_0^2/c^2)^{-\frac{1}{2}}$ as usual. The expression (11.78) refers to a one-electron target. Since in this approximation the binding to the target has no influence, the cross section capture from a target containing Z_B electrons is just $Z_B \sigma_c$. The photoionization cross section is well known and for relativistic energies tabulations have been given by Hultberg *et al.* (1967) and others (Anholt and Gould 1986). In the limit $\gamma \gg 1$, it is found that

$$\sigma_c = \frac{4\pi}{c^8} \frac{Z_A^5}{\gamma}. \tag{11.79}$$

While the approximation of ignoring the target binding is reasonable for small Z_B, it becomes progressively less accurate as Z_B increases, and for $Z_A = 1$, $Z_B = 100$ it has been estimated by Deco and Rivarola (1989) that at high energies the ratio of the true cross section to that provided by (11.78) is about 7. The relativistic generalizations of the formulae (4.101) and (4.104) are (in the impact parameter formalism)

$$\frac{d\sigma}{d\mathbf{k}} = \frac{1}{2}\sum_{\mu\mu'}\sum_{j=1}^{L}\int |M_{\mathrm{if}}^{(j)\mu\mu'}(\mathbf{b})|\,d\mathbf{b}, \tag{11.80}$$

where $M_{\mathrm{if}}^{j(\mu\mu')}(\mathbf{b})$ is the matrix element of the dipole coupling for the emission of a photon with wave vector $\mathbf{k}$ and polarization vector $\boldsymbol{\varepsilon}^j$. The indices μ and μ' represent the spin of the electron in the initial and final states. In the frame of reference centred on B, we have

$$M_{\mathrm{if}}^{(j)\mu\mu'}(\mathbf{b}) = -\mathrm{i}c\left(\frac{2\pi}{kc}\right)^{\frac{1}{2}}\int_{-\infty}^{\infty}(2\pi)^{-\frac{3}{2}}\langle\Psi_j^{\mu-}(\mathbf{r},t)|\mathrm{e}^{-\mathrm{i}\mathbf{k}\cdot\mathbf{r}}\,\boldsymbol{\alpha}\cdot\boldsymbol{\varepsilon}^j|\Psi_i^{\mu+}(\mathbf{r},t)\rangle\,dt. \tag{11.81}$$

With the aid of this formula studies of radiative capture have been made in the approximation in which strong potential Born wave functions are employed for $\Psi_{\mathrm{i,f}}$ (Hino and Watanabe 1987) and also employing relativistic CDW wave functions (Deco and Rivarola 1989). In the latter case a simple formula is obtained in the limit $\gamma \gg 1$ for the total cross section; it is

$$\sigma_{\mathrm{Rc}} = \sigma_{\mathrm{c}}F_1F_2 \tag{11.82}$$

where σ_{c} is given by (11.79) and

$$F_1 = \left[\frac{2c^2}{3\pi(Z_B^2 + c^2)}\left(\frac{Z_B}{c}\right)^3 + \left\{\frac{c^2}{(Z_B^2 + c^2)} + 1\right\}\frac{1}{\pi}\frac{Z_B}{c} + 1\right] \tag{11.83}$$

$$F_2 = |N(\nu_A)|^2|N(\nu_B)|^2 \tag{11.84}$$

with $N(\nu) = exp(\pi\nu/2)\Gamma(1 - \mathrm{i}\nu)$ and $\nu_A = Z_A/v_0$, $\nu_B = Z_B/v_0$. The modification of σ_{c} by the factors F_1 and F_2 increases as Z_B increases and for $Z_B = 20$, 50, and 100 we have $F_1 = 1.09$, 1.23, and 1.44, and $F_2 = 1.56$, 2.61, and 4.74, respectively. While the expression (11.82) has not yet been subjected to a detailed experimental test, the basic result that large deviations from the simple formulae (11.79) can be expected for large values of the charge on the target nucleus is undoubtedly correct.

11.5 ELECTRON–POSITRON PAIR PRODUCTION AND ELECTRON CAPTURE

Free pair creation

As we remarked at the beginning of this chapter, pair production in which the electron is captured by one of the colliding ions is an increasingly important charge changing mechanism in ion–ion collisions at very high energies. Because of its importance in cosmic ray physics, the creation of *free* electron–positron pairs in the collisions of charged particles has been studied theoretically since 1934, when Williams, and also von Weizsächer, obtained the approximate expression, valid for large γ,

$$\sigma = \frac{28}{27\pi} \frac{\alpha^2 Z_A^2 Z_B^2 [\log(\gamma/4)]^3}{c^2} \tag{11.85}$$

for the cross section for electron–positron pair production in the collision between two fully stripped ions of charge Z_A and Z_B (Williams 1934, 1935; von Weizsächer 1934). In the Williams–Weizsächer method, the electromagnetic pulse caused by the passage of the projectile is expressed in terms of a spectrum of virtual photons. At extreme relativistic energies, the field is nearly transverse and the photons are almost on the energy shell, which allows the probability for pair production during collision to be related to that for the production of a pair by a real photon. The Williams–Weizsächer method is described in the texts of Heitler (1954) and of Landau and Lifshitz (1971), and more recent developments of the method have been discussed by Bertulani and Baur (1988) (see also Brodsky *et al.* 1971). The motion of the colliding ions can always be accurately represented by the straight line classical trajectory, and in this case the lowest-order Feynman diagrams contain two photons. These second-order diagrams have been evaluated numerically by Bottcher and Strayer (1989) and the corresponding cross section in the large γ limit is of the form

$$\sigma = \frac{3.4\alpha^2 Z_A^2 Z_B^2 (\log \gamma)^3}{c^2}. \tag{11.86}$$

The cross sections (11.85) and (11.86) increase without limit as $\gamma \to \infty$. This is an indication that the perturbation theory employed will eventually break down as γ increases, the calculated probability exceeding unity for small values of the impact parameter. Nevertheless, the formula (11.86) correctly indicates that in U + U collisions the pair creation cross section will exceed capture cross sections for $\gamma > 10$ and approach the ionization cross section for $\gamma > 10^4$.

In the second-order perturbation theory used by Botcher and Strayer, the electron and positron are described by free-particle wave functions with no allowance for distortion by the Coulomb field of the target. This effect is taken into account in the method of Becker *et al.* (1986). The heavy particle motion is again described by a straight line trajectory and pair creation is viewed as the ionization of a negative energy electron during the collision. The probability amplitude is then of the same form as that for excitation or ionization, (11.69). We have

$$\mathscr{A}_{ji}(\mathbf{b}, E_e, E_P) = Z_A \gamma \int_{-\infty}^{\infty} \exp[i(E_e + E_p)t] \langle \phi_j^e(\mathbf{r}) | L \frac{1}{x'} | \phi_i^p(\mathbf{r}) \rangle \, dt \qquad (11.87)$$

where $\phi_j^e(\mathbf{r})$ and $\phi_i^p(\mathbf{r})$ are the Coulomb–Dirac wave functions for an electron of energy E_e and for a positron, described as an electron of energy $-E_p$. The indices i (and j) stand collectively for the partial-wave quantum numbers, and the double differential probability becomes

$$\frac{dP(b)}{dE_p \, dE_e} = \sum_{i,j} |\mathscr{A}_{ji}(\mathbf{b}, E_e, E_p)|^2. \qquad (11.88)$$

The corresponding total cross section for proton–uranium collisions is shown in Fig. 11.5, where it is seen to be in reasonable agreement with the results of the Williams–Weizsächer model for $E_L \simeq 10^4$ MeV/a.m.u., with some disagreement at higher energies possibly due to numerical difficulties.

Pair production with electron capture

The final state of the electron produced in pair creation can be a bound state. If the electron is captured into a bound state of the target B, the probability amplitude will again be given by (11.87), but this time $\phi_j^p(\mathbf{r})$ will be a bound state Dirac hydrogenic wave function. Capture into the K-shell is the most likely process and this has been calculated by Becker *et al.* (1987) and Becker (1987). In subsequent work, the partial-wave analysis associated

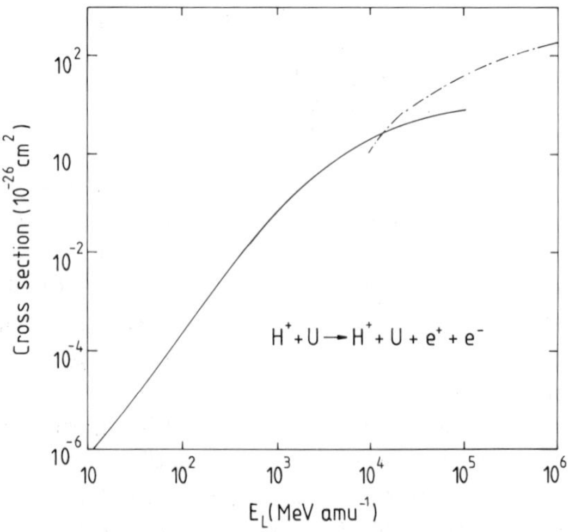

Fig. 11.5. The total cross section for electron–positron pair creation in proton–uranium collisions (after Becker *et al.* 1986): ——— calculated from (11.87) by Becker *et al.* (1986); —·—·— calculated from the Williams–Weizsäcker formula (11.85).

with the use of exact Dirac wave functions has been avoided by using Darwin or Sommerfeld–Maue wave functions (Becker *et al.* 1987, Deco and Rivarola 1988*a*, *b*), and Deco and Grün (1989*b*) have discussed the distortion of the electron and positron wave functions by the field of the projectile A. The capture cross section again increases logarithmically with γ and is proportional to $Z_A^2 Z_B^5$. As an illustration, the cross section for capture by S^{16+} ions in collision with targets of charge Z_B at an energy of 16 GeV/a.m.u. has already been shown in Fig. 11.1, and for $Z_B \gtrsim 80$ capture with the production of positrons is seen to be the most important process.

It should be noted that the cross section for the production of muon pairs (or of pairs of heavier particles) can be treated in the same way as for electron–positron pair creation; however the cross sections are too small to be of importance in the energy region which is at present the subject of experiment.

APPENDIX A

A NOTE ON UNITS AND CONSTANTS

Unless otherwise stated, theoretical quantities and formulae in this book have been given in atomic units in which the mass of the electron (m), magnitude of the electronic charge (e), and Planck's constant divided by 2π ($\hbar$) all have the value unity.

Measured cross sections are generally displayed in units of cm^2. Usually the energy of the incident particle is stated for the laboratory system of coordinates, in which the target projectile is at rest before the collision. Since the ion–atom cross sections depend primarily on the relative velocity v_0, rather than the incident energy, cross sections are generally shown as a function of laboratory energy measured in electron–volts (eV), or kilo-electron-volts (keV), divided by the mass of the incident particle in atomic mass units (a.m.u.). The relationship, valid at non-relativistic energies,

$$E_L = 25v_0^2 \text{ keV/a.m.u.} \tag{A.1}$$

should be noted, where v_0 is the relative velocity of the colliding particles in atomic units. When v_0 is comparable with c, the velocity of light, the corresponding relationship is

$$E_L = 50(\gamma - 1)c^2 \text{ keV/a.m.u.} \tag{A.2}$$

where $\gamma = (1 - v_0^2/c^2)^{-\frac{1}{2}}$ and both v_0 and c are expressed in atomic units.

Tables A.1 and A.2 summarize the units and constants used in this book.

Table A.1. Atomic units.

Quantity	Unit	Significance	Value
Mass	m	The mass of the electron	9.10953×10^{-31} kg
Charge	e	The magnitude of the electronic charge	1.60219×10^{-19} C
Length	a_0	Bohr radius of atomic hydrogen	5.2918×10^{-11} m
Velocity	u	Electronic velocity in the first Bohr orbit of atomic hydrogen	2.18769×10^{6} ms^{-1}

Table *continued*

Table A.1. Atomic units—*continued.*

Quantity	Unit	Significance	Value
Time	a_0/u	Time required for an electron in the first Bohr orbit of atomic hydrogen to travel the distance a_0	2.41889×10^{-17} s
Energy	$e^2/(4\pi\varepsilon_0 a_0)$	Twice the ionization potential of atomic hydrogen (with infinite nuclear mass)	4.35981×10^{-18} J (27.2116 eV)
Angular momentum	$\hbar$	Planck's constant divided by 2π	1.05459×10^{-34} Js

Table A.2. Atomic constants and conversion factors.

		Atomic units	SI units
Planck's constant	h	2π	6.62618×10^{-34} Js
Velocity of light	c	137.036	2.99792×10^{8} m s^{-1}
Magnitude of the electronic charge	e	1	1.60219×10^{-19} C
Mass of the electron	m	1	9.10953×10^{-31} kg
Mass of the proton	$M(M_P)$	1836.15	1.67265×10^{-27} kg
Ionization potential of atomic hydrogen (infinite nuclear mass)	I	$\frac{1}{2}$	2.17991×10^{-18} J (= 13.6058 eV)
Permitivity of free space	ε_0	$1/4\pi$	8.85419×10^{-12} Fm^{-1}
Permeability of free space	μ_0	$4\pi/c^2$	1.25664×10^{-6} H m^{-1}
1 electron volt	eV	0.036749	1.60219×10^{-19} J
1 Angström	Å	1.8897	10^{-10} m

APPENDIX B

CENTRE-OF-MASS AND LABORATORY COORDINATES

As in Chapter 2, let us consider the elastic scattering of an atom A of mass M_A by a target atom B of mass M_B. The *laboratory system* is the frame in which the target atom B is at rest *before* the collision, while the *centre-of-mass system* is the frame in which the centre of mass of A and B is at rest throughout the collision.

In the centre-of-mass system, the projectile A and the target B move with respect to the centre of mass with equal and opposite momentum, $\mathbf{P}_A = -\mathbf{P}_B = \mathbf{P}$, where $\mathbf{P}$ is the relative momentum which, before the collision, will be taken to be parallel to the Z-axis. if $\mathbf{R}_A$ and $\mathbf{R}_B$ are the position vectors of A and B with respect to the centre of mass as origin, and $\mathbf{R}$ is the position vector of A with respect to B, $\mathbf{R} = \mathbf{R}_A - \mathbf{R}_B$, and we have

$$\mathbf{R}_A = \frac{M_B}{M_A + M_B}\mathbf{R}, \qquad \mathbf{R}_B = -\frac{M_A}{M_A + M_B}\mathbf{R}. \tag{B.1}$$

Since

$$\mathbf{P}_A = M_A \dot{\mathbf{R}}_A, \qquad \mathbf{P}_B = M_B \dot{\mathbf{R}}_B \tag{B.2}$$

we have that

$$\mathbf{P} = \mathbf{P}_A = -\mathbf{P}_B = \mu \dot{\mathbf{R}} \tag{B.3}$$

where μ is the reduced mass

$$\mu = \frac{M_A M_B}{M_A + M_B}. \tag{B.4}$$

If $W(R)$ is the interaction potential between A and B, the Hamiltonian of the system is

$$\begin{aligned} H = T + W &= \frac{P_A^2}{2M_A} + \frac{P_B^2}{2M_B} + W(R) \\ &= \frac{1}{2\mu} P^2 + W(R). \end{aligned} \tag{B.5}$$

It follows that the classical equations of motion (and the Schrödinger equation) for the system in the centre-of-mass frame are the same as for a particle of mass μ with a position vector $\mathbf{R}$ with respect to a fixed centre of force.

Before the collision,

$$\mathbf{P} = \mathbf{p} = \mu \mathbf{v}_0 \tag{B.6}$$

where $\mathbf{v}_0$ is the relative velocity of A with respect to B, and the kinetic energy is $T = \frac{1}{2}\mu v_0^2$.

After the collision the two atoms A and B emerge in the centre-of-mass frame with equal and opposite momenta $\mathbf{p}'_A = -\mathbf{p}'_B = \mathbf{p}'$. Since we are dealing with elastic collisions in which the internal energies of A and B do not change, the magnitudes of $\mathbf{p}$ and $\mathbf{p}'$ must be the same: $p = p'$. The direction of $\mathbf{p}'$ can be described by polar angles (θ, ϕ), taking the incident direction $\hat{\mathbf{v}}_0$ as the polar axis, that is the Z-axis. The angle θ is the angle of scattering in the centre-of-mass system.

Transformation to the laboratory system

In the laboratory system of coordinates in which the origin is taken to be at the initial position of B, the centre of mass moves with the constant velocity $\mathbf{v}_c$ where

$$\mathbf{v}_c = \frac{M_A \mathbf{v}_0}{M_A + M_B}. \tag{B.7}$$

Let $\mathbf{v}_A$ be the final velocity of A in the centre-of-mass system and $\mathbf{v}'_A$ be the same quantity in the laboratory system. Then $\mathbf{v}_A$ and $\mathbf{v}'_A$ are related by

$$\mathbf{v}_A = \mathbf{v}'_A - \mathbf{v}_c. \tag{B.8}$$

The angle between the incident direction $\hat{\mathbf{v}}_0$ and $\mathbf{v}_A$ is the scattering angle, θ, in the centre-of-mass system, while the angle between $\mathbf{v}_0$ and $\mathbf{v}'_A$ is the angle of scattering, θ_L, in the laboratory system. By taking components at right angles and parallel to $\hat{\mathbf{v}}_0$ we obtain

$$\begin{aligned} v_A \cos\theta &= v'_A \cos\theta_L - v_c \\ v_A \sin\theta &= v'_A \sin\theta_L \end{aligned} \tag{B.9}$$

from which

$$\tan\theta_L = \frac{\sin\theta}{\cos\theta + \tau} \tag{B.10}$$

where

$$\tau = \frac{v_c}{v_A}.$$

Since

$$v_{\mathrm{A}} = \frac{p_{\mathrm{A}}}{M_{\mathrm{A}}} = \frac{M_{\mathrm{B}}}{M_{\mathrm{A}} + M_{\mathrm{B}}} v_0$$

and using (B.7) we have that

$$\tau = \frac{M_{\mathrm{A}}}{M_{\mathrm{B}}}. \tag{B.11}$$

As the plane of scattering is the same in both coordinate systems the azimuthal angle, ϕ, is unaltered in transforming from the centre-of-mass to the laboratory system ($\phi_{\mathrm{L}} = \phi$).

Cross sections

The probability that scattering has occurred cannot depend on the coordinate frame from which the event is viewed so that the total cross section σ is the same in all coordinate systems. It follows that if

$$\left.\frac{\mathrm{d}\sigma(\theta, \phi)}{\mathrm{d}\Omega}\right|_{\mathrm{cm}} \quad \text{and} \quad \left.\frac{\mathrm{d}\sigma(\theta_{\mathrm{L}}, \phi)}{\mathrm{d}\Omega}\right|_{\mathrm{L}}$$

are the cross sections in the centre-of-mass and laboratory systems, respectively, then

$$\sigma = \int_0^{2\pi} \mathrm{d}\phi \int_{-1}^{+1} \mathrm{d}(\cos\theta) \left.\frac{\mathrm{d}\sigma(\theta, \phi)}{\mathrm{d}\Omega}\right|_{\mathrm{cm}} = \int_0^{2\pi} \mathrm{d}\phi \int_{-1}^{+1} \mathrm{d}(\cos\theta_{\mathrm{L}}) \left.\frac{\mathrm{d}\sigma(\theta_{\mathrm{L}}, \phi)}{\mathrm{d}\Omega}\right|_{\mathrm{L}} \tag{B.12}$$

from which we obtain the relationship

$$\frac{\mathrm{d}\sigma_{\mathrm{L}}(\theta_{\mathrm{L}}, \phi)}{\mathrm{d}\Omega} = \frac{\mathrm{d}\sigma(\theta, \phi)}{\mathrm{d}\Omega}\left|\frac{\mathrm{d}(\cos\theta)}{\mathrm{d}(\cos\theta_{\mathrm{L}})}\right|. \tag{B.13}$$

Evaluating $|\mathrm{d}(\cos\theta)/\mathrm{d}(\cos\theta_{\mathrm{L}})|$ with the help of (B.10) we find that

$$\left.\frac{\mathrm{d}\sigma(\theta_{\mathrm{L}}, \phi)}{\mathrm{d}\Omega}\right|_{\mathrm{L}} = \frac{(1 + \tau^2 + 2\tau\cos\theta)^{\frac{3}{2}}}{|1 + \tau\cos\theta|} \left.\frac{\mathrm{d}\sigma(\theta, \phi)}{\mathrm{d}\Omega}\right|_{\mathrm{cm}}, \qquad \tau = \frac{M_{\mathrm{A}}}{M_{\mathrm{B}}}. \tag{B.14}$$

Although many experiments are carried out by scattering a beam of particles from a stationary target, in other types of experiment the scattering of one beam of particles by another is studied. These crossed-beam experiments in which two beams, inclined at a certain angle, interact are neither in the laboratory or centre-of-mass frames, but, following similar arguments to those given above, the measured quantities can be transformed into the centre-of-mass system to allow for easy comparison with theoretical calculations.

APPENDIX C

THE WKB APPROXIMATION TO THE PHASE SHIFTS

In this appendix we shall obtain a semi-classical approximation for the phase shifts η_l. We shall start by first deriving the WKB (Wentzel–Kramers–Brillouin) approximation to the one-dimensional Schrödinger equation,†

$$\left[-\frac{\hbar^2}{2m}\frac{d^2}{dx^2} + V(x) - E\right]\psi(x) = 0, \qquad -\infty < x < \infty. \tag{C.1}$$

A solution of (C.1) is sought in the form

$$\psi(x) = \exp\left[\frac{i\,S(x)}{\hbar}\right]. \tag{C.2}$$

Substitution into the Schrödinger equation shows that $S(x)$ satisfies

$$-i\hbar\frac{d^2S}{dx^2} + \left(\frac{dS}{dx}\right)^2 = F(x), \tag{C.3}$$

where

$$F(x) = 2m[E - V(x)]. \tag{C.4}$$

By expanding $S(x)$ in powers of $\hbar$,

$$S(x) = S_0(x) + \hbar S_1(x) + \hbar^2 S_2(x) + \cdots, \tag{C.5}$$

and substituting into (C.3) the following hierarchy of equations is obtained by equating the coefficients of each power of $\hbar$:

$$\left(\frac{dS_0}{dx}\right)^2 = F(x) \tag{C.6a}$$

$$-i\frac{d^2S_0}{dx^2} + 2\frac{dS_0}{dx}\frac{dS_1}{dx} = 0, \tag{C.6b}$$

$$-i\frac{d^2S_1}{dx^2} + \left(\frac{dS_1}{dx}\right)^2 + 2\frac{dS_0}{dx}\frac{dS_2}{dx} = 0 \tag{C.6c}$$

$$\cdots$$

† Atomic units are *not* employed in equations (C.1) to (C.15).

On integrating (C.6a), S_0 is found to be

$$S_0(x) = \pm \int F^{\frac{1}{2}}(x)\,\mathrm{d}x. \tag{C.7}$$

The next term $S_1(x)$ is obtained by using S_0 in (C.6b) and integrating

$$S_1(x) = \mathrm{i}\,\log|F(x)|^{\frac{1}{4}}. \tag{C.8}$$

By examining (C.6c) it can be shown that the third and higher terms in the series (C.5) can be neglected if the condition (Bransden and Joachain 1989, p. 391)

$$\left|\frac{\hbar}{2} F^{-\frac{3}{2}}(x) \frac{\mathrm{d}F}{\mathrm{d}x}\right| = \left|\hbar m F^{-\frac{3}{2}}(x) \frac{\mathrm{d}V(x)}{\mathrm{d}x}\right| \ll 1 \tag{C.9}$$

is satisfied. Under nearly classical conditions $\hbar m(\mathrm{d}V(x)/\mathrm{d}x)$ is small compared with $F^{\frac{3}{2}}(x)$, so that $S_2(x)$ and higher terms can be neglected, except near the classical turning points at which $F(x) = 0$. If (C.9) is satisfied, and if we are away from the classical turning points, the most general solution for the case $E > V(x)$. $F(x)$ positive (the classically allowed region) is

$$\psi_+(x) = |F(x)|^{-\frac{1}{4}}\left\{A \exp\left[\frac{\mathrm{i}}{\hbar}\int^x \mathrm{d}x' F(x')^{\frac{1}{2}}\right] + B \exp\left[-\frac{\mathrm{i}}{\hbar}\int^x \mathrm{d}x' F(x')^{\frac{1}{2}}\right]\right\}, \tag{C.10}$$

and that for $E < V(x)$, $F(x) < 0$ (the classically forbidden region) is

$$\psi_-(x) = |F(x)|^{-\frac{1}{4}}\left\{C \exp\left[\frac{1}{\hbar}\int^x \mathrm{d}x' |F(x')|^{\frac{1}{2}}\right] + D \exp\left[-\frac{1}{\hbar}\int^x \mathrm{d}x' |F(x')|^{\frac{1}{2}}\right]\right\}, \tag{C.11}$$

where A, B, C, D are arbitrary constants.

Now let us consider the case with one turning point at $x = x_0$, such that $F(x) < 0$ for $x < x_0$ and $F(x) > 0$ for $x > x_0$. Since $\psi(x)$ must be everywhere finite, we must set $D = 0$ and the physical solution in the region $x < x_0$ can be written

$$\psi_-(x) = |F(x)|^{-\frac{1}{4}} C \exp\left[\frac{1}{\hbar}\int_{x_0}^x \mathrm{d}x' |F(x')|^{\frac{1}{2}}\right]. \tag{C.12}$$

The lower limit of the integral has been taken to be $x = x_0$ for convenience. We now have $\psi_-(x) \to 0$ as $x \to -\infty$. The problem is to find the values of A and B in (C.10) which determine the corresponding solution $\psi_+(x)$ in the region $x > x_0$. Such a 'connection formula' can be found by noting that near the turning point $F(x)$ can be approximated in the form

$$F(x) \approx A^2(x - x_0), \tag{C.13}$$

where

$$A^2 = \left.\frac{\mathrm{d}F}{\mathrm{d}x}\right|_{x=x_0}.$$

The Schrödinger equation in the vicinity of the turning point is, in this approximation,

$$\left[\frac{\mathrm{d}^2}{\mathrm{d}x^2} + \frac{A^2}{\hbar^2}(x - x_0)\right]\psi(x) = 0. \tag{C.14}$$

This equation can be solved exactly in terms of Bessel functions of order $\pm\frac{1}{3}$ (Airy functions). In the asymptotic region where $|x - x_0|$ is large, these solutions go over into the WKB approximations and it is then straightforward to verify that the particular WKB solution in the region $x > x_0$ which joins the solution vanishing as $x \to -\infty$, (C.12) is

$$\psi_+(x) = N[F(x)]^{-\frac{1}{4}} \cos\left[\int_{x_0}^{x} \frac{1}{\hbar}|F(x)|^{\frac{1}{2}}\,\mathrm{d}x' - \frac{\pi}{4}\right]. \tag{C.15}$$

where N is a constant. The detailed derivation of the connection formula can be found in Bransden and Joachain (1989) or in Bransden (1983).

We shall now revert to the use of atomic units with $\hbar = 1$ and write (C.15) as

$$\psi_+(x) = N[F(x)]^{-\frac{1}{4}} \cos\left[\int_{x_0}^{x} |F(x')|^{\frac{1}{2}}\,\mathrm{d}x' - \frac{\pi}{4}\right], \tag{C.16}$$

where

$$F(x) = k^2 - 2mV(x) \tag{C.17}$$

and

$$k^2 = 2mE. \tag{C.18}$$

The WKB approximation for the one-dimensional Schrödinger equation cannot be applied immediately to the radial equation for potential scattering (cf. (2.37))

$$\frac{\mathrm{d}^2U_l(R)}{\mathrm{d}R^2} + \left\{k^2 - 2\mu W(R) - \frac{l(l+1)}{R^2}\right\}U_l(R) = 0, \qquad 0 \le R < \infty \tag{C.19}$$

because of the different boundary conditions. For the one-dimensional case $\psi_-(x) \to 0$ as $x \to -\infty$, while for the radial function the condition $U_l(R) = 0$ at $R = 0$ must be imposed. Langer (1937) showed that this problem could be overcome by making substitutions which mapped the radial coordinate, defined for $0 \leq R < \infty$, onto a variable x defined on the whole real axis $-\infty < x < \infty$. Let us set

$$G_l(x) = R^{-\frac{1}{2}} U_l(R), \qquad \text{with } x = \log R, \tag{C.20}$$

in which case from (C.9), $G_l(x)$ satisfies

$$\frac{\mathrm{d}^2 G_l(x)}{\mathrm{d}x^2} + \tilde{F}(x) G_l(x) = 0, \tag{C.21}$$

where

$$\tilde{F}(x) = \mathrm{e}^{2x}\{k^2 - 2\mu W(\mathrm{e}^x) - \mathrm{e}^{-2x}(l + \tfrac{1}{2})^2\}. \tag{C.22}$$

Since the point at $R = 0$ is transformed to the point at $x = -\infty$, the boundary condition to be placed on $G_l(x)$ is that

$$\lim_{x \to -\infty} G_l(x) = 0. \tag{C.23}$$

The WKB approximation to (C.21) with the boundary condition (C.23) is now of the same form as we studied for the one-dimensional problem, and comparing with (C.15), the WKB approximation for $G_l(x)$ is seen to be

$$G_l(x) = N|\tilde{F}(x)|^{-\frac{1}{4}} \cos\left[\int_{x_0}^{x} |\tilde{F}(x')|^{\frac{1}{2}}\,\mathrm{d}x' - \frac{\pi}{4}\right]. \tag{C.24}$$

Using (C.20) and returning to the variable R, we have

$$G_l(R) = N|\kappa(R)|^{-\frac{1}{2}} \cos\left[\int_{R_0}^{R} |\kappa(R)|\,\mathrm{d}k' - \frac{\pi}{4}\right] \tag{C.25}$$

where

$$\kappa^2(R) = k^2 - 2\mu W(R) - \frac{(l + \frac{1}{2})^2}{R^2} \tag{C.26}$$

and where R_0 is the largest root of the equation $\kappa^2(R) = 0$. It is seen that in (C.26) the effective potential $\{2\mu W(R) + l(l+1)/R^2\}$ has been replaced by $\{2\mu W(R) + (l + \frac{1}{2})^2/R\}$. The change from $l(l+1)$ to $(l + \frac{1}{2})^2$ is known as the *Langer modification.*

For $R \to \infty$, $\kappa^2(R) \to k^2$, so we can write for large R

$$\int_{R_0}^{R} \kappa(R')\,\mathrm{d}R' \to \int_{R_0}^{\infty} \{\kappa(R') - k\}\,\mathrm{d}R' + k(R - R_0). \tag{C.27}$$

Thus the asymptotic form of $U_l(R)$ is (from (C.25))

$$U_l(R) \sim \cos\left[\int_{R_0}^{\infty} (\kappa(R') - k)\,\mathrm{d}R' + k(R - R_0) - \frac{\pi}{4}\right]. \tag{C.28}$$

On comparing with the standard asymptotic form (see Chapter 2)

$$U_l(R) \sim \sin(kR - \tfrac{1}{2}l\pi + \eta_l) \tag{C.29}$$

the phase shift of η_l can be identified as

$$\eta_l = \int_{R_0}^{\infty} [\kappa(R') - k]\,\mathrm{d}R + (l + \tfrac{1}{2})\frac{\pi}{2} - kR_0. \tag{C.30}$$

Since $\eta_l = 0$ when $W = 0$, it is easy to see that an alternative form for η_l is

$$\eta_l = \int_{R_0}^{\infty} \kappa(R)\,\mathrm{d}R - \int_{R_1}^{\infty} \kappa_0(R)\,\mathrm{d}R \tag{C.31}$$

where

$$\kappa_0^2(R) = k^2 - \frac{(l + \frac{1}{2})^2}{R^2} \tag{C.32}$$

and R_1 is the outermost zero of $\kappa_0(R)$.

When the angular momentum is large the turning points R_0 and R_1 are both approximately equal to the impact parameter $b = (l + \frac{1}{2})/k$. Expanding $\kappa(R)$ in powers of the potential, we find:

$$\kappa(R) - \kappa_0(R) \approx -\frac{1}{2k} W(R)\left[1 - \frac{(l + \frac{1}{2})^2}{k^2 R}\right]^{-\frac{1}{2}} \tag{C.33}$$

and

$$\eta_l \approx -\frac{1}{2k}\int_b^{\infty} \frac{W(R)}{[1 - (l + \frac{1}{2})^2/k^2R^2]^{\frac{1}{2}}}\,\mathrm{d}\kappa. \tag{C.34}$$

This approximation for η_l, which is accurate both for large l and for large k, was introduced by Massey and Mohr (1933).

In the above discussion it has been assumed that the effective potential $[2\mu W(R) + (l + \frac{1}{2})^2/R^2]$ has only one classical turning point at $R = R_0$. If the effective potential has three turning points at $R = R_0$, R_1, and R_2, as in the situation illustrated in Fig. 2.5(b), additional contributions to η_l arise. These can be calculated by applying the WKB approximation in each of the regions $R > R_0$, $R_1 < R < R_0$, $R_2 < R < R_1$, and $0 < R < R_2$. Provided the turning points are well separated, connection formulae can be employed to connect the approximate solutions in each interval to find the solution obeying the physical boundary conditions (Curtis and Power 1964; Livingston 1966; Berry 1966; Connor 1969). The result for the phase shifts has been given in (2.71).

APPENDIX D

ADIABATIC DIATOMIC MOLECULAR ORBITALS FOR ONE-ELECTRON SYSTEMS

In the important case in which one electron moves in the field of two fixed charges Z_A, Z_B, the Schrödinger equation is separable in prolate spheroidal coordinates and the orbitals and eigenvalues can be calculated exactly. The wave functions for this system, $\bar{\psi}_j(\bar{\mathbf{s}}_g, R)$, satisfy the Schrödinger equation in the body-fixed coordinate frame

$$\left(-\tfrac{1}{2}\nabla^2_{\bar{\mathbf{s}}_g} - \frac{Z_A}{x} - \frac{Z_B}{r}\right)\bar{\psi}_j(\bar{\mathbf{s}}_g, R) = E_j(R)\bar{\psi}_j(\bar{\mathbf{s}}_g, R) \tag{D.1}$$

where (see Chapters 3 and 4), x and r are the distances of the electron from the fixed charges at A and B, respectively, $\bar{\mathbf{s}}_g$ is the position vector of the electron with respect to the mid-point of the internuclear line AB, which is taken to be the Z-axis, and R is the internuclear distance which enters as a parameter.

The prolate spheroidal coordinates in which (D.1) is separable can be defined as (Morse and Feshbach 1953),

$$\begin{aligned} \xi &= \frac{x+r}{R} & \quad 1 \le \xi < \infty \\ \eta &= \frac{x-r}{R} & \quad -1 \le \eta \le 1 \\ \phi &= \tan^{-1}\left(\frac{\bar{y}_g}{\bar{x}_g}\right) & \quad 0 \le \phi < 2\pi \end{aligned} \tag{D.2}$$

where $(\bar{x}_g, \bar{y}_g, \bar{z}_g)$ are the Cartesian coordinates of $\bar{\mathbf{s}}_g$. The inverse relations to (D.2) are

$$\begin{aligned} \bar{x}_g &= \frac{R}{2}[(\xi^2-1)(1-\eta^2)]^{\frac{1}{2}}\cos\phi \\ \bar{y}_g &= \frac{R}{2}[(\xi^2-1)(1-\eta^2)]^{\frac{1}{2}}\sin\phi \\ \bar{z}_g &= \frac{R}{2}[\xi\eta]. \end{aligned} \tag{D.3}$$

Substituting

$$\bar{\psi}_j(\bar{\mathbf{s}}_g, R) = X(\xi)Y(\eta)\Phi(\phi) \quad \text{(D.4)}$$

into (D.1) leads to the set of equations:

$$\left[\frac{d^2}{d\phi^2} - m^2\right]\Phi(\phi) = 0 \quad \text{(D.5)}$$

$$\left[\frac{d}{d\xi}(\xi^2 - 1)\frac{d}{d\xi} - \frac{m^2}{\xi^2 - 1} + R(Z_A + Z_B)\xi - p^2(\xi^2 - 1) + A\right]X(\xi) = 0 \quad \text{(D.6)}$$

$$\left[\frac{d}{d\eta}(1 - \eta^2)\frac{d}{d\eta} - \frac{m^2}{1 - \eta^2} - R(Z_A - Z_B)\eta - p^2(1 - \eta^2) - A\right]Y(\eta) = 0 \quad \text{(D.7)}$$

where m^2 and A are separation constants, and $p^2 = -R^2E_j(R)/2$. The solution of (D.5) is, with $m \geq 0$,

$$\Phi(\phi) = (2\pi)^{-\frac{1}{2}} \exp(\pm im\phi) \quad \text{(D.8)}$$

where, in order that $\Phi(\phi)$ should be single valued, m must be a positive integer, or zero. Solutions of equations (D.6) and (D.7) can be obtained in the form of series expansions, both for discrete and continuum states (Greenland and Greiner, 1976). A convenient discussion of these solutions, together with references to earlier work, has been given by Greenland (1982).

The solutions of (D.6) and (D.7) must be finite and single valued everywhere and in the case of bound states $X(\xi)$ must vanish at infinity

$$X(\xi) \to 0, \qquad \xi \to \infty. \quad \text{(D.9)}$$

These boundary conditions lead to an eigenvalue problem which allows the determination of the allowed values of both p^2 and A. For a given value of p^2, (D.6) will possess eigenfunctions $X_{km}(\xi)$ belonging to eigenvalues $A \equiv A_1(k, m; p^2)$. The label k can be taken to be the number of nodes in the function $X_{km}(\xi)$. Similarly solving (D.7), we can obtain for each value of p^2 eigenfunctions $Y_{qm}(\eta)$ belonging to eigenvalues $A \equiv A_2(q, m; p^2)$, where q is the number of nodes in $Y_{qm}(\eta)$. Since the separation constant A must be the same in both equations (D.6) and (D.7), the condition

$$A_1(k, m; p^2) = A_2(q, m; p^2) \quad \text{(D.10)}$$

must be satisfied. For given k, q, and m, (D.10) determines the allowed value of p^2 and hence the energy $E_j(R)$, and the different energy levels can be labelled by the set of quantum numbers $j \equiv (k, q, m)$. The component of the orbital angular momentum along the internuclear line (in atomic units) is $\pm m$ and this is a conserved quantity since AB is an axis of symmetry. As explained in Chapter 4, levels with $m = 0, 1, 2$ are, by convention, known as $\sigma, \pi, \delta, \ldots$ levels.

In the limit $R \to 0$, the energy $E_j(R)$ must become equal to one of the energy levels of the united atom with charge $Z = Z_A + Z_B$. Correspondingly in the limit $R \to \infty$, $E_j(R)$ must become equal to one of the energy levels of either the atom $(A + e^-)$ or of the atom $(B + e^-)$. In finding the correlation between these united atom and separated atom limits, account must be taken of the Wigner–von Neumann non-crossing rule (a simple proof of which may be found in Bransden and Joachain (1983)) which states that the energy eigenvalues $E_{kqm}(R)$ and $E_{k'q'm'}(R)$ cannot become equal at any value of R, and therefore cannot cross, if the states $j = (k, q, m)$ and $j = (k', q', m')$ are of the same symmetry. To enumerate the symmetries of the system we have to find which operators commute with the Hamiltonian. One such operator is L_z so that levels with different values of m may cross. In addition to this geometrical symmetry, there is a further dynamical symmetry arising from the fact that the Schrödinger equation is separable. The operator concerned, which commutes with both H and L_z, is the separation operator $\mathbf{A}$ defined as

$$\mathbf{A} = -\mathbf{L}^2 - \frac{R}{4}\left(\frac{\partial^2}{\partial \bar{x}_g^2} + \frac{\partial^2}{\partial \bar{y}_g^2}\right) - R(Z_A \cos\theta_x - Z_B \cos\theta_r) \qquad \text{(D.11)}$$

where θ_x and θ_r are the angles made between $\mathbf{x}$ and the internuclear line, and $\mathbf{r}$ and the internuclear line, respectively. The separation constant A is an eigenvalue of the operator $\mathbf{A}$, so that the symmetry of each state can be defined by the values of m and A. Potential energy curves for two states with the same values of m and A cannot cross. If the system is homonuclear $Z_A = Z_B$, and, as we have already seen in Chapter 4, parity is also conserved; states of even and odd parity being known as gerade or ungerade states, respectively. States with the same values of m, A, and the parity cannot cross. The consequences of these considerations can be summarized as follows:

(1) for homonuclear systems states of different parity are allowed to cross;

(2) states with different values of m are allowed to cross;

(3) two states with the same value of m may cross if $k \neq k'$ and $q \neq q'$.

Rule (3) can be shown as follows. We first note that since equations (D.6) and (D.7) can be put in the form of one-dimensional Schrödinger equations for motion in a potential well, the discrete eigenvalues are non-degenerate (see for example Bransden and Joachain (1989)) and, in addition, the nth eigenvalue has $n - 1$ nodes. Now consider two states j, j' with quantum numbers $j = kqm$ and $j' = kq'm$, so that each state has the same values of m

and k. Let the two energy curves E_j and $E_{j'}$ cross at $R = R_0$ so that

$$E_j(R_0) = E_{j'}(R_0). \qquad \text{(D.12)}$$

Since the eigenvalues of (D.6) are non-degenerate, $X_j = X_{j'}$, from which we see that the separation constants A_j and $A_{j'}$ must be equal. Turning to (D.7), we notice that since both $E_j = E_{j'}$ and $A_j = A_{j'}$, and in view of the fact that the eigenvalues of (D.7) are non-degenerate, $Y_j = Y_{j'}$. This shows that $q = q'$, in which case the two wave functions ψ_j and $\psi_{j'}$ are identical. It follows that two levels cannot cross if $m = m'$ and $k = k'$, even if $q \neq q'$. Similarly two levels cannot cross if $m = m'$ and $q = q'$, even if $k \neq k'$. If, on the other hand, $k \neq k'$ and $q \neq q'$, the two separation constants A_j and $A_{j'}$ are different and a crossing may occur.

The united atom limit

To connect the quantum numbers (kqm) of a level j with the spherical quantum numbers (nlm) in the united atom limit $R \to 0$, we rely on the fact that the number of nodes in the wave function is independent of R (for R finite). This follows from the fact that ψ_j is a continuous function of R, so that the number of nodes can only change if two coalesce. If this happens both ψ_j and $\mathrm{d}\psi_j/\mathrm{d}R$ must be zero at some point, and in turn this can only occur if ψ_j is identically zero for all R.

From (D.2), we see that in the limit $R \to 0$, $x \to r$ and

$$\xi \to \frac{2r}{R}, \qquad \eta \to \cos\theta, \qquad \text{(D.13)}$$

where (r, θ, ϕ) are the spherical polar coordinates of the electron. Equation (D.6) becomes, in this limit, the equation for the radial hydrogenic wave functions $R_{nl}(r)$ for an ion with nuclear charge $Z_A + Z_B$. Since $R_{nl}(r)$ has $n - l - 1$ nodes, we have that

$$k = n - l - 1. \qquad \text{(D.14)}$$

Similarly, equation (D.7) becomes the equation for the associated Legendre function $P_l^m(\cos\theta)$, which has $l - m$ nodes, from which we see that

$$q = l - m. \qquad \text{(D.15)}$$

From (D.14) and (D.15), the quantum numbers (n, l, m) of the united atom are related to the spheroidal quantum numbers (k, q, m) by

$$\begin{aligned} n &= k + q + m + 1 \\ l &= q + m \\ m &= m. \end{aligned} \qquad \text{(D.16)}$$

For small value of R, Power (1973*a*) has obtained the following expression for the energy $E_j(R)$

$$E_j(R) = -\frac{(Z_A + Z_B)^2}{2n^2} - \frac{2Z_A Z_B[l(l+1) - 3m^2](Z_A + Z_B)^2 R^2}{n^3 l(l+1)(2l+3)(2l+1)(2l-1)} + O(R^3). \tag{D.17}$$

The separated atom limit

In the limit $R \to \infty$, if the electron forms an ion centred on A, $x \ll r$ and $r \simeq R$, so that $\eta \to -1$. In this case, the spheroidal coordinates become parabolic coordinates (λ_A, μ_A, ϕ) where

$$\xi \to 1 + \frac{\lambda_A}{R}, \qquad \eta \to -1 + \frac{\mu_A}{R}, \qquad \phi \to \phi. \tag{D.18}$$

In the region of nucleus A, $\eta \approx -1$ and $\mu_A \ll R$. The coordinates λ_A, μ_A, ϕ can also be expressed in terms of the spherical coordinates (x, θ_x, ϕ) centred on A by

$$\lambda_A = x(1 - \cos\theta_x), \qquad \mu_A = x(1 + \cos\theta_x), \qquad \phi = \phi. \tag{D.19}$$

In the same way if the system separates to the atom $(B + e^-)$, the spheroidal coordinates become parabolic coordinates (λ_B, μ_B, ϕ) where

$$\xi \to 1 + \frac{\lambda_B}{R}, \qquad \eta \to 1 - \frac{\mu_B}{R}, \qquad \phi \to \phi. \tag{D.20}$$

In the region of nucleus B, $\eta \approx 1$ and $\mu_B \ll R$. In terms of the spherical coordinates (r, θ_r, ϕ) centred on B

$$\lambda_B = r(1 + \cos\theta_r), \qquad \mu_B = r(1 - \cos\theta_r), \qquad \phi = \phi. \tag{D.21}$$

In the limit $R \to \infty$, the Schrödinger equation separates in terms of the parabolic coordinates. If the electron separates to the centre B, the wave function for the atom $(B + e^-)$ can be written as (Bethe and Saltpeter 1957)

$$\psi(\lambda_B, \mu_B, \phi) = f_1(\lambda_B) f_2(\mu_B)(2\pi)^{-\frac{1}{2}} \exp(\pm im\phi) \tag{D.22}$$

where $f_1(\lambda_B)$, $f_2(\mu_B)$ satisfy

$$\begin{aligned}\frac{d}{d\lambda_B}\left(\lambda_B \frac{df_1(\lambda_B)}{d\lambda_B}\right) - \left\{\frac{m^2}{4\lambda_B} + \tfrac{1}{2}|E|\lambda_B + C\right\} f_1(\lambda_B) &= 0 \\ \frac{d}{d\mu_B}\left(\mu_B \frac{df_2(\mu_B)}{d\mu_B}\right) - \left\{\frac{m^2}{4\mu_B} + \tfrac{1}{2}|E|\mu_B + (1 - C)\right\} f_2(\mu_B) &= 0\end{aligned} \tag{D.23}$$

where C is a separation constant.

The states can be labelled by $(\bar{K}_B, K_B, m)$, where $\bar{K}_B$, K_B are the number of nodes in $f_1(\lambda_B)$ and $f_2(\mu_B)$, respectively; m is the magnetic quantum number as before.

If n_B is the principal quantum number of a level of the atom $(B + e^-)$ then

$$n_B = \bar{K}_B + K_B + m + 1 \tag{D.24}$$

and

$$E_{n_B} = -\frac{Z_B^2}{2n_B^2}. \tag{D.25}$$

The atomic levels can be labelled by the set of quantum numbers (n_B, K_B, m) where, for a given n_B, m can take the values $0 \le m \le (n_B - 1)$ and K_B can take the $n_B - m$ values $0, 1, \ldots, n_B - m - 1$. For fixed values of n_B, K_B, and m, the parabolic wave function is a linear combination of the spherical wave functions differing in l_B belonging to the degenerate level E_{n_B}.

Using perturbation theory, the energy of the system when it separates to the atom $(B + e^-)$ in the limit $R \to \infty$, is

$$E_{n_B, K_B, m} \approx -\frac{Z_B^2}{2n_B^2} - \frac{Z_A}{R} + O\left(\frac{1}{R^2}\right). \tag{D.26}$$

An exactly similar set of results to (D.22)–(D.26) can be obtained if the system separates to the atom $(A + e^-)$. If the system is homogeneous, then the wave function is a symmetrical, or antisymmetrical, combination (with $n_A = n_B$, $\lambda_A = \lambda_B = \lambda$)

$$\psi_{g,u} = f_1(\lambda)[f_2(\mu_A) \pm f_2(\mu_B)](2\pi)^{-\frac{1}{2}} \exp(\pm im\phi). \tag{D.27}$$

Correlation between the united and separated atom limits

The correlation between the united and separated atom limits has been discussed by Gershtein and Krivchenkov (1961), Power (1973*a*), and Greenland (1982) who has corrected certain defects in the earlier work. The rules are obtained by noting the order in which states are filled taking into account the non-crossing rules and the conservation of the number of nodes k of the function $X(\xi)$. The rules in the form given by Greenland are as follows, for the case $Z_A > Z_B$ and starting with a state of the united atom with quantum numbers (n, l, m).

1. Note that $k = n - l - 1$.

2. Define n_2 as

$$n_2 = \text{Ent}\left[\frac{Z_B n - Z_A(k + m + 1)}{Z_A + Z_B}\right]$$

where Ent denotes the integer part of the expression in square brackets (positive or negative).

3. Then if $n_2 < 0$, the system separates to $(\mathrm{A} + \mathrm{e}^-)$ with $K_\mathrm{A} = l - m$.
4. If $n_2 \geq 0$ but $n_2 \leq \{Z_\mathrm{B}(n-1) - Z_\mathrm{A}(k+m+1)\}/(Z_\mathrm{A} + Z_\mathrm{B})$, then the system separates to $(\mathrm{A} + \mathrm{e}^-)$ with $K_\mathrm{A} = l - m - 1 - n_2$.
5. Otherwise the system separates to $(\mathrm{B} + \mathrm{e}^-)$ and $K_\mathrm{B} = n_2$.

The inverse rules are as follows, again for the case $Z_\mathrm{A} > Z_\mathrm{B}$.

1. If the separated atom is $(\mathrm{A} + \mathrm{e}^-)$, then in the united atom limit we have, if $K_\mathrm{A} < n_\mathrm{A}(Z_\mathrm{A} - Z_\mathrm{B})/Z_\mathrm{A}$

$$n = n_\mathrm{A} \qquad \text{and} \qquad l = K_\mathrm{A} + m$$

otherwise

$$n = l + n_\mathrm{A} - K_\mathrm{A} - m, \qquad l = K_\mathrm{A} + m + 1 + \mathrm{Ent}\left[K_\mathrm{A} - \frac{n_\mathrm{A}(Z_\mathrm{A} - Z_\mathrm{B})}{Z_\mathrm{A}}\right].$$

2. If the separated atom is $(\mathrm{B} + \mathrm{e}^-)$ then

$$l = n + m + K_\mathrm{B} - n_\mathrm{B}$$

and

$$n = \begin{cases} K_\mathrm{B} + \dfrac{n_\mathrm{B} Z_\mathrm{A}}{Z_\mathrm{B}} & \text{if } n_\mathrm{B}/Z_\mathrm{B} \text{ is an integer} \\[2ex] K_\mathrm{B} + 1 + \mathrm{Ent}\left[\dfrac{n_\mathrm{B} Z_\mathrm{A}}{Z_\mathrm{B}}\right] & \text{otherwise.} \end{cases}$$

For the homonuclear case $Z_\mathrm{A} = Z_\mathrm{B}$, $n_\mathrm{A} = n_\mathrm{B} = N$, $K_\mathrm{A} = K_\mathrm{B} = K$ and it is found that

$$K = \mathrm{Ent}[\tfrac{1}{2}(l - m)]$$

and

$$N = n - l + m + K.$$

Numerical solutions

Numerical procedures to obtain solutions of equations (D.6) and (D.7) have been discussed by, among others, Bates and Reid (1968), Bates and Carson (1956), Peek (1965), Madsen and Peek (1971), and Ponomarev and Puzynina (1967, 1968, 1970). A computer program has been published by Power (1973*b*) which allows the efficient determination of the eigenenergies $E_j(R)$. Using Power's program as input, Salin (1978) has made available a program which determines the corresponding wave functions, and, in addition, calculates the matrix elements of the radial and rotational coupling required by the perturbed stationary state model of ion–atom scattering (see Chapter 4).

APPENDIX E

EXCHANGE MATRIX ELEMENTS WITH A SLATER BASIS

In this appendix we are concerned with the evaluation of integrals occurring in the impact parameter coupled channel model employing a two-centre basis with plane-wave translation factors. The one-electron problem will be discussed first, in which case the required matrix elements are defined by (5.85). It will be assumed that the basis functions $\phi_j(\mathbf{r})$ and $\chi_j(\mathbf{x})$ are expressed in linear combinations of Slater functions having the general form

$$F_{plm}(\lambda, \mathbf{r}) = r^{p+l} \exp(-\lambda r) Y^{\pm}_{l|m|}(\hat{\mathbf{r}}) \tag{E.1}$$

where $Y^{\pm}_{l|m|}$ are the real spherical harmonics defined by (5.86). For the particular case in which the functions $\phi_j(\mathbf{r})$ and $\chi_k(\mathbf{x})$ are hydrogenic wave functions of bound states this representation is exact, but if effective potentials are employed, or for pseudostates, it is approximate but convenient. With the functions (E.1) the analytic evaluation of the direct interaction matrix elements H_{jk} and $\bar{H}_{jk}$ presents no difficulty, but the exchange and overlap matrix elements K_{jk}, $\bar{K}_{jk}$, and S_{jk} cannot be completely reduced to analytic expressions and are time consuming to compute. Some techniques for evaluating these integrals efficiently will be outlined in this appendix, followed by a brief discussion of the additional integrals that arise for systems of two (or more) electrons. The integrals that are needed when triple centred expansions are employed can be treated along similar lines, but are more complicated and will not be considered here (Anderson and Antal 1975; Lin *et al.* 1982).

Apart from a normalization factor, the functions $r^l Y^{\pm}_{l|m|}(\hat{\mathbf{r}})$ can be expressed as the sums of products of r_α ($\alpha = 1, 2, 3$), the Cartesian components of $\hat{\mathbf{r}}$ (as shown in Table E.1), and for a one-electron system the exchange and overlap matrix elements can be expressed in terms of integrals of the general form

$$I(n_A l_A p, n_B l_B q) = \int x^{n_A + l_A - 2}\, e^{-\lambda_A x}(x_1^i x_2^j x_3^k)\, e^{i\mathbf{v}\cdot\mathbf{s}_g}\, r^{n_A + l_A - 2}\, e^{-\lambda_B r}\, (r_1^u r_2^v r_3^w)\, d\mathbf{s}_g \tag{E.2}$$

where x_α ($\alpha = 1, 2, 3$) are the Cartesian components of $\mathbf{x}$, $i + j + k = l_A$, and $u + v + w = l_B$. The labels of p and q stand for (i, j, k) and (u, v, w), respectively, and n_A, n_B are integers with $n_A, n_B \geq 1$. We recollect that $\mathbf{s}_g$ is the position vector of the electron with respect to the mid-point of the

Table E.1. The spherical harmonics in real form for $l \leq 2$.

l	$\lvert m \rvert$	$\pm$	$r^l Y^{\pm}_{l\lvert m\rvert}(\hat{r})$
0	0	+	$(1/4\pi)^{\frac{1}{2}}$
1	0	+	$(3/4\pi)^{\frac{1}{2}} r_3$
	1	+	$(3/4\pi)^{\frac{1}{2}} r_1$
	1	−	$(3/4\pi)^{\frac{1}{2}} r_2$
2	0	+	$(5/16\pi)^{\frac{1}{2}}(2r_3^2 - r_1^2 - r_2^2)$
	1	+	$(15/4\pi)^{\frac{1}{2}} r_1 r_3$
	1	−	$(15/4\pi)^{\frac{1}{2}} r_2 r_3$
	2	+	$(15/16\pi)^{\frac{1}{2}}(r_1^2 - r_2^2)$
	2	−	$(15/4\pi)^{\frac{1}{2}} 2r_1 r_2$

interionic line and that

$$\mathbf{x} = \mathbf{s}_g - \tfrac{1}{2}\mathbf{R}, \qquad \mathbf{r} = \mathbf{s}_g + \tfrac{1}{2}\mathbf{R}. \tag{E.3}$$

The general integral (E.2) can be obtained from a basic integral $J(\mathbf{a}, \mathbf{b})$ where

$$J(\mathbf{a}, \mathbf{b}) = \int \frac{e^{-\lambda_A x}}{x} \frac{e^{-\lambda_B r}}{r} e^{i\{\mathbf{v}\cdot\mathbf{s}_g + \mathbf{a}\cdot\mathbf{x} + \mathbf{b}\cdot\mathbf{r}\}} d\mathbf{s}_g \tag{E.4}$$

by parametric differentiation with respect to λ_A, λ_B and the components of the vectors **a** and **b**. We have

$$\begin{aligned} I(n_A l_A p, n_B l_B q) = \lim_{\substack{\mathbf{a}\to 0 \\ \mathbf{b}\to 0}} &\left(-\frac{\partial}{\partial \lambda_A}\right)^{n_A - 1} \left(-\frac{\partial}{\partial \lambda_B}\right)^{n_B - 1} \left(-i\frac{\partial}{\partial a_1}\right)^{i} \\ &\times \left(-i\frac{\partial}{\partial a_2}\right)^{j} \left(-i\frac{\partial}{\partial a_3}\right)^{k} \\ &\times \left(-i\frac{\partial}{\partial b_1}\right)^{u} \left(-i\frac{\partial}{\partial b_2}\right)^{v} \left(-i\frac{\partial}{\partial b_3}\right)^{w} J(\mathbf{a}, \mathbf{b}). \end{aligned} \tag{E.5}$$

Method A: Fourier transforms

The Fourier transform method to evaluate J, and hence I, was introduced by Sin Fai Lam (1967*a*, *b*) and later by Shakeshaft (1975*b*). The Fourier transform

$$\frac{e^{-\mu y}}{y} = \frac{1}{2\pi^2} \int \frac{e^{i\mathbf{k}\cdot\mathbf{y}}}{k^2 + \mu^2} d\mathbf{k} \tag{E.6}$$

is used to express each of the factors $r^{-1}\exp(-\lambda_B r)$, $x^{-1}\exp(-\lambda_A x)$ in the integrand of the basic integral (E.4), to obtain

$$J(\mathbf{a}, \mathbf{b}) = \frac{1}{4\pi^4} \int \mathrm{d}\mathbf{s}_\mathrm{g} \int \mathrm{d}\mathbf{k} \int \mathrm{d}\mathbf{K} \frac{\exp[\mathrm{i}\{(\mathbf{k} + \mathbf{a})\cdot\mathbf{x} + (\mathbf{K} + \mathbf{b})\cdot\mathbf{r} + \mathbf{v}\cdot\mathbf{s}_\mathrm{g}\}]}{(k^2 + \lambda_\mathrm{A}^2)(K^2 + \lambda_\mathrm{B}^2)}. \tag{E.7}$$

With the help of (E.3) the integration over $\mathbf{s}_\mathrm{g}$ can be effected, providing a factor $(2\pi)^2\delta(\mathbf{k} + \mathbf{K} + \bar{\mathbf{v}})$ where

$$\bar{\mathbf{v}} = \mathbf{v} + \mathbf{a} + \mathbf{b}. \tag{E.8}$$

By using this delta function to integrate over $\mathbf{K}$, $J(\mathbf{a}, \mathbf{b})$ becomes

$$J(\mathbf{a}, \mathbf{b}) = \frac{2}{\pi} \exp\left[-\mathrm{i}\frac{(2\mathbf{a} + \mathbf{v})\cdot\mathbf{R}}{2}\right] \int \frac{\exp(-\mathrm{i}\mathbf{k}\cdot\mathbf{R})}{(k^2 + \lambda_\mathrm{A}^2)(|\mathbf{k} + \bar{\mathbf{v}}|^2 + \lambda_\mathrm{B}^2)} \mathrm{d}\mathbf{k}. \tag{E.9}$$

The Feynman identity

$$\frac{1}{\alpha\beta} = \int_0^1 \frac{1}{[\alpha y + \beta(1 - y)]^2} \mathrm{d}y \tag{E.10}$$

can be used to combine the two factors in the denominator of (E.9), giving

$$J(\mathbf{a}, \mathbf{b}) = \exp\left[-\mathrm{i}\frac{(2\mathbf{a} + \mathbf{v})\cdot\mathbf{R}}{2}\right] \frac{2}{\pi} \int_0^1 \frac{\exp(-\mathrm{i}\mathbf{k}\cdot\mathbf{R})}{[A^2 + |\mathbf{k} + \mathbf{B}|^2]^2} \mathrm{d}y \tag{E.11}$$

where

$$A^2 = \lambda_\mathrm{A}^2 y + \lambda_\mathrm{B}^2(1 - y) + \bar{v}^2 y(1 - y) \tag{E.12}$$

and

$$\mathbf{B} = \bar{\mathbf{v}}(1 - y). \tag{E.13}$$

The integration over $\mathbf{k}$ can now be carried out using the Fourier transform

$$\mathrm{e}^{-\mu y} = \frac{\mu}{\pi^2} \int \frac{\mathrm{e}^{\mathrm{i}\mathbf{k}\cdot\mathbf{y}}}{(k^2 + \mu^2)^2} \mathrm{d}\mathbf{k}. \tag{E.14}$$

It is found that

$$J(\mathbf{a}, \mathbf{b}) = \exp\left[-\mathrm{i}\frac{(2\mathbf{a} + \mathbf{v})\cdot\mathbf{R}}{2}\right] (2\pi) \int_0^1 \frac{\exp(-AR + \mathrm{i}\mathbf{B}\cdot\mathbf{R})}{A} \mathrm{d}y$$

$$= 2\pi \int_0^1 \frac{\exp(-AR)}{A}$$

$$\times \exp\left[-\mathrm{i}\left\{\curvearrowright y\mathbf{a}\cdot\mathbf{R} + (1 + y)\mathbf{b}\cdot\mathbf{R} + \frac{(1 - 2y)\mathbf{v}\cdot\mathbf{R}}{2}\right\}\right] \mathrm{d}y. \tag{E.15}$$

The evaluation of $I(100, 100) \equiv J(0, 0)$, proceeds by setting **a** and **b** equal to zero in (E.15) and integrating over the parameter y using a numerical integration rule. A low-order Gauss–Legendre formula is generally sufficient.

The general integral $I(n_A l_A p, n_B l_B q)$ can be expressed in terms of a one-parameter integral by carrying out the differentiations indicated in (E.5) before setting **a** and **b** to zero. The explicit expressions become very complicated as n_A, l_A, n_B, and l_B increase; however they can be produced automatically from a computer program of Noble (1980), which uses the REDUCE symbolic manipulation language. A second program in this package evaluates the one-dimensional integrals over y to produce numerical tables of the standard integrals (E.2).

Method B: the differential equation method

When the relative motion of the ions A and B is described by a straight line trajectory traversed at a constant velocity $\mathbf{v} = \mathbf{v}_0$, Cheshire (1967) and Chatterjee *et al.* (1967) were able to show that the integrals $I(n_A l_A p, n_B l_B q)$, considered as functions of t, satisfied simple systems of first-order differential equations which could be solved numerically, using standard methods, to obtain the integrals for all the required values of t.

We shall start by redefining a basic integral $J(\gamma_A, \gamma_B)$ as

$$J(\gamma_A, \gamma_B) = \int \frac{e^{-\lambda_A x}}{x} \frac{e^{-\lambda_B r}}{r} e^{i\mathbf{v}_0 \cdot \mathbf{s}_g} \, d\mathbf{s}_g \tag{E.16}$$

where the definitions of **x** and **r** are modified so that

$$\mathbf{x} = \mathbf{s}_g - \tfrac{1}{2}\mathbf{R} + \gamma_A, \qquad \mathbf{r} = \mathbf{s}_g + \tfrac{1}{2}\mathbf{R} + \gamma_B \tag{E.17}$$

where γ_A and γ_B are constant vectors. These vectors play a similar role to **a** and **b** and differentiation with respect to their components introduces the angular factors into the integrands. In the final formula γ_A and γ_B are set equal to zero.

By introducing the Fourier transforms of $x^{-1} \exp(-\lambda_A x)$ and $r^{-1} \exp(-\lambda_B r)$, $J(\gamma_A, \gamma_B)$ can be represented in a similar form to (E.9):

$$J(\gamma_A, \gamma_B) = \exp[-i\mathbf{v}_0 \cdot (\gamma_B + \tfrac{1}{2}\mathbf{R})] \frac{2}{\pi} \int \frac{\exp[-i\mathbf{k} \cdot (\mathbf{R} + \gamma_B - \gamma_A)]}{(k^2 + \lambda_A^2)(|\mathbf{k} + \mathbf{v}_0|^2 + \lambda_B^2)} d\mathbf{k}. \tag{E.18}$$

Differentiating with respect to t, and recollecting $\dot{\mathbf{R}} = \mathbf{v}$ where for a straight line trajectory $\mathbf{v} = \mathbf{v}_0$ is constant, we have that

$$\frac{dJ(\gamma_A, \gamma_B)}{dt} = -i \frac{v_0^2}{2} J(\gamma_A, \gamma_B) + \exp[-i\mathbf{v}_0 \cdot (\gamma_B + \tfrac{1}{2}\mathbf{R})]$$
$$\times \frac{2}{\pi} \int \frac{(-i\mathbf{k} \cdot \mathbf{v}_0) \exp[-i\mathbf{k} \cdot (\mathbf{R} + \gamma_B - \gamma_A)]}{(k^2 + \lambda_A^2)(|\mathbf{k} + v_0|^2 + \lambda_B^2)} d\mathbf{k}. \tag{E.19}$$

Noting that

$$\frac{2\mathbf{k}\cdot\mathbf{v}_0}{(k^2+\lambda_A^2)(|\mathbf{k}+\mathbf{v}_0|^2+\lambda_B^2)} = \frac{\lambda_A^2-\lambda_B^2-v_0^2}{(k^2+\lambda_A^2)(|\mathbf{k}+\mathbf{v}_0|^2+\lambda_B^2)} + \frac{1}{k^2+\lambda_A^2} - \frac{1}{|\mathbf{k}+\mathbf{v}_0|^2+\lambda_B^2} \tag{E.20}$$

we see that

$$\left[\mathrm{i}\frac{\mathrm{d}}{\mathrm{d}t} - \tfrac{1}{2}(\lambda_A^2-\lambda_B^2)\right] J(\boldsymbol{\gamma}_A, \boldsymbol{\gamma}_B) = \frac{1}{\pi}\exp\left[-\mathrm{i}\mathbf{v}_0\cdot(\boldsymbol{\gamma}_B+\tfrac{1}{2}\mathbf{R})\right]\int \mathrm{d}\mathbf{k}\left(\frac{1}{k^2+\lambda_A^2} - \frac{1}{|\mathbf{k}+\mathbf{v}_0|^2+\lambda_B^2}\right)\exp(-\mathrm{i}\mathbf{k}\cdot\mathbf{R}') \tag{E.21}$$

where

$$\mathbf{R}' = \mathbf{R} + \boldsymbol{\gamma}_B - \boldsymbol{\gamma}_A. \tag{E.22}$$

Using (E.6) both the terms on the right-hand side of (E.21) can be integrated and we obtain the first-order differential equation

$$QJ(\boldsymbol{\gamma}_A, \boldsymbol{\gamma}_B) = G_{11}(\boldsymbol{\gamma}_A, \boldsymbol{\gamma}_B) \tag{E.23}$$

where

$$Q = \mathrm{i}\frac{\mathrm{d}}{\mathrm{d}t} - \tfrac{1}{2}(\lambda_A^2-\lambda_B^2) \tag{E.24}$$

and

$$G_{11}(\boldsymbol{\gamma}_A, \boldsymbol{\gamma}_B) = \frac{2\pi}{R'}\left[\exp\left(-\lambda_A R' - \mathrm{i}\mathbf{v}_0\cdot\boldsymbol{\gamma}_B - \mathrm{i}\mathbf{v}_0\cdot\frac{\mathbf{R}}{2}\right) - \exp\left(-\lambda_B R' - \mathrm{i}\mathbf{v}_0\cdot\boldsymbol{\gamma}_A + \mathrm{i}\mathbf{v}_0\cdot\frac{\mathbf{R}}{2}\right)\right]. \tag{E.25}$$

To complete the evaluation of $I(100, 100) \equiv J(0, 0)$, $\boldsymbol{\gamma}_A$ and $\boldsymbol{\gamma}_B$ are set equal to zero and the equation (E.23) integrated subject to the boundary condition $\mathbf{J} \to 0$, $t \to -\infty$.

Systems of equations can be obtained for the integrals with $n_A, n_B > 1$ or with $l_A, l_B > 0$. For instance, using

$$I(n_A 00, n_B 00) = \left(-\frac{\partial}{\partial\lambda_A}\right)^{n_A-1}\left(-\frac{\partial}{\partial\lambda_B}\right)^{n_B-1} J(\boldsymbol{\gamma}_A, \boldsymbol{\gamma}_B) \tag{E.26}$$

we find for $l_A = l_B = 0$ and $n_A, n_B \leq 2$,

$$\begin{aligned}
QI(100, 100) &= G_{11} \\
QI(200, 100) &= -\lambda_A I(100, 100) + G_{21} \\
QI(100, 200) &= \lambda_B I(100, 100) + G_{12} \\
QI(200, 200) &= -\lambda_A I(100, 200) + \lambda_B I(200, 100) + G_{22}
\end{aligned} \tag{E.27}$$

where

$$G_{n_A n_B} = \left[\left(-\frac{\partial}{\partial \lambda_A}\right)^{n_A - 1}\left(-\frac{\partial}{\partial \lambda_B}\right)^{n_B - 1} G_{11}\right]. \tag{E.28}$$

To introduce components of the vectors **x** and **r** into the integrand to build up the angular factors $Y^{\pm}_{lm}$, parametric differentiation by the components of $\boldsymbol{\gamma}_A$ or $\boldsymbol{\gamma}_B$ can be employed. As a simple example, the operator $T^A_i = \partial/\partial\lambda_A(\lambda_A^{-1}\,\partial/\partial\gamma_{A_i})$ will introduce the component x_i into the integrand of $I(200, 100)$

$$I(21(i), 100) = T^A_i I(200, 100). \tag{E.29}$$

By operating throughout (E.27) by T^A_i, a new set of equations can be obtained containing the integrals $I(21(i), 100)$. In general, products of the operators T^A_i, T^B_i applied to $I(n_A 00, n_B 00)$ produce linear combinations of integrals $I(n_A l_A p, n_B l_B q)$, which satisfy simple sets of equations like (E.27).

Method C: prolate spheroidal coordinates

In an approach to the evaluation of the one-electron exchange integrals employed by McCarroll (1961) and subsequently elaborated by Lovell and McElroy (1965), Msezane and Gallaher (1975, 1977), and Gallaher and Msezane (1974), the family of integrals $I(n_A l_A p, n_B l_B q)$ are first expressed in prolate spheroidal coordinates (ξ, μ, ϕ). We have

$$x = \tfrac{1}{2}R(\xi + \mu), \qquad r = \tfrac{1}{2}R(\xi - \mu) \tag{E.30}$$

and ϕ is the azimuthal of $\mathbf{s}_g$ with respect to the plane of scattering containing the vectors $\mathbf{v}\ (=\dot{\mathbf{R}})$ and $\mathbf{R}$. The ranges of ξ, μ, and ϕ are

$$1 \leq \xi \leq \infty, \qquad -1 \leq \mu \leq 1, \qquad 0 \leq \phi \leq 2\pi \tag{E.31}$$

and the volume element is

$$\mathrm{d}\mathbf{s}_g = \tfrac{1}{8}R^3(\xi^2 - \mu^2)\,\mathrm{d}\xi\,\mathrm{d}\mu\,\mathrm{d}\phi. \tag{E.32}$$

Consider the basic integral

$$I(100, 100) = \int \frac{\mathrm{e}^{-\lambda_A x}}{x}\frac{\mathrm{e}^{-\lambda_B y}}{y}\,\mathrm{e}^{\mathrm{i}\mathbf{v}\cdot\mathbf{s}_g}\,\mathrm{d}\mathbf{s}_g. \tag{E.33}$$

The scalar product $\mathbf{v}\cdot\mathbf{s}_g$ can be expressed as

$$\mathbf{v}\cdot\mathbf{s}_g = vs_g(\cos\theta\cos\delta + \sin\theta\sin\delta\cos\phi) \tag{E.34}$$

where θ is the angle between $\mathbf{s}_g$ and $\mathbf{R}$, and δ is the angle between $\mathbf{v}$ and $\mathbf{R}$. We note that if a straight line trajectory is employed $v = v_0$, $\cos\delta = v_0 t/R$ and $\sin\delta = b/R$. In terms of ξ, μ, and ϕ,

$$\mathbf{v}\cdot\mathbf{s}_g = vR[\xi\mu\cos\delta + \{(\xi^2-1)(1-\mu^2)\}^{\frac{1}{2}}\sin\delta\cos\phi] \tag{E.35}$$

$$I(100, 100) = \frac{R}{2}\int_1^\infty d\xi \int_{-1}^{+1} d\mu \int_0^{2\pi} d\phi \exp\Bigg[-\xi(\lambda_A+\lambda_B)\frac{R}{2} - \mu(\lambda_A-\lambda_B)\frac{R}{2}$$
$$+ iv\frac{R}{2}\{\xi\mu\cos\delta + \{(\xi^2-1)(1-\mu^2)\}^{\frac{1}{2}}\sin\delta\cos\phi\Bigg]. \tag{E.36}$$

Using Bessel's integral

$$J_0(\alpha) = \frac{1}{2\pi}\int_0^{2\pi} e^{i\alpha\cos\phi}\, d\phi \tag{E.37}$$

the integration over ϕ can be carried and $I(100, 100)$ reduces to

$$I(100, 100) = \pi R \int_1^\infty d\xi \int_{-1}^{+1} d\mu$$
$$\times \exp\left[-\xi(\lambda_A+\lambda_B)\frac{R}{2} - \mu(\lambda_A-\lambda_B)\frac{R}{2} + iv\frac{R}{2}\cos\delta\xi\mu\right]$$
$$\times J_0\left\{\frac{vR}{2}(\lambda^2-1)^{\frac{1}{2}}(1-\mu^2)^{\frac{1}{2}}\sin\delta\right\}. \tag{E.38}$$

The integration over μ can be effected by employing the result:

$$\int_{-1}^{+1} e^{iT\cos\chi\mu} J_0\{T\sin\chi(1-\mu^2)^{\frac{1}{2}}\} P_l(\mu)\, d\mu = 2i^l P_l(\cos\chi) j_l(T). \tag{E.39}$$

In the present case

$$T\cos\chi = \frac{R}{2}[v\xi\cos\delta + i(\lambda_A-\lambda_B)]$$

$$T\sin\chi = \frac{R}{2}v(\xi^2-1)^{\frac{1}{2}}\sin\delta \tag{E.40}$$

and we obtain, with $l = 0$,

$$I(100, 100) = 2\pi R\int_1^\infty e^{-\xi(\lambda_A+\lambda_B)R/2} j_0(T)\, d\xi \tag{E.41}$$

where from (E.40)

$$T^2 = \tfrac{1}{4}R^2[v^2(\xi^2 - \sin^2\delta) - (\lambda_A - \lambda_B)^2 + 2i(\lambda_A - \lambda_B)v\xi\cos\delta] \quad \text{(E.42)}$$

The one-dimensional integral (E.37) must be computed numerically.

The integrals $I(n_A00, n_B00)$ can also be expressed in terms of one-dimensional integrals over ξ by operating with $(-\partial/\partial\lambda_A)^{n_A-1}(-\partial/\partial\lambda_B)^{n_B-1}$. The integrals can be simplified by using the recursion relations for spherical Bessel functions. When l_A or l_B are non-zero the angular functions are more easily introduced by working in the body-fixed coordinate frame, in which case the components of the vectors **x** and **r** can be expressed in terms of ξ, μ, and ϕ using the results

$$\cos\theta_A = \frac{\mathbf{x}\cdot\mathbf{r}}{xR} = \frac{\xi\mu + 1}{\xi + \mu}$$
$$\cos\theta_B = \frac{\mathbf{r}\cdot\mathbf{R}}{rR} = \frac{\xi\mu - 1}{\xi + \mu}. \quad \text{(E.43)}$$

Having introduced the angular factors expressed in terms of ξ, μ, and ϕ, a reduction to one-dimensional integrals is again possible, although the resulting formulae are very complicated, and it is often preferable to integrate over both ξ and μ numerically. Having computed matrix elements in the body-fixed frame, the appropriate rotation can be made to obtain the corresponding matrix elements in the space-fixed frame.

Two-electron matrix elements For systems of two or more electrons the basis functions can be expressed in terms of products of one-electron functions which in turn are combinations of Slater functions of the form (E.1). With one exception, all the integrals required can be reduced to products of integrals of one-electron form. The exceptional case is the two-electron two-centre exchange integral containing the Coulomb interaction between two electrons. For basis functions with zero angular momentum the integrals required are of the form

$$M(n_An_B, n'_An'_B) = \int d\mathbf{s}_{g_1} \int d\mathbf{s}_{g_2}\, e^{-\lambda_C x_2}\, e^{-\lambda_D r_2} \frac{e^{i\mathbf{v}\cdot(\mathbf{s}_{g1}-\mathbf{s}_{g2})}}{|\mathbf{s}_{g_1} - \mathbf{s}_{g_2}|} e^{-\lambda_A x_1}\, e^{-\lambda_B r_1}$$
$$\times\, x_1^{n_A-2} r_1^{n_B-2} x_2^{n_C-2} r_2^{n_D-2} \quad \text{(E.44)}$$

where $\mathbf{s}_{g_1}$, $\mathbf{s}_{g_2}$ are the position vectors of the two electrons with respect to the mid-point of the interionic line AB and

$$\mathbf{x}_i = \mathbf{s}_{g_i} - \tfrac{1}{2}\mathbf{R}, \qquad \mathbf{r}_i = \mathbf{s}_{g_i} + \tfrac{1}{2}\mathbf{R}, \qquad i = 1, 2. \quad \text{(E.45)}$$

The most direct way of evaluating these integrals (Green *et al.* 1965) is to employ spheroidal coordinates, as in Method C for one-electron integrals, and to use the Neumann expansion of $|\mathbf{s}_{g_1} - \mathbf{s}_{g_2}|^{-1}$, namely

$$\frac{1}{|\mathbf{s}_{g_1} - \mathbf{s}_{g_2}|} = \frac{4}{R} \sum_{L=0}^{\infty} \sum_{m=-L}^{L} v_L^m P_l^{|m|}(\xi_<) Q_l^{|m|}(\xi_>) P_l^{|m|}(\mu_1) P_l^{|m|}(\mu_2)\, \mathrm{e}^{im(\phi_1 - \phi_2)} \tag{E.46}$$

where

$$v_L^m = (-1)^m \frac{(l - |m|)!}{(l + |m|)!}. \tag{E.47}$$

In (E.42), the spheroidal coordinates of electrons 1 and 2 are (ξ_1, μ_1, ϕ_1) and (ξ_2, μ_2, ϕ_2), respectively, and $\xi_<, \xi_>$ are the lesser and greater of ξ_1 and ξ_2.

After expressing $\mathbf{v}\cdot(\mathbf{s}_{g_1} - \mathbf{s}_{g_2})$ in a similar form to (E.31), it is found that the integrals over ϕ_1 and ϕ_2 are of the form

$$\int_0^{2\pi} \mathrm{d}\phi\, \mathrm{e}^{\pm \mathrm{i}(\phi - q\cos\phi)} = (2\pi)(-1)^m i^{|m|} J_{|m|}(q) \tag{E.48}$$

while the integrals over μ_1 and μ_2 can be achieved with a generalization of (E.35), that is

$$\int_{-1}^{+1} \mathrm{e}^{\mathrm{i}T\cos\chi\mu} J_m\{T \sin\chi(1 - \mu^2)^{\frac{1}{2}}\} P_l^m(\mu)\, \mathrm{d}\mu = 2\mathrm{i}^l P_l^m(\cos\chi) j_l(T). \tag{E.49}$$

The resulting expression for $M(11\ 11)$ is a *double* integral over ξ_1 and ξ_2, the integrand of which contains a double sum of the form $(\sum_{l=0}^{\infty} \sum_{M=0}^{L})$. Notwithstanding the relatively rapid convergence of the series over L, the computation of even the basic integral $M(11\ 11)$ is very time consuming, since the double integration must be evaluated numerically. It is almost certainly more rapid to represent the Slater functions in terms of Gaussian functions, after which the methods of Appendix F can be employed.

APPENDIX F

MATRIX ELEMENTS WITH A GAUSSIAN BASIS

Although basis sets of Gaussian functions had been shown by Boys (1950) to be efficient in solving molecular structure problems, the first use of Gaussian orbitals together with plane-wave translational factors in charge exchange appears to date from the work of Dose and Semini (1974). In this approach the one-electron basis functions $\phi_j(\mathbf{r})$ and $\chi_k(\mathbf{x})$ are expressed as linear combinations of functions of the general form

$$G_{plm}(\lambda, \mathbf{r}) = r^{2p+l} \exp(-\lambda r^2) Y^{\pm}_{l|m|}(\hat{\mathbf{r}}). \tag{F.1}$$

The great advantage of these functions is that all the matrix elements encountered, including the difficult two-electron exchange term, can be evaluated analytically. The corresponding disadvantage is that it is difficult to represent the long-range parts of atomic wave functions by Gaussians. For example, to obtain reasonable accuracy Dose and Semini required four functions to represent the ground-state wave function of atomic hydrogen. This in turn implies that the basis sets must be larger than for those composed of Slater functions and that the number of integrals required is also larger.

Formation of a Gaussian basis

There are a number of ways in which a Gaussian basis can be obtained. Perhaps the best method is to diagonalize the atomic Hamiltonians centred on the heavy particles A and B in terms of a primitive Gaussian set to obtain wave functions representing both real states and pseudostates. The screening parameters λ in the primitive Gaussian set have to be pre-chosen, and how this should be done has been discussed by Gramlich *et al.* (1989).

An alternative procedure to the direct diagonalization of these internal Hamiltonians is to first obtain a Slater basis and subsequently to represent each atomic orbital in terms of Gaussians, either by least squares fitting or by using the transforms

$$\frac{e^{-\alpha r}}{r} = \left(\frac{1}{\pi}\right)^{\frac{1}{2}} \int_0^\infty \lambda^{-\frac{1}{2}} \exp\left(-\frac{\alpha^2}{4\lambda} - \lambda r^2\right) d\lambda \tag{F.2}$$

and

$$r^{n-1} e^{-\alpha r} = \left(\frac{1}{\pi}\right)^{\frac{1}{2}} \int_0^\infty \lambda^{-\frac{1}{2}} \left(-\frac{\partial}{\partial \alpha}\right)^n \exp\left(-\frac{\alpha^2}{4\lambda} - \lambda r^2\right) d\lambda. \tag{F.3}$$

One-electron integrals

Let us consider a basis of terms G_{lm} defined by (F.1) with $p = 0$ for a one-electron system in which the interactions are Coulomb. The case of effective potentials V_{Ae}, V_{Be} will be discussed a little later. There are three types of integrals: overlap, kinetic energy, and potential energy.

Overlap integrals The general form is

$$I(l_{\mathrm{A}}p, l_{\mathrm{B}}q) = \int (x_1^i x_2^j x_3^k)\, \mathrm{e}^{\mathrm{i}\mathbf{v}\cdot\mathbf{s}_{\mathrm{g}}}\, \mathrm{e}^{-\lambda_{\mathrm{A}}x^2}\, (r_1^u r_2^v r_3^w)\, \mathrm{e}^{-\lambda_{\mathrm{B}}r^2}\, \mathrm{d}\mathbf{s}_{\mathrm{g}} \tag{F.4a}$$

with

$$\mathbf{x} = \mathbf{s}_{\mathrm{g}} + \frac{\mathbf{R}}{2}; \qquad \mathbf{r} = \mathbf{s}_{\mathrm{g}} - \frac{\mathbf{R}}{2} \tag{F.4b}$$

where (see Appendix E) x_α and r_β are the components of $\mathbf{x}$ and $\mathbf{r}$, $i + j + k = l_{\mathrm{A}}$ and $u + v + w = l_{\mathrm{B}}$, while p and q stand collectively for ijk and uvw, respectively. Introducing a basic integral

$$J(\mathbf{a}, \mathbf{b}) = \int \mathrm{e}^{-\lambda_{\mathrm{A}}x^2}\, \mathrm{e}^{-\lambda_{\mathrm{B}}r^2}\, \mathrm{e}^{\mathrm{i}\{\mathbf{v}\cdot\mathbf{s}_{\mathrm{g}} + \mathbf{a}\cdot\mathbf{x} + \mathbf{b}\cdot\mathbf{r}\}}\, \mathrm{d}\mathbf{s}_{\mathrm{g}} \tag{F.5}$$

the general overlap integral (F.4) can be obtained by parametric differentiation

$$\begin{aligned} I(l_{\mathrm{A}}p, l_{\mathrm{B}}q) = \lim_{\substack{\mathbf{a}\to 0\\ \mathbf{b}\to 0}} &\left(-\mathrm{i}\frac{\partial}{\partial a_1}\right)^i \left(-\mathrm{i}\frac{\partial}{\partial a_2}\right)^j \left(-\mathrm{i}\frac{\partial}{\partial a_3}\right)^k \\ &\times \left(-\mathrm{i}\frac{\partial}{\partial b_1}\right)^u \left(-\mathrm{i}\frac{\partial}{\partial b_2}\right)^v \left(-\mathrm{i}\frac{\partial}{\partial b_3}\right)^w J(\mathbf{a}, \mathbf{b}). \end{aligned} \tag{F.6}$$

The basic integral $J(\mathbf{a}, \mathbf{b})$ can be expressed as

$$J(\mathbf{a}, \mathbf{b}) = \int \exp[-As_{\mathrm{g}}^2 - 2\mathbf{B}\cdot\mathbf{s}_{\mathrm{g}} - C]\, \mathrm{d}\mathbf{s}_{\mathrm{g}} \tag{F.7}$$

where

$$A = \lambda_{\mathrm{A}} + \lambda_{\mathrm{B}} \tag{F.8}$$

$$2\mathbf{B} = (\lambda_{\mathrm{B}} - \lambda_{\mathrm{A}})\mathbf{R} - \mathrm{i}(\mathbf{v} + \mathbf{a} + \mathbf{b}) \tag{F.9}$$

and

$$C = \frac{(\lambda_{\mathrm{A}} + \lambda_{\mathrm{B}})R^2}{4} + \mathrm{i}\,\frac{(\mathbf{a} - \mathbf{b})\cdot\mathbf{R}}{2}. \tag{F.10}$$

By using the standard integral

$$\int_{-\infty}^{\infty} \exp(-Ax^2 - 2Bx - C)\,\mathrm{d}x = \left(\frac{\pi}{A}\right)^{\frac{1}{2}} \mathrm{e}^{-C}\,\mathrm{e}^{B^2/A} \tag{F.11}$$

to integrate over each of the Cartesian components of $\mathbf{s}_g$, it is found that

$$J(\mathbf{a}, \mathbf{b}) = \left(\frac{\pi}{A}\right)^{\frac{3}{2}} \exp\left[-C + \frac{\mathbf{B}^2}{A}\right]. \tag{F.12}$$

The differentiation indicated in (F.6) can be carried out without difficulty to obtain the integrals $I(l_A p, l_B q)$. It is indeed possible to find general expressions for I (and the other integrals to be discussed below) in the form of finite sums over certain simple auxiliary functions (Errea *et al.* 1979) or, alternatively, it is not difficult to find recursion formulae to generate the integrals for higher values of p and q. In fact the standard methods for the calculation of molecular integrals with a Gaussian basis, reviewed by Saunders (1983), can be taken over for the generation of integrals required for collision processes which contain the translation factor.

Kinetic energy integrals Integrals containing the electronic kinetic energy operator can be generated from the basic integral

$$K(\mathbf{a}, \mathbf{b}) = \int \mathrm{e}^{-\lambda_A x}\,\mathrm{e}^{\mathrm{i}\mathbf{a}\cdot\mathbf{x}}\,\mathrm{e}^{\mathrm{i}\mathbf{v}\cdot\mathbf{s}_g}[\nabla_r^2\{\mathrm{e}^{-\lambda_B r^2}\,\mathrm{e}^{\mathrm{i}\mathbf{b}\cdot\mathbf{r}}\}]\,\mathrm{d}\mathbf{s}_g. \tag{F.13}$$

From the result

$$\nabla_r^2\{\mathrm{e}^{-\lambda_B r^2}\,\mathrm{e}^{\mathrm{i}\mathbf{b}\cdot\mathbf{r}}\} = [-(6\lambda_B + b^2) - 4\lambda_B \mathrm{i}(\mathbf{b}\cdot\mathbf{r}) + 4\lambda_B r^2]\,\mathrm{e}^{-\lambda_B r^2}\,\mathrm{e}^{\mathrm{i}\mathbf{b}\cdot\mathbf{r}} \tag{F.14}$$

we see that

$$K(\mathbf{a}, \mathbf{b}) = -(6\lambda_B + b^2)J(\mathbf{a}\cdot\mathbf{b}) - 4\lambda_B b\frac{\partial}{\partial b}J(\mathbf{a}, \mathbf{b}) + 4\lambda_B\left(-\frac{\partial}{\partial\lambda_B}\right)J(\mathbf{a}, \mathbf{b}) \tag{F.15}$$

where $J(\mathbf{a}, \mathbf{b})$ is given by (F.12). The general kinetic energy integral can be obtained from $K(\mathbf{a}, \mathbf{b})$ by parametric differentiation with respect to the components of **a** and **b**.

The nuclear attraction integrals The exchange integrals containing the Coulomb interaction between the electron and the centre can be obtained from a basic integral

$$V(\mathbf{a}, \mathbf{b}) = \int \mathrm{e}^{-\lambda_A x^2}\,\mathrm{e}^{-\lambda_B r^2}\left(\frac{1}{r}\right)\mathrm{e}^{\mathrm{i}\{\mathbf{v}\cdot\mathbf{s}_g + \mathbf{a}\cdot\mathbf{x} + \mathbf{b}\cdot\mathbf{r}\}}\,\mathrm{d}\mathbf{s}_g. \tag{F.16}$$

Changing the variable of integration from $\mathbf{s}_g$ to $\mathbf{r}$, we have

$$V(\mathbf{a}, \mathbf{b}) = \int \frac{1}{r} \exp(-A'r^2 - 2\mathbf{B}' \cdot \mathbf{r} - C')\, \mathrm{d}\mathbf{r} \tag{F.17}$$

where

$$\begin{aligned} A' &= A = \lambda_A + \lambda_B \\ 2\mathbf{B}' &= -\lambda_B 2\mathbf{R} - \mathrm{i}(\mathbf{v}_0 + \mathbf{a} + \mathbf{b}) \\ C' &= \lambda_A R^2 + \mathrm{i}\left(\frac{\mathbf{v}_0 \cdot \mathbf{R}}{2} + \mathbf{a} \cdot \mathbf{R}\right). \end{aligned} \tag{F.18}$$

Integrating over angles in (F.17) we find

$$V(\mathbf{a}, \mathbf{b}) = 2\pi \int_0^\infty \frac{\sin 2B'r}{B'} \exp(-A'r^2 - C')\, \mathrm{d}r. \tag{F.19}$$

The integral over r is given by

$$\int_0^\infty \sin 2B'r\, \mathrm{e}^{-A'r^2}\, \mathrm{d}r = \frac{B'}{A'} \mathrm{e}^{-B'^2/A'^2}\, {}_1F_1\left(\frac{1}{2}, \frac{3}{2}, \frac{B'^2}{A'}\right) \tag{F.20}$$

so that

$$V(\mathbf{a}, \mathbf{b}) = \frac{2\pi}{A'} \exp\left(-C' - \frac{B'^2}{A'^2}\right) {}_1F_1\left(\frac{1}{2}, \frac{3}{2}, \frac{B'^2}{A'}\right). \tag{F.21}$$

The confluent hypergeometric function in (F.20) is related to the error function of complex argument, since

$$\begin{aligned} \Phi(z) &= \frac{2}{\sqrt{\pi}} \int_0^z \exp(-t^2)\, \mathrm{d}t \\ &= \frac{2z}{\sqrt{\pi}}\, {}_1F_1(\tfrac{1}{2}, \tfrac{3}{2}, -z^2) \end{aligned} \tag{F.22}$$

and this function can be evaluated numerically with the help of standard computer packages.

The general integrals including the angular factors can be obtained in the usual way from (F.21) by parametric differentiation with respect to the components of **a** or of **b**. As in the case of the overlap and kinetic energy integrals, explicit formulae can be deduced for the general integral and recursion relations can be derived.

Effective potentials If an effective potential between the active electron and an ion core is represented by a sum of terms of the form $\sum_{i=1}^{N} a_i r^{i-2}$

$\times \exp(-\lambda_i r)$, the nuclear attraction integrals can be evaluated by fitting the potential to Gaussian functions, or by employing the transformations (E.3).

Two-electron exchange integrals

The 'difficult' two-electron exchange integral can also be evaluated analytically if a Gaussian base is used. The basic integral is

$$T(\mathbf{a}, \mathbf{b}; \mathbf{c}, \mathbf{d}) = \int \mathrm{d}\mathbf{s}_{g_1} \int \mathrm{d}\mathbf{s}_{g_2}\, \mathrm{e}^{-\lambda_A x_1^2}\, \mathrm{e}^{-\lambda_B r_1^2}\, \mathrm{e}^{-\lambda_c x_2^2}\, \mathrm{e}^{-\lambda_d r_2^2}$$

$$\times \frac{1}{|\mathbf{s}_{g_1} - \mathbf{s}_{g_2}|}\, \mathrm{e}^{\mathrm{i}[\mathbf{v}\cdot(\mathbf{s}_{g1} - \mathbf{s}_{g2}) + \mathbf{a}\cdot\mathbf{x}_1 + \mathbf{b}\cdot\mathbf{r}_1 + \mathbf{c}\cdot\mathbf{x}_2 + \mathbf{d}\cdot\mathbf{r}_2]}, \quad \text{(F.23a)}$$

with

$$\mathbf{x}_i = \mathbf{s}_{g_i} - \frac{\mathbf{R}}{2}; \qquad \mathbf{r}_i = \mathbf{s}_{g_i} + \frac{\mathbf{R}}{2}. \qquad \text{(F.23b)}$$

The general integral containing the components of $\mathbf{x}_1$, $\mathbf{x}_2$; $\mathbf{r}_1$, $\mathbf{r}_2$ can be found by differentiating with respect to the components of **a**, **b**, **c**, and **d**. By employing the Fourier transform

$$\frac{1}{|\mathbf{s}_{g_1} - \mathbf{s}_{g_2}|} = \frac{1}{2\pi^2} \int \mathrm{e}^{\mathrm{i}\mathbf{k}\cdot(\mathbf{s}_{g1} - \mathbf{s}_{g2})}\, k^{-2}\, \mathrm{d}\mathbf{k}, \qquad \text{(F.24)}$$

into the integrand of (F.23), and using the result (F.12), we find immediately that

$$T(\mathbf{a}, \mathbf{b}, \mathbf{c}, \mathbf{d}) = \frac{1}{2\pi^2} \int \frac{1}{k^2} \left(\frac{\pi^2}{A_1 A_2}\right) \exp\left[-C_1 - C_2 + \left(\mathbf{B}_1 - \frac{\mathrm{i}\mathbf{k}}{2}\right)^2 + \left(\mathbf{B}_2 + \frac{\mathrm{i}\mathbf{k}}{2}\right)^2\right] \mathrm{d}\mathbf{k} \qquad \text{(F.25)}$$

where

$$\begin{aligned}
A_1 &= \lambda_A + \lambda_B, \qquad A_2 = \lambda_C + \lambda_D \\
2\mathbf{B}_1 &= (\lambda_B - \lambda_A)\mathbf{R} - \mathrm{i}(\mathbf{v} + \mathbf{a} + \mathbf{b}) - \mathrm{i}\mathbf{k} \\
2\mathbf{B}_2 &= (\lambda_D - \lambda_C)\mathbf{R} + \mathrm{i}(\mathbf{v} - \mathbf{c} - \mathbf{d}) + \mathrm{i}\mathbf{k} \\
C_1 &= \frac{(\lambda_A + \lambda_B)R^2}{4} + \mathrm{i}\,\frac{(\mathbf{a} - \mathbf{b})\cdot R}{2} \\
C_2 &= \frac{(\lambda_C + \lambda_D)R^2}{4} + \mathrm{i}\,\frac{(\mathbf{c} - \mathbf{d})\mathbf{R}}{2}.
\end{aligned} \qquad \text{(F.26)}$$

Extracting the remaining integral over **k**:

$$T(\mathbf{a}, \mathbf{b}, \mathbf{c}, \mathbf{d}) = \left(\frac{\pi^2}{A_1 A_2}\right)^{\frac{3}{2}} \exp\left(-C_1 - C_2 + \frac{\mathbf{B}_1^2}{A_1} + \frac{\mathbf{B}_2^2}{A_2}\right)\mathscr{I}, \tag{F.27}$$

where

$$\mathscr{I} = \frac{1}{2\pi^2}\int \mathrm{d}\mathbf{k}\, \frac{1}{k^2} \exp(-\alpha k^2 + \mathrm{i}\boldsymbol{\beta}\cdot\mathbf{k}), \tag{F.28}$$

with

$$\alpha = \frac{1}{4A_1} + \frac{1}{4A_2}, \qquad \boldsymbol{\beta} = \mathbf{B}_2 - \mathbf{B}_1. \tag{F.29}$$

The integral $\mathscr{I}$ is easily evaluated. We have, integrating over angles,

$$\mathscr{I} = \frac{2}{\pi}\int_0^\infty \mathrm{e}^{-\alpha k^2} \frac{\sin \beta k}{\beta k}\,\mathrm{d}k, \tag{F.30}$$

and using the standard result

$$\int_0^\infty \mathrm{e}^{-\alpha k^2} k^{-1} \sin \beta k \,\mathrm{d}k = \frac{\pi}{2}\Phi\left(\frac{\beta}{2\alpha}\right), \tag{F.31}$$

we find that

$$\mathscr{I} = \frac{1}{\beta}\Phi\left(\frac{\beta}{2\alpha}\right). \tag{F.32}$$

Employing (F.32) in equation (F.27) completes the evaluation of $T(\mathbf{a}, \mathbf{b}, \mathbf{c}, \mathbf{d})$. Once again a general formula for the integrals with non-zero angular momentum can be obtained.

APPENDIX G

GREEN'S FUNCTIONS IN THE IMPACT PARAMETER MODEL

The problem is to construct integral equations for the solutions of the time-dependent Schrödinger equation (10.6). Let us write equation (10.6) in the form

$$\left(\mathrm{i}\frac{\partial}{\partial t} - H_{\mathrm{B}}\right)\tilde{\psi}_i(\mathbf{r}, t) = U_{\mathrm{A}}\tilde{\psi}_i(\mathbf{r}, t) \tag{G.1}$$

where H_{B} is defined in (10.9) and U is given by

$$U_{\mathrm{A}}(\mathbf{r}, t) = V_{\mathrm{Ae}}(x) + \frac{Z_{\mathrm{A}}}{R(t)}. \tag{G.2}$$

A formal solution of (G.1) can be written as

$$\tilde{\psi}_i = \Phi_i^{\mathrm{B}} + g_{\mathrm{B}}U_{\mathrm{A}}\tilde{\psi}_i \tag{G.3}$$

where Φ_i^{B} is one of the unperturbed functions satisfying (10.8) and defined by (10.12) and (10.13), while g_{B} is the inverse operator

$$g_{\mathrm{B}} = \frac{1}{\mathrm{i}\dfrac{\partial}{\partial t} - H_{\mathrm{B}}}. \tag{G.4}$$

In configuration space the unit operator $\delta(\mathbf{r} - \mathbf{r}')\delta(t - t')$ can be expressed in terms of the complete set of functions $\Phi_j^{\mathrm{BT}}(\mathbf{r}, t)$, which are also solutions of (10.8), and which are defined by (10.15) and (10.16). We have

$$\delta(\mathbf{r} - \mathbf{r}')\delta(t - t') = \frac{1}{2\pi}\sum_k \Phi_k^{\mathrm{BT}}(\mathbf{r}, t)[\Phi_k^{\mathrm{BT}}(\mathbf{r}', t')]^* \int_{-\infty}^{\infty} \mathrm{e}^{-\mathrm{i}\omega\tau}\,\mathrm{d}\omega \tag{G.5}$$

where $\tau = t - t'$ and the sum runs over both the discrete and continuum eigenfunctions of H_{B}. Thus $g_{\mathrm{B}}U\tilde{\psi}_i$ can be written explicitly as

$$g_{\mathrm{B}}U\tilde{\psi}_i = \int \mathrm{d}\mathbf{r}' \int_{-\infty}^{\infty} \mathrm{d}t' g_{\mathrm{B}}(\mathbf{r}, t; \mathbf{r}', t')U(r', t')\tilde{\psi}_i(\mathbf{r}', t') \tag{G.6}$$

where

$$g_{\mathrm{B}}(\mathbf{r}, t; \mathbf{r}', t') = \frac{1}{2\pi}\int_{-\infty}^{\infty} \sum_k \frac{1}{\mathrm{i}\dfrac{\partial}{\partial t} - H_{\mathrm{B}}} \Phi_k^{\mathrm{BT}}(\mathbf{r}, t)[\Phi_k^{\mathrm{BT}}(\mathbf{r}', t')]^* \,\mathrm{e}^{-\mathrm{i}\omega\tau}\,\mathrm{d}\omega$$

$$= \frac{1}{2\pi}\int_{-\infty}^{\infty} \sum_k \frac{\mathrm{e}^{-\mathrm{i}\omega\tau}}{\omega} \Phi_k^{\mathrm{BT}}(\mathbf{r}, t)[\Phi_k^{BT}(\mathbf{r}', t')]^* \,\mathrm{d}\omega. \tag{G.7}$$

The integral over ω can be written as

$$I^{\pm}(\tau) = \lim_{\varepsilon \to 0^+} \int_{-\infty}^{\infty} \frac{\mathrm{e}^{-\mathrm{i}\omega\tau}}{\omega \pm \mathrm{i}\varepsilon}\,\mathrm{d}\omega. \tag{G.8}$$

The addition of $\pm \mathrm{i}\varepsilon$ to the denominator is to avoid the singularity at $\omega = 0$. Corresponding to the choice $\omega \to \omega + \mathrm{i}\varepsilon$ or $\omega \to \omega - \mathrm{i}\varepsilon$ two different Green's functions can be obtained, which will be denoted by g_{B}^+ and g_{B}^-, respectively.

To evaluate $I^+(\tau)$, the integration along the real axis can be closed by an infinite semicircle in the lower half-plane if $\tau > 0$ and the upper half-plane if $\tau < 0$. Since $\exp(-\mathrm{i}\omega\tau)$ vanishes on these semicircles, we have from Cauchy's theorem

$$I^+(\tau) = \begin{cases} -2\pi\mathrm{i} & \tau > 0 \\ 0 & \tau < 0 \end{cases} \tag{G.9}$$

because for $\tau > 0$ the pole at $\omega = -\mathrm{i}\varepsilon$ is encircled, while for $\tau < 0$ the pole is excluded from the contour. Proceeding in a similar manner, we find

$$I^-(\tau) = \begin{cases} 0 & \tau > 0 \\ 2\pi\mathrm{i} & \tau < 0. \end{cases} \tag{G.10}$$

Using these results in (G.7), we have†

$$g_{\mathrm{B}}^{\pm}(\mathbf{r}', t'; \mathbf{r}, t) = \mp \mathrm{i}\theta(\pm\tau) \sum_k \Phi_k^{\mathrm{BT}}(\mathbf{r}, t)[\Phi_k^{\mathrm{BT}}(\mathbf{r}', t')]^*. \tag{G.11}$$

Because of the step functions, $\theta(\pm\tau)$, $g_{\mathrm{B}}^+ \to 0$ as $t \to -\infty$ and $g_{\mathrm{B}}^- \to 0$ as $t \to \infty$. It follows that a solution of (G.1) satisfying the boundary condition,

$$\tilde{\psi}_i(\mathbf{r}, t) \to \Phi_i^{\mathrm{B}}(\mathbf{r}, t), \qquad \text{as } t \to -\infty, \tag{G.12}$$

† We have chosen to write $g_{\mathrm{B}}^{\pm}$ in terms of the time-reversed functions Φ_j^{BT} but, as the derivation makes clear, it is equally possible to employ the functions Φ_j^{B}, so that $g_{\mathrm{B}}^{\pm}(\mathbf{r}, t'; \mathbf{r}', t) = \mp \mathrm{i}\theta(\pm\tau)$ $\sum_k \Phi_k^{\mathrm{B}}(\mathbf{r}, t)\,[\Phi_k^{\mathrm{B}}(\mathbf{r}', t')]^*$.

is

$$\tilde{\psi}_i = \Phi_i^{\rm B} + g_{\rm B}^{+} U_{\rm A} \tilde{\psi}_i$$

$$= \Phi_i^{\rm B} + \lim_{\varepsilon \to 0^+} \frac{1}{{\rm i}\dfrac{\partial}{\partial t} + {\rm i}\varepsilon - H_{\rm B}} U_{\rm A} \tilde{\psi}_i. \tag{G.13}$$

The corresponding time-reversed equation to (G.13) is

$$\tilde{\psi}_i^{\rm T} = \Phi_i^{\rm BT} + g_{\rm B}^{-} U_{\rm A} \tilde{\psi}_i^{\rm T}, \tag{G.14}$$

and $\tilde{\psi}_i^{\rm T}$ satisfies the boundary condition

$$\tilde{\psi}_i^{\rm T} \to \Phi_i^{\rm BT}, \qquad \text{as } t \to +\infty. \tag{G.15}$$

The expression for the Green's functions $g_{\rm A}^{-}$ given in (10.49), and the relation $g_{\rm A}^{+} = g_{\rm A}^{-\dagger}$, can be obtained in exactly the same way as for $g_{\rm B}^{\pm}$. In operator form

$$g_{\rm A}^{\pm} = \lim_{\varepsilon \to 0^+} \frac{1}{{\rm i}\dfrac{\partial}{\partial t} \pm {\rm i}\varepsilon - H_{\rm A}}. \tag{G.16}$$

The explicit form of $g_{\rm A}^{\pm}$ is

$$g_{\rm A}^{\pm}(\mathbf{r}, t; r', t') = \mp {\rm i}\theta(\tau) \sum_m {\rm X}_m^{\rm AT}(\mathbf{r}, t)[{\rm X}_m^{\rm A}(\mathbf{r}'t')]^*. \tag{G.17}$$

The corresponding integral equations for $\tilde{\psi}_n^{\rm T}$ is

$$\tilde{\psi}_n^{\rm (T)} = {\rm X}_n^{\rm A(T)} + g_{\rm A}^{+(-)} U_{\rm B} \tilde{\psi}_n^{\rm (T)} \tag{G.18}$$

where

$$U_{\rm B} = V_{\rm Be} + \frac{Z_{\rm B}}{R}. \tag{G.19}$$

Integral equations for $g_{\rm A}^{\pm}$ and $g_{\rm B}^{\pm}$

It is often useful to find integral equations satisfied by Green's functions of the kind we have been considering. Let us take the case of the functions $g_{\rm B}^{\pm}$

$$g_{\rm B}^{\pm} = \lim_{\varepsilon \to 0^+} \frac{1}{{\rm i}\dfrac{\partial}{\partial t} \pm {\rm i}\varepsilon - H_{\rm B}}. \tag{G.20}$$

If V_0 is an arbitrary potential a new Green's function can be defined as

$$\mathscr{G}_{\rm B}^{\pm} = \lim_{\varepsilon \to 0^+} \frac{1}{{\rm i}\dfrac{\partial}{\partial t} \pm {\rm i}\varepsilon - H_{\rm B} + V_0}. \tag{G.21}$$

Applying the operator identity

$$\frac{1}{P} = \frac{1}{Q} + \frac{1}{Q}(Q - P)\frac{1}{P} \tag{G.22}$$

by setting $P = \mathrm{i}(\partial/\partial t) \pm \mathrm{i}\varepsilon - H_{\mathrm{B}}$ and $Q = (\partial/\partial t) \pm \mathrm{i}\varepsilon - H_{\mathrm{B}} - V_0$, $g_{\mathrm{B}}^{\pm}$ is found to satisfy the following integral equation

$$g_{\mathrm{B}}^{\pm} = \mathscr{G}_{\mathrm{B}}^{\pm} + \mathscr{G}_{\mathrm{B}}^{\pm} V_0 g_{\mathrm{B}}^{\pm}. \tag{G.23}$$

By iteration, a series expansion for $g_{\mathrm{B}}^{\pm}$ can be obtained as

$$g_{\mathrm{B}}^{\pm} = \mathscr{G}_{\mathrm{B}}^{\pm} + \mathscr{G}_{\mathrm{B}}^{\pm} V_0 \mathscr{G}_{\mathrm{B}}^{\pm} + \mathscr{G}_{\mathrm{B}}^{\pm} V_0 \mathscr{G}_{\mathrm{B}}^{\pm} V_0 \mathscr{G}_{\mathrm{B}}^{\pm} + \cdots. \tag{G.24}$$

This is, of course, a formal manipulation and convergence of the series is by no means guaranteed; each case requires separate investigation.

If the choice $V_0 = V_{\mathrm{Be}}$ is made, $\mathscr{G}_{\mathrm{B}}^{\pm}$ reduces to the free-particle Green's function, modified by the Coulomb phase due to the interaction $Z_{\mathrm{A}}/R(t)$. We have

$$\mathscr{G}_{\mathrm{B}}^{\pm} \to g_0^{\pm} = \lim_{\varepsilon \to 0^+} \frac{1}{\mathrm{i}\dfrac{\partial}{\partial t} \pm \mathrm{i}\varepsilon - H_0} \tag{G.25}$$

with $H_0 = -\frac{1}{2}\nabla_r^2 - Z_{\mathrm{A}}/R$. The eigenfunctions of H_0 are plane waves modified by a Coulomb phase, and the explicit form of $g_0^{\pm}$ is

$$g_0^{\pm}(\mathbf{r}, t; \mathbf{r}'t') = \mp \mathrm{i}\theta(\pm\tau)(2\pi)^{-3} \int \mathrm{e}^{\mathrm{i}\mathbf{k}\cdot(\mathbf{r}-\mathbf{r}')}\, \mathrm{e}^{-\mathrm{i}k^2(t-t')/2 - \mathrm{i}[\sigma(t)-\sigma(t')]}\, \mathrm{d}\mathbf{k} \tag{G.26}$$

where $\sigma(t)$ is defined by (10.14). The integration over $\mathbf{k}$ can be performed (Joachain 1975, p. 301) and $g_0^{\pm}$ obtained in the closed form:

$$g_0^{\pm}(\mathbf{r}, t; \mathbf{r}', t') = \mp \mathrm{i}\theta(\pm\tau)\left[\frac{1}{2\pi\mathrm{i}|\tau|}\right]^{\frac{3}{2}} \exp\left[\frac{\mathrm{i}R^2}{2|\tau|}\right] \exp[-\mathrm{i}\{\sigma(t) - \sigma(t')\}] \tag{G.27}$$

where $R = |\mathbf{r} - \mathbf{r}'|$.

The Green's function $g_{\mathrm{B}}^{\pm}$ can be expressed in terms of $g_0^{\pm}$, as

$$\begin{aligned} g_{\mathrm{B}}^{\pm} &= g_0^{\pm} + g_0^{\pm} V_{\mathrm{B}} g_{\mathrm{B}}^{\pm} \\ &= g_0^{\pm} + g_0^{\pm} V_{\mathrm{B}} g_0^{\pm} + g_0^{\pm} V_{\mathrm{B}} \mathrm{g}_0^{\pm} V_{\mathrm{B}} g_0^{\pm} + \cdots. \end{aligned} \tag{G.28}$$

A similar expansion can be written down for $g_{\mathrm{A}}^{\pm}$.

SELECTED BIBLIOGRAPHY OF BOOKS AND REVIEW ARTICLES

Anholt, R. and Gould, H. (1985). Relativistic heavy ion–atom collisions. *Adv. At. Mol. Phys.*, **22**, 315.

Basu, P., Mukherjee, S. C., Sural, D. P. (1979). Electron capture processes in ion–atom collisions. *Physics Repts.*, **42C**, 145.

Belkić, Dž., Gayet, R., and Salin, A. (1979). Electron capture in high energy ion–atom collisions. *Phys. Repts.*, **56**, 279.

Bransden, B. H. (1972). The theory of charge exchange. *Repts. Prog. Phys.*, **35**, 949.

Bransden, B. H. (1979). The theory of fast heavy particle collisions. *Adv. At. Mol. Phys.*, **15**, 263.

Bransden, B. H. (1983). *Atomic collision theory*, (2nd edn). Benjamin Cummings, Reading, Mass.

Bransden, B. H. and Dewangan, D. P. (1988). High energy charge transfer. *Adv. At. Mol. Phys.*, **25**, 343.

Bransden, B. H. and Janev, R. K. (1983). Electron capture in collisions of hydrogen atoms with fully stripped ions. *Adv. At. Mol. Phys.*, **19**, 1.

Bransden, B. H. and Joachain, C. J. (1983). *Physics of atoms and molecules.* Longman, London.

Briggs, J. S. (1976). The excitation of inner shells in slow atomic collisions. *Repts. Prog. Phys.*, **39**, 217.

Child, M. (1974). *Molecular collision theory.* Academic Press, New York.

Crothers, D. S. F. (1981). Non adiabatic charge transfer. *Adv. At. Mol. Phys.*, **17**, 55.

Eichler, J. (1990). Theory of relativistic ion–atom collisions. *Phys. Repts.*, **193**, 165.

Fritsch, W. and Lin, C. D. (1991). The semiclassical close-coupling description of atomic collisions: Recent developments and results. *Phys. Repts.*, **202**, 1.

Gilbody, H. B. (1986). Measurements of charge transfer and ionization in collisions involving hydrogen atoms. *Adv. At. Mol. Phys.*, **22**, 18.

Janev, R. K. and Presnyakov, L. P. (1981). Collision processes of multiply charged ions with atoms. *Phys. Repts.*, **70**, 1.

Janev, R. K. and Winter, H. (1978). State-selection electron capture in atom–highly charged ion collisions. *Phys. Repts.*, **117**, 267.

Janev, R. K., Presnyakov, L. P., and Shevelko, V. P. (1985). *Physics of highly charged atoms.* Springer-Verlag, Berlin.

Joachain, C. J. (1983). *Quantum collision theory* (3rd edn). North-Holland, Amsterdam.

Kimura, N. and Lane, N. F. (1989). The low energy, heavy particle collisions – a close coupling treatment. *Adv. At. Mol. Phys.*, **26**, 79.

Lin, C. D. and Richard, P. (1981). Inner shell vacancy production in ion–atom collisions. *Adv. At. Mol. Phys.*, **17**, 273.

Mapleton, R. A. (1972). *Theory of charge exchange.* Wiley, New York.

Massey, H. S. W. and Gilbody, H. B. (1974). *Electronic and ionic impact phenomena vol. 4 : recombination and fast collisions of heavy particles.* Oxford University Press.

Moiseiwitsch, B. L. (1989). Electron capture at relativistic energies. *Adv. At. Mol. Phys.*, **36**, 51.

Nikitin, E. E. and Umanskii, S. Ya. (1984). *Theory of slow atomic collisions.* Springer-Verlag, Berlin.

Pollack, E. and Hahn, Y. (1985). Electron capture by simple ions. *Adv. At. Mol. Phys.*, **22**, 243.

Shakeshaft, R. and Spruch, L. (1979). Mechanisms for charge transfer (or for capture of any light particle) at asymptotically high impact velocities. *Rev. Mod. Phys.* **51**, 369.

REFERENCES

Aaron, R., Amado, R. D., and Lee, B. W. (1961). *Phys. Rev.*, **121**, 319.

Abramov, V. A., Baryshnikov, F. F., and Lisitsa, V. S. (1979). *Soviet Phys. JETP*, **46**, 469.

Abrines, R. and Percival, I. C. (1966). *Proc. Phys. Soc.*, **88**, 861, 873.

Afrosimov, V. V., Leiko, G. A., and Panov, M. N. (1975). *Sov. Phys. JETP*, **40**, 661.

Allan, R. J. (1986). *J. Phys. B*, **19**, 321.

Allan, R. J. and Hanssen, J. (1985). *J. Phys. B*, **18**, 1981.

Allison, S. K. (1958). *Rev. Mod. Phys.*, **30**, 1137.

Alston, S. (1983). *Phys. Rev. A*, **33**, 2959.

Alston, S. (1989*a*). *Nucl. Inst. and Meths. B*, **43**, 19.

Alston, S. (1989*b*). *Phys. Rev. A*, **40**, 4907.

Amundsen, P. A. and Aashamer, K. (1981). *J. Phys. B*, **14**, 4047.

Anderson, D. G. M. and Antal, J. (1975). *J. Phys. B*, **8**, 1507.

Anderson, D. G. M., Antal, M. J., and McElroy, M. B. (1974). *J. Phys. B*, **7**, L118.

Anderson, L. H., Hvelplund, P., Knudsen, H., Moller, S. P., Elsener, K., Rensfelt, K. G., and Uggerhoj, E. (1986). *Phys. Rev. Lett.* **57**, 2147.

Angel, G. C., Sewell, E. C., Dunn, K. F., and Gilbody, H. B. (1978). *J. Phys. B*, **11**, L297.

Anholt, R. (1985). *Phys. Rev.*, A**31**, 3579.

Anholt, R. and Eichler, J. (1985). *Phys. Rev.* **31**, 3505.

Anholt, R. and Gould, H. (1986). *Adv. Atom. Mol. Phys.*, **22**, 315.

Anholt, R., Meyerhof, W. E., Stoller, C. L., Morenzoni, E., Andriamonje, S. A., Molitoris, J. D., Baker, O. K., Hoffman, D. H. H., Bowman, H., Xu, J.-S., Xu, Z.-Z., Frankel, K., Murphy, D., Grove, K., and Rasmussen, J. O. (1984). *Phys. Rev.*, A**30**, 2234.

Ashburn, J. R., Cline, R. A., Stone, C. D., van der Burgt, B. J. M., Westerveld, W. B., and Risley, J. S. (1989). *Phys. Rev. A*, **40**, 4885.

Ashburn, J. R., Cline, R. A., Stone, C. D., van der Burgt, B. J. M., Westerveld, W. B., and Riley, J. S. (1990). *Phys. Rev. A*, **41**, 2407.

Ast, H., Lüdde, H. J., and Dreizler, R. M. (1988). *J. Phys. B*, **21**, 4143.

Austern, N., Iseri, Y., Kamimura, M., Kawai, M., Rawitscher, G., and Yahiro, M. (1987). *Phys. Repts.*, **154**, 126.

Balian, R., Encrenaz, P., and Lequeux, J. (eds) (1974). *Atomic and molecular physics and the interstellar matter.* North-Holland, Amsterdam.

Band, Y. B. (1974). *J. Phys. B*, **7**, 2557.

Bandrauk, A. D. (1972). *Mol. Phys.*, **24**, 661.

Bang, J. and Hansteen, J. M. (1959). *K. Dan. Vidensk, Selsk. Mat-Fys. Medd.*, **31**, No. 13.

Bárány, A., Astner, G., Cederquist, H., Danared, H., Hult, S., Hvelplund, P., Johnson, A., Knudsen, H., Liljeby, L., and Rensfelt, K.-G. (1985). *Nucl. Inst. and Meths*, **B9**, 397.
Bárány, A., Danard, H., Cederquist, H., Hvelplund, P., Knudsen, H., Pedersen, J. O. K., Cock, C. L., Tunnell, L. N., Waggoner, W., and Giesce, J. P. (1986). *J. Phys. B*, **19**, L427.
Barat, M. (1983). In *Fundamental processes in energetic atomic collisions* (ed. H. O. Lutz, J. S. Briggs, and H. Kleinpoppen), p. 389, Plenum, New York.
Barat, M. and Lichten, W. (1972). *Phys. Rev. A*, **6**, 211.
Bardsley, J. N. (1974). *Case Stud. At. Phys.*, **4**, 299.
Barnett, C. F. and Reynolds, H. K. (1958). *Phys. Rev.*, **109**, 358.
Bassel, R. H. and Gerjuoy, E. (1960). *Phys. Rev.*, **117**, 749.
Bassichis, W. H., Reading, J. F., and Scheerbaum, R. R. (1975). *Phys. Rev.*, C**11**, 316.
Bassichis, W. H., Strayer, M. R., Reading, J. F., and Scheerbaum, R. (1978). *Phys. Rev.*, C**18**, 632.
Bates, D. R. (1958). *Proc. Roy. Soc.*, A**247**, 294.
Bates, D. R. (1960). *Proc. Roy. Soc.*, A**257**, 22.
Bates, D. R. (1962). *Disc. Faraday Soc.*, **33**, 7.
Bates, D. R. (1988). *Proc. Roy. Soc.*, A**274**, 294.
Bates, D. R. and Boyd, A. H. (1962). *Proc. Phys. Soc.*, **80**, 1301.
Bates, D. R. and Carson, T. R. (1956). *Proc. Roy. Soc.*, A**234**, 207.
Bates, D. R. and Crothers, D. S. F. (1970). *Proc. Roy. Soc.*, A**315**, 465.
Bates, D. R. and Dalgarno, A. (1952). *Proc. Phys. Soc.*, A**65**, 919.
Bates, D. R. and Dalgarno, A. (1953). *Proc. Phys. Soc.*, A**66**, 972.
Bates, D. R. and Griffing, G. (1953). *Proc. Phys. Soc.*, A**66**, 961.
Bates, D. R. and Holt, A. R. (1966). *Proc. Roy. Phys. Soc.*, A**292**, 168.
Bates, D. R. and Reid, R. H. G. (1968). *Adv. At. Mol. Phys.*, **4**, 13.
Bates, D. R. and Lynn, N. (1959). *Proc. Roy. Soc.*, A**253**, 141.
Bates, D. R. and McCarroll, R. (1958). *Proc. Roy. Soc.*, A**245**, 175.
Bates, D. R. and McCarroll, R. (1962). *Adv. in Phys.*, **11**, 39.
Bates, D. R. and Moiseiwitsch, B. L. (1954). *Proc. Phys. Soc.*, A**67**, 805.
Bates, D. R. and Sprevak, D. (1970). *J. Phys. B*, **3**, 1483.
Bates, D. R. and Williams, D. A. (1964). *Proc. Roy. Soc.*, **83**, 425.
Bates, D. R., Johnson, H. C., and Stewart, I. (1964). *Proc. Phys. Soc.*, **84**, 517.
Baye, D. (1970). *Bull. Acad. r. Belg. Cl. Sci.*, **56**, 1274.
Baye, D. and Heenan, P.-H. (1973). *J. Phys. B*, **6**, 105.
Baye, D., Biron, F., and Demeur, M. (1970). *Bull. Acad. r. Belg. Cl. Sci.*, **56**, 1238.
Bayfield, J. E. (1969). *Phys. Rev.*, **185**, 105.
Bayfield, J. E. and Khayrallah, G. A. (1975*a*). *Phys. Rev. A*, **11**, 920.
Bayfield, J. E. and Khayrallah, G. A. (1975*b*). *Phys. Rev. A*, **12**, 869.
Becker, R. L., Ford, A. L., and Reading, J. R. (1984). *Phys. Rev. A*, **29**, 3111.
Becker, U. (1987). *J. Phys. B*, **20**, 6563.
Becker, U., Grün, N. and Scheid, W. (1987). *J. Phys. B*, **20**, 2075.
Becker, U., Grün, N., and Scheid, W. (1986). *J. Phys. B*, **19**, 1347.
Begum, S., Bransden, B. H., and Coleman, J. P. (1973). *J. Phys. B*, **6**, 837.
Belkić, Dž. (1977). *J. Phys. B*, **10**, 3075.
Belkić, Dž. (1978). *J. Phys. B*, **11**, 3529.
Belkić, Dž. (1980). *J. Phys. B*, **13**, L589.

Belkić, Dž. (1988). *Europhys. Lett.*, **7**, 323.
Belkić, Dž. and Gayet, R. (1977). *J. Phys. B*, **10**, 1911.
Belkić, Dž., Gayet, R., and Salin, A. (1979). *Phys. Repts.* **56**, 279.
Belkić, Dž., Gayet, R., Hanssen, J., and Salin, A. (1986*a*). *J. Phys. B*, **19**, 2945.
Belkić, Dž., Saini, S., and Taylor, H. S. (1986*b*). *Z. Phys. D*, **3**, 59.
Belkić, Dž., Saini, S., and Taylor, H. S. (1987*a*). *Phys. Rev. A*, **35**, 1991.
Belkić, Dž., Saini, S., and Taylor, H. S. (1987*b*). *Phys. Rev. A*, **36**, 1601.
Bell, K. L. and Kingston, A. E. (1974). *Adv. At. Mol. Phys.*, **10**, 53.
Bell, K. L. and Kingston, A. E. (1976). In *Atomic processes and applications* (ed. P. G. Burke and B. L. Moiseiwitsch), p. 493. North-Holland, Amsterdam.
Bell, K. L. and Kingston, A. E. (1978). *J. Phys. B*, **11**, 1259.
Belyaev, V. A., Brezhnev, B. G., and Erastov, E. M. (1967). *Soviet Phys. JETP*, **25**, 77.
Berkner, K. H., Pyle, A. V., Stearns, J. W., and Warren, J. C. (1968). *Phys. Rev.*, **166**, 44.
Berrington, K. A., Bransden, B. H., and Coleman, J. P. (1973). *J. Phys. B*, **6**, 436.
Berry, M. V. (1966). *Proc. Phys. Soc.*, **89**, 479.
Berry, M. V. and Mount, K. E. (1972). *Rep. Prog. Phys.*, **35**, 315.
Berson, I. Ya. (1968). *Izv. Akad. Nauk Latv. SSR, Ser. Fiz. Tekh.* **4**, 47.
Bertulani, C. A. and Baur, G. (1985). *Phys. Rept.*, **163**, 299.
Bethe, H. A. and Saltpeter, E. E. (1957). *Quantum mechanics of one- and two electron atoms.* Springer-Verlag, Berlin.
Bienstock, S., Dalgarno, A., and Heil, T. G. (1984). *Phys. Rev. A*, **29**, 2238.
Bliman, S. Hitz, D., Jacquot, B., Harel, C., and Salin, A. (1983). *J. Phys. B*, **16**, 2849.
Boellard, A. (1984). FOM report 58.248.
Bohr, N. and Lindhardt, J. (1954). *K. Dan. Vid. Sel. Mat. Phys. Medd.*, **28** No. 7.
Bottcher, C. (1982). *Phys. Rev. Lett.*, **48**, 85.
Bottcher, C. (1985). *Nucl. Inst. and Meths*, **B11**, 7.
Bottcher, C. and Strayer, M. R. (1987). *Ann. Phys. (N.Y.)*, **175**, 64.
Bottcher, C. and Strayer, M. R. (1989). *Phys. Rev. D*, **39**, 1330.
Bottcher, C., Strayer, M. R., Umar, A. S., and Oberacher, V. E. (1988). *Phys. Rev. C*, **37**, 2487.
Boys, S. F. (1950). *Proc. Roy. Soc.*, A**200**, 542.
Brandt, D. (1983). *Phys. Rev. A*, **27**, 1314.
Bransden, B. H. (1965). *Adv. At. Mol. Phys.*, **1**, 85.
Bransden, B. H. (1969). In *Lectures in theoretical physics: X1-C atomic collisions* (ed. S. Geltman, K. T. Mahanthappa, and V. B. Brittan) p. 139. Gordon and Breach, New York.
Bransden, B. H. (1983). *Atomic collision theory* (2nd edn). Benjamin-Cummings, Rending, Mass.).
Bransden, B. H. and Cheshire, I. M. (1963). *Proc. Phys. Soc.*, **81**, 820.
Bransden, B. H. and Coleman, J. P. (1972). *J. Phys. B*, **5**, 537.
Bransden, B. H. and Dewangan, D. P. (1979). *J. Phys. B*, **12**, 1372.
Bransden, B. H. and Issa, M. (1975). *J. Phys. B*, **8**, 1088.
Bransden, B. H. and Joachain, C. J. (1983). *Physics of atoms and molecules.* Longman, London.
Bransden, B. H. and Joachain, C. J. (1989). *Introduction to quantum mechanics.* Longman, London and Wiley, New York.
Bransden, B. H. and Noble, C. J. (1979). *Phys. Lett.*, **70**A, 404.
Bransden, B. H. and Noble, C. J. (1981). *J. Phys. B*, **14**, 1849.

Bransden, B. H. and Noble, C. J. (1982). *J. Phys. B*, **15**, 451.
Bransden, B. H. and Sin Fai Lam, L. T. (1966). *Proc. Phys. Soc.*, **87**, 653.
Bransden, B. H. and Stelbovics, A. T. (1984). *J. Phys. B*, **17**, 1877.
Bransden, B. H., Dalgarno, A., and King, W. M. (1954). *Proc. Phys. Soc. A*, **67**, 1075.
Bransden, B. H., Coleman, J. P., and Sullivan, J. (1972). *J. Phys. B*, **5**, 546.
Bransden, B. H., Dewangan, D. P., and Noble, C. J. (1979). *J. Phys. B*, **12**, 3563.
Bransden, B. H., Newby, C. W., and Noble, C. J. (1980). *J. Phys. B*, **13**, 4245.
Bransden, B. H., Scott, T., Shingal, R., and Roychoudhury, R. K. (1981). *J. Phys. B*, **15**, 4605.
Bransden, B. H., Noble, C. J., and Chandler, J. (1983). *J. Phys. B*, **16**, 4191.
Bransden, B. H., Ermolaev, A. M., and Shingal, R. (1984). *J. Phys. B*, **17**, 4515.
Brauner, M., Briggs, J. S., and Klar, H. (1989). *J. Phys. B*, **22**, 2265.
Briggs, J. S. (1976). *Rept. Prog. Phys.*, **39**, 217.
Briggs, J. S. (1977). *J. Phys. B*, **10**, 3075.
Briggs, J. S. (1983). In *Fundamental processes in energetic atomic collisions* (ed. H. O. Lutz, J. S. Briggs, and H. Kleinpoppen) p. 421. Plenum, New York.
Briggs, J. S. (1985). In *Semi-classical descriptions of atomic and nuclear collisions* (ed. J. Bang and J. de Boer) p. 183. North Holland, Amsterdam.
Briggs, J. S. (1986). *J. Phys. B*, **19**, 2703.
Briggs, J. S. and Dettmann, K. (1974). *Phys. Rev. Lett.*, **33**, 1125.
Briggs, J. S. and Macek, J. H. (1973). *J. Phys. B*, **6**, 982, 2484.
Briggs, J. S. and Taulbjerg, K. (1975). *J. Phys. B*, **8**, 1909.
Briggs, J. S. and Taulbjerg, K. (1978). In *Structure and collisions of ions and atoms* (ed. A. Selin) p. 105. Springer-Verlag, Berlin.
Brinkman, H. C. and Kramers, H. A. (1930). *Proc. Acad. Scien. Amsterdam*, **33**, 973.
Broad, J. T. (1978). *Phys. Rev. A*, **18**, 1012.
Brodsky, J. S., Kinoshita, T., and Terazawa, H. (1971). *Phys. Rev. D*, **4**, 1532.
Brower, M. C. and Pipkin, F. M. (1989). *Phys. Rev. A*, **39**, 3323.
Burden, F. R. and Wilson, R. M. (1972). *Adv. in Phys.*, **21**, 825.
Burgdörfer, J. and Taulbjerg, K. (1986). *Phys. Rev. A*, **33**, 2959.
Burke, P. G. and Mitchell, J. F. B. (1973). *J. Phys. B*, **6**, 320.
Burke, P. G. and Webb, T. G. (1970). *J. Phys. B*, **3**, L131.
Burke, P. G., Eissner, W. B., Hummer, D. G., and Percival, I. C. (eds) (1983). *Atoms in astrophysics.* Plenum, London.
Burns, W. D. and Crothers, D. S. F. (1976). *J. Phys. B*, **9**, 2479.
Butler, S. E. and Dalgarno, A. (1980). *Astr. J.*, **241**, 838.
Butler, S. E., Guberman, S. L., and Dalgarno, A. (1977). *Phys. Rev. A*, **16**, 500.
Butler, S. E., Heil, T. G., and Dalgarno, A. (1980). *Ast. J.*, **241**, 442.
Bykhovskii, V. K. and Nikitin, E. E. (1965). *Sov. Phys. JETP*, **21**, 10003.
Byron, F. W. and Joachain, C. J. (1977). *Phys. Rept.*, **34C**, 233.
Callaway, J. and Bartling, J. Q. (1966). *Phys. Rev.*, **150**, 69.
Callaway, J. and Bayer, E. (1965). *Phys. Rev. A*, **140**, 1072.
Callaway, J. and Duggan, G. (1966). *Phys. Letts.*, **21**, 295.
Callaway, J. and Wooten, J. W. (1973). *Phys. Lett.*, **45**A, 85.
Callaway, J. and Wooten, J. W. (1974). *Phys. Rev.*, A**9**, 1929.
Callaway, J., McDowell, M. R. C., and Morgan, L. A. (1975). *J. Phys. B*, **8**, 2181.
Chan, F. T. and Eichler, J. K. M. (1979). *Phys. Rev. A*, **20**, 1841.
Chatterjee, G., Bhattacharya, D. M., and Sil, N. C. (1967). *Ind. J. Phys.*, **41**, 934.

Cheshire, I. M. (1963). *Proc. Phys. Soc.*, **82**, 113.
Cheshire, I. M. (1964). *Proc. Phys. Soc.*, **84**, 89.
Cheshire, I. M. (1967). *Proc. Phys. Soc. A*, **92**, 862.
Cheshire, I. M. (1968). *J. Phys. B*, **1**, 428.
Cheshire, I. and Sullivan, E. (1967). *Phys. Rev.*, **16**, 4.
Cheshire, I. M., Gallaher, D. F., and Taylor, A. J. (1970). *J. Phys. B*, **3**, 813.
Chew, G. F. (1950). *Phys. Rev.*, **80**, 196.
Chew, G. F. and Wick, G. C. (1952). *Phys. Rev.*, **85**, 636, 686.
Chidichimo-Frank, U. C. and Piacentini, R. D. (1974). *J. Phys. B*, **7**, 548.
Child, M. S. (1971). *Mol. Phys.*, **20**, 171.
Child, M. S. (1974). *Molecular collision theory.* Academic Press, London and New York.
Chong, Y. P. and Fite, W. L. (1977). *Phys. Rev. A*, **16**, 933.
Ćirić, D., Dijkkamp, D., Vlieg, E., and de Heer, F. J. (1985*a*). *J. Phys. B*, **18**, 4745.
Ćirić, D., Dijkkamp, D., Vlieg, E, and de Heer, F. J. (1985*b*). *J. Phys. B*, **18**, L17.
Cohen, J. S. (1985). *J. Phys. B*, **18**, 1754.
Coleman, J. P. and McDowell, M. R. C. (1965). *Proc. Phys. Soc.*, **85**, 1097.
Colgrave, R. K. and Stephens, D. B. L. (1968). *J. Phys. B*, **1**, 428.
Connor, J. N. L. (1969). *Mol. Phys.*, **16**, 525.
Cooper, D. L., Kirby, K., and Dalgarno, A. (1984). *Cand. J. Phys.*, **62**, 1622.
Crandall, D. H. and Jaecks, D. H. (1971). *Phys. Rev. A*, **4**, 2271.
Crandall, D. H., Phaneuf, R. A., and Meyer, F. W. (1979). *Phys. Rev. A*, **19**, 504.
Crawford, H. J. (1979). Lawrence Berkeley Lab. Report No. 8807.
Crooks, J. B. and Rudd, M. E. (1970). *Phys. Rev. Letts.*, **25**, 159.
Crothers, D. S. F. (1971). *Adv. Phys.*, **20**, 405.
Crothers, D. S. F. (1981). *Adv. Atom. Mol. Phys.*, **17**, 55.
Crothers, D. S. R. (1982). *J. Phys. B*, **15**, 2061.
Crothers, D. S. F. (1985). *J. Phys. B*, **18**, 2879, 2893.
Crothers, D. S. F. (1989). *Proc. Roy. Irish Acad.*, **89**A, 91.
Crothers, D. S. F. and Dunseath, K. M. (1987). *J. Phys. B*, **20**, 4115.
Crothers, D. S. F. and Dunseath, K. M. (1990). *J. Phys. B*, **23**, L365.
Crothers, D. S. F. and Hughes, J. G. (1978*a*). *Proc. Roy. Soc.*, A**359**, 345.
Crothers, D. S. F. and Hughes, J. G. (1978*b*). *Phil. Trans. Roy. Soc.* (*Lond.*), A**292**, 539.
Crothers, D. S. F. and McCann, J. F. (1983). *J. Phys. B*, **16**, 3229.
Crothers, D. S. F. and McCann, J. F. (1984). *J. Phys. B*, **17**, L177.
Crothers, D. S. F. and McCann, J. F. (1985). *J. Phys. B*, **18**.
Crothers, D. S. F. and McCann, J. F. (1987). *J. Phys. B*, **20**, L19.
Crothers, D. S. F. and Todd, N. R. (1978). *J. Phys. B*, **11**, L663.
Crothers, D. S. F. and Todd, N. R. (1980*a*). *J. Phys. B*, **13**, 547.
Crothers, D. S. F. and Todd, N. R. (1980*b*). *J. Phys. B*, **13**, 2277.
Crothers, D. S. F. and Todd, N. R. (1981). *J. Phys. B*, **14**, 2233, 2251.
Curtis, C. F. and Power, R. S. (1964). *J. Chem. Phys.*, **40**, 2145.
Daley, H. L. and Perel, J. (1969). VI ICPEAC Abs. p. 1051.
Dalgarno, A. (1954). *Proc. Phys. Soc.*, A**67**, 1010.
Dalgarno, A. and McDowell, M. R. C. (1956). *Proc. Phys. Soc.*, A**69**, 615.
Dalgarno, A. and Yadav, H. N. (1983). *Proc. Phys. Soc.*, A**66**, 173.
Daniele, R. (1979). *J. Chem. Phys.*, **70**, 3462.
Datta, S. K., Crothers, D. S. F., and McCarroll, R. (1990). *J. Phys. B*, **23**, 479.

Davies, J. P. and Thorson, W. R. (1978). *Cand. J. Phys.*, **56**, 996.
Deb, N. C. (1988). *Phys. Rev. A*, **38**, 1202.
Deb, N. C. and Crothers, D. S. F. (1989). *J. Phys. B*, **22**, 3725.
Deb, N. C. and Crothers, D. S. F. (1990). *J. Phys. B*, **23**, L799.
Decker, F. and Eichler, J. K. M. (1989*a*). *J. Phys. B*, **22**, L95.
Decker, F. and Eichler, J. K. M. (1989*b*). *J. Phys. B*, **22**, 3023.
Deco, G. R. and Grün, N. (1989*a*). *J. Phys. B*, **22**, 1357.
Deco, G. R. and Grün, N. (1989*b*). *J. Phys. B*, **22**, 3709.
Deco, G. R. and Rivarola, R. (1985). *J. Phys. B*, **18**, 2283.
Deco, G. R. and Rivarola, R. (1986). *J. Phys. B*, **19**, 1759.
Deco, G. R. and Rivarola, R. (1987*a*). *J. Phys. B*, **20**, 317.
Deco, G. R. and Rivarola, R. (1987*b*). *J. Phys. B*, **20**, 3853.
Deco, G. R. and Rivarola, R. (1988*a*). *J. Phys. B*, **21**, 1224.
Deco, G. R. and Rivarola, R. (1988*b*). *J. Phys. B*, **21**, 1861.
Deco, G. R. and Rivarola, R. (1989). *Phys. Rev. A*, **39**, 5451.
Deco, G. R., Maidagan, J. M., and Rivarola, R. D. (1984). *J. Phys. B*, **17**, L707.
Deco, G. R., Fainstein, P. D., and Rivarola, R. D. (1986*a*). *J. Phys. B*, **19**, 213.
Deco, G. R., Piacentini, R. D., and Rivarola, R. D. (1986*b*). *J. Phys. B*, **19**, 3727.
Deco, G. R., Hanssen, J., and Rivarola, R. D. (1986*c*). *J. Phys. B*, **19**, L635.
Deco, G. R., Momberger, K., and Grün, N. (1990). *J. Phys. B*, **23**, 2091.
Delos, J. B. (1981*a*). *Rev. Mod. Phys.*, **53**, 287.
Delos, J. B. (1981*b*). *Phys. Rev. A*, **23**, 230.
Delos, J. B. and Thorson, W. R. (1972). *Phys. Rev. A*, **6**, 720.
Demkov, Yu.N. (1963). *Soviet Phys. JETP*, **45**, 195.
Dettmann, K. (1971). *Z. Phys.*, **244**, 86.
Dettmann, K. and Liebfried, G. (1966). *Phys. Rev.*, **14**, 1271.
Dettmann, K. and Liebfried, G. (1969). *Z. Phys.*, **21**, 1.
Dettmann, K., Harrison, K. G., and Lucas, M. W. (1974). *J. Phys. B*, **7**, 269.
Dewangan, D. P. (1975). *J. Phys. B*, **8**, L119.
Dewangan, D. P. (1977). *J. Phys. B*, **10**, 1053.
Dewangan, D. P. (1988). In *Electronic and atomic collisions* (ed. H. B. Gilbody, W. R. Newell, F. H. Read, and A. C. H. Smith) p. 425. Elsevier, Amsterdam.
Dewangan, D. P. and Bransden, B. H. (1988). *J. Phys. B*, **21**, L353.
Dewangan, D. P. and Chakraborty, H. S. (1989). *J. Phys. B*, **22**, L415.
Dewangan, D. P. and Eichler, J. K. M. (1985). *J. Phys. B*, **18**, L65.
Dewangan, D. P. and Eichler, J. K. M. (1986). *J. Phys. B*, **19**, 2939.
Dewangan, D. P. and Eichler, J. K. M. (1987). *Nucl. Inst. and Meths*, B**23**, 160.
Diercksen, G. H. F. and Wilson, S. (eds) (1983). *Methods in computational molecular physics.* Reidel, Dordrecht.
Dijkkamp, D., Ćirić, D., and de Heer, F. J. (1985*a*). *Phys. Rev. Lett.*, **54**, 1004.
Dijkkamp, D., Ćirić, D., Vlieg, E., de Boer, A., and de Heer, E. J. (1985*b*). *J. Phys. B*, **18**, 4763.
Dodd, L. R. and Greider, K. R. (1966). *Phys. Rev.*, **146**, 671, 675.
Dose, V. and Semini, C. (1974). *Helv. Phys. Acta*, **47**, 307, 609.
Drisko, R. M. (1955). Ph.D. thesis, Carnegie Institute of Technology.
Dubé, L. J. and Briggs, J. (1981). *J. Phys. B*, **14**, 4595.
Dubé, L. J., Mensour, B., Dewangan, D. P., and Chakraborty, H. S. (1990). *J. Phys. B*, **23**, L711.

DuBois, R. D. and Toburen, L. H. (1985). *Phys. Rev. A*, **31**, 3603.
Dunseath, K. M., Crothers, D. S. F., and Ishihara, T. (1988). *J. Phys. B*, **21**, L461.
Dyson, F. J. (1949). *Phys. Rev.*, **75**, 486.
Edmonds, A. R. (1960). *Angular momentum in quantum mechanics.* Princeton University Press, Princeton.
Eichenauer, D., Grün, N., and Scheid, W. (1981). *J. Phys. B*, **14**, 3929.
Eichler, J. K. M. (1981). *Phys. Rev. A*, **23**, 498.
Eichler, J. K. M. (1985). *Phys. Rev. A*, **32**, 112.
Eichler, J. K. M. (1987). *Phys. Rev. A*, **35**, 3248.
Eichler, J. K. M. (1988). *Phys. Rev. A*, **37**, 287(E).
Eichler, J. K. M. (1990). *Phys. Repts.*, **193**, 165.
Eichler, J. K. M., Wille, U., Fastrap, B., and Taulbjerg, K. (1976). *Phys. Rev. A*, **14**, 707.
Ellison, R. and Borowitz, S. (1964). *Atomic collision processes.* North-Holland, Amsterdam.
Erdelyi, A., Magnus, W., Oberhettinger, F., and Tricomi, E. G. (1953). *Higher transcendental functions*, Vol. 1. McGraw-Hill, New York.
Ermolaev, A. M. (1984). *J. Phys. B*, **17**, 1069.
Ermolaev, A. M. (1985). In *The physics of ionized gases* (ed. M. M. Popović and P. Krstić) p. 275. World Scientific, Singapore.
Ermolaev, A. M. (1987). In *Atomic physics with positrons* (ed. J. W. Humberston and E. A. G. Armour) p. 393. Plenum, New York.
Ermolaev, A. M. (1988*a*). *Hyperfine Interactions*, **44**, 375.
Ermolaev, A. M. (1988*b*). *J. Phys. B*, **21**, 81.
Ermolaev, A. M. (1990*a*). *J. Phys. B*, **23**, L45.
Ermolaev, A. M. (1990*b*). *Phys. Lett. A*, **149**, 151.
Ermolaev, A. M. and Bransden, B. H. (1984). *J. Phys. B*, **17**, 1083.
Ermolaev, A. M. and Hewitt, R. (1985). *Proceedings of the International Conference on the Physics of Highly Ionised Atoms* (Oxford) (ed. J. D. Silver and N. J. Peacock) p. 487. North-Holland, Amsterdam.
Ermolaev, A. M. and McDowell, M. R. C. (1987). *J. Phys. B*, **20**, L379.
Ermolaev, A. M., Hewitt, R. N., and McDowell, M. R. C. (1987*a*). *J. Phys. B*, **20**, 3125.
Ermolaev, A. M., Hewitt, R. N., Shingal, R., and McDowell, M. R. C. (1987*b*). *J. Phys. B*, **20**, 4507.
Errea, L. F., Méndez, L., and Riera, A. (1979). *J. Phys. B*, **12**, 69.
Errea, L. F., Méndez, L., and Riera, A. (1982). *J. Phys. B*, **15**, 101.
Errea, L. F., Gomez-Llorente, J. M., Méndez, L., and Riera, A. (1985). *Phys. Rev. A*, **32**, 2158.
Errea, L. F., Gomez-Llorente, J. M., Méndez, L., and Riera, A. (1987*a*). *Phys. Rev. A*, **35**, 4060.
Errea, L. F., Gomez-Llorente, J. M., Méndez, L., and Riera, A. (1987*b*). *J. Phys. B*, **20**, 6089.
Errea, L. F., Méndez, L., and Riera, A. (1989*a*). *Phys. Rev. A*, **39**, 2404.
Errea, L. F., Méndez, L., and Riera, A. (1989*b*). *Chem. Phys. Lett.*, **164**, 261.
Errea, L. F., Gomez-Llorente, J. M., Méndez, L., and Riera, A. (1989*c*). *Z. Phys. D*, **14**, 229.
Errea, L. F., Méndez, L., and Riera, A. (1990). *Europhys. Lett.*, **13**, 43.
Eu, B. C. (1984). *Semiclassical theories of molecular scattering.* Springer-Verlag, Berlin.
Faddeev, C. D. (1960). *Soviet Phys. JETP*, **12**, 1014.

Fainstein, P. D. and Rivarola, R. D. (1987). *J. Phys. B*, **20**, 1285.
Fainstein, P. D., Ponce, V. H., and Rivarola, R. D. (1988). *J. Phys. B*, **21**, 287.
Fainstein, P. D., Ponce, V. H., and Rivarola, R. D. (1989). *J. Phys. B*, **22**, L559.
Feagin, J. M., Briggs, J. S., and Reeve, T. M. (1984). *J. Phys. B*, **17**, 1057.
Ferguson, A. F. (1961). *Proc. Roy. Soc.*, **264**, 540.
Feshbach, H. (1958). *Ann. Phys.*, **5**, 357.
Feshbach, H. (1962). *Ann. Phys.*, **19**, 287.
Firsov, O. B. (1951). *Zh. Eksperim. i. Teor. Fiz.*, **21**, 1001.
Fitchard, E., Ford, A. L., and Reading, J. F. (1977). *Phys. Rev. A*, **16**, 1325.
Fite, L. W., Stebbings, R. F., Hummer, D. F., and Brackman, R. T. (1960). *Phys. Rev.*, **119**, 663.
Fite, W. L., Smith, A. C. H., and Stebbings, R. F. (1962). *Proc. Roy. Soc. A*, **268**, 527.
Flannery, M. R. (1969*a*). *J. Phys. B*, **2**, 913, 1044.
Flannery, M. R. (1969*b*). *Phys. Rev.*, **183**, 231, 241.
Flannery, M. R. (1970). *J. Phys. B*, **3**, 306.
Flannery, M. R. and McCann, K. J. (1973). *Phys. Rev. A*, **8**, 2915.
Flannery, M. R. and McCann, K. J. (1974). *J. Phys. B*, **7**, 840, 1349, 1558.
Ford, A. L. and Reading, J. F. (1988). *J. Phys. B*, **21**, L685.
Ford, A. L. and Reading, J. F. (1990*a*). *J. Phys. B*, **23**, 2567.
Ford, A. L. and Reading, J. F. (1990*b*). *J. Phys. B*, **23**, 3131.
Ford, A. L., Fitchard, E., and Reading, J. F. (1972). *Phys. Rev. A*, **16**, 133.
Ford, A. L., Reading, J. F., and Becker, R. L. (1977). *Phys. Rev. A*, **16**, 133.
Ford, A. L., Reading, J. F., and Becker, R. L. (1982). *J. Phys. B*, **15**, 3257.
Ford, K. W. and Wheeler, J. A. (1959). *Ann. Phys. (N.Y.)*, **7**, 259, 287.
Fricke, B., Rashid, K., Bertoncini, P., and Wahl, A. C. (1975). *Phys. Rev. Lett.*, **34**, 243.
Fritsch, W. (1982). *J. Phys. B*, **15**, L389.
Fritsch, W. (1983). Private communication to Grün and Scheid (1983).
Fritsch, W. (1984). *Phys. Rev. A*, **30**, 1135.
Fritsch, W. and Lin, C. D. (1982*a*). *J. Phys. B*, **15**, 1255.
Fritsch, W. and Lin, C. D. (1982*b*). *J. Phys. B*, **15**, L281.
Fritsch, W. and Lin, C. D. (1982*c*). *Phys. Rev. A*, **26**, 762.
Fritsch, W. and Lin, C. D. (1983*a*). *Phys. Rev. A*, **27**, 3361.
Fritsch, W. and Lin, C. D. (1983*b*). Abs. Contr. Papers XIII Int. Conf. on Physics of Electronic and Atomic Collisions (ed. J. Eichler, W. Fritsch, I. V. Hertel, N. Stolterfoht, and V. Willie) p. 502. ICPEAC, Berlin.
Fritsch, W. and Lin, C. D. (1983*c*). *J. Phys. B*, **16**, 1595.
Fritsch, W. and Lin, C. D. (1984*a*). *J. Phys. B*, **17**, 3271.
Fritsch, W. and Lin, C. D. (1984*b*). *Phys. Rev. A*, **29**, 3039.
Fritsch, W. and Lin, C. D. (1986). *J. Phys. B*, **19**, 2683.
Fritsch, W. and Lin, C. D. (1987). *Phys. Lett. A*, **123**, 128.
Fritsch, W. and Lin, C. D. (1988). *Phys. Rev. Lett.*, **61**, 690.
Fritsch, W. and Lin, C. D. (1990). *Phys. Rev. A*, **41**, 4776.
Fritsch, W. and Wille, U. (1971). *J. Phys. B*, **12**, L645.
Fröman, N. and Fröman, P. O. (1965). *JWKB approximation – contributions to the theory*. North-Holland, Amsterdam.
Gallaher, D. F. and Msezane, A. (1974). *J. Phys. B*, **7**, 127.
Gallaher, D. F. and Wilets, L. (1968). *Phys. Rev.*, **169**, 139.
Gargaud, M. (1981). *J. Phys. B*, **14**, 2263.

Gargaud, M. and McCarroll, R. (1985). *J. Phys. B*, **18**, 463.
Gargaud, M. and McCarroll, R. (1988). *J. Phys. B*, **21**, 513.
Gargaud, M., Hanssen, J., McCarroll, R., and Valiron, P. (1981). *J. Phys. B*, **14**, 2259.
Gargaud, M., McCarroll, R., and Valiron, P. (1987). *J. Phys. B*, **20**, 1555.
Gargaud, M., McCarroll, R., and Opradolce, L. (1988). *J. Phys. B*, **21**, 521.
Garrett, B. C. and Truhlar, D. G. (1981). In *Theoretical chemistry: advances and perspectives* (ed. D. Henderson) Vol. 6A, p. 215. Academic Press, New York.
Gaussorgues, C., Le Sech, C., Masnou-Seews, F., McCarroll, R., and Riera, A. (1975*a*). *J. Phys. B*, **8**, 239.
Gaussorgues, C., Le Sech, C., Masnou-Seews, F., McCarroll, R., and Riera, A. (1975*b*). *J. Phys. B: At. Mol. Phys.*, **8**, 253.
Gaussorgues, C., Piancentini, R. D., and Salin, A. (1975*c*). *Comp. Phys. Comm.*, **10**, 223.
Gayet, R. (1972). *J. Phys. B*, **5**, 483.
Gayet, R. (1989). *J. de Physique C1*, Suppl. 1, **50**, 53.
Gealy, M. W. and van Zyl, B. (1987). *Phys. Rev. A*, **36**, 3091.
Gershtein, S. S. and Krivchenkov, V. D. (1961). *Sov. Phys. JETP*, **13**, 1044.
Gibson, D. K. and Reid, I. D. (1986). *J. Phys. B*, **19**, 3265.
Gilbody, H. B. (1986). *Adv. Atom. Molec. Phys.*, **22**, 143.
Gilbody, H. B. and Ireland, J. V. (1963). *Proc. Roy. Soc.*, A**277**, 137.
Gilbody, H. B. and Ryding, G. (1966). *Proc. Roy. Soc.*, A**291**, 438.
Goffe, J. N., Shah, M. B., and Gilbody, H. B. (1979). *J. Phys. B*, **12**, 3763.
Gonzáles, A. D. and Miraglia, J. E. (1984). *Phys. Rev. A*, **30**, 2292.
Gramlich, K., Grün, N., and Scheid, W. (1986). *J. Phys. B*, **19**, 1457.
Gramlich, K., Grün, N., and Scheid, W. (1989). *J. Phys. B*, **22**, 2567.
Green, T. A. (1965). *Proc. Phys. Soc.*, **86**, 1017.
Green, T. A. (1981*a*). *Phys. Rev. A*, **23**, 519.
Green, T. A. (1981*b*). *Phys. Rev. A*, **23**, 532.
Green, T. A., Stanley, H. I., and Chiang, Y. C. (1965). *Helv. Phys. Act.*, **38**, 109.
Green, T. A., Shipsey, E. J., and Browne, J. C. (1982). *Phys. Rev. A*, **25**, 1364.
Greenland, P. T. (1978). *J. Phys. B*, **11**, L191.
Greenland, P. T. (1982). *Phys. Repts.*, **81**, 131.
Greenland, P. T. and Greiner, W. (1976). *Theor. Chim. Acta (Berlin)*, **42**, 273.
Greiner, W. (1990). *Relativistic quantum mechanics.* Springer-Verlag, Berlin.
Grozdanov, T. P. (1980). *J. Phys. B*, **13**, 3835.
Grün, N. and Scheid, W. (1983). *J. Phys. B*, **16**, L425.
Grün, N., Mühlkens, A., and Scheid, W. (1982). *J. Phys. B*, **15**, 4043.
Hahn, Y. (1989). *Phys. Rev. A*, **40**, 2950.
Hanssen, J., Gayet, R., Harel, C., and Salin, A. (1984). *J. Phys. B*. **17**, L323.
Hardie, D. J. W. and Olson, R. E. (1983). *J. Phys. B*, **16**, 1983.
Harel, C. and Salin, A. (1980). *J. Phys. B*, **13**, 785.
Harrison, K. G. and Lucas, M. W. (1970). *Phys. Rev. A*, **33**, 142.
Hatton, G. J., Lane, N. F., and Winter, T. G. (1979). *J. Phys. B*, **12**, L571.
Haywood, C. A. (1959). *Proc. Phys. Soc.*, **73**, 201.
Hazi, A. and Taylor, H. S. (1970). *Phys. Rev. A*, **1**, 1109.
Heil, T. G., Butler, S. E., and Dalgarno, A. (1981). *Phys. Rev. A*, **23**, 1100.
Heil, T. G., Butler, S. E., and Dalgarno, A. (1983). *Phys. Rev. A*, **27**, 2365.
Heinz, U., Müller, B., and Greiner, W. (1981). *Phys. Rev.*, **23**, 562.
Heitler, W. (1954). *The quantum theory of radiation.* Oxford University Press.

Helbig, H. F. and Everhart, E. (1965). *Phys. Rev. A*, **140**, 715.
Heller, E. J. (1973). Thesis, Harvard University.
Heller, E. J., Reinhardt, W. P., and Yamani, H. A. (1973*a*). *J. Comp. Phys.*, **13**, 536.
Heller, E. J., Rescigno, T. N., and Reinhardt, W. P. (1973*b*). *Phys. Rev. A*, **8**, 2946.
Henderson, G. H. (1923). *Proc. Roy. Soc.*, A**102**, 496.
Henderson, G. H. (1925). *Proc. Roy. Soc.*, A**109**, 157.
Henne, A., Toepfer, A., Lüdde, H. J., and Dreizler, R. M. (1986). *J. Phys. B*, **19**, L361.
Hill, J., Geddes, J., and Gilbody, H. B. (1979*a*). *J. Phys. B*, **12**, L341.
Hill, J., Geddes, J., and Gilbody, H. B. (1979*b*). *J. Phys. B*, **12**, 2875.
Hino, K. and Watanabe, T. (1987). *Phys. Rev. A*, **36**, 581.
Hippler, R. (1988). In *Coherence in atomic collision physics* (ed. H. J. Beyer, K. Blum, and R. Hippler) p. 137. Plenum, New York.
Hippler, R. (1991*a*). *Phys. Rev. A*, **43**, 2587.
Hippler, R. (1991*b*). *Z. Phys. D*, **18**, 61.
Hippler, R., Madeheim, H., Harbich, W., Kleinpoppen, H., and Lütz, H. O. (1988). *Phys. Rev. A*, **38**, 1662.
Hirsch, G., Bruno, S. P. J., Buenker, R. J., and Peyeremhoff, S. D. (1980). *J. Chem. Phys.*, **45**, 335.
Holt, A. R. (1969). *J. Phys. B*, **2**, 1253.
Holt, A. R. and Moiseiwitsch, B. L. (1968). *J. Phys. B*, **1**, 36.
Horbatsch, M. (1984). *J. Phys. B*, **17**, 2591.
Horsdal-Pedersen, E., Cocke, C. L., and Stocki, M. (1983*a*). *Phys. Rev. Lett.*, **50**, 1910.
Horsdal-Pedersen, E., Cocke, C. L., Rasmussen, J. L., Varghese, S. L., and Waggmer, W. (1983*b*). *J. Phys. B*, **16**, 1799.
Houver, J. C., Fayeton, J., and Barat, M. (1974). *J. Phys. B*, **7**, 1358.
Hughes, J. G. and Crothers, D. S. F. (1977). *J. Phys. B*, **10**, L605.
Hughes, R. H., Stokes, E. D., Choe, S.-S., and King, T. J. (1971). *Phys. Rev. A*, **4**, 1453.
Hultberg, S., Nagel, B., and Olsson, P. (1967). *Ark. Fys.*, **38**, 1.
Humphries, W. J. and Moiseiwitsch, B. L. (1984). *J. Phys. B*, **17**, 2655.
Hunter, G. and Kuriyan, M. (1977*a*). *Proc. Roy. Soc.*, A**353**, 575.
Hunter, G. and Kuriyan, M. (1977*b*). *Proc. Roy. Soc.*, A**358**, 321.
Huq, M. S., Havener, C. C., and Phaneuf, R. A. (1989). *Phys. Rev. A*, **40**, 1811.
Hvelplund, P., Heinemeier, J., Horsdal-Pedersen, E., and Simpson, F. R. (1976). *J. Phys. B*, **7**, 491.
Hylleraas, E. A. (1931). *Z. Phys.*, **71**, 739.
Iwai, T., Kaneko, Y., Kimura, M., Kobayashi, N., Ohtami, S., Okuno, K., Takagi, S., Tawara, H., and Tswabuchi, S. (1982). *Phys. Rev. A*, **26**, 105.
Jackson, J. D. and Schiff, H. (1953). *Phys. Rev.*, **89**, 359.
Jacobsen, J. C. (1926). *Nature*, **117**, 858.
Jaecks, D. W., van Zyl, B., and Geballe, R. (1965). *Phys. Rev. A*, **137**, 340.
Jain, A., Lin, C. D., and Fritsch, W. (1987*a*). *Phys. Rev. A*, **35**, 3180.
Jain, A., Lin, C. D., and Fritsch, W. (1987*b*). *Phys. Rev. A*, **36**, 2041.
Jain, A., Lin, C. D., and Fritsch, W. (1988). *Phys. Rev. A*, **37**, 3611(E).
Jakubassa-Amundsen, D. H. and Amundsen, P. A. (1980). *Z. Phys. A*, **29**, 203.
Janev, R. K., Belkić, Dž., and Bransden, B. H. (1983). *Phys. Rev. A*, **28**, 1293.
Janev, R. K. and Winter, H. (1985). *Phys. Repts.*, **117**, 265.
Joachain, C. J. (1975). *Quantum collision theory*. North-Holland, Amsterdam.

Joachain, C. J. and Post, D. E. (eds) (1983). *Atomic and molecular physics of controlled thermonuclear fusion.* Plenum, London.

Johnson, R. (1983). *Phys. Rev. A*, **28**, 1460.

Kadota, K., Dykkamp, D., van der Woude, R. L., de Boer, A., Pan Guang Yan, and de Heer, F. J. (1982). *J. Phys. B*, **15**, 3275.

Kaneko, Y., Iwai, T., Ohtani, S., Okuno, K., Kobayashi, N., Tsurubuchi, S., Kimura, M., Tawara, H., and Takagi, S. (1981). Proceedings of the 12th International Conference on the Physics of Electronic and Atomic Collisions, Abstracts Vol. II, p. 696. North Holland, Amsterdam.

Karnakov, B. M. (1982). *Sov. Phys. JETP*, **55**, 816.

Kemble, E. C. (1937). *The fundamental principles of quantum mechanics.* McGraw-Hill, New York.

Kimura, M. (1981). Thesis, University of Alberta.

Kimura, M. (1985). *Phys. Rev. A*, **31**, 2158.

Kimura, M. (1988). *J. Phys. B*, **21**, L19.

Kimura, M. and Lane, N. F. (1989). *Adv. Atom. Mol. Phys.*, **26**, 79.

Kimura, M. and Lin, C. D. (1985*a*). *Phys. Rev. A*, **31**, 590.

Kimura, M. and Lin, C. D. (1985*b*). *Phys. Rev. A*, **32**, 1357.

Kimura, M. and Lin, C. D. (1986). *Phys. Rev. A*, **34**, 176.

Kimura, M. and Olson, R. E. (1984). *J. Phys. B*, **17**, L713.

Kimura, M. and Thorson, W. R. (1981*a*). *Phys. Rev. A*, **24**, 1780.

Kimura, M. and Thorson, W. R. (1981*b*). *Phys. Rev. A*, **24**, 3019.

Kimura, M. and Thorson, W. R. (1981*c*). 12th International Conference on the Physics of Electronic and Atomic Collisions, 1981, Abstracts, p. 638. North-Holland, Amsterdam.

Kimura, M. and Thorson, W. R. (1989); see Kimura and Lane (1989).

Kimura, M., Sato, H., and Olsen, R. E. (1983). *Phys. Rev. A*, **28**, 2085.

Kirschbaum, C. L. and Wilets, L. (1980). *Phys. Rev. A*, **21**, 834.

Knudsen, S. W. and Thorson, W. R. (1970). *Cand. J. Phys.*, **48**, 313.

Knudsen, H., Haugen, H. K., and Hvelplund, P. (1981). *Phys. Rev. A*, **23**, 597.

Knudsen, H., Andersen, L. H., Hvelplund, P., Astner, G. Cederquist, H., Donared, H., Liljeby, L., and Rensfelt, K.-G. (1984). *J. Phys. B*, **17**, 3545.

Kocbach, L. (1976). *Z. Phys. A*, **279**, 233.

Kondow, T., Chong, Y. P., Fite, W. L., and Givnius, R. J. (1974). *Phys. Rev. A*, **10**, 1167.

Kramer, P. J. (1972). *Phys. Rev. A*, **6**, 2125.

Kronig, R. de L. (1930). *Band spectra and molecular structure.* Cambridge University Press.

Kulander, K. C., Sandhya Devi, K. R., and Koonin, S. E. (1982). *Phys. Rev. A*, **25**, 2968.

Kvale, T. J., Seely, D. G., Blankenship, D. M., Redd, E., Guy, T. J., Kimura, M., Rille, E., Peacher, J. L., and Park, J. T. (1985). *Phys. Rev. A*, **32**, 1369.

Landau, L. D. (1932). *Phys. Z. Soviet Union*, **2**, 46.

Landau, L. D. and Lifshitz, E. M. (1958). *Quantum mechanics.* Pergamon Press, Oxford.

Landau, L. D. and Lifshitz, E. M. (1960). *Mechanics.* Pergamon Press, Oxford.

Landau, L. D. and Lifshitz, E. M. (1971). *Relativistic quantum theory.* Pergamon Press, Oxford.

Langer, R. E. (1937). *Phys. Rev.*, **51**, 669.

Lichten, W. (1967). *Phys. Rev.*, **164**, 131.
Lin, C. D., Soong, S. C., and Tunnell, L. N. (1978). *Phys. Rev. A*, **17**, 1646.
Lin, C. D., Winter, T. G., and Fritsch, W. (1982). *Phys. Rev. A*, **25**, 2395.
Livingston, P. W. (1966). *J. Chem. Phys.*, **45**, 601.
Lockwood, G. T. and Everhart, E. (1962). *Phys. Rev.*, **125**, 567.
Lovell, S. E. and McElroy, M. B. (1965). *Proc. Phys. Soc. A*, **283**, 100.
Lüdde, H. J. and Dreizler, R. M. (1981). *J. Phys. B*, **14**, 2191.
Lüdde, H. J. and Dreizler, R. M. (1982*a*). *J. Phys. B*, **15**, 2703.
Lüdde, H. J. and Dreizler, R. M. (1982*b*). *J. Phys. B*, **15**, 2713.
Lüdde, H. J. and Dreizler, R. M. (1983). *J. Phys. B*, **16**, 3973.
Lüdde, H. J. and Dreizler, R. M. (1985). *J. Phys. B*, **18**, 113.
Lüdde, H. J. and Dreizler, R. M. (1989). *J. Phys. B*, **22**, 3243.
Lüdde, H. J., Ast, H. and Dreizler, R. M. (1987). *Phys. Lett.*, **125**A, 197.
Lüdde, H. J., Ast, H., and Dreizler, R. M. (1988). *J. Phys. B*, **21**, 4131.
Macek, J. (1970). *Phys. Rev. A*, **1**, 235.
Macek, J. (1988). *Phys. Rev. A*, **37**, 2765.
Macek, J. H. and Alston, S. (1982). *Phys. Rev. A*, **26**, 250.
Macek, J. and Shakeshaft, R. (1980). *Phys. Rev. A*, **22**, 1441.
Macek, J. H. and Taulbjerg, K. (1981). *Phys. Rev. Lett.*, **46**, 170.
Macek, J. H. and Taulbjerg, K. (1989*a*). *Phys. Rev. A*, **39**, 6064.
Macek, J. M. and Taulbjerg, K. (1989*b*). *J. Phys. B*, **22**, L523.
Maciás, A. and Riera, A. (1982). *Phys. Repts.*, **90**, 299.
Madsen, M. M. and Peek, J. M. (1971). *At. Data*, **2**, 171.
Magnus, W. (1954). *Comm. Proc. Appl. Math.*, **7**, 649.
Maidagan, J. M. and Rivarola, R. D. (1984). *J. Phys. B*, **17**, 2477.
Mandal, C. R., Shyamal Daata, and Mukherjee, S. C. (1981). *Phys. Rev. A*, **24**, 3044.
Manson, S. T., Toburen, L. H., Madison, D. H., and Stolterfoht, N. (1975). *Phys. Rev. A*, **12**, 60.
Mapleton, R. A. (1961). *J. Maths. Phys.*, **2**, 482.
Mapleton, R. A. (1964). *Proc. Phys. Soc.*, **83**, 895.
Mapleton, R. A. (1967). *Proc. Phys. Soc.*, **91**, 868.
Mapleton, R. A. (1972). *Theory of charge exchange*. Wiley, New York.
Martin, P. J., Blankenship, D. M., Kvale, T. J., Redd, E., Peacher, J. L., and Park, J. T. (1981). *Phys. Rev. A*, **23**, 3357.
Martir, M. H., Ford, A. L., Reading, J. F., and Becker, R. L. (1982). *J. Phys. B*, **15**, 1729.
Maruhn-Rezwani, V., Grün, N., and Scheid, W. (1979). *Phys. Rev. Lett.*, **43**, 512.
Massey, H. S. W. (1971). *Electronic and ionic impact phenomena*, Vol. 3. Oxford University Press.
Massey, H. S. W. and Mohr, C. B. O. (1933). *Proc. Roy. Soc.*, A**141**, 434.
Massey, H. S. W. and Smith, R. A. (1933). *Proc. Roy. Soc.*, A**142**, 142.
McCann, J. F. (1985). *J. Phys. B*, **18**, L569.
McCarroll, R. (1961). *Proc. Roy. Soc.*, A**264**, 547.
McCarroll, R. and McElroy, M. B. (1962). *Proc. Roy. Soc.*, A**266**, 422.
McCarroll, R. and Salin, A. (1968). *J. Phys. B*, **1**, 163.
McCarroll, R. and Valiron, P. (1979). *Astron. Ast.*, **78**, 177.
McCarroll, R., Piacentini, R. D., and Salin, A. (1970). *J. Phys. B*, **3**, 1336.
McClure, G. W. (1966). *Phys. Rev.*, **148**, 47.

McCullough, R. W., Goffe, T. V., Shah, M. B., Lennon, M., and Gilbody, H. B. (1982). *J. Phys. B*, **15**, 111.

McDowell, M. R. C. and Coleman, J. P. (1970). *Introduction to the theory of ion–atom collisions.* North-Holland, Amsterdam.

McDowell, M. R. C. and Ferendeci, A. M. (eds) (1980). *Atomic and molecular processes in controlled thermonuclear fusion.* Plenum, London.

McDowell, M. R. C. and Janev, R. K. (1984). *Phys. Rev. A*, **17**, 2295.

McElroy, M. B. (1963). *Proc. Roy. Soc.*, A**272**, 542.

McGuire, J. H. and Weaver, L. (1977). *Phys. Rev. A*, **16**, 41.

McGuire, J. H., Stockli, M., Cocke, C. L., Horsdal-Pedersen, E., and Sil, N. C. (1984). *Phys. Rev. A*, **30**, 89.

McGuire, J. H., Müller, A., Schuch, B., Gosh, W., and Salzborn, E. (1987). *Phys. Rev. A*, **35**, 2479.

McKenzie, M. L. and Olson, R. E. (1987). *Phys. Rev. A*, **35**, 2863.

McLaughlin, B. M. and Bell, K. L. (1983). *J. Phys. B*, **16**, 3797.

McLaughlin, B. M. and Bell, K. L. (1985). *J. Phys. B*, **18**, L771.

Meckbach, W., Focke, P. J., Goni, A. R., Suarez, S., Macek, J., and Mendez, M. F. (1986). *Phys. Rev. Lett.*, **57**, 1587.

Mehler, G., Soff, G., Rumrich, K., and Greiner, W. (1989). *Z. Phys. D*, **13**, 193.

Melchert, F., Rink, K., Salzborn, E., and Grün, N. (1987). *J. Phys. B*, **20**, L223.

Men, F. K., Kimura, M., and Olson, R. E. (1986). *Phys. Rev. A*, **33**, 3800.

Merzbacher, E. (1970). *Quantum mechanics*, (2nd edn). Wiley, New York.

Meyer, F. W., Howald, A. M., Havener, C. C., and Phaneuf, R. A. (1985). *Phys. Rev. A*, **32**, 3310.

Meyerhof, W. E. (1988). Quoted by Toshima and Eichler (1988*b*).

Meyerhof, W. E., Anholt, R., Eichler, J., and Salop, A. (1978). *Phys. Rev. A*, **17**, 108.

Meyerhof, W. E., Anholt, R., Eichler, J., Gould, H., Munger, C. L., Alonso, J., Thieberger, P., and Wegner, H. E. (1985). *Phys. Rev. A*, **32**, 3291.

Miraglia, J. E., Piancentini, R. D., Rivarola, R. D., and Salin, A. (1981). *J. Phys. B*, **14**, L197.

Mittleman, M. H. (1964). *Proc. Phys. Sci.*, **84**, 453.

Mittleman, H. H. (1969). *Phys. Rev.*, **188**, 221.

Moiseiwitsch, B. L. (1980). *Adv. Atom. Mol. Phys.*, **16**, 281.

Moiseiwitsch, B. L. (1985). *Phys. Repts*, **118**, 133.

Moiseiwitsch, B. L. (1986). *J. Phys. B*, **19**, 3733.

Moiseiwitsch, B. L. (1987). *J. Phys. B*, **20**, L171 and 4111.

Moiseiwitsch, B. L. (1988). *J. Phys. B*, **21**, 603.

Moiseiwitsch, B. L. (1989). *Phys. Rev. A*, **39**, 5609.

Moiseiwitsch, B. L. and Stockman, S. G. (1980). *J. Phys. B*, **13**, 2975, 4607.

Momberger, K., Grün, N., Scheid, W., and Becker, U. (1990). *J. Phys. B*, **23**, 22935.

Morgan, T. J., Geddes, J., and Gilbody, H. B. (1973). *J. Phys. B*, **6**, 2118.

Morgan, T. J., Geddes, J., and Gilbody, H. B. (1974). *J. Phys. B*, **7**, 142.

Morgan, T. J., Stone, J., and Mayo, R. (1980). *Phys. Rev. A*, **22**, 1460.

Morrison, M. G. and Öpik, U. (1978). *J. Phys. B*, **11**, 473.

Morrison, M. G. and Öpik, U. (1984). *J. Phys. B*, **17**, 857.

Morse, P. and Feshbach, H. (1953). *Methods of theoretical physics.* McGraw-Hill, New York.

Mott, N. F. and Massey, H. S. W. (1965). *Theory of atomic collisions*, (3rd edn). Oxford University Press.
Msezane, A. and Gallaher, D. F. (1975). *J. Phys. B*, **6**, 2334.
Msezane, A. and Gallaher, D. F. (1977). *J. Phys. B*, **8**, 112.
Mukherjee, S. C., Sil, N. C., and Basu, D. (1979). *J. Phys. B*, **12**, 1259.
Müller, B. (1989). In *Physics of highly ionised atoms*, (ed. R. Marrus). Plenum, New York.
Müller, B. and Greiner, W. (1973). *Phys. Lett.*, **47B**, 5.
Müller, B. and Greiner, W. (1976). *Z. Naturf.*, **31a**, 1.
Murray, G. A., Stone, J., Mayo, M., and Morgan, T. J. (1982). *Phys. Rev. A*, **25**, 1805.
Newby, C. W. (1985). *J. Phys. B*, **18**, 1781.
Newman, J. H., Cogan, J. D., Ziegler, D. L., Nitz, D. E., Rundel, R. D., Smith, K. A., and Stebbings, R. F. (1982). *Phys. Rev. A*, **25**, 2976.
Nikitin, E. E. (1962*a*). *Opt. Spectrosc. USSR*, **13**, 431.
Nikitin, E. E. (1962*b*). *Disc. Faraday Soc.*, **33**, 14.
Nikitin, E. E. (1963). *Izv. Akad. Nauk. SSSR Ser. Fiz.*, **27**, 996.
Nikitin, E. E. (1970). *Adv. in Quantum Chemistry*, **5**, 135.
Nikitin, E. E. and Umanskii, S. Yu. (1984). *Theory of slow atomic collisions*. Springer-Verlag, Berlin.
Noble, C. J. (1980). *Com. Phys. Comm.*, **19**, 327.
Nutt, W. L., McCullough, R. N., Brady, K., Shah, M. B., and Gilbody, H. B. (1978). *J. Phys. B*, **11**, 1457.
Okubo, S. and Feldman, D. (1960). *Phys. Rev.*, **117**, 292.
Oldham, W. J. B. (1967). *Phys. Rev.*, **161**, 1.
Olson, R. E. (1978). *J. Phys. B*, **11**, L227.
Olson, R. E. (1981). *Phys. Rev. A*, **24**, 1726.
Olson, R. E. (1983). *Phys. Rev. A*, **27**, 1871.
Olson, R. E. and Gay, T. J. (1988). *Phys. Rev. Lett.*, **61**, 302.
Olson, R. E. and Salop, A. (1976). *Phys. Rev. A*, **14**, 576.
Olson, R. E. and Salop, A. (1977). *Phys. Rev. A*, **16**, 531.
Olson, R. E., and Salop, A., Phanneuf, R. A., and Meyer, F. W. (1977). *Phys. Rev. A*, **16**, 1867.
Olson, R. E., Shipsey, E. J., and Browne, J. C. (1978). *J. Phys. B.*, **11**, 699.
Omidvar, K. (1967). *Phys. Rev.*, **153**, 121.
Omidvar, K. (1975). *Phys. Rev. A*, **12**, 911.
Oppenheimer, J. R. (1928). *Phys. Rev.*, **31**, 349.
Opradolce, L., Benmeuraiem, L., McCarroll, R., and Piacentini, R. D. (1988). *J. Phys. B*, **21**, 503.
Park, J. T. (1982). In *Physics of electronic and atomic collisions* (ed. S. Datz), p. 109. North-Holland, Amsterdam.
Park, J. T. (1983). *Adv. At. Mol. Phys.*, **19**, 67.
Park, J. T. and Schowengerdt, F. D. (1969). *Phys. Rev.*, **185**, 152.
Park, J. T., Aldag, J. E., and George, J. M. (1975). *Phys. Lett.*, **34**, 1253.
Park, J. T., Aldag, J. E., George, J. M., and Peacher, J. L. (1976). *Phys. Rev. A*, **14**, 608.
Park, J. T., Aldag, J. E., George, J. M., Peacher, J. L., and McGuire, J. H. (1977). *Phys. Rev. A*, **15**, 508.
Park, J. T., Aldag, J. E., Peacher, J. L., and George, J. M. (1978). *Phys. Rev. Lett.*, **40**, 1646.

Pauli, M. and Trautmann, D. (1978). *J. Phys. B*, **11**, 667.

Pauly, H. (1979). In *Atom–molecule collisons – a guide for the experimentalist* (ed. R. B. Bernstein). p. 111. Plenum Press, New York and London.

Peach, G. (1982). *Comm. At. Mol. Phys.*, **11**, 101.

Peach, G., Willis, S. L., and McDowell, M. R. C. (1985). *J. Phys. B*, **18**, 3921.

Peart, B., Rinn, K., and Dolder, K. T. (1983). *J. Phys. B*, **16**, 1461.

Peek, J. M. (1965). *J. Chem. Phys.*, **43**, 4003.

Phaneuf, R. A., Meyer, F. W., and McKnight, R. H. (1978). *Phys. Rev. A*, **17**, 534.

Phaneuf, R. A., Alvaraz, I., Meyer, F. W., and Crandell, D. H. (1982). *Phys. Rev. A*, **26**, 1892.

Piacentini, R. D. and Salin, A. (1974). *J. Phys. B*, **7**, 1666.

Pilar, F. L. (1968). *Elementary quantum chemistry*. McGraw-Hill, New York.

Pivovar, L. I., Novikov, M. T., and Tubaev, V. M. (1962). *Sov. Phys. JETP*, **15**, 1035.

Pluvinage, P. (1980). *Ann. Phys.* (*Paris*), **5**, 145.

Ponce, V. H. (1979). *J. Phys. B*, **12**, 3731.

Ponomarev, L. J. and Puzynina, T. P. (1967). *Soviet Phys. JETP*, **25**, 846.

Ponomarev, L. J. and Puzynina, T. P. (1968). *USSR Comp. Math & Math. Phys.*, **8**, 94.

Ponomarev, L. J. and Puzynina, T. P. (1970). *J. Inst. Nucl. Res.*, Dubna Report HINR-14-5040.

Potvliege, R. M., Furtado, F., and Joachain, C. J. (1987). *J. Phys. B*, **20**, 1771.

Power, J. D. (1973*a*). *Phil. Trans. Roy. Soc.*, A**274**, 55.

Power, J. D. (1973*b*). QCPE Program 233, Quantum Chemistry Program Exchanges, Chemistry Dept., Indiana University, USA.

Pradhan, T. (1957). *Phys. Rev.*, **10**, 1250.

Presnyakov, L. P. and Ulanstev, A. D. (1975). *Sov. J. Quan. Elect.*, **4**, 1320.

Puckett, L. J. and Martin, D. W. (1970). *Phys. Rev. A*, **1**, 1432.

Rankin, J. and Thorson, W. R. (1978). *Phys. Rev. A*, **18**, 1990.

Rapp, D. and Dinwiddie, D. (1972). *J. Chem. Phys.*, **57**, 4919.

Reading, J. F. (1973). *Phys. Rev. A*, **8**, 3262.

Reading, J. F. and Ford, A. L. (1979). *J. Phys. B*, **12**, 1367.

Reading, J. F. and Ford, A. L. (1980). *Phys. Rev. A*, **21**, 124.

Reading, J. F. and Ford, A. L. (1987*a*). *Phys. Rev. Lett.*, **58**, 543.

Reading, J. F. and Ford, A. L. (1987*b*). *J. Phys. B*, **20**, 3747.

Reading, J. F., Ford, A. L., and Fitchard, A. E. G. (1976). *Phys. Rev. Lett.*, **36**, 573.

Reading, J. F., Ford, A. L., Swafford, G. L., and Fitchard, A. (1979). *Phys. Rev. A*, **20**, 130.

Reading, J. F., Ford, A. L., and Becker, R. L. (1981). *J. Phys. B*, **14**, 1995.

Reading, J. F., Ford, A. L., and Becker, R. L. (1982). *J. Phys. B*, **15**, 625.

Reading, J. F., Ford, A. L., Smith, J. S., and Becker, R. L. (1984). In *Electronic and atomic collisions* (eds. J. Eichler, I. V. Hertel, and N. Stolterfoht) p. 201. Elsevier, Amsterdam.

Reading, J. F., Ford, A. L., and Fang, X. (1989). *Phys. Rev. Lett.*, **62**, 245.

Reinhold, C. O. and Falcon, C. A. (1986). *Phys. Rev. A*, **33**, 3859.

Reinhold, C. O. and Falcon, C. A. (1988*a*). *J. Phys. B*, **21**, 1829.

Reinhold, C. O. and Falcon, C. A. (1988*b*). *J. Phys. B*, **21**, 2473.

Reinhold, C. O. and Olson, R. E. (1989*a*). *J. Phys. B*, **22**, L39.

Reinhold, C. O. and Olson, R. E. (1989*b*). *Phys. Rev. A*, **39**, 3861.

Reinhold, C. O. and Schulz, D. R. (1989). *J. Phys. B*, **22**, L565.

Riera, A. (1984). *Phys. Rev. A*, **30**, 2304.
Riley, M. E. (1973). *Phys. Rev. A*, **8**, 742.
Riley, M. E. and Green, T. A. (1971). *Phys. Rev. A*, **4**, 619.
Rinn, K., Melchert, F., and Salzborn, E. (1985). *J. Phys. B*, **18**, 3783.
Rinn, K., Melchert, F., and Salzborn, E. (1987). *J. Phys. B*, **20**, L797.
Rinn, K., Melchert, F., Rink, K., and Salzborn, E. (1986). *J. Phys. B*, **19**, 3717.
Rivarola, R. D. (1985). *Nucl. Inst. and Meths*, A**240**, 508.
Rivarola, R. D. (1989). *Nucl. Inst. and Meths*, B**42**, 491.
Rivarola, R. D., Deco, G. R., and Maidagan, J. M. (1985). *Nucl. Inst. and Meths*, **B10/11**, 222.
Rosen, W. and Zener, C. (1932). *Phys. Rev.*, **40**, 502.
Rosenthal, H. (1971). *Phys. Rev. Lett.*, **27**, 635.
Rotenberg, M. (1962). *Ann. Phys.*, **19**, 262.
Roy, P. K., Saha, B. C., and Sil, N. C. (1980). *J. Phys. B*, **13**, 3401.
Rudd, M. E. and Jorgenson, T. (1963). *Phys. Rev.*, **131**, 666.
Rudd, M. E. and Macek, J. (1972). *Case Stud. At. Phys.*, **3**, 47.
Rudd, M. E. and Madison, D. H. (1976). *Phys. Rev. A*, **14**, 128.
Rudd, M. E., Toburen, L. H., and Stolterfoht, N. (1976). *At. Data and Nucl. Data Tables*, **18**, 413.
Russek, A. (1966). Private communication to Wilets and Gallaher (1966).
Rutherford, E. (1924). *Phil. Mag. 6th Ser.*, **47**, 276.
Ryufuku, H. (1982). *Phys. Rev. A*, **25**, 733.
Ryufuku, H. and Watanabe, T. (1978). *Phys. Rev. A*, **18**, 2005.
Ryufuku, H. and Watanabe, T. (1979). *Phys. Rev. A*, **19**, 1538, 1828.
Saha, H. P., Maiti, N., and Sil, N. C. (1980). *J. Phys. B*, **13**, 327.
Sakurai, J. J. (1967). *Advanced quantum mechanics.* Addison Wesley, Reading.
Salin, A. (1969). *J. Phys. B*, **5**, 979.
Salin, A. (1978). *Comp. Phys. Comm.*, **14**, 121.
Salin, A. (1987). *Phys. Rev. A*, **36**, 5471.
Salin, A. (1991). *Comp. Phys. Comm.*, **62**, 58.
Salop, A. and Olson, R. E. (1976). *Phys. Rev. A*, **13**, 1312.
Salzborn, E. (1989). *J. de Physique Coll.*, C1, Suppl. 1, **50**, 207.
Sandhya Devi, K. R. and Garcia, J. D. (1983). *J. Phys. B*, **16**, 2837.
Sandhya Devi, K. R. and Garcia, J. D. (1984). *Phys. Rev. A*, **30**, 600.
Sato, H. and Kimura, M. (1983). *Phys. Lett.*, **96**A, 286.
Saunders, V. R. (1983). In *Methods in computational molecular physics*, (eds. G. H. F. Diercksen and S. Wilson). Reidel, Dordrecht.
Schaudt, K. J., Kwong, N. H., and Garcia, J. D. (1991). *Phys. Rev. A*, **43**, 2294.
Schiff, H. (1984). *Cand. J. Phys.*, **32**, 393.
Schinke, R. and Krüger, H. (1976). *J. Phys. B*, **9**, 2469.
Schiwietz, G. (1990). *Phys. Rev. A*, **42**, 296.
Schiwietz, G. and Fritsch, W. (1987). *J. Phys. B*, **20**, 5463.
Schlachter, A. S., Berkner, K. H., Graham, W. G., Pyle, R. V., Stearms, J. W., and Tanis, J. A. (1981). *Phys. Rev. A*, **24**, 1110.
Schneiderman, S. M. and Russek, A. (1969). *Phys. Rev.*, **181**, 311.
Schryber, U. (1967). *Helv. Phys. Acta*, **40**, 1023.
Schwab, W., Baptista, G. B., Justiniano, E., Schuch, R., Vogt, H., and Weber, E. W. (1987). *J. Phys. B*, **20**, 2825.

Seim, W., Müller, A., and Salzborn, E. (1980). *Phys. Lett.*, **80A**, 20.
Seim, W., Müller, A., Wirkner-Bott, I., and Salzborn, E. (1981). *J. Phys. B*, **14**, 3475.
Shah, M. B. and Gilbody, H. B. (1978). *J. Phys. B*, **11**, 121.
Shah, M. B. and Gilbody, H. B. (1981). *J. Phys. B*, **14**, 2361.
Shah, M. B. and Gilbody, H. B. (1982). *J. Phys. B*, **15**, 413.
Shah, M. B. and Gilbody, H. B. (1983). *J. Phys. B*, **16**, L449.
Shah, M. B. and Gilbody, H. B. (1985). *J. Phys. B*, **18**, 899.
Shah, M. B., Goffe, T. V., and Gilbody, H. B. (1978). *J. Phys. B*, **11**, L233.
Shah, M. B., Elliott, D. S., and Gilbody, H. B. (1985). *J. Phys. B*, **18**, 4245.
Shah, M. B., Elliott, D. S., and Gilbody, H. B. (1987). *J. Phys. B*, **20**, 2481.
Shah, M. B., Elliott, D. S., McCallion, P. and Gilbody, H. B. (1988). *J. Phys. B.*, **21**, 2455.
Shah, M. B., McCallion, P., and Gilbody, H. B. (1989). *J. Phys. B*, **22**, 3037.
Shakeshaft, R. (1974*a*). *Phys. Rev. A*, **10**, 1907.
Shakeshaft, R. (1974*b*). *J. Phys. B*, **7**, 1059.
Shakeshaft, R. (1975*a*). *J. Phys. B*, **8**, 1114.
Shakeshaft, R. (1975*b*). *J. Phys. B*, **8**, L134.
Shakeshaft, R. (1976). *Phys. Rev. A*, **14**, 1626.
Shakeshaft, R. (1978*a*). *Phys. Rev. A*, **17**, 1011.
Shakeshaft, R. (1978*b*). *Phys. Rev. A*, **18**, 1930.
Shakeshaft, R. (1979). *Phys. Rev. A*, **20**, 779.
Shakeshaft, R. and Spruch, L. (1973). *Phys. Rev. A*, **8**, 206.
Shakeshaft, R. and Spruch, L. (1979). *Rev. Mod. Phys.*, **51**, 369.
Shampine, L. F. and Gordon, M. K. (1975). *Computer solution of ordinary differential equations: the initial value problem.* Freeman, San Francisco.
Shimakura, N., Inouye, H., Koike, F., and Watanabe, T. (1981). *J. Phys. B.* **14**, 2203.
Shingal, R. (1988*a*). *J. Phys. B*, **21**, 125.
Shingal, R. (1988*b*). *J. Phys. B*, **21**, 2065.
Shingal, R. (1990*a*). Preprint.
Shingal, R. (1990*b*). *Bull. Amer. Phys. Soc.*, **35**, 1778.
Shingal, R. and Bransden, B. H. (1987). *J. Phys. B*, **20**, L127.
Shingal, R. and Bransden, B. H. (1990). *J. Phys. B*, **23**, 1203.
Shingal, R. and Lin, C. D. (1991*a*). *J. Phys. B*, **24**, 251.
Shingal, R. and Lin, C. D. (1991*b*). *J. Phys. B*, **24**, 963.
Shingal, R., Bransden, B. H., and Flower, D. R. (1985). *J. Phys. B*, **18**, 2485.
Shingal, R., Bransden, B. H., Ermolaev, A. M., Flower, D. R., Newby, C. M., and Noble, C. J. (1986*a*). *J. Phys. B*, **19**, 309.
Shingal, R., Noble, C. J., Bransden, B. H., and Flower, D. R. (1986*b*). *J. Phys. B*, **19**, 3951.
Shingal, R., Noble, C. J., and Bransden, B. H. (1987). *J. Phys. B*, **19**, 793.
Shingal, R., Bransden, B. H., and Flower, D. R. (1989). *J. Phys. B*, **22**, 855.
Shipsey, E. J., Redman, L. T., Browne, J. C., and Olson, R. E. (1978). *Phys. Rev. A*, **18**, 1916.
Shipsey, E. J., Green, T. A., and Browne, J. C. (1983). *Phys. Rev. A*, **27**, 821.
Sidis, V., Kubach, C., and Fussen, D. (1981). *Phys. Rev. Lett.*, **47**, 1280.
Sidis, V., Kubach, C., and Fussen, D. (1983). *Phys. Rev. A*, **27**, 2431.
Simony, P. R. and McGuire, J. H. (1981). *J. Phys. B*, **14**, L747.
Sin Fai Lam, L. T. (1967*a*). *Proc. Phys. Soc.*, **92**, 67.

Sin Fai Lam, L. T. (1967*b*). Thesis, University of Durham.
Skinner, B. G. (1961). *Proc. Phys. Soc.*, **77**, 151.
Skogvall, B. and Schiwietz, G. (1990). *Phys. Rev. Lett.*, **65**, 3265.
Slim, H., Heck, L., Bransden, B. H., and Flower, D. R. (1990). *J. Phys. B*, **23**, L611.
Slim, H., Heck, L., Bransden, B. H., and Flower, D. R. (1991*a*). *J. Phys. B*, **24** (1683).
Slim, H., Heck, L., Bransden, B. H., and Flower, D. R. (1991*b*). *J. Phys. B*, **24** (in press).
Smith, F. J. (1964). *Proc. Phys. Soc.*, **84**, 889.
Smith, F. J. (1966*a*). *Proc. Phys. Soc.*, **92**, 866.
Smith, F. J. (1966*b*). *Planet. Space Sci.*, **14**, 929.
Smith, F. J. (1969). *Phys. Rev.*, **179**, 111.
Sobel'man, I. I. (1972). *An introduction to the theory of atomic spectra*, (trans. T. F. J. Le Vierge). Pergamon, Oxford.
Spruch, L. (1969). In *Lectures in theoretical physics: XI-C atomic collisions* (eds. S. Geltman, K. T. Mahanthappa, and V. B. Britten), p. 77. Gordon and Breach, New York.
Spruch, L. (1978). *Phys. Rev. A*, **18**, 2016.
Stary, C., Lüdde, H. J., and Dreizler, R. M. (1990). *J. Phys. B*, **23**, 263.
Stebbings, R. F., Young, R. A., Oxley, C. L., and Ehrhardt, H. (1965). *Phys. Rev. A*, **138**, 1312.
Stelbovics, A. T. (1989). *J. Phys. B*, **22**, L159.
Stelbovics, A. T. and Slim, H. A. (1986). *Phys. Rev. A*, **9**, 3993.
Stich, W., Lüdde, H. J., and Dreizler, R. M. (1983). *Phys. Lett.*, **99A**, 5.
Stich, W., Lüdde, H. J., and Dreizler, R. M. (1985). *J. Phys. B*, **18**, 1195.
Stier, P. M. and Barnett, C. F. (1956). *Phys. Rev.*, **103**, 896.
Stolterfoht, N. (1979). In *The physics of ionised gases* (ed. R. K. Janev), p. 93. Institute of Physics, Belgrade.
Storm, D. (1974). *Phys. Rev. A*, **10**, 1008.
Storn, D. and Rapp, D. (1976). *Phys. Rev. A*, **14**, 193.
Stueckelberg, E. C. G. (1932). *Helv. Phys. Acta*, **5**, 361.
Sullivan, J., Coleman, J. P., and Bransden, B. H. (1972). *J. Phys. B*. **5**, 2061.
Swafford, G. L., Reading, J. F., Ford, A. L., and Fitchard, E. (1977). *Phys. Rev. A*, **16**, 1329.
Takayanagi, K. (1963). *Progr. Theor. Phys.* (*Kyoto*), Suppl. **25**, 43.
Tan, J., Lin, C. D., and Kimura, M. (1987). *J. Phys. B*, **20**, L91.
Tanis, J. A., Bernstein, E. M., Clark, M. W., Graham, W. G., McFarland, R. H., Morgan, T. J., Johnson, B. B., Jones, K. W., and Merm, M. (1985). *Phys. Rev. A*, **31**. 4040.
Taulbjerg, K., Barrachina, R. O., and Macek, J. H. (1990). *Phys. Rev. A*, **41**, 207.
Tawara, H., Kato, T., and Nakai, Y. (1983). IPPJ-AM30, Nagoya, p. 21.
Temkin, A. (1962). *Phys. Rev. Lett.*, **6**, 354.
Temkin, A. and Bhatia, A. H. (1964). *Rev. Mod. Phys.*, **36**, 1050.
Temkin, A. and Sullivan, E. (1963). *Phys. Rev.*, **129**, 1250.
Thomas, L. H. (1927). *Proc. Roy. Soc.*, A**114**, 561.
Thorson, W. R. and Delos, J. B. (1978*a*). *Phys. Rev. A*, **18**, 117.
Thorson, W. R. and Delos, J. B. (1978*b*). *Phys. Rev. A*, **18**, 135.
Thorson, W. R., Kimura, M., Choi, J. H., and Knudsen, S. K. (1981). *Phys. Rev. A*, **24**, 1768.
Toburen, L. H., Nakai, M. Y., and Langley, R. A. (1968). *Phys. Rev.*, **171**, 114.

Toepfer, A., Henne, A., Lüdde, H. J., Horbatsch, M., and Dreizler, R. M. (1987). *Phys. Lett.*, A**126**, 11.
Toshima, N. and Eichler, J. (1988*a*). *Phys. Rev. Lett.*, **60**, 573.
Toshima, N. and Eichler, J. (1988*b*). *Phys. Rev. A*, **38**, 2305.
Toshima, N. and Eichler, J. (1989). *Phys. Rev. A*, **40**, 125.
Toshima, N. and Eichler, J. (1990). *Phys. Rev. A*, **41**, 5221.
Toshima, N. and Ishihara, T. (1989). *Phys. Rev. A*, **40**, 638.
Toshima, N., Ishihara, T., and Eichler, B. (1987). *Phys. Rev. A*, **36**, 2659.
Toshima, N., Ishihara, T., Ohsaki, A., and Watanabe, T. (1989). *Phys. Rev. A*, **40**, 2192.
Vaaben, J. and Taulbjerg, K. (1981). *J. Phys. B.*, **14**, 1815.
van Hemert, M. C., van Dishoeck, E. F., van der Hart, J. A., and Koike, F. (1985). *Phys. Rev. A*, **31**, 2227.
Varghese, S. L., Waggoner, W., and Corke, C. L. (1984). *Phys. Rev. A*, **29**, 2453.
Vogt, H., Schuch, R., Justiniano, E., Shulz, M., and Schwab, W. (1986). *Phys. Rev. Lett.*, **57**, 2256.
Vučić, S., Potvliege, R. M., and Joachain, C. J. (1987). *J. Phys. B*, **20**, 3157.
Watts, M. F., Dunn, K. F., and Gilbody, H. B. (1978). *J. Phys. B*, **11**, L227.
Watts, M. F., Dunn, K. F., and Gilbody, H. B. (1986). *J. Phys. B*, **19**, L355.
von Weizsäcker, C. F. (1934). *Z. Phys.*, **88**, 612.
Welsh, L. M., Berkner, K. H., Kaplan, S. N., and Pyle, R. V. (1967). *Phys. Rev.*, **158**, 85.
West, B., Lane, N. F., and Cohen, J. S. (1982). *Phys. Rev. A*, **26**, 3164.
Weyl, H. (1910). *Math. Annal.*, **68**, 220.
Whittaker, E. T. and Watson, G. N. (1927). *Modern analysis* (4th edn). Cambridge University Press.
Wilcox, R. M. (1967). *J. Math. Phys.*, **8**, 962.
Wilets, L. and Gallagher, D. F. (1966). *Phys. Rev.*, **147**, 13.
Wilets, L. and Wallace, S. J. (1968). *Phys. Rev.*, **169**, 84.
Wille, U. (1981). *Phys. Lett. A*, **82**, 389.
Wille, U. (1982). *Z. Phys. A*, **308**, 3.
Williams, E. J. (1934). *Phys. Rev.*, **45**, 729.
Williams, E. J. (1935). *Dan. Vidensk, Selsk. Mat-Fys. Medd.*, **13**, No. 4.
Williams, J. F. (1967). *Phys. Rev.*, **157**, 97.
Winter, T. G. (1982). *Phys. Rev. A*, **25**, 697.
Winter, T. G. (1986). *Phys. Rev. A*, **33**, 3842.
Winter, T. G. (1987). *Phys. Rev. A*, **35**, 3799.
Winter, T. G. (1988). *Phys. Rev. A*, **37**, 4656.
Winter, T. G. (1991). *Phys. Rev. A*, **44**, 4353.
Winter, T. G. and Hatton, G. J. (1978). *Phys. Rev. A*, **21**, 793.
Winter, T. G. and Hatton, G. J. (1980). *Phys. Rev. A*, **21**, 793.
Winter, T. G. and Lane, N. F. (1985). *Phys. Rev.*, **31**, 2698.
Winter, T. G. and Lin, C. D. (1974). *Phys. Rev. A*, **10**, 2141.
Winter, T. G. and Lin, C. D. (1984). *Phys. Rev. A*, **29**, 567, 3071.
Winter, T. G., Hatton, G. F., and Lane, N. F. (1980). *Phys. Rev. A*, **22**, 930.
Wirkner-Bott, I., Seim, W., Müller, A., Kester, P., and Salzborn, E. (1981). *J. Phys. B*, **14**, 3987.

Wittkower, A. B., Ryding, G., and Gilbody, H. B. (1966). *Proc. Phys. Soc. A*, **89**, 547.

Yamani, A. J. and Reinhardt, W. P. (1975). *Phys. Rev. A*, **11**, 1444.

Zener, C. (1932). *Proc. Roy. Soc.*, A**137**, 696.

Zimmerman, I. H. (1972). Thesis, University of Washington.

Zygelman, B., Dalgarno, A., Kimura, M., and Lane, N. F. (1989). *Phys. Rev. A*, **40**, 2340.

INDEX